STABLE ADAPTIVE CONTROL AND ESTIMATION FOR NONLINEAR SYSTEMS

Adaptive and Learning Systems for Signal Processing, Communications, and Control

Editor: Simon Haykin

Beckerman / ADAPTIVE COOPERATIVE SYSTEMS

Chen and Gu / CONTROL-ORIENTED SYSTEM IDENTIFICATION: An $\mathcal{H}_\infty$ Approach

Cherkassky and Mulier / LEARNING FROM DATA: Concepts, Theory, and Methods

Diamantaras and Kung / PRINCIPAL COMPONENT NEURAL NETWORKS: Theory and Applications

Haykin / UNSUPERVISED ADAPTIVE FILTERING: Blind Source Separation

Haykin / UNSUPERVISED ADAPTIVE FILTERING: Blind Deconvolution

Haykin and Puthussarypady / CHAOTIC DYNAMICS OF SEA CLUTTER

Hrycej / NEUROCONTROL: Towards an Industrial Control Methodology

Hyvärinen, Karhunen, and Oja / INDEPENDENT COMPONENT ANALYSIS

Kristić, Kanellakopoulos, and Kokotović / NONLINEAR AND ADAPTIVE CONTROL DESIGN

Mann / INTELLIGENT IMAGE PROCESSING

Nikias and Shao / SIGNAL PROCESSING WITH ALPHA-STABLE DISTRIBUTIONS AND APPLICATIONS

Passino and Burgess / STABILITY ANALYSIS OF DISCRETE EVENT SYSTEMS

Sánchez-Peña and Sznaier / ROBUST SYSTEMS THEORY AND APPLICATIONS

Sandberg, Lo, Fancourt, Principe, Katagiri, and Haykin / NONLINEAR DYNAMICAL SYSTEMS: Feedforward Neural Network Perspectives

Spooner, Maggiore, Ordóñez, and Passino / STABLE ADAPTIVE CONTROL AND ESTIMATION FOR NONLINEAR SYSTEMS: Neural and Fuzzy Approximator Techniques

Tao and Kokotović / ADAPTIVE CONTROL OF SYSTEMS WITH ACTUATOR AND SENSOR NONLINEARITIES

Tsoukalas and Uhrig / FUZZY AND NEURAL APPROACHES IN ENGINEERING

Van Hulle / FAITHFUL REPRESENTATIONS AND TOPOGRAPHIC MAPS: From Distortion- to Information-Based Self-Organization

Vapnik / STATISTICAL LEARNING THEORY

Werbos / THE ROOTS OF BACKPROPAGATION: From Ordered Derivatives to Neural Networks and Political Forecasting

Yee and Haykin / REGULARIZED RADIAL BIAS FUNCTION NETWORKS: Theory and Applications

STABLE ADAPTIVE CONTROL AND ESTIMATION FOR NONLINEAR SYSTEMS

Neural and Fuzzy Approximator Techniques

Jeffrey T. Spooner
Sandia National Laboratories

Manfredi Maggiore
University of Toronto

Raúl Ordóñez
University of Dayton

Kevin M. Passino
The Ohio State University

WILEY-INTERSCIENCE

A JOHN WILEY & SONS, INC., PUBLICATION

This text is printed on acid-free paper. ∞

Published simultaneously in Canada.

For ordering and customer service, call 1-800-CALL-WILEY.

Library of Congress Cataloging-in-Publication Data is available.

ISBN 0-471-41546-4

10 9 8 7 6 5 4 3 2 1

To our families

Contents

Preface

A key issue in the design of control systems has long been the robustness of the resulting closed-loop system. This has become even more critical as control systems are used in high consequence applications in which certain process variations or failures could result in unacceptable losses. Appropriately, the focus on this issue has driven the design of many robust nonlinear control techniques that compensate for system uncertainties.

At the same time neural networks and fuzzy systems have found their way into control applications and in sub-fields of almost every engineering discipline. Even though their implementations have been rather ad hoc at times, the resulting performance has continued to excite and capture the attention of engineers working on today's "real-world" systems. These results have largely been due to the ease of implementation often possible when developing control systems that depend upon fuzzy systems or neural networks.

In this book we attempt to merge the benefits from these two approaches to control design (traditional robust design and so called "intelligent control" approaches). The result is a control methodology that may be verified with the mathematical rigor typically found in the nonlinear robust control area while possessing the flexibility and ease of implementation traditionally associated with neural network and fuzzy system approaches. Within this book we show how these methodologies may be applied to state feedback, multi-input multi-output (MIMO) nonlinear systems, output feedback problems, both continuous and discrete-time applications, and even decentralized control. We attempt to demonstrate how one would apply these techniques to real-world systems through both simulations and experimental settings.

This book has been written at a first-year graduate level and assumes some familiarity with basic systems concepts such as state variables and stability. The book is appropriate for use as a text book and homework problems have been included

Organization of the Book

This book has been broken into four main parts. The first part of the book is dedicated to background material on the stability of systems, optimization, and properties of fuzzy systems and neural networks. In Chapter 1 a brief introduction to the control philosophy used throughout the book is presented. Chapter 2 provides the necessary mathematical background for the book (especially needed to understand the proofs), including stability and convergence concepts and methods, and definitions of the notation we will use. Chapter 3 provides an introduction to the key concepts from neural networks and fuzzy systems that we need. Chapter 4 provides an introduction to the basics of optimization theory and the optimization techniques that we will use to tune neural networks and fuzzy systems to achieve the estimation or control tasks. In Chapter 5 we outline the key properties of neural networks and fuzzy systems that we need when they are used as approximators for unknown nonlinear functions.

The second part of the book deals with the state-feedback control problem. We start by looking at the non-adaptive case in Chapter 6 in which an introduction to feedback linearization and backstepping methods are presented. It is then shown how both a direct (Chapter 7) and indirect (Chapter 8) adaptive approach may be used to improve both system robustness and performance. The application of these techniques is further explained in Chapter 9, which is dedicated to implementation issues.

In the third part of the book we look at the output-feedback problem in which all the plant state information is not available for use in the design of the feedback control signals. In Chapter 10, output-feedback controllers are designed for systems using the concept of uniform complete observability. In particular, it is shown how the separation principle may be used to extend the approaches developed for state-feedback control to the output-feedback case. In Chapter 11 the output-feedback methodology is developed for adaptive controllers applicable to systems with a great degree of uncertainty. These methods are further explained in Chapter 12 where output-feedback controllers are designed for a variety of case studies.

The final part of the book addresses miscellaneous topics such as discrete-time control in Chapter 13 and decentralized control in Chapter 14. Finally, in Chapter 15 the methods studied in this book will be compared to conventional adaptive control and to other "intelligent" adaptive control methods (e.g., methods based on genetic algorithms, expert systems, and planning systems).

Acknowledgments

The authors would like to thank the various sponsors of the research that formed the basis for the writing of this textbook. In particular, we would like to thank the Center for Intelligent Transportation Systems at The Ohio

State University, Litton Corp., the National Science Foundation, NASA, Sandia National Laboratories, and General Electric Aircraft Engines for their support throughout various phases of this project.

This manuscript was prepared using LaTeX. The simulations and many of the figures throughout the book were developed using MATLAB.

As mentioned above, the material in this book depends critically on conventional robust adaptive control methods, and in this regard it was especially influenced by the excellent books of P. Ioannou and J. Sun, and S. Sastry and M. Bodson (see Bibliography). As outlined in detail in the "For Further Study" section of the book, the methods of this book are also based on those developed by several colleagues, and we gratefully acknowledge their contributions here. In particular, we would like to mention: J. Farrell, H. Khalil, F. Lewis, M. Polycarpou, and L-X. Wang. Our writing process was enhanced by critical reviews, comments, and support by several persons including: A. Bentley, Y. Diao, V. Gazi, T. Kim, S. Kohler, M. Lau, Y. Liu, and T. Smith. We would like to thank B. Codey, S. Paracka, G. Telecki, and M. Yanuzzi for their help in producing and editing this book. Finally, we would like to thank our families for their support throughout this entire project.

Jeff Spooner
Manfredi Maggiore
Raúl Ordóñez
Kevin Passino

March, 2002

Chapter 1

Introduction

1.1 Overview

The goal of a control system is to enhance automation within a system while providing improved performance and robustness. For instance, we may develop a cruise control system for an automobile to release drivers from the tedious task of speed regulation while they are on long trips. In this case, the output of the plant is the sensed vehicle speed, y, and the input to the plant is the throttle angle, u, as shown in Figure 1.1. Typically, control systems are designed so that the plant output follows some reference input (the driver-specified speed in the case of our cruise control example) while achieving some level of "disturbance rejection." For the cruise control problem, a disturbance would be a road grade variation or wind. Clearly we would want our cruise controller to reduce the effects of such disturbances on the quality of the speed regulation that is achieved.

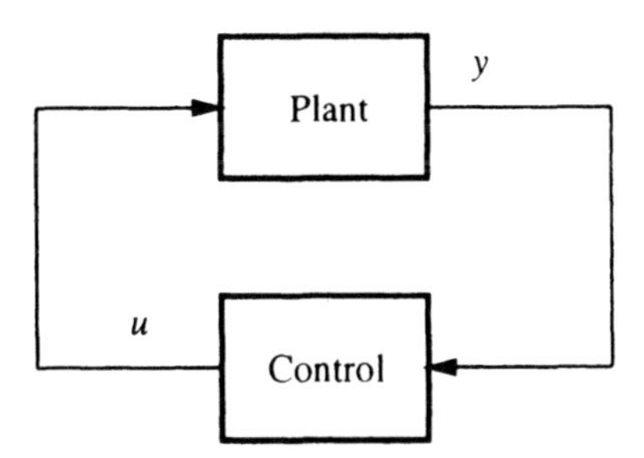

Figure 1.1. Closed loop control.

In the area of "robust control" the focus is on the development of controllers that can maintain good performance even if we only have a poor model of the plant or if there are some plant parameter variations. In the area of adaptive control, to reduce the effects of plant parameter variations, robustness is achieved by adjusting (i.e., adapting) the controller on-line.

For instance, an adaptive controller for the cruise control problem would seek to achieve good speed tracking performance even if we do not have a good model of the vehicle and engine dynamics, or if the vehicle dynamics change over time (e.g., via a weight change that results from the addition of cargo, or due to engine degradation over time). At the same time it would try to achieve good disturbance rejection. Clearly, the performance of a good cruise controller should not degrade significantly as your automobile ages or if there are reasonable changes in the load the vehicle is carrying.

We will use adaptive mechanisms within the control laws when certain parameters within the plant dynamics are unknown. An adaptive controller will thus be used to improve the closed-loop system robustness while meeting a set of performance objectives. If the plant uncertainty cannot be expressed in terms of unknown parameters, one may be able to reformulate the problem by expressing the uncertainty in terms of a fuzzy system, neural network, or some other parameterized nonlinearity. The uncertainty then becomes recast in terms of a new set of unknown parameters that may be adjusted using adaptive techniques.

1.2 Stability and Robustness

Often, when given the challenge of designing a control system for a particular application, one is provided a model of the plant that contains the dominant dynamic characteristics. The engineer responsible for the design of a control system may then proceed to formulate a control algorithm assuming that when the model is controlled to within specifications, then the true plant will also be controlled within specifications. This approach has been successfully applied to numerous systems. More often, however, the controller may need to be adjusted slightly when moving from the design model to the actual implementation due to a mismatch between the model and true system. There are also cases when a control system performs well for a particular operating region, but when tested outside that region, performance degrades to unacceptable levels.

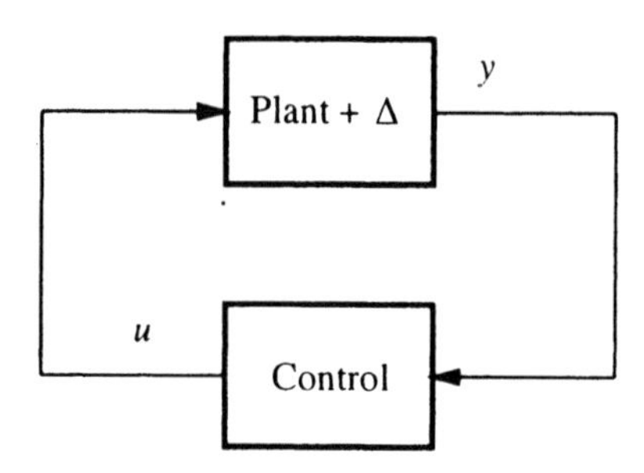

Figure 1.2. Robust control of a plant with unmodeled dynamics.

These issues, among others, are addressed by robust control design. When developing a robust control design, the focus is often on maintaining stability even in the presense of unmodeled dynamics or external disturbances applied to the plant. Figure 1.2 shows the situation in which the controller must be designed to operate given any possible plant variation Δ. Unmodeled dynamics are typically associated with every control problem in which a controller is designed based upon a model. This may be due to any one of a number of reasons:

- It may be the case that only a nominal set of parameters are available for the control design. If a controller is to be incorporated into a mass-produced product, for example, it may not be practical to measure the exact parameter values for each plant so that a controller can be customized to each particular system.
- It may not be cost effective to produce a model that exactly (or even closely) represents the plant's dynamics. It may be possible to spend fewer resources on a robust control design using an incomplete model than developing a high fidelity model so that traditional non-robust techniques may be used.

Hence, the approach in robust control is to accept a priori that there will be model uncertainty, and try to cope with it.

The issue of robustness has been studied extensively in the control literature. When working with linear systems, one may define phase and gain margins which quantify the range of uncertainty a closed-loop system may withstand before becoming unstable. In the world of nonlinear control design, we often investigate the stability of a closed-loop system by studying the behavior of a Lyapunov function candidate. The Lyapunov function candidate is a mathematical function designed to provide a simplified measure of the control objectives allowing complex nonlinear systems to be analyzed using a scalar differential equation. When a controller is designed that drives the Lyapunov function to zero, the control objectives are met. If some system uncertainty tends to drive the Lyapunov candidate away from zero, we will often simply add an additional stabilizing term to the control algorithm that dominates the effect of the uncertainty, thereby making the closed-loop system more robust.

We will find that by adding a static term in the control law that simply dominates the plant uncertainty, it is often easy to simply stabilize an uncertain plant, however, driving the system error to zero may be difficult if not impossible. Consider the case when the plant is defined by

$$\dot{x} = \theta x + u, \tag{1.1}$$

where $x \in \mathsf{R}$ is the plant state that we wish to drive to the point $x = 1$, $u \in \mathsf{R}$ is the plant input, and θ is an unknown constant. Since θ is unknown, one may not define a static controller that causes $x = 1$ to be a stable equilibrium point. In order for $x = 1$ to be a stable equilibrium point, it is necessary that $\dot{x} = 0$ when $x = 1$, so $u(x) = -\theta$ when $x = 1$. Since θ is unknown, however, we may not define such a controller.

In this case, the best that a static nonlinear controller may do is to keep x bounded in some region around $x = 1$. If dynamics are included in the nonlinear controller, then it turns out that one may define a control system that does drive $x \to 1$ even if θ is unknown. In this book we will use the approach of adaptive control to help us define such a nonlinear dynamic controller that will stabilize a certain class of nonlinear uncertain systems.

1.3 Adaptive Control: Techniques and Properties

An adaptive controller can be designed so that it estimates some uncertainty within the system, then automatically designs a controller for the estimated plant uncertainty. In this way the control system uses information gathered on-line to reduce the model uncertainty, that is, to figure out exactly what the plant is at the current time so that good control can be achieved. Considering the system defined by (1.1), an adaptive controller may be defined so that an estimate of θ is generated, which we will denote $\hat{\theta}$. If θ were known, then including a term $-\theta x$ in the control law would cancel the effects of the uncertainty. If $\hat{\theta} \to \theta$ over time, then including the term $-\hat{\theta}x$ in the control law would also cancel the effects of the uncertainty over time. This approach is referred to as indirect adaptive control.

1.3.1 Indirect Adaptive Control Schemes

An indirect approach to adaptive control is made up of an approximator (often referred to as an "identifier" in the adaptive control literature) that is used to estimate unknown plant parameters and a "certainty equivalence" control scheme in which the plant controller is defined ("designed") assuming that the parameter estimates are their true values. The indirect adaptive approach is shown in Figure 1.3. Here the adjustable approximator is used to model some component of the system. Since the approximation is used in the control law, it is possible to determine if we have a good estimate of the plant dynamics. If the approximation is good (i.e., we know how the plant should behave), then it is easy to meet our control objectives. If, on the other hand, the plant output moves in the wrong direction, then we may assume that our estimate is incorrect and should be adjusted accordingly.

As an example of an indirect adaptive controller, consider the cruise control problem where we have an approximator that is used to estimate the vehicle mass and aerodynamic drag. Assume that the vehicle dynamics

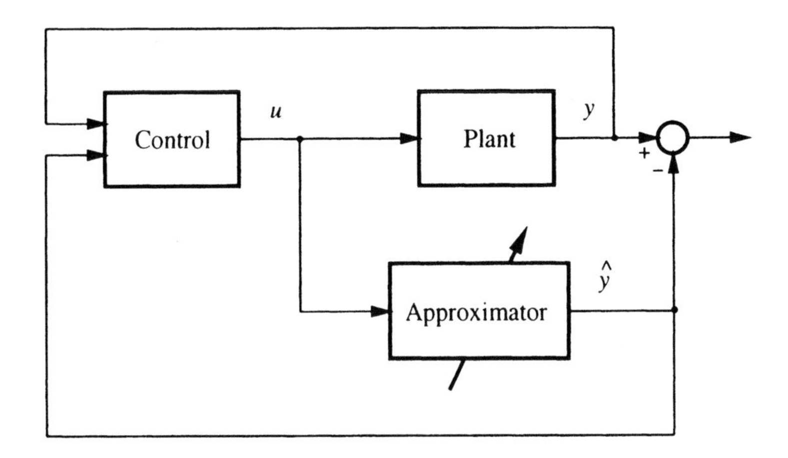

Figure 1.3. Indirect adaptive control.

may be approximated by

$$m\dot{x} = -\rho x^2 + u, \tag{1.2}$$

where m is the vehicle mass, ρ is the coefficient of aerodynamic drag, x is the vehicle velocity, and u is the plant input. Assume that an approximator has been defined so that estimates of the mass and drag are found such that $\hat{m} \to m$ and $\hat{\rho} \to \rho$. Then the control law

$$u = \hat{\rho}x^2 + \hat{m}\nu(t)$$

may be used so that $\dot{x} = \nu(t)$ when $\hat{m} = m$ and $\hat{\rho} = \rho$. Here $\nu(t)$ may be considered a new control input that is defined to drive x to any desired value.

Later in this book, we will learn how to define an approximator for $\hat{m}$ and $\hat{\rho}$ in the above example that allows us to drive x to some desired velocity. We will also find that the indirect approach remains stable when $\hat{m}(0) \neq m$ and $\hat{\rho}(0) \neq \rho$ though the initial parameter values may affect the transient performance of the closed-loop system.

1.3.2 Direct Adaptive Control Schemes

Yet another approach to adaptive control is shown in Figure 1.4. Here the adjustable approximator acts as a controller. The adaptation mechanism is then designed to adjust the approximator causing it to match some unknown nonlinear controller that will stabilize the plant and make the closed-loop system achieve its performance objectives.

Note that we call this scheme "direct" since there is a direct adjustment of the parameters of the controller without identifying a model of the plant.

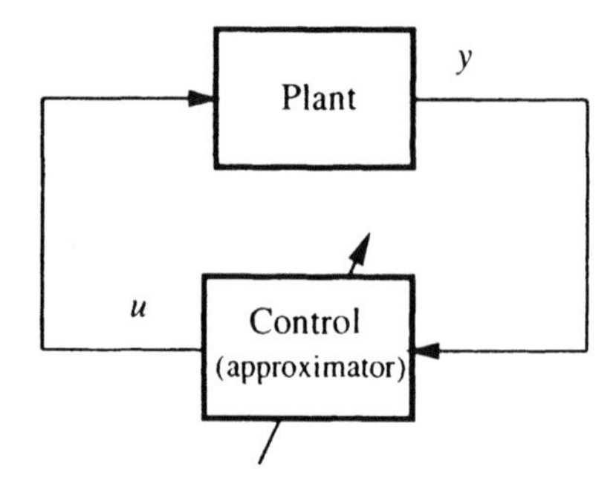

Figure 1.4. Direct adaptive control.

Direct adaptive control, while a somewhat less popular approach (at least in the neural/fuzzy adaptive control literature), will be considered each time we consider an indirect scheme in this book. Part of the reason we give a relatively equal treatment to direct adaptive schemes is that in several implementations we have found them to work more effectively than their indirect adaptive counterparts.

1.4 The Role of Neural Networks and Fuzzy Systems

In this section we outline how neural networks and fuzzy systems can be used as the "approximator" in the adaptive schemes outlined in the previous section. Then we discuss the advantages of using neural networks or fuzzy systems as approximators in adaptive systems.

1.4.1 Approximator Structures and Properties

Neural networks are parameterized nonlinear functions. Their parameters are, for instance, the weights and biases of the network. Adjustment of these parameters results in different shaped nonlinearities. Typically, the adjustment of the neural network parameters is achieved by a gradient descent approach on an error function that measures the difference between the output of the neural network and the output of the actual system (function). That is, we try to adjust the neural network to serve as an approximator for an unknown function that we only know by how it specifies output values for the given input values (i.e., the training data). Or, viewed another way, we seek to adjust the neural network so that it serves as an "interpolator" for the input-output data so that if it is presented with input data, it will produce an output that is close to the actual output that the function (system) would create.

Due to the wide range of roles that the neural network can play in adaptive schemes we will simply call them "approximators," and below we will focus on their properties and advantages. It is important to note, however, that neural networks are not unique in their ability to serve as approximators. There are conventional approximator structures such as polynomials. Moreover, there is the possibility of using a fuzzy system as an approximator structure as we discuss next.

Historically, fuzzy controllers have stirred a great deal of excitement in some circles since they allow for the simple inclusion of heuristic knowledge about how to control a plant rather than requiring exact mathematical models. This can sometimes lead to good controller designs in a very short period of time. In situations where heuristics do not provide enough information to specify all the parameters of the fuzzy controller a priori, researchers have introduced adaptive schemes that use data gathered during the on-line operation of the controller, and special adaptation heuristics, to automatically learn these parameters.

Hence, fuzzy systems have served not only their originally intended function of providing an approach to nonadaptive control, but also in adaptive controllers where, for example, the membership functions are adjusted. Fuzzy systems are indeed simply nonlinear functions that are parameterized by, for example, membership function parameters. In fact, in some situations they are mathematically identical to a certain class of radial basis function neural networks. It is then not surprising that we can use fuzzy systems as approximators in the same way that we can use neural networks. It is possible, however, that the fuzzy system can offer an additional advantage in that it may be easier to incorporate heuristic knowledge about how the input-output map for which you are gathering data from should be shaped. In some situations this can lead to better convergence properties (simply because it may be easier to initialize the shape of the nonlinearity implemented by the approximator).

In this book we will provide some insights into how to pick an approximator (e.g., based on physical considerations); however, the question of which approximator is best to use is still an open research issue. In our discussions on approximator properties, when we refer to an "approximator structure," we mean the nonlinear function that is tuned by the parameters of the approximator. The "size" of the approximator is some measure of the complexity of the mapping it implements (e.g., for a neural network it might be the total number of parameters used to adjust the network). Another feature that we will use to distinguish among different approximators is whether they are "linear in their parameters." For instance, when only certain parameters in a neural network are adjusted, these may be ones that enter in a linear fashion. Clearly, linear in the parameter approximators are a special case of nonlinear in the parameter approximators and hence they can be more limited in what functions that they can approximate.

We will study approximators (neural or fuzzy) that satisfy the "universal approximation property." If an approximator possesses the universal approximation property, then it can approximate any continuous function on a closed and bounded domain with as much accuracy as desired (however, most often, to get an arbitrarily accurate approximation you have to be willing to increase the size the the approximator structure arbitrarily). It turns out that some approximator structures provide much more efficient parameterized nonlinearities in the sense that to get definite improvement in approximation accuracy they only have to grow in size in a linear fashion. Other approximator structures may have to grow exponentially to achieve small increases in approximation accuracy. However, it is important to note that the inclusion of physical domain knowledge may help us to avoid prohibitive increases in the size of the approximator.

The "approximation error" is some suitably defined measure (e.g., the maximum distance between the two functions over their domains) of the error between the function you are trying to approximate (e.g., the plant) and the function implemented by the approximator. The "ideal approximation error" (also known as the "representation error") is the minimum error that would result from the best choice of the approximator parameters (i.e., the "ideal parameters"). For a class of neural networks it can be shown that the ideal approximation error has definite decreases with an increase in the size of the approximator (i.e., it decreases at a certain rate with a linear increase in the size of the neural network); however, in this case you must adjust the parameters that enter in a nonlinear fashion and there are no general guarantees for current algorithms that you will find the ideal parameters. Linear in the parameter approximators provide no such guarantees of reduction of the ideal approximation error; however, when one incorporates physical domain knowledge, experience with applications shows that increases in approximator accuracy can often be found with reasonable increases in the size of the approximator.

1.4.2 Benefits for Use in Adaptive Systems

First, for comparison purposes it is useful to point out that we can broadly think of many conventional adaptive estimation and control approaches for linear systems as techniques that use linear approximation structures for systems with known model order (of course, this is for the state feedback case and ignores the results for plants where the order is not assumed known). Most often, in these cases, the problems are set up so that the linear approximator (e.g., a linear model with tunable parameters) can perfectly represent the underlying unknown function that it is trying to approximate (e.g., the plant model). However, it may take a certain "persistency of excitation" to achieve perfect approximation and conditions for this were derived for adaptive estimation and control.

Regardless, thinking along these lines, linear robust adaptive control studies how to tune linear approximators when it is not possible to perfectly approximate the unknown function with a linear map. In this sense, it becomes clear why there is such a strong reliance of the methods of on-line approximation based control via neural or fuzzy systems on conventional robust control of linear systems. While the universal approximation property guarantees that our approximators can represent the unknown function, for practical reasons we have to limit their size so a finite approximation error arises and must be dealt with; on-line approximation approaches deal with it in similar (or the same) ways to how it is dealt with in linear robust control.

Now, while there is a strong connection to the conventional robust adaptive control approaches, the on-line approximation based approach allows you to go further since it does not restrict the unknown function to be linear. In this way, it provides a logical extension to create nonlinear robust control schemes where there is no need to assume that the *plant* is a linear parameterization of *known* nonlinear functions (as in the early work on adaptive feedback linearization [192] and the more recently developed systematic approach of adaptive backstepping [115]).

It is interesting to note, however, that while there are strong connections to conventional adaptive schemes, there is an additional interesting characteristic of the resulting adaptive systems in that if designed properly they can implement something that is more similar to the way we think of "learning" than conventional adaptive schemes. Some on-line approximation based schemes (particularly some that are implemented with approximators that have basis functions with "local support" like radial basis function neural networks and fuzzy systems) achieve local adjustments to parameters so that only local adjustments to the tuned nonlinearity take place. In this case, if designed properly, the controller can be taught one operating condition, then learn a different operating condition, and later return to the first operating condition with a controller that is already properly tuned for that region. Another way to think of this is that since we are tuning nonlinear functions that have an ability to be tuned locally (something a simple linear map cannot do since if you change a parameter it affects the shape of the map over the whole space) they can remember past tuning to a certain extent.

To summarize, in many ways, the advantages of using neural networks or fuzzy systems arise as practical rather than theoretical benefits in the sense that we could avoid their use all together and simply use some conventional approximator structure (e.g., a polynomial approximator structure). The practical benefits of neural networks or fuzzy systems are the following:

- They offer forms of nonlinearities (e.g., the neural network) that are universal approximators (hence more broadly applicable to many applications) and that offer reduced ideal approximation error for only a linear increase in the number of parameters.

- They offer convenient ways to incorporate heuristics on how to initialize the nonlinearity (e.g., the fuzzy system).

In addition, to help demonstrate the practical nature of the approaches we introduce in this book, there will be an experimental component where we discuss several laboratory implementations of the methods.

1.5 Summary

The general control philosophy used within this book may be summarized as follows:

1. We use concepts and techniques from robust control theory,

2. Adaptive approaches are used to compensate for unknown system characteristics, and

3. When a system uncertainty may be characterized by a function, the problem is reformulated in terms of fuzzy systems or neural networks to extend the applicability of the adaptive approaches.

We will use the traditional controller development and analysis approaches used in robust, adaptive, and nonlinear control, with the mathematical flexibility provided by fuzzy systems and neural networks, to develop a powerful approach to solving many of today's challenging real-world control problems.

Overall, while we understand that many people do not read introductions to books, we tried to make this one useful by giving you a broad view of the lines of reasoning that we use, and by explaining what benefits the methods may provide to you.

Part I

Foundations

Chapter 2

Mathematical Foundations

2.1 Overview

Engineers have applied knowledge gained in certain areas of science in order to develop control systems. Physics is needed in the development of mathematical models of dynamical systems so that we may analyze and test our adaptive controllers. Throughout this book, we will assume that a mathematical model of the system is provided so we will not cover the physics required to develop the model. We do, however, require an understanding of background material from mathematics, and thus it is the primary focus of this chapter. In particular, mathematical foundations are presented in this chapter to establish the notation used in this book and to provide the reader with the background necessary to construct adaptive systems and analyze their resulting dynamical behavior. Here, we overview some ideas from vector, matrix, and signal norms and properties; function properties; and stability and boundedness analysis.

The reader who already understands all these topics should quickly skim this chapter to get familiar with the notation. For the reader who is unfamiliar with all or some of these topics, or for those in need of a review of these topics, we recommend doing a variety of the examples throughout the chapter and some of the homework problems at the end of it.

2.2 Vectors, Matrices, and Signals: Norms and Properties

Norms measure the size of elements in a set $\mathcal{S}$. In general, given two elements $x, y \in \mathcal{S}$, a **norm**, denoted by $\|\cdot\|$ (or $|\cdot|$), is a real valued function which must satisfy

1. $\|x\| \geq 0$, and $\|x\| = 0$ if and only if (iff) $x = 0$.

2. $\|ax\| = |a|\|x\|$ for any $a \in \mathsf{R}$, where R is the set of real numbers.

3. $\|x + y\| \leq \|x\| + \|y\|$.

The third relationship is called the **triangle inequality**. If $\|x - y\| = 0$ we say that x and y are the same element in $\mathcal{S}$.

2.2.1 Vectors

Given a vector $x \in \mathsf{R}^n$ (the Euclidean space) with elements defined by $x = [x_1, \ldots, x_n]^\top$ (where $\top$ denotes transpose), its ***p*-norm** is defined as

$$|x|_p = \left(\sum_{i=1}^{n} |x_i|^p \right)^{1/p}, \tag{2.1}$$

where $p \in [1, \infty)$. If a is a scalar, then $|a|$ denotes the absolute value of a. We also define the **∞-norm** as

$$|x|_\infty = \max_{1 \le i \le n} \{|x_i|\}. \tag{2.2}$$

When we use a vector norm without an explicit subscript, to be concrete we adopt the convention in this book that it is the 2-norm (also called the Euclidean norm) although in most cases any other p-norm or ∞-norm would also be valid. Notice that the 2-norm may be written in vector notation as $|x| = \sqrt{x^\top x}$.

Example 2.1 Consider the 1-norm, and real n-vectors $x = [x_1, \ldots, x_n]^\top$, and $y = [y_1, \ldots, y_n]^\top$. Here, we show that all three properties of a norm hold for the $p = 1$ case. Notice that since $|x|_1 = |x_1| + |x_2| + \cdots + |x_n|$, clearly $|x|_1 \ge 0$, and $|x|_1 = 0$ iff $x = 0$ so that the first property of the norm is satisfied. Next, notice that since for any two scalars a and b, $|ab| = |a||b|$, we know that $|ax|_1 = |ax_1| + |ax_2| + \cdots + |ax_n| = |a||x_1| + |a||x_2| + \cdots + |a||x_n| = |a||x|_1$ and the second property of norms is satisfied. For the triangle inequality notice that $|x + y|_1 = |x_1 + y_1| + |x_2 + y_2| + \cdots + |x_n + y_n|$, and since for any two scalars a and b, $|a + b| \le |a| + |b|$, we know that $|x_i + y_i| \le |x_i| + |y_i|$ for each i, $1 \le i \le n$. Hence, we have $|x + y|_1 \le |x_1| + |y_1| + |x_2| + |y_2| + \cdots + |x_n| + |y_n| = |x|_1 + |y|_1$ so that the triangle inequality holds. For practice, show that all the properties of a norm hold for the $p = 2$ case. $\triangle$

We also note that each of the vector norms are similar in that we may define constants $a, b \in \mathsf{R}$ such that $a|x|_q \le |x|_p \le b|x|_q$ for any $p, q \in [1, \infty]$. For instance, if $x \in \mathsf{R}^n$,

$$|x|_\infty \le |x|_2 \le \sqrt{n}|x|_\infty,$$

and

$$|x|_\infty \le |x|_1 \le n|x|_\infty.$$

Example 2.2 Let $x = [x_1, \ldots, x_n]^\top$. To see that $|x|_\infty \leq |x|_1 \leq n|x|_\infty$ note first that $|x|_\infty \leq |x|_1$ since the sum of all the absolute values of the elements of x must be larger than the largest element of x since this element is contained in the sum. Next, it is easy to see that $|x|_1 \leq n|x|_\infty$ since if you take the largest element of x, and sum it up n times (equivalent to multiplying by n), this must be larger than the sum of the absolute values of the elements of x. For practice, show that $|x|_\infty \leq |x|_2 \leq \sqrt{n}|x|_\infty$. △

Cauchy's inequality is given by

$$|x^\top y| \leq |x||y|. \tag{2.3}$$

Equality holds in Cauchy's inequality iff $x = \alpha y$ for a scalar $\alpha \in \mathsf{R}$. This, and other relationships such as the one discussed in the next example will be useful in the study of adaptive systems.

Example 2.3 We may use the definition of vector norms to show that

$$-|x|^2 \pm x^\top y \leq -\frac{|x|^2}{2} + \frac{|y|^2}{2} \tag{2.4}$$

for any $x, y \in \mathsf{R}^n$. First, notice that

$$\frac{1}{2}|x \mp y|^2 = \frac{|x|^2}{2} \mp x^\top y + \frac{|y|^2}{2} \geq 0.$$

From this inequality, we rearrange terms to get the desired result. △

2.2.2 Matrices

Given a matrix $A \in \mathsf{R}^{m \times n}$ with elements

$$A = \begin{bmatrix} a_{1,1} & \cdots & a_{1,n} \\ \vdots & & \vdots \\ a_{m,1} & \cdots & a_{m,n} \end{bmatrix},$$

(also denoted $A = [a_{ij}]$) and matrix $B \in \mathsf{R}^{n \times q}$ then the transpose satisfies

$$(AB)^\top = B^\top A^\top. \tag{2.5}$$

Also, if $n = m = q$ then a matrix A is said to be a **symmetric matrix** if $A = A^\top$. The **trace operator** is defined by

$$\mathrm{tr}[A] = \sum_{i=1}^{n} a_{i,i} \tag{2.6}$$

and has the property

$$\mathrm{tr}[AB] = \mathrm{tr}[BA]. \tag{2.7}$$

Induced Norms

The **induced p-norm** is defined as

$$\|A\|_p = \sup_{x \neq 0} \frac{|Ax|_p}{|x|_p}, \tag{2.8}$$

(where A is $m \times n$ and x is $n \times 1$) which may be expressed as

$$\|A\|_p = \sup_{|x|=1} |Ax|_p, \tag{2.9}$$

where $p \in [1, \infty)$. The operator sup, called the supremum, gives the least upper bound of its argument over its subscript. Some of the more commonly used matrix norms are

$$\|A\|_1 = \max_{1 \leq j \leq n} \sum_{i=1}^{m} |a_{ij}|, \tag{2.10}$$

$$\|A\|_2 = \sqrt{\lambda_{\max}(A^\top A)}, \tag{2.11}$$

and

$$\|A\|_\infty = \max_{1 \leq i \leq m} \sum_{j=1}^{n} |a_{ij}|, \tag{2.12}$$

where $\lambda_{\max}(A)$ is the maximum eigenvalue of A (refer to [74]). In this book when we use the notation $\|A\|$ (i.e., without a subscript) to be concrete we will be referring to the case where $p = 2$, although in most cases any p-norm or ∞-norm will work also.

Consider $m \times n$ matrices A and B. The induced p-norm of a $m \times n$ matrix satisfies the axioms of a norm on $\mathsf{R}^{m \times n}$, so the triangle inequality

$$\|A + B\| \leq \|A\| + \|B\| \tag{2.13}$$

holds. In addition, it is useful to note that

$$\|Ax\| \leq \|A\||x| \tag{2.14}$$

and

$$\|AB\| \leq \|A\|\|B\| \tag{2.15}$$

for matrices A and B and vector x. Also,

$$\max_{i,j} |a_{i,j}| \leq \|A\|_2 \leq \sqrt{mn} \max_{i,j} |a_{i,j}|. \tag{2.16}$$

Example 2.4 Let

$$A = \begin{bmatrix} 2 & 1 \\ 0 & 3 \end{bmatrix}. \tag{2.17}$$

In this case, $\|A\|_1 = 4$, $\|A\|_2 = \frac{1}{2}(\sqrt{26} + \sqrt{2}) = 3.26$, and $\|A\|_\infty = 3$. If $x = [1, 1]^\top$ then notice that $4.24 = \sqrt{18} = \|Ax\|_2 \leq \|A\|_2 |x|_2 = 3.26(\sqrt{2}) = 4.61$. Notice that

$$3 = \max_{i,j} |a_{i,j}| \leq \|A\|_2 \leq \sqrt{mn} \max_{i,j} |a_{i,j}| = 6.$$

△

Positive Definite Matrices

We will use the properties of positive definite matrices throughout the analysis in this book. A real symmetric $n \times n$ matrix P is said to be **positive semidefinite** (denoted $P \geq 0$, which is *not* an element-wise inequality) if $x^\top P x \geq 0$ for all $x \neq 0$, while it is said to be **positive definite** (denoted $P > 0$) if $x^\top P x > 0$ for all $x \neq 0$. Given a real symmetric matrix P, then $P > 0$ ($P \geq 0$) iff all its eigenvalues are positive (nonnegative). This provides a convenient way to test for positive definiteness (semidefiniteness). Since the determinant of P, $\det(P) = \lambda_1 \cdots \lambda_n$, where the λ_i are eigenvalues, we know that $\det(P) > 0$ if $P > 0$. As an example, note that if $D = [d_{ij}]$ is a diagonal matrix (i.e., $d_{ij} = 0$, $i \neq j$) with $d_{ii} > 0$ ($d_{ii} \geq 0$) then D is positive definite (positive semidefinite); hence, the identity matrix is a positive definite matrix.

There are other ways to test if a matrix is positive definite. For instance, given a square matrix $P \in \mathbb{R}^{n \times n}$, a **leading principle submatrix** is defined by

$$P_i = \begin{bmatrix} p_{11} & \cdots & p_{1i} \\ \vdots & & \vdots \\ p_{i1} & \cdots & p_{ii} \end{bmatrix} \tag{2.18}$$

for any $i = 1, \ldots, n$. If the leading principal submatrices $P_1, \ldots, P_n$ all have positive determinants, then $P > 0$. Next, we outline some useful properties of positive definite matrices.

If $P > 0$ then the matrix inverse satisfies $P^{-1} > 0$. If P^{-1} exists and $P \geq 0$ then $P > 0$. If A and B are $n \times n$ positive semidefinite (definite) matrices, then the matrix $P = \lambda A + \mu B$ is also positive semidefinite (definite) for all $\lambda > 0$ and $\mu > 0$. If A is an $n \times n$ positive semidefinite matrix and C is an $m \times n$ matrix, then the matrix $P = CAC^\top$ is positive semidefinite. If an $n \times n$ matrix A is positive definite, and an $m \times n$ matrix C has rank m, then $P = CAC^\top$ is positive definite. An $n \times n$ positive definite and symmetric matrix P can be written as $P = CC^\top$ where C is a square invertible matrix. If an $n \times n$ matrix P is positive semidefinite and symmetric, and its rank is m, then it can be written as $P = CC^\top$, where C is an $n \times m$ matrix of full rank. Given $P > 0$, one can factor $P = U^\top DU$ where D is a diagonal matrix and U is a **unitary matrix** (if we let U^* denote the complex conjugate transpose of U, then U is called a unitary matrix if $U^* = U^{-1}$). This implies that we may express $P = P^{\top/2}P^{1/2}$ where $P^{1/2} = D^{1/2}U$.

We may use positive definite matrices to define the vector norm

$$|x|_{[P]}^2 = x^\top Px, \tag{2.19}$$

where $P > 0$ is a real symmetric positive definite matrix.

Example 2.5 In this example we show how to use properties of positive definite matrices to show that $|x|_{[P]}$ is a vector norm. Clearly, $|x|_{[P]} \geq 0$ and $|x|_{[P]} = 0$ iff $x = 0$. Also, it is clear that $|ax|_{[P]} = |a||x|_{[P]}$ for any $a \in R$ so the first two properties of a norm are satisfied. Next, we need to show that $|x+y|_{[P]} \leq |x|_{[P]} + |y|_{[P]}$. The triangle inequality may be established by first noting that

$$|x+y|_{[P]} = \sqrt{|x^\top Px + 2x^\top Py + y^\top Py|}. \tag{2.20}$$

Since $x^\top Py = x^\top P^{\top/2}P^{1/2}y = (P^{1/2}x)^\top(P^{1/2}y)$, we know that $|x^\top Py| \leq \sqrt{x^\top Px}\sqrt{y^\top Py}$ so

$$\begin{aligned} |x+y|_{[P]} &\leq \sqrt{\left[\sqrt{x^\top Px} + \sqrt{y^\top Py}\right]^2} \quad (2.21)\\ &= |x|_{[P]} + |y|_{[P]}. \quad (2.22)\end{aligned}$$

△

Next, note that if P is $n \times n$ and symmetric then for any real n-vector x, the **Rayleigh-Ritz inequality**

$$\lambda_{\min}(P)x^\top x \leq x^\top Px \leq \lambda_{\max}(P)x^\top x \tag{2.23}$$

holds where $\lambda_{\min}(P)$ and $\lambda_{\max}(P)$ are the smallest and largest eigenvalues of P. Also, if $P \geq 0$ then

$$||P||_2 = \lambda_{\max}(P).$$

If $P > 0$, then

$$||P^{-1}||_2 = 1/\lambda_{\min}(P),$$

and the trace $tr[P]$ is bounded by

$$||P||_2 \leq tr[P] \leq n||P||_2. \tag{2.24}$$

2.2.3 Signals

Norms may also be defined for a signal $x(t) : \mathsf{R}^+ \to \mathsf{R}^n$ to quantify its magnitude. Here $\mathsf{R}^+ = [0, \infty)$ is the set of positive real numbers so $x(t)$ is simply a vector function whose n elements vary with time. The p-norm for a continuous signal is defined as

$$\|x\|_p = \left(\int_0^\infty |x(t)|^p dt \right)^{1/p}, \tag{2.25}$$

where $p \in [1, \infty)$. If $x(t) = e^{-t}$, what is $\|x\|_2$? If $x(t)$ is a vector quantity, then $|\cdot|$ represents the vector 2-norm in R^n. Additionally,

$$\|x\|_\infty = \sup_{t \in \mathsf{R}^+} |x(t)|. \tag{2.26}$$

The **supremum** operator gives the least upper bound of its argument, and hence $\sup_{t\in\mathsf{R}^+} |x(t)|$ is the least upper bound of the signal over all values of time $t \geq 0$ (inf denotes **infimum** and it is the greatest lower bound). For example, if $x(t) = \sin(t)$, $t \geq 0$, then $\sup_{t\in\mathsf{R}^+} |x(t)| = 1$ and if $x(t) = 2 - 2e^{-t}$, $t \geq 0$, then $\sup_{t\in\mathsf{R}^+} |x(t)| = 2$.

The functional space over which the signal norm exists is defined by

$$\mathcal{L}_p = \{x(t) \in \mathsf{R}^n : \|x\|_p < \infty\}. \tag{2.27}$$

for $p \in [1, \infty]$, that is, $\mathcal{L}_p$ is the set of all vector functions in R^n for which the p-norm is well defined (finite). In other words, we say that a signal $x \in \mathcal{L}_p$ if $\|x\|_p$ exists. We define $\mathcal{L}_\infty$ in a similar way. Hence, as an example, $x(t) = \sin(t) \in \mathcal{L}_\infty$ but $\sin(t) \notin \mathcal{L}_2$. It is also easy to see that $e^{-t} \in \mathcal{L}_2$ and that $e^t \notin \mathcal{L}_\infty$. If scalar functions $x(t), y(t) \geq 0$, $t \geq 0$, are defined such that $x(t) \leq y(t)$, $t \geq 0$, and $y(t) \in \mathcal{L}_p$, then $x(t) \in \mathcal{L}_p$ for all $p \in [1, \infty)$. As an example, since $e^{-2t} \leq e^{-t}$ and $e^{-t} \in \mathcal{L}_2$ we immediately know that $e^{-2t} \in \mathcal{L}_2$. Also, if $x(t) \in \mathcal{L}_1 \cap \mathcal{L}_\infty$ then $x(t) \in \mathcal{L}_p$ for all $p \in [1, \infty)$.

If $x(k)$ is a sequence, then the signal norm becomes

$$\|x\|_p = \left(\sum_{k=0}^{\infty} |x(k)|^p \right)^{1/p}, \tag{2.28}$$

and we define

$$\ell_p = \{x(k) \in \mathsf{R}^n : \|x\|_p < \infty\} \tag{2.29}$$

to be the space of discrete time signals over which the norm exists ($\|x\|_\infty$ and ℓ_∞ are defined in an analogous manner).

The following inequalities for signals will be useful:

1. **Hölder's Inequality:** If scalar time functions $x \in \mathcal{L}_p$ and $y \in \mathcal{L}_q$ for $p, q \in [1, \infty]$ and $1/p + 1/q = 1$, then $xy \in \mathcal{L}_1$ and $\|xy\|_1 \leq \|x\|_p \|y\|_q$. When $p = q = 2$, this reduces to the Schwartz inequality.

2. **Minkowski Inequality:** If scalar time functions $x, y \in \mathcal{L}_p$ for $p \in [1, \infty]$, then $x + y \in \mathcal{L}_p$ and $||x + y||_p \leq ||x||_p + ||y||_p$.

3. **Young's Inequality:** For scalar time functions $x(t) \in \mathsf{R}$ and $y(t) \in \mathsf{R}$, it holds that
$$2xy \leq \frac{1}{\epsilon}x^2 + \epsilon y^2$$
for any $\epsilon > 0$.

4. **Completing the Square:** For scalar time functions $x(t) \in \mathsf{R}$ and $y(t) \in \mathsf{R}$,
$$-x^2 + 2xy = -x^2 + 2xy - y^2 + y^2 \leq y^2.$$

Example 2.6 Given $x, y : \mathsf{R}^+ \to \mathsf{R}^n$, then if $x \in \mathcal{L}_p$ and $y \in \mathcal{L}_\infty$ for some $p \in [0, \infty)$, then $y^\mathsf{T} x \in \mathcal{L}_p$. This may be shown as follows: Since $y \in \mathcal{L}_\infty$, there exists some finite $c > 0$ such that $\sup_{t \geq 0}\{|y|\} \leq c$. By definition,
$$||y^\mathsf{T} x||_p \leq \left(\int_0^\infty c^p |x(\tau)|^p d\tau\right)^{1/p} \leq c||x||_p < \infty \tag{2.30}$$
so that $y^\mathsf{T} x \in \mathcal{L}_p$. △

Next, we examine how to quantify the effects of linear systems on the sizes of signals. The relationship between input $u(t)$ and output $y(t)$ of a linear time-invariant causal system may be expressed as
$$y(t) = \int_{-\infty}^{\infty} g(t - \tau)u(\tau)d\tau, \tag{2.31}$$
where $g(t)$ is the impulse response of the system transfer function $G(s)$, and s is the complex variable used in the Laplace transform representation $y(s) = G(s)u(s)$. We may define the following system norms
$$||G||_2 = \sqrt{\frac{1}{2\pi}\int_{-\infty}^{\infty} |G(j\omega)|^2 \, d\omega} \tag{2.32}$$
and
$$||G||_\infty = \sup_\omega |G(j\omega)| \tag{2.33}$$
and we note that
$$\begin{aligned} ||y||_2 &= ||G||_\infty ||u||_2 && (2.34)\\ ||y||_\infty &= ||G||_2 ||u||_2 && (2.35)\\ ||y||_p &\leq ||g||_1 ||u||_p && (2.36) \end{aligned}$$
and $||y||_\infty = ||g||_1 ||u||_\infty$. For example, if $G(s) = 1/(s + 1)$ then $||G||_\infty = 1$; if the input to $G(s)$ is $u(t) = e^{-t}$ what is $||y||_2$?

2.3 Functions: Continuity and Convergence

In this section we overview some properties of functions and summarize some useful convergence results.

2.3.1 Continuity and Differentiation

We begin with basic definitions of continuity.

Definition 2.1: A function $f : D \to \mathbb{R}^n$ is **continuous at a point** $x \in D \subseteq \mathbb{R}^n$ if for each $\epsilon > 0$, there exists a $\delta(\epsilon, x)$ such that for all $y \in D$ satisfying $|x - y| < \delta(\epsilon, x)$, then $|f(x) - f(y)| < \epsilon$. A function $f : D \to \mathbb{R}^n$ is **continuous** on D if it is continuous at every point in D.

As an example, the function $f(x) = \sin(x)$ is continuous. However, the function defined by $f(x) = 1$, $x \geq 0$, $f(x) = 0$, $x < 0$, is not continuous. This is the unit step function that has a discontinuity at $x = 0$. It is not continuous since if we pick $x = 0$ and $\epsilon = \frac{1}{2}$, then there does not exist a $\delta > 0$ such that for all $y \in D = \mathbb{R}$ satisfying $|y| < \delta(\epsilon, x)$, $|f(x) - f(y)| = |0 - f(y)| < \epsilon$. In particular, such a δ does not exist since for $y > 0$, $f(y) = 1$.

Definition 2.2: A function $f : D \to \mathbb{R}^n$ is **uniformly continuous** on $D \subseteq \mathbb{R}^n$ if for each $\epsilon > 0$, there exists a $\delta(\epsilon)$ (depending only on ϵ) such that for all $x, y \in D$ satisfying $|x - y| < \delta(\epsilon)$, then $|f(x) - f(y)| < \epsilon$.

The difference between uniform continuity and continuity is the lack of dependence of δ on x in uniform continuity. As an example, note that the function $f(x) = 1/x$ is continuous over the open interval $(0, \infty)$, but it is *not* uniformly continuous within that interval. What happens if we consider the interval $[\epsilon_1, \infty)$ instead, where ϵ_1 is a small, positive number? A scalar function f with $f, \dot{f} \in \mathcal{L}_\infty$ is uniformly continuous on $[0, \infty)$. The unit step function discussed above is not uniformly continuous.

Definition 2.3: A function $f : [0, \infty) \to \mathbb{R}$ is **piecewise continuous** on $[0, \infty)$ if f is continuous on any finite interval $[a, b] \subset [0, \infty)$ except at a finite number of points on each of these intervals.

Note that the unit step function and a finite frequency square wave are both piecewise continuous.

Definition 2.4: A function $f : D \to \mathbb{R}^m$ is said to be **Lipschitz continuous** (or simply Lipschitz) if there exists a constant $L > 0$ (which is sometimes called the Lipschitz constant) such that $|f(x) - f(y)| \leq L|x - y|$ for all $x, y \in D$ where $D \subseteq \mathbb{R}^n$.

Intuitively, Lipschitz continuous functions have a finite slope at all points on their domain. Clearly, the unit step function is not Lipschitz continuous but a sine wave is. Note that if $f : D \to \mathsf{R}$ is a Lipschitz function, then f is uniformly continuous. To see this, note that if f is Lipschitz with constant L, then given any $\epsilon > 0$, we may choose $\delta = \epsilon/L$. If $x, y \in D$ such that $|x - y| < \delta$, then $|f(x) - f(y)| < L\left(\frac{\epsilon}{L}\right) = \epsilon$, and therefore f is uniformly continuous on D. As practice, show that if $f(x) = e^{-2x}$, f is Lipschitz continuous. What is the Lipschitz constant in this case?

If $f : D \to \mathsf{R}$ has a derivative at x, then f is continuous at x. However, continuity of a point does not ensure that the derivative exists at that point. A function $f : \mathsf{R}^n \to \mathsf{R}^m$ is continuously differentiable if the first partial derivatives of the components of $f(x)$ with respect to the components of x are continuous functions of x.

The **Jacobian matrix** of $f : \mathsf{R}^n \to \mathsf{R}^m$ is defined as

$$\frac{\partial f}{\partial x} = \begin{bmatrix} \frac{\partial f_1}{\partial x_1} & \cdots & \frac{\partial f_1}{\partial x_n} \\ \vdots & & \vdots \\ \frac{\partial f_m}{\partial x_1} & \cdots & \frac{\partial f_m}{\partial x_n} \end{bmatrix}. \tag{2.37}$$

If $f(x) = [x_1^2 + x_2^2, 2x_1^3 + 3x_2^4]^\top$, find $\frac{\partial f}{\partial x}$. For a scalar function $f(x, y)$ that depends on x and y, the **gradient** with respect to x is defined as

$$\nabla_x f(x, y) = \left[\frac{\partial f}{\partial x}\right] = \left[\frac{\partial f}{\partial x_1}, \frac{\partial f}{\partial x_2}, \ldots, \frac{\partial f}{\partial x_n}\right]. \tag{2.38}$$

If f is only a function of x we will often use the notation $\nabla f(x)$. As an example, let $x = [x_1, x_2]^\top$ and $f(x) = x_1^2 + x_2^2$. We have $\nabla f(x) = [2x_1, 2x_2]$ which is a row vector.

Next, to specify the mean value theorem we define a line segment between points $a, b \in D = \mathsf{R}^n$ as

$$L(a, b) = \{x \in D : x = \gamma a + (1 - \gamma)b \quad \text{for some } \gamma \in [0, 1]\}, \tag{2.39}$$

where $D \subseteq \mathsf{R}^n$ is a **convex set** (D is convex set if for every $x, y \in D$, and every scalar $\alpha \in [0, 1]$, we have $\alpha x + (1 - \alpha)y \in D$). The **mean value theorem** says that if a function $f : \mathsf{R}^n \to \mathsf{R}^m$ is differentiable at each point $x \in D$ where $D \subseteq \mathsf{R}^n$, and $x, y \in D$ such that the line segment $L(x, y) \in D$, then there exists some $z \in L(x, y)$ such that

$$f(y) - f(x) = \left.\frac{\partial f(x)}{\partial x}\right|_{x=z} (y - x). \tag{2.40}$$

This may be rearranged so that

$$\left.\frac{\partial f(x)}{\partial x}\right|_{x=z} = \frac{f(y) - f(x)}{y - x}.$$

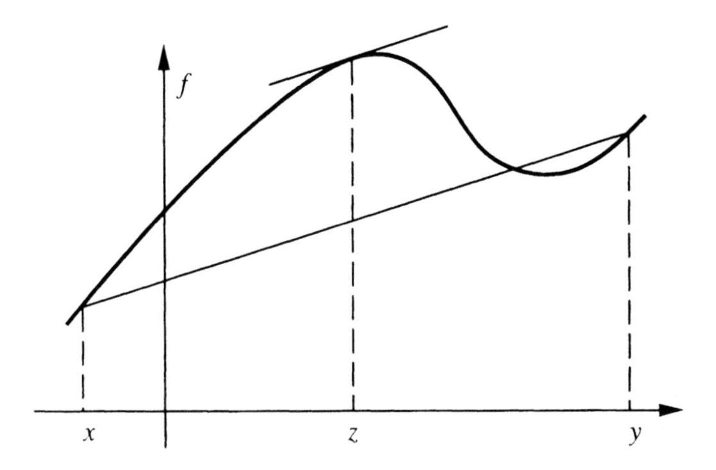

Figure 2.1. The mean value theorem.

From this form, we see that there exists some z such that the slope $\partial f / \partial x$ is equal to the mean slope between the points x and y as shown in Figure 2.1.

This completes the ideas we need from continuity and differentiability. Next, we discuss some basic ideas about convergence.

2.3.2 Convergence

A function $f(t)$ that is bounded from below and is nonincreasing has a limit as $t \to \infty$; but it may not be the case that $f(t) \in \mathcal{L}_\infty$ (e.g., consider $f(t) = 1/t$). If, in addition, $f(0)$ is finite, then $f(t) \leq f(0)$, $t \geq 0$, and $f(t) \in \mathcal{L}_\infty$; for example, consider $f(t) = e^{-t}$. Knowing that $\lim_{t\to\infty} \dot{f}(t) = 0$ does not imply that $f(t)$ has a limit as $t \to \infty$. Also, if $\lim_{t\to\infty} f(t) = c$ for some $c \in \mathrm{R}$ this does not imply that $\dot{f}(t) \to 0$ as $t \to \infty$. These last two statements are included to make sure that when you analyze convergence you do not over-generalize some conclusions and conclude that there is convergence when there may not be.

Barbalat's lemma may be used to show that signals converge to zero. Barbalat's lemma says that if $x(t)$ is a uniformly continuous signal and $\lim_{t\to\infty} \int_0^t x(\tau)d\tau$ exists and is finite, then $x(t) \to 0$ as $t \to \infty$. As an example, if $x(t) = e^{-t}$ then $\lim_{t\to\infty} \int_0^t e^{-\tau}d\tau$ exists and is finite (simple integration shows this) so that $x(t) \to 0$ as $t \to \infty$. On the other hand if $x(t) = \cos(t)$ then $\lim_{t\to\infty} \int_0^t \cos(\tau)d\tau$ does not exist so we cannot conclude that $x(t) \to 0$ as $t \to \infty$.

From Barbalat's lemma we also know that if $f, \dot{f} \in \mathcal{L}_\infty$, and $f \in \mathcal{L}_p$ for some $p \in [0, \infty)$, then $|f(t)| \to 0$ as $t \to \infty$. Another way to say this is that if a signal is uniformly continuous and an $\mathcal{L}_p$ signal, then it converges to zero. Also, if $\dot{f} \in \mathcal{L}_\infty$, and $f \in \mathcal{L}_2$, then $|f(t)| \to 0$ as $t \to \infty$. Notice that in this case if you are willing to use the 2-norm you do not need to assume that $f \in \mathcal{L}_\infty$. Barbalat's lemma for a discrete-time sequence is simplified since if $f \in \ell_p$ for some $p \in [0, \infty)$, then $|f(k)| \to 0$ as $k \to \infty$.

Example 2.7 Suppose that for the scalar ordinary differential equation

$$\dot{x} = f(x)$$

with f Lipschitz continuous in x (so we know that it possesses a unique solution $x(t, x_0)$ for a given initial condition $x(0) = x_0$) you are given a function $V : \mathsf{R} \to \mathsf{R}$ with $V(x) = ax^2 \geq 0$ and $\dot{V}(x) = -bx^2 \leq 0$ for some $a, b > 0$. Since $V(x)$ is bounded from below ($V(x) \geq 0$) and is nonincreasing, it has a limit, so $V \in \mathcal{L}_\infty$. From this, and the definition of $V(x)$ we see that $x \in \mathcal{L}_\infty$, and thus $\dot{V} \in \mathcal{L}_\infty$. Then, since f is Lipschitz continuous, $\dot{x} \in \mathcal{L}_\infty$. Notice that

$$\int_0^t \dot{V}(x(\tau, x_0))d\tau = V(x(t, x_0)) - V(x(0, x_0)). \tag{2.41}$$

Hence, we know that

$$\begin{aligned} V(x(t, x_0)) &= V(x(0, x_0)) + \int_0^t \dot{V}(x(\tau, x_0))d\tau \\ &= V(x(0, x_0)) - \int_0^t bx^2(\tau)d\tau. \end{aligned} \tag{2.42}$$

Since $\int_0^t \dot{V}(x(\tau, x_0))d\tau < \beta$ for some finite $\beta > 0$ and any t, then $x \in \mathcal{L}_2$ and from Barbalat's lemma we obtain that $\lim_{t \to \infty} x(t) = 0$.
△

2.4 Characterizations of Stability and Boundedness

Suppose that a nonautonomous (time-varying) dynamical system may be expressed as

$$\dot{x} = f(t, x), \tag{2.43}$$

where $x \in \mathsf{R}^n$ is an n dimensional vector and $f : \mathsf{R}^+ \times D \to \mathsf{R}^n$ with $D = \mathsf{R}^n$ or $D = B_h$ for some $h > 0$, where

$$B_h = \{x \in \mathsf{R}^n : |x| < h\}$$

is a ball centered at the origin with a radius of h. If $D = \mathsf{R}^n$ then we say that the dynamics of the system are defined globally, whereas if $D = B_h$

they are only defined locally. We will not consider systems whose dynamics are defined over disjoint subspaces of $\mathbb{R}^n$. It is assumed that $f(t, x)$ is piecewise continuous in t and Lipschitz in x for existence and uniqueness of state solutions. As an example, the linear system

$$\dot{x}(t) = Ax(t)$$

fits the form of (2.43) with $D = \mathbb{R}^n$.

Assume that for every x_0 the initial value problem

$$\dot{x}(t) = f(t, x(t)), \quad x(t_0) = x_0 \tag{2.44}$$

possesses a unique solution $x(t, t_0, x_0)$. We call $x(t, t_0, x_0)$ a solution to (2.43) if

1. $x(t_0, t_0, x_0) = x_0$
2. $\frac{d}{dt}x(t, t_0, x_0) = f(t, x(t, t_0, x_0))$.

You should think of $x(t, t_0, x_0)$ as a trajectory that is indexed by t_0 and x_0. Different t_0 and x_0 in general result in different solutions $x(t, t_0, x_0)$ of the ordinary differential equation (2.43). Hence, the ordinary differential equation can be thought of as a generator of system trajectories. The theory that we develop in this section actually applies to any set of trajectories (under mild assumptions), not just the ones generated by the ordinary differential equation (2.43). As an example of the generality of the approach, in Chapter 13 we will explain how with only slight modifications the theory applies to a wide class of nonlinear discrete time systems.

A point $x_e \in \mathbb{R}^n$ is called an equilibrium point of (2.43) if $f(t, x_e) = 0$ for all $t \geq 0$. An equilibrium point x_e is an isolated equilibrium point if there exists an $\rho > 0$ such that the ball around x_e,

$$B_\rho(x_e) = \{x \in \mathbb{R}^n : |x - x_e| < \rho\}, \tag{2.45}$$

contains no other equilibrium points besides x_e.

Example 2.8 Consider the system defined by

$$\begin{aligned} \dot{x}_1 &= x_1 x_2 \\ \dot{x}_2 &= -x_2 \end{aligned} \tag{2.46}$$

For this system, $x = [x_1, x_2]^\top = 0$ is not an isolated equilibrium point since $x_2 = 0$ and $x_1 = a$ for any $a \in \mathbb{R}$ is also an equilibrium point. △

As is standard we will assume (unless otherwise stated) that the equilibrium of interest is an isolated equilibrium located at the origin of $\mathbb{R}^n$

for $t \geq 0$. Studying the equilibrium of the origin results in no loss of generality since if $x_e \neq 0$ is an equilibrium of (2.43) for $t \geq t_0$ and we let $\bar{x}(t) = x(t) - x_e$ and $\tau = t + t_0$, then $\bar{x} = 0$ is an equilibrium of the transformed system $\dot{\bar{x}}(\tau) = f(t + t_0, \bar{x}(t + t_0) + x_e)$. An example of how to translate an equilibrium in this manner is given in Example 2.10.

2.4.1 Stability Definitions

Stability is a property of systems that we often witness around us. For example, it can refer to the ability of an airplane or ship to maintain its planned flight trajectory or course after displacement by wind or waves. In mathematical studies of stability we begin with a model of the dynamics of the system (e.g., airplane or ship) and investigate if the system possesses a stability property. Of course, with this approach we can only ensure that the *model* possesses (or does not possess) a stability property. In a sense, the conclusions we reach about stability will only be valid about the actual physical system to the extent that the model we use to represent the physical system is valid (i.e., accurate).

While we have a general intuitive notion of how a stable system behaves, next we will show a wide range of precise (and standard) mathematical characterizations of stability and boundedness.

Definition 2.5: The equilibrium $x_e = 0$ of (2.43) is said to be **stable** (in the sense of Lyapunov) if for every $\epsilon > 0$ and any $t_0 \geq 0$ there exists a $\delta(\epsilon, t_0) > 0$ such that $|x(t, t_0, x_0)| < \epsilon$ for all $t \geq t_0$ whenever $|x_0| < \delta(\epsilon, t_0)$ and $x(t, t_0, x_0) \in B_h(x_e)$ for some $h > 0$.

That is, the equilibrium is stable if when the system (2.43) starts close to x_e, then it will stay close to it. Note that stability is a property of an equilibrium, not a system. Often, however, we will refer to a system as being stable if all its equilibrium points are stable. Also, notice that according to this definition, stability in the sense of Lyapunov is a "local property." It is a local property since if x_e is stable for some small h, then $x(t, t_0, x_0) \in B_{h'}(x_e)$, for some $h' > h$.

Next, notice that the definition of stability is for a single equilibrium $x_e \in \mathsf{R}$ but actually such an equilibrium is a trajectory of points that satisfy the differential equation in (2.43). That is, the equilibrium is a solution to the differential equation, $x(t, t_0, x_e) = x_e$ for $t \geq 0$. We call any set such that when the initial condition of (2.43) starts in the set and stays in the set for all $t \geq 0$, an **invariant set**. As an example, if $x_e = 0$ is an equilibrium, then the set containing only the point x_e is an invariant set for (2.43) (of course, in general, an invariant set may have many more points in it). With only slight modifications, all the stability definitions in this section, and analysis approaches in the next section, are easy to extend to be valid for invariant sets rather than just equilibria.

Definition 2.6: An equilibrium that is not stable is called **unstable**.

Hence, if an equilibrium is unstable, there does not exist an $h > 0$ such that it is stable. Clearly, a single system can contain both stable and unstable equilibria.

Example 2.9 As an example, suppose that

$$\dot{x}(t) = ax(t),$$

where $a > 0$ is a fixed constant. In this case, $x_e = 0$ is an isolated equilibrium. Using ideas from calculus, the solution to the ordinary differential equation is easy to find as

$$x(t, x_0) = x_0 e^{at}.$$

We use $x(t, x_0)$ to denote a solution that does not depend on t_0. Notice that for every $\epsilon > 0$ you can pick, there exists no $\delta > 0$ such that $|x(t, x_0)| < \epsilon$, since no matter which δ you pick if $|x_0| < \delta$, the solutions $x(t, x_0) \to \infty$ as $t \to \infty$ so long as $x_0 \neq 0$. Because of this we conclude that $x_e = 0$ is an unstable equilibrium point. △

Generally, we try to design adaptive systems in this book so that they do not exhibit instabilities. In fact, we often seek to construct adaptive systems that possess even "stronger" stability properties such as the ones we provide next.

Definition 2.7: If in Definition 2.5, δ is independent of t_0, that is, if $\delta = \delta(\epsilon)$, then the equilibrium x_e is said to be **uniformly stable**.

If in (2.43) f does not depend on time (i.e., $f(x)$), then x_e being stable is equivalent to it being uniformly stable. Of course, uniform stability is also a local property. Next, we introduce a very commonly used form of stability.

Definition 2.8: The equilibrium $x_e = 0$ of (2.43) is said to be **asymptotically stable** if it is stable and for every $t_0 \geq 0$ there exists $\eta(t_0) > 0$ such that

$$\lim_{t \to \infty} |x(t, t_0, x_0)| = 0$$

whenever $|x_0| < \eta(t_0)$.

That is, it is asymptotically stable if when it starts close to the equilibrium it will converge to it. Asymptotic stability is also a local property. It is a "stronger" stability property since it requires that the solutions to the ordinary differential equation converge to zero in addition to what is required for stability in the sense of Lyapunov. See Figure 2.2.

Definition 2.9: The equilibrium $x_e = 0$ of (2.43) is said to be **uniformly asymptotically stable** if it is uniformly stable and for every $\epsilon > 0$ and and $t_0 \geq 0$, there exist a $\delta_0 > 0$ independent of t_0 and ϵ, and a $T(\epsilon) > 0$ independent of t_0, such that $|x(t, t_0, x_0) - x_e| \leq \epsilon$ for all $t \geq t_0 + T(\epsilon)$ whenever $|x_0 - x_e| < \delta(\epsilon)$.

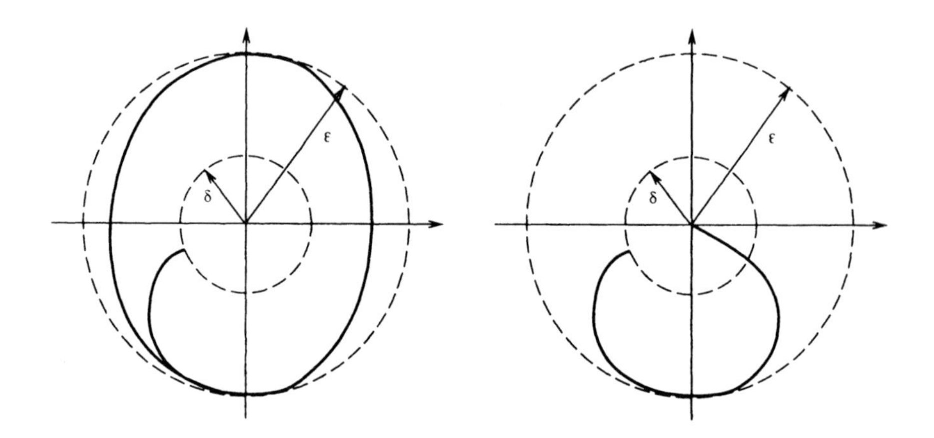

Figure 2.2. Stable and asymptotically stable equilibrium points.

Again, if in (2.43) f does not depend on time (i.e., $f(x)$), then x_e being asymptotically stable is equivalent to it being uniformly asymptotically stable. Of course, uniform asymptotic stability is also a local property.

Example 2.10 Consider

$$\dot{x}(t) = -a(x(t) - b), \tag{2.47}$$

where $a, b > 0$. Notice that $x_e = b$ is an isolated equilibrium. We study stability of the origin so we must translate the equilibrium to the origin. To translate the equilibrium to the origin we let $\bar{x}(t) = x(t) - b$ for $t \geq 0$ and notice that

$$\dot{\bar{x}}(t) = \dot{x}(t) = -a(\bar{x}(t) + b - b) = -a\bar{x}(t) = \bar{f}(t, \bar{x}(t)) \tag{2.48}$$

and $\bar{x}_e = 0$ is an equilibrium of this new system. The solution to this differential equation is

$$\bar{x}(t, x_0) = \bar{x}_0 e^{-at} \tag{2.49}$$

for $t \geq 0$. Notice that $|\bar{x}(t, x_0)| \to 0$ as $t \to \infty$ for all $\bar{x}_0 \in \mathsf{R}^n$ so that $\bar{x}_e = 0$ of (2.48) is asymptotically stable. What conclusions can be drawn about the equilibrium $x_e = b$ of (2.47)? It holds the *same* stability properties, since translation of the equilibrium does not change its stability properties. To explicitly see this, note that the solution to (2.47) is

$$x(t, x_0) = (x_0 - b)e^{-at} + b$$

for $t \geq 0$. Notice that $x(t, x_0) \to b$ as $t \to \infty$ for all $x_0 \in \mathsf{R}$. If $x_0 < b$ then $x(t, x_0)$ increases monotonically to b while if $x_0 > b$, then $x(t, x_0)$ decreases monotonically to b. We see that the equilibrium $x_e = b$ is an asymptotically stable equilibrium point as we expect. $\triangle$

Definition 2.10: The set $\mathcal{X}_d \subset \mathsf{R}^n$ of all $x_0 \in \mathsf{R}^n$ such that $|x(t, t_0, x_0)| \to 0$ as $t \to \infty$ is called the **domain of attraction** of the equilibrium $x_e = 0$ of (2.43).

Sometimes, if such an $\mathcal{X}_d \subset \mathsf{R}^n$ is *known* for a system, then it is said to possess a "regional" stability property to contrast with the local cases just discussed (and exponential stability below), or the global one to be discussed next.

Definition 2.11: The equilibrium $x_e = 0$ is said to be **asymptotically stable in the large** if $\mathcal{X}_d = \mathsf{R}^n$.

That is, an equilibrium is asymptotically stable in the large if no matter where the system starts, its state converges to the equilibrium asymptotically. Notice that this is a global property as opposed to the earlier stability definitions that characterized local properties. This means that for asymptotic stability in the large, the local property of asymptotic stability holds for $B_h(x_e)$ with $h = \infty$ (i.e., on the whole state space). As an example, notice that the equilibrium $x_e = b$ in Example 2.10 has a domain of attraction $\mathcal{X}_d = \mathsf{R}$ so in this case x_e is asymptotically stable in the large.

Definition 2.12: The equilibrium $x_e = 0$ is said to be **exponentially stable** if there exists an $\alpha > 0$ and for every $\epsilon > 0$ there exists a $\delta(\epsilon) > 0$ such that

$$|x(t, t_0, x_0)| \leq \epsilon e^{-\alpha(t - t_0)}, \tag{2.50}$$

whenever $|x_0| < \delta(\epsilon)$ and $t \geq t_0 \geq 0$. The constant α is sometimes called the **rate of convergence**.

Exponential stability is sometimes said to be a "stronger" form of stability since in its presence we know that system trajectories decrease exponentially to zero. It is a local property, but we next define its global version.

Definition 2.13: The equilibrium point $x_e = 0$ is **exponentially stable**

in the large if there exists $\alpha > 0$ and for any $\beta > 0$ there exists $\epsilon(\beta) > 0$ such that

$$|x(t, t_0, x_0)| \leq \epsilon(\beta)e^{-\alpha(t-t_0)}, \tag{2.51}$$

whenever $|x_0| < \beta$ and $t \geq t_0 \geq 0$.

As an example, consider the equilibrium $x_e = b$ of Example 2.10. In this case we had translated the equilibrium and found that the solution for the translated system was given in (2.49). Notice that we can pick $t_0 = 0$, $\alpha = a$, and $\epsilon(\beta) = \beta$ and we see that $\bar{x}_e = 0$ is exponentially stable in the large so that $x_e = b$ is also exponentially stable in the large. It is, in fact, the case that every linear time invariant system that is asymptotically stable is also exponentially stable in the large (and vice versa).

2.4.2 Boundedness Definitions

Next, we introduce some standard definitions of boundedness. Notice that each of these is a global property of a system in the sense that they apply to trajectories (solutions) of the system that can be defined over all of the state space.

Definition 2.14: A solution $x(t, t_0, x_0)$ is **bounded** if there exists a $\beta > 0$, that may depend on each solution, such that

$$|x(t, t_0, x_0)| < \beta \tag{2.52}$$

for all $t \geq t_0 \geq 0$. A system is said to possess **Lagrange stability** if for each $t_0 \geq 0$ and $x_0 \in \mathbb{R}^n$, the solution $x(t, t_0, x_0)$ is bounded.

Notice that if an equilibrium is asymptotically stable in the large or exponentially stable in the large then the system for which the equilibrium is defined is also Lagrange stable (but not necessarily vice versa). Also, if an equilibrium is stable, it does not imply that the system for which the equilibrium is defined is Lagrange stable (notice you must be careful in saying whether a system or an equilibrium is stable) since there may be a way to pick x_0 such that it is near an unstable equilibrium and $x(t, t_0, x_0) \to \infty$ as $t \to \infty$.

As an example, consider the system in (2.47) that has a solution

$$x(t, x_0) = (x_0 - b)e^{-at} + b.$$

We see that since b is finite, the maximum value $x(t, x_0)$ achieves is at $t = 0$ or $t \to \infty$. Hence, we can pick

$$\beta = \max\{x_0, b\}$$

as a bound on each trajectory that is dependent on the trajectory since it depends on the initial condition. Hence, the system has bounded solutions and possesses Lagrange stability.

Definition 2.15: The solutions $x(t, t_0, x_0)$ are **uniformly bounded** if for any $\alpha > 0$ and $t_0 \geq 0$, there exists a $\beta(\alpha) > 0$ (independent of t_0) such that if $|x_0| < \alpha$, then $|x(t, t_0, x_0)| < \beta(\alpha)$ for all $t \geq t_0 \geq 0$.

This is yet another type of boundedness property. If the solutions are uniformly bounded then they are bounded and the system is Lagrange stable; however, the fact that a system is Lagrange stable does not mean it is uniformly stable (why?).

As an example, consider the system in (2.47). Notice that if we pick

$$\beta = \max\{\alpha, b\}$$

then $|x(t, x_0)| < \beta(\alpha)$ if $|x_0| < \alpha$ so that the solutions to the system in (2.47) are uniformly bounded.

Definition 2.16: The solutions $x(t, t_0, x_0)$ are said to be **uniformly ultimately bounded** if there exists some $B > 0$, and if corresponding to any $\alpha > 0$ and $t_0 > 0$ there exists a $T(\alpha) > 0$ (independent of t_0) such that $|x_0| < \alpha$ implies that $|x(t, t_0, x_0)| < B$ for all $t \geq t_0 + T(\alpha)$.

Hence, a system is said to be uniformly ultimately bounded if eventually all trajectories end up in a B-neighborhood of the origin. As an example, suppose that the solutions to some ordinary differential equation are

$$x(t, x_0) = x_0 e^{-at} + 0.1\cos(t).$$

In this case, we can choose $B = 0.11$ and we know that since the $x_0 e^{-at}$ term will eventually die out, there will exist a $T(\alpha)$ (which is generally larger for smaller α) no matter how big x_0 is ($|x_0| < \alpha$) such that $|x(t, x_0)| < B$ for $t > T(\alpha)$.

2.5 Lyapunov's Direct Method

A. M. Lyapunov invented two methods to analyze stability. In his first method (called the indirect method) he showed that if you linearize a system about an equilibrium point, certain conclusions about local stability properties can be made (e.g., if the eigenvalues of the linearized system are in the left half plane then the equilibrium is stable but if one is in the right half plane it is unstable).

In his second method (called the direct method) the stability results for an equilibrium $x_e = 0$ of (2.43) depend on the existence of an appropriate "Lyapunov function" $V : D \to \mathsf{R}$ where $D = \mathsf{R}^n$ for global results (e.g., asymptotic stability in the large) and $D = B_h$ for some $h > 0$, for local results (e.g., stability in the sense of Lyapunov or asymptotic stability). If V is continuously differentiable with respect to its arguments then the derivative of V with respect to t along the solutions of (2.43) is

$$\dot{V}(t,x)\Big|_{(2.43)} = \frac{\partial V}{\partial t} + \frac{\partial V}{\partial x} f(t,x).$$

Using the subscript on $\dot{V}$ is sometimes cumbersome so we will at times omit it with the understanding that the derivative of V is taken along the solutions of the differential equation that we are studying. As an example, suppose that in Equation (2.43) f does not depend on t, and let $V(x) = x^\top P x$ where $x \in \mathbb{R}^n$ and $P = P^\top$. Then,

$$\dot{V}(x) = \frac{\partial V}{\partial x} f(t,x) = \dot{x}^\top P x + x^\top P \dot{x}.$$

Now, notice that $\dot{x}^\top P x$ is a scalar and the transpose of a scalar is the same as the scalar so $\dot{V}(x) = 2x^\top P \dot{x}$.

While at times it can be difficult to construct an appropriate Lyapunov function, which limits the applicability of the method somewhat, the method does allow us to avoid finding the explicit solution to the nonlinear differential equation in Equation (2.43) (which, for some nonlinear ordinary differential equations, can be very difficult or impossible). Notice, for example, that in all the examples we presented in the last section to illustrate the various stability and boundedness properties, we explicitly solved a (simple) differential equation, and then showed that the system possessed the stability property via mathematical analysis of the solution trajectories. Using Lyapunov's direct method, we can make conclusions about possibly infinite sets of solutions to a differential equation (indexed, e.g., by the intial condition) without explicitly finding the solutions. For very complex nonlinear systems this can offer a significant advantage when investigating its qualitative properties.

2.5.1 Preliminaries: Function Properties

Before we introduce Lyapunov's direct method we need the following definitions:

Definition 2.17: A function $\gamma : D \to \mathbb{R}$ is said to be **monotone increasing** on $D \subseteq \mathbb{R}$ if for every $x, y \in D$ with $x \leq y$, then $\gamma(x) \leq \gamma(y)$. If for every $x, y \in D$ with $x < y$, $\gamma(x) < \gamma(y)$, then γ is said to be **strictly increasing**.

As an example, if $\gamma(x) = ax + b$ where $a, b > 0$, then γ is strictly increasing for all $x \in D = \mathbb{R}$. If, however, $\gamma(x) = ax^2 + b$ where $a, b > 0$ then it is not strictly increasing on all of $D = \mathbb{R}$.

Definition 2.18: A continuous function $\gamma : D \to \mathsf{R}^+$ is said to belong to **class** $\mathcal{K}$ (denoted by $\gamma \in \mathcal{K}$) if it is strictly increasing on $D = [0, r)$ for some $r \in \mathsf{R}$ (or on $D = [0, \infty)$), and $\gamma(0) = 0$. A continuous function $\gamma : \mathsf{R}^+ \to \mathsf{R}^+$ is said to belong to **class** $\mathcal{K}_\infty$ if $\gamma \in \mathcal{K}$ with γ defined on $D = [0, \infty)$ and $\gamma(x) \to \infty$ as $x \to \infty$.

As an example, the function $\gamma(x) = ax^2$ where $a > 0$, is strictly increasing on $[0, \infty)$, $\gamma \in \mathcal{K}$, and $\gamma \in \mathcal{K}_\infty$.

Definition 2.19: A continuous function $\beta : D \times \mathsf{R}^+ \to \mathsf{R}^+$ is said to belong to class-$\mathcal{KL}$ if $\beta(r, s) \in \mathcal{K}$ for each fixed s and $\beta(r, s)$ is decreasing with respect to s for each fixed r with $\beta(r, s) \to 0$ as $s \to \infty$.

Definition 2.20: A continuous function $V(t, x) : \mathsf{R}^+ \times B_h \to \mathsf{R}$ ($V(t, x) : \mathsf{R}^+ \times \mathsf{R}^n \to \mathsf{R}$) is said to be **positive definite** if $V(t, 0) = 0$ for $t \geq 0$ and there exists a function $\gamma \in \mathcal{K}$ defined on $[0, h)$ such that $V(t, x) \geq \gamma(|x|)$ for all $t \geq 0$ and $x \in B_h$ for some $h > 0$ ($x \in \mathsf{R}^n$). $V(t, x)$ is said to be **negative definite** if $-V(t, x)$ is positive definite. A continuous function $V(t, x) : \mathsf{R}^+ \times B_h \to \mathsf{R}$ ($V(t, x) : \mathsf{R}^+ \times \mathsf{R}^n \to \mathsf{R}$) is said to be **positive semidefinite** if $V(t, 0) = 0$ for $t \geq 0$ and $V(t, x) \geq 0$ for all $t \geq 0$ and $x \in B_h$ for some $h > 0$ ($x \in \mathsf{R}^n$). For **negative semidefinite** replace "$V(t, x) \geq 0$" with "$V(t, x) \leq 0$" in the definition of positive semidefinite.

As an example, let $V(x) = px^2$ where $x \in \mathsf{R}$ and $p \in \mathsf{R}$ are scalars with $p > 0$. Notice that $V(0) = 0$, and $\gamma(x) = \alpha x^2$, $0 < \alpha < p$, has $\gamma \in \mathcal{K}$ and $V(x) \geq \gamma(|x|)$ so that $V(x)$ is positive definite.

Sometimes it is convenient to use the fact that a continuous function $w(x) : B_h \to \mathsf{R}$ is positive (negative) definite if and only if $w(0) = 0$ and $w(x) > 0$ ($w(x) < 0$) for all $x \in B_h - \{0\}$. Of course, a continuous function $w(x) : \mathsf{R}^n \to \mathsf{R}$ is positive (negative) definite if and only if $w(0) = 0$ and $w(x) > 0$ ($w(x) < 0$) for all $x \in \mathsf{R}^n - \{0\}$. Also, a continuous function $V(t, x) : \mathsf{R}^+ \times B_h \to \mathsf{R}$ is positive (negative) definite if and only if there exists a positive (negative) definite function $w(x)$ defined on B_h such that $V(t, 0) = 0$ for all $t \geq 0$ and $V(t, x) \geq w(x)$ for all $x \in B_h$ and $t \geq 0$. Similarly, if we replace "B_h" by R^n.

Definition 2.21: A continuous function $V(t, x) : \mathsf{R}^+ \times B_h \to \mathsf{R}$ ($V(t, x) : \mathsf{R}^+ \times \mathsf{R}^n \to \mathsf{R}$) is said to be **decrescent** if there exists a function $\gamma \in \mathcal{K}$ defined on $[0, r)$ for some $r > 0$ (defined on $[0, \infty)$) such that $V(t, x) \leq \gamma(|x|)$ for all $t \geq 0$ and $x \in B_h$ for some $h > 0$ ($x \in \mathsf{R}^n$).

Note that a continuous function $V(t, x) : \mathsf{R}^+ \times B_h \to \mathsf{R}$ ($V(t, x) : \mathsf{R}^+ \times \mathsf{R}^n \to \mathsf{R}$) is decrescent if and only if there exists a positive definite function on B_h (on R^n), such that $|V(t, x)| \leq w(x)$ for all $x \in B_h$ ($x \in \mathsf{R}^n$) and

$t \geq 0$. Also, any time independent function that is positive or negative definite is decrescent.

Definition 2.22: A continuous function $V(t, x) : R^+ \times R^n \to R$ is said to be **radially unbounded** if $V(t, 0) = 0$ for $t \geq 0$ and there exists a function $\gamma \in \mathcal{K}_\infty$ such that $V(t, x) \geq \gamma(|x|)$ for all $t \geq 0$ and $x \in R^n$.

Also, note that a continuous function $w(x) : R^n \to R$ is said to be radially unbounded if $w(0) = 0$, $w(x) > 0$ for all $x \in R^n - \{0\}$, and $w(x) \to \infty$ as $|x| \to \infty$. Hence, a continuous function $V(t, x) : R^+ \times R^n \to R$ is said to be radially unbounded if $V(t, 0) = 0$ for $t \geq 0$ and there exists a radially unbounded function $w(x)$ such that $V(t, x) \geq w(x)$ for all $t \geq 0$ and $x \in R^n$.

Example 2.11 Suppose we define $V(x) = x^\top P x$ where $P = P^\top$ and $P > 0$ is positive definite. Let $\gamma_1(y) = \lambda_{\min}(P)y^2$ and $\gamma_2(y) = \lambda_{\max}(P)y^2$. Notice that $V(x)$ is positive definite, decrescent, and radially unbounded, since $\gamma_1, \gamma_2 \in \mathcal{K}_\infty$, and

$$\gamma_1(|x|) \leq x^\top P x \leq \gamma_2(|x|) \tag{2.53}$$

(by the Rayleigh-Ritz inequality in (2.23)). △

2.5.2 Conditions for Stability

Let $x_e = 0$ be an isolated equilibrium point of (2.43). Assume that a unique solution exists to the differential equation in (2.43) on $x \in B_h$ for some $h > 0$ for local results, or on $x \in R^n$ for global results. Below, we let $V : R^+ \times B_h \to R$ for some $h > 0$ (for local results) or $V : R^+ \times R^n \to R$ for global results be a continuously differentiable function (i.e., it has continuous first order partial derivatives with respect to x and t).

Lyapunov's direct method provides for the following ways to test for stability. The first two are strictly for local properties while the last two have local and global versions.

- **Stable:** If $V(t, x)$ is continuously differentiable, positive definite, and $\dot{V}(t, x) \leq 0$, then $x_e = 0$ is stable.

- **Uniformly stable:** If $V(t, x)$ is continuously differentiable, positive definite, decrescent, and $\dot{V}(t, x) \leq 0$, then $x_e = 0$ is uniformly stable.

- **Uniformly asymptotically stable:** If $V(t, x)$ is continuously differentiable, positive definite, and decrescent, with negative definite $\dot{V}(t, x)$, then $x_e = 0$ is uniformly asymptotically stable (uniformly asymptotically stable in the large if all these properties hold globally).

Or, said another way for the local case: if there exists a continuously differentiable $V(t,x)$ and $\gamma_1, \gamma_2, \gamma_3 \in \mathcal{K}$ defined on $[0,r)$ for some $r > 0$, such that

$$\gamma_1(|x|) \leq V(t,x) \leq \gamma_2(|x|) \tag{2.54}$$

$$\dot{V}(t,x) \leq -\gamma_3(|x|) \tag{2.55}$$

for all $t \geq 0$ and $x \in B_h$ for some $h > 0$, then $x_e = 0$ is uniformly asymptotically stable.

Similarly, we can state the global case as: if there exists a continuously differentiable $V(t,x)$ and $\gamma_1, \gamma_2, \gamma_3 \in \mathcal{K}$ defined on $[0,\infty)$ where $\gamma_3 \in \mathcal{K}_\infty$, and Equations (2.54) and (2.55) hold for all $x \in \mathsf{R}^n$ and $t \geq 0$, then $x_e = 0$ is uniformly asymptotically stable in the large.

In addition, the **LaSalle-Yoshizawa theorem** tells us that if there exists a continuously differentiable $V(t,x)$ and $\gamma_1, \gamma_2 \in \mathcal{K}_\infty$ such that (2.54) holds for all $x \in \mathsf{R}^n$ and $t \geq 0$, and

$$\dot{V}(t,x) \leq -W(x) \leq 0$$

for all $x \in \mathsf{R}^n$ and $t \geq 0$, where W is a continuous function (i.e., positive semidefinite), then the solutions of (2.43) are uniformly bounded and

$$\lim_{t\to\infty} W(x(t)) = 0.$$

If, in addition, $W(x)$ is positive definite, then $x_e = 0$ is uniformly asymptotically stable in the large.

- **Exponentially stable:** If there exists a continuously differentiable $V(t,x)$ and $c, c_1, c_2, c_3 > 0$ such that

$$c_1|x|^c \leq V(t,x) \leq c_2|x|^c \tag{2.56}$$

$$\dot{V}(t,x) \leq -c_3|x|^c \tag{2.57}$$

for all $x \in B_h$ and $t \geq 0$, then $x_e = 0$ is exponentially stable. If there exists a continuously differentiable $V(t,x)$ and Equations (2.56) and (2.57) hold for some $c, c_1, c_2, c_3 > 0$ for all $x \in \mathsf{R}^n$ and $t \geq 0$, then $x_e = 0$ is exponentially stable in the large.

Example 2.12 As an example, consider $\dot{x} = -x^3$ which has an equilibrium $x_e = 0$. Choose $V(x) = \frac{1}{2}x^2$, $\gamma_1(y) = \frac{1}{2}y^2$, and $\gamma_2(y) = y^2$, so that $\gamma_1, \gamma_2 \in \mathcal{K}_\infty$, and (2.54) holds so that V is positive definite, decrescent, and radially unbounded. Notice that $\dot{V}(x) = x\dot{x} = -x^4 \leq 0$

so $x_e = 0$ is uniformly stable. However, also note that $\dot{V}(x) < 0$ if $x \neq 0$ and $\dot{V}(x) = 0$ for $x = 0$ so that $\dot{V}(x)$ is negative definite and hence $x_e = 0$ is uniformly asymptotically stable in the large. It is interesting to note that $x_e = 0$ of $\dot{x} = -x^3$ is *not* exponentially stable. △

Example 2.13 Consider

$$\dot{x}_1 = -x_1 - x_2 \tag{2.58}$$

$$\dot{x}_2 = x_1 - x_2 \tag{2.59}$$

which has a state $x = [x_1, x_2]^\top$ and an equilibrium $x_e = 0$. Let $V(x) = x_1^2 + x_2^2$, which has $V(x) = 0$, and is continuously differentiable. If we pick $\gamma_1(y) = \gamma_2(y) = y^2$ (which are defined on $[0, \infty)$), $\gamma_1, \gamma_2 \in \mathcal{K}$, and both (2.54) and (2.56) hold (in (2.56) pick $c_1 = c_2 = 1$) on all $x \in \mathbb{R}^n$ so V is positive definite and decrescent. Notice that

$$\begin{aligned} \dot{V} &= 2x_1(-x_1 - x_2) + 2x_2(x_1 - x_2) &(2.60)\\ &= -2x_1^2 - 2x_2^2. &(2.61) \end{aligned}$$

Choose $\gamma_3(y) = 2y^2$ so $\gamma_3 \in \mathcal{K}$, and $\gamma_3 \in \mathcal{K}_\infty$, and we see that $\dot{V}(x) \leq -\gamma_3(|x|)$ for all $x \in \mathbb{R}^n$ so that the equilibrium $x_e = 0$ is uniformly asymptotically stable in the large and also exponentially stable in the large. △

The last example studies stability of a simple two-dimensional linear time-invariant system. There are, in fact, many stability results for the general n-dimensional case for linear time invariant systems and some of these are outlined in Section 2.7.

Finally, note that in stability analysis it is sometimes convenient to use a **Lyapunov-like function** that satisfies all but some properties of a Lyapunov function, then combine the analysis with other properties of the system to conclude convergence of some signals. For instance, later in our stability proofs for adaptive systems we will augment our analysis with boundedness concepts to prove properties of asymptotic tracking.

2.5.3 Conditions for Boundedness

Suppose that there exists a specified function $V(t, x)$ defined on $|x| \geq R$ (where R may be large) and $t \geq 0$ that is continuously differentiable (i.e., it has continuous first order partial derivatives with respect to x and t). Assume that unique solutions exist to the underlying differential equation over all of $\mathbb{R}^n$.

- **Uniform boundedness:** If there exists a continuously differentiable $V(t,x)$ and $\gamma_1, \gamma_2 \in \mathcal{K}_\infty$ such that

$$\gamma_1(|x|) \leq V(t,x) \leq \gamma_2(|x|) \tag{2.62}$$

$$\dot{V}(t,x) \leq 0 \tag{2.63}$$

for all $|x| \geq R$ and $t \geq 0$ then the solutions to the differential equation are uniformly bounded. Notice that this is less restrictive than the LaSalle-Yoshizawa theorem for uniform boundedness since we only need $\dot{V}(t,x) \leq 0$ for all $|x| \geq R$ for some R, not on all R^n.

- **Uniform ultimate boundedness:** If there exists a continuously differentiable $V(t,x)$, $\gamma_1, \gamma_2 \in \mathcal{K}_\infty$, and $\gamma_3 \in \mathcal{K}$ defined on $[0,\infty)$ such that

$$\gamma_1(|x|) \leq V(t,x) \leq \gamma_2(|x|) \tag{2.64}$$

$$\dot{V}(t,x) \leq -\gamma_3(|x|) \tag{2.65}$$

for all $|x| \geq R$ and $t \geq 0$ then the solutions to the differential equation are uniformly ultimately bounded.

Example 2.14 As an example, consider the system

$$\dot{x} = f(t,x), \tag{2.66}$$

where there are known class $\mathcal{K}$ functions γ_1, γ_2 such that

$$\gamma_1(|x|) \leq V(t,x) \leq \gamma_2(|x|) \tag{2.67}$$

$$\dot{V} \leq -k_1 V + k_2 \tag{2.68}$$

and $k_1, k_2 > 0$. We wish to find some γ_3 such that $\dot{V} \leq -\gamma_3(|x|)$ when $|x| \geq R$, proving that the trajectory $x(t)$ is uniformly ultimately bounded.

Choose some ϵ such that $0 < \epsilon < 1$. Then

$$\begin{aligned} \dot{V} &\leq -\epsilon k_1 V - (1-\epsilon)k_1 V + k_2 \\ &\leq -\epsilon k_1 \gamma_1(|x|) - (1-\epsilon)k_1 V + k_2. \end{aligned} \tag{2.69}$$

Choosing $\gamma_3 = \epsilon k_1 \gamma_1(|x|)$ we see that

$$\dot{V} \leq -\gamma_3(|x|) - (1-\epsilon)k_1 V + k_2. \tag{2.70}$$

Now if $|x| \geq R$ where

$$R = \gamma_1^{-1}\left(\frac{k_2}{(1-\epsilon)k_1}\right),$$

then $(1-\epsilon)k_1 V \geq (1-\epsilon)k_1\gamma_1(R) = k_2$. Thus

$$\dot{V} \leq -\gamma_3(|x|) \tag{2.71}$$

for all $|x| \geq R$ so the solutions of (2.66) are uniformly ultimately bounded. △

Notice that we do not automatically get the explicit value of B in Definition 2.16; all we know from the theorem is that its value exists. Often, however, it is possible from the application at hand to determine the explicit value of B.

The following lemma may be helpful in determining the ultimate bound when provided a differential inequality satisfying Equations (2.67) and (2.68).

Lemma 2.1: *If $V(t,x)$ is positive definite and $\dot{V} \leq -k_1 V + k_2$ where $k_1 > 0$ and $k_2 \geq 0$ are bounded constants, then*

$$V(t,x) \leq \frac{k_2}{k_1} + \left(V(0) - \frac{k_2}{k_1}\right) e^{-k_1 t}$$

for all t.

Proof: Let $\dot{\eta} = -k_1\eta + k_2$ and choose $\eta(0) = V(0) \geq 0$ so

$$\eta = \frac{k_2}{k_1} + \left(V(0) - \frac{k_2}{k_1}\right) e^{-k_1 t}.$$

Since $\dot{V} \leq \dot{\eta}$ (V decreases at least as quickly as η) and $V(0) = \eta(0)$, we find $V(t,x) \leq \eta(t)$ for all t, which completes the proof. ■

If $\dot{V} \leq -k_1 V + k_2$, then Lemma 2.1 may be used to show that as $t \to \infty$ we find $|V| \leq k_2/k_1$. Moreover, if $\gamma_1(|x|) \leq V(t,x)$, then

$$\lim_{t\to\infty} |x| \leq \gamma_1^{-1}\left(\frac{k_2}{k_1}\right). \tag{2.72}$$

2.6 Input-to-State Stability

In this section we overview a few concepts from the study of input-to-state stability. We start with definitions, then provide results that will be useful in our later analysis.

2.6.1 Input-to-State Stability Definitions

In the following we will introduce the basic notions of input-to-state stability and input-to-state practical stability (also referred to as compact input-to-state stability) which are very useful in the study of the stability properties of interconnected systems. Consider the dynamical system

$$\dot{x} = f(x, u), \tag{2.73}$$

where $x \in \mathrm{R}^n$, $u \in \mathrm{R}^m$, f is locally Lipschitz in x and u, and u, representing the input of the system, is a piece-wise continuous and bounded function of t.

Definition 2.23: System (2.73) is said to be **input-to-state stable** if, for any initial condition $x(t_0)$ and any bounded input $u(t)$, the solution $x(t)$ satisfies

$$|x(t)| \leq \beta(|x(t_0)|, t - t_0) + \gamma\left(\sup_{t_0 \leq \tau \leq t} |u(\tau)|\right) \tag{2.74}$$

for all $t \geq t_0$, where β is class $\mathcal{KL}$, and γ is class $\mathcal{K}$.

When dealing with uncertainties, it is often useful to define a more general notion of input-to-state stability.

Definition 2.24: System (2.73) is said to be **input-to-state practically stable** if, for any initial condition $x(t_0)$ and any bounded input $u(t)$, the solution $x(t)$ satisfies

$$|x| \leq \beta(|x(t_0)|, t - t_0) + \gamma\left(\sup_{t_0 \leq \tau \leq t} |u(\tau)|\right) + d \tag{2.75}$$

for all $t \geq t_0$, where β is class $\mathcal{KL}$, γ is class $\mathcal{K}$, and d is a nonnegative constant.

2.6.2 Conditions for Input-to-State Stability

The following provides useful characterizations of input-to-state stability properties, and a uniform ultimate boundedness property for interconnected systems, in terms of Lyapunov functions.

- **Input-to-State Stability:** System (2.73) is input-to-state stable if, and only if, there exists a continuously differentiable function V such that

$$\gamma_1(|x|) \leq V(x) \leq \gamma_2(|x|) \tag{2.76}$$

$$\frac{\partial V}{\partial x} f(x, u) \leq -\gamma_3(|x|), \quad \forall |x| \geq \psi(|u|) > 0 \quad , \tag{2.77}$$

  where γ_1, γ_2 are class $\mathcal{K}_\infty$and γ_3, ψ are class $\mathcal{K}$.

- **Input-to-State Practical Stability:** System (2.73) is input-to-state practically stable if, and only if, there exists a continuously differentiable function V and constants $c > 0$, $d \geq 0$ such that

$$\gamma_1(|x|) \leq V(x) \leq \gamma_2(|x|) \tag{2.78}$$

$$\tfrac{\partial V}{\partial x} f(x,u) \leq -cV + \psi(|u|) + d, \tag{2.79}$$

where γ_1, γ_2, and ψ are class $\mathcal{K}_\infty$.

- **Uniform Ultimate Boundedness:** Consider $\dot{x} = f(x,y)$ and $\dot{y} = g(x,y)$, where f and g are locally Lipschitz, $x \in \mathsf{R}^n$ and $y \in \mathsf{R}^m$. If there exist continuously differentiable functions $V_x : \mathsf{R}^n \to \mathsf{R}$ and $V_y : \mathsf{R}^m \to \mathsf{R}$ with $\gamma_{x1}(|x|) \leq V_x \leq \gamma_{x2}(|x|)$ and $\gamma_{y1}(|y|) \leq V_y \leq \gamma_{y2}(|y|)$ such that

$$\dot{V}_x \leq 0 \quad \text{when } V_x \geq V_r \tag{2.80}$$

$$\dot{V}_y \leq -\gamma_{y3}(|y|), \quad \forall |y| \geq \psi(|x|), \tag{2.81}$$

where $\gamma_{x1}, \gamma_{x2}, \gamma_{y1}, \gamma_{y2}$ are class-$\mathcal{K}_\infty$, γ_{y3} and ψ are class-$\mathcal{K}$, and $V_r \geq 0$, then x and y are uniformly ultimately bounded. To see why this is the case note the following: From (2.80) we find $V_x \leq \max(V_x(0), V_r)$ so

$$\gamma_{x1}(|x|) \leq \max(V_x(0), V_r).$$

Thus $|x| \leq d$ for all t, where $d = \gamma_{x1}^{-1} \circ \max(V_x(0), V_r)$. If $|y| \geq \psi(d)$, then $\dot{V}_y \leq -\gamma_{y3}(|y|) \leq 0$. Thus if $V_y \geq \gamma_{y2} \circ \psi(d)$ (which implies $|y| \geq \psi(d)$), then $\dot{V}_y \leq 0$ so V_y is bounded. Thus $V_y \leq \max(V(0), \gamma_{y2} \circ \psi(d))$ so

$$|y| \leq \gamma_{y1}^{-1} \circ \max(V_y(0), \gamma_{y2} \circ \psi(d))$$

for all t.

As a simple example for input-to-state stability, consider the scalar ordinary differential equation

$$\dot{x} = -ax + bu, \tag{2.82}$$

where $a > 0$, so that when $u = 0$ the origin is an exponentially stable in the large equilibrium. Is this system input-to-state stable? Choose $V(x) = \frac{1}{2}x^2$, $\gamma_1 = \gamma_2 = \frac{1}{2}x^2$. Note that

$$\frac{\partial V}{\partial x} f(x,u) = x(-ax + bu) = -ax^2 + bxu. \tag{2.83}$$

Choose $\psi(|u|) = |u|$. Now we must show that when $|x| \geq \psi(|u|)$ we can find an appropriate γ_3. Note that

$$-ax^2 + bxu \leq -ax^2 + |b||x||u| \leq -ax^2 + |b|x^2 = -(a - |b|)x^2 \tag{2.84}$$

so that if $\gamma_3 = (a - |b|)x^2$, and $a > |b|$, then (2.82) is input-to-state stable.

As another simple example, for the interconnected system case, consider the two-dimensional ordinary differential equation

$$\begin{aligned}\dot{x} &= -ax \\ \dot{y} &= -dy + cx,\end{aligned}$$

where $a > 0$ and $d > 0$. We can think of the $x-$subsystem as generating trajectories to input to the $y-$subsystem. Choose $V_x = \gamma_{x1} = \gamma_{x2} = \frac{1}{2}x^2$, and $V_y = \gamma_{y1} = \gamma_{y2} = \frac{1}{2}y^2$. Choose $\psi(|x|) = |x|$. Can we find a $\gamma_{y3}(|y|)$ and V_r? Notice that for any $V_r \geq 0$, $\dot{V}_x = -ax^2$ and that when $|y| \geq |x|$

$$\dot{V}_y = -dy^2 + cxy \leq -dy^2 + |c||x||y| \leq -dy^2 + |c|y^2 = -(d - |c|)y^2.$$

Hence, choosing $\gamma_{y3} = (d - |c|)y^2$, and $d > |c|$, then $|x|$ and $|y|$ are uniformly ultimately bounded. Intuitively, we wee that if the $x-$subsystem generates a bounded input to an input-to-state stable $y-$subsystem, we find that the $y-$subsystem will generate bounded trajectories.

2.7 Special Classes of Systems

Here, we explain how certain analysis and results hold when we restrict our attention to autonomous (time-invariant) or linear time-invariant systems.

2.7.1 Autonomous Systems

If we assume that in (2.43) f does not depend explicitly on time t, then

$$\dot{x}(t) = f(x(t)) \tag{2.85}$$

is the system under consideration and several simplifications are possible; in particular, some sufficient conditions for asymptotic stability exist that are sometimes easier to satisfy than the previous ones. First, note that for (2.85) we only need a Lyapunov function that does not depend on time, and since all positive definite functions are automatically decrescent we can ignore the need for $V(x)$ to be decrescent in all the stability conditions considered. Second, recall that for (2.85) uniform stability is equivalent to stability in the sense of Lyapunov and uniform asymptotic stability (in the large) is equivalent to asymptotic stability (in the large). Moreover, some invariance theorems due to LaSalle hold. Next, we overview a special case of his more general invariance theorem that proves to be useful in the construction of asymptotically stable adaptive systems.

We will call a set $\Omega \subset \mathsf{R}^n$ invariant with respect to (2.85) if every solution $x(t, x_0)$ of (2.85) with $x(0, x_0) \in \Omega$ has $x(t, x_0) \in \Omega$ for all $t \geq 0$. Assume that (2.85) possesses unique solutions for all $x_0 \in D \subset \mathsf{R}^n$ where D contains the origin. Suppose that there exists a continuously differentiable, positive definite, and radially unbounded function $V(x) : D \to \mathbb{R}^+$ with $\dot{V}(x) \leq 0$ on D. If the origin is the only invariant subset of

$$E = \left\{x \in D : \dot{V}(x) = 0\right\}$$

with respect to (2.85), then the equilibrium $x_e = 0$ of (2.85) is asymptotically stable. Also, if in this case, $D = \mathbb{R}^n$ then the equilibrium $x_e = 0$ of (2.85) is asymptotically stable in the large.

Example 2.15 As an example, consider

$$\dot{x}(t) = -\mathrm{sgn}(x(t))x^2(t) = f(x), \tag{2.86}$$

where we define $\mathrm{sgn}(x) = 1$ if $x \geq 0$ and $\mathrm{sgn}(x) = -1$ if $x < 0$. In this case, f is Lipschitz continuous so a unique solution exists for each x_0. Also, $x_e = 0$ is an isolated equilibrium of (2.86). Choose $V(x) = \frac{1}{2}x^2$, which is positive definite and radially unbounded on $D = \mathbb{R}$. Notice that

$$\dot{V}(x) = x\dot{x} = -\mathrm{sgn}(x)x^3 = -x^2(x\mathrm{sgn}(x)) = -x^2|x| \leq 0$$

for all $x \in \mathbb{R}$. Notice, also that

$$E = \left\{x \in D : \dot{V}(x) = 0\right\} = \{0\}$$

so the origin is the only nonempty subset of E so clearly it can be the only invariant subset of E, so $x_e = 0$ is asymptotically stable in the large.

As another approach to study convergence suppose we use Barbalat's lemma. Notice that since $\dot{V}(x) \leq 0$ for all $x \in \mathbb{R}$, $\dot{V}(x) \in \mathcal{L}_\infty$ for all $x \in \mathbb{R}$. Also, since $V(x(t, x_0)) \geq 0$ (i.e., it is bounded from below) and is nonincreasing ($\dot{V}(x) \leq 0$) it has a limit so $V(x(t, x_0)) \in \mathcal{L}_\infty$ and hence $x(t, x_0) \in \mathcal{L}_\infty$ for all $x_0 \in \mathbb{R}$. Also, since the system in (2.86) is Lipschitz continuous, $\dot{x} \in \mathcal{L}_\infty$. In general,

$$V(x(t, x_0)) = V(x_0) + \int_0^t \dot{V}(x(\tau, x_0))d\tau$$

and since $V(x(t, x_0)) \in \mathcal{L}_\infty$, there exists a $\beta > 0$ such that

$$\int_0^t |x(\tau, x_0)|^3 d\tau = V(x_0) - V(x(t, x_0)) < \beta.$$

This implies that $x \in \mathcal{L}_3$ so that by Barbalat's lemma $\lim_{t \to \infty} x(t) = 0$.

Yet another approach to stability analysis for this system is to use the LaSalle-Yoshizawa theorem and note that $W(x) = x^2|x|$ which is positive semidefinite so $\lim_{t\to\infty} W(x(t)) = \lim_{t\to\infty} x^2|x| = 0$ and this can only happen if $x \to 0$. Of course, since $W(x)$ is also positive definite we can conclude that $x_e = 0$ is uniformly asymptotically stable in the large. In this simple example we obtain the same stability result from both parts of the LaSalle-Yoshizawa theorem; in general this will not be the case. △

2.7.2 Linear Time-Invariant Systems

In the case where (2.85) is a linear system we can obtain additional stability results. In particular, consider the linear time invariant ordinary differential equation

$$\dot{x}(t) = Ax(t), \tag{2.87}$$

where $x \in \mathsf{R}^n$. The equilibrium $x_e = 0$ being asymptotically stable in the large is *equivalent* to the following three statements:

1. All eigenvalues of A are in the open left half plane (A is **Hurwitz**).

2. The equilibrium $x_e = 0$ is exponentially stable in the large.

3. For every $n \times n$ matrix Q such that $Q = Q^\top$ and $Q > 0$, the **Lyapunov matrix equation**

   $$A^\top P + PA = -Q$$

   has a unique solution matrix P such that $P = P^\top$ and $P > 0$.

Notice that if we know that $x_e = 0$ of (2.87) is asymptotically stable in the large, then we know that if we are given $P > 0$, there is a unique associated $Q > 0$. We will use this fact later in some stability proofs. For illustration of the above ideas consider Example 2.13 and note that testing stability of a linear system via simple examination of the eigenvalues is particularly attractive as widely available computational tools can be employed for finding eigenvalues.

Special results related to boundedness also hold when it is assumed that the system is linear and time-invariant. For instance, consider the dynamical system

$$\dot{x} = Ax + bu, \tag{2.88}$$

where A is a Hurwitz matrix. Then $|x(t)| \leq \psi(t, |u|)$ for all t, where $\psi : \mathsf{R}^+ \times \mathsf{R}^+ \to \mathsf{R}$ is bounded for any bounded u and nonincreasing with respect to $|u|$ for each fixed t. To see this, note that the solution $x(t)$ of the linear differential equation (2.88) is given by the convolution integral

$$x(t) = e^{At}x(0) + \int_0^t e^{A(t-\tau)}bu(\tau)d\tau. \tag{2.89}$$

Since A is assumed to be Hurwitz, we have that $|e^{At}| \leq c_1 e^{-c_2 t}$ for some positive scalars c_1 and c_2. By using this inequality in (2.89) we get

$$|x(t)| \leq c_1 e^{-c_2 t}|x(0)| + \int_0^t c_1 e^{-c_2(t-\tau)}|u(\tau)|d\tau. \tag{2.90}$$

By noting that (2.90) is bounded for any bounded $|u|$ and nonincreasing with respect to $|u|$ we obtain the desired result by setting

$$\psi(t, |u|) = c_1 e^{-c_2 t}|x(0)| + \int_0^t c_1 e^{-c_2(t-\tau)}|u(\tau)|d\tau. \tag{2.91}$$

The following example provides a simple method to find values for c_1 and c_2 defined above so that $|e^{At}| \leq c_1 e^{-c_2 t}$.

Example 2.16 Consider the system defined by $\dot{x} = Ax$ where A is Hurwitz. Given that the solution of this unforced system is $x(t) = e^{At}x(0)$, we want to find constants c_1 and c_2 which satisfy $|e^{At}| \leq c_1 e^{-c_2 t}$. If we let $V = x^\top Px$ where P is a symmetrix positive definite matrix, then

$$\dot{V} = x^\top(PA + A^\top P)x. \tag{2.92}$$

Now choose P to satisfy the Lyapunov matrix equation $PA + A^\top P = -I$ so that

$$\dot{V} = -x^\top x \leq -\frac{V}{\lambda_{\max}(P)}. \tag{2.93}$$

Here we have used the Rayleigh-Ritz inequality (2.23). Solving the above differential inequality, we find

$$V \leq V(0)e^{-k_1 t}, \tag{2.94}$$

where $k_1 = 1/\lambda_{\max}(P)$. Since

$$|x|^2 \leq \frac{1}{\lambda_{\min}(P)}V \leq \frac{V(0)}{\lambda_{\min}(P)}e^{-k_1 t}, \tag{2.95}$$

we may let $c_1 = \sqrt{x^\top(0)Px(0)/\lambda_{\min}(P)}$ and $c_2 = 1/(2\lambda_{\max}(P))$. △

2.8 Summary

Within this chapter we have presented various mathematical tools that have been found to be useful in the construction of adaptive systems. In particular, we have covered the following topics:

- Vectors, vector norms, and their properties.
- Matrices, induced norms, and their properties
- Positive definite matrices and their properties.
- Signals, signal norms, and their properties.
- Continuity, differentiability, Barbalat's lemma and other convergence properties.
- Stability definitions: stable in the sense of Lyapunov, uniformly stable, asymptotically stable (in the large), exponentially stable (in the large).
- Boundedness definitions: Lagrange stability, uniform boundedness, uniform ultimate boundedness.
- Lyapunov's direct method for stability and boundedness analysis (including results for all the stability and boundedness definitions).
- The LaSalle-Yoshizawa theorem and a special case of LaSalle's invariance theorem.
- Input-to-state stability definitions and analysis.

This provides a list of the main topics covered in this chapter. You should make sure that you understand all these topics before proceeding to any later chapter (unless perhaps you are not at all concerned about proving stability of the various adaptive schemes).

2.9 Exercises and Design Problems

Exercise 2.1 (Norms, Norm Properties) Prove that if $x \in \mathrm{R}^n$, $|x|_\infty$ is a norm. Prove that $|x|_\infty \leq |x|_2 \leq \sqrt{n}|x|_\infty$.

Exercise 2.2 (Positive Definite Matrices) If

$$P = \begin{bmatrix} 10 & 1 \\ 1 & 20 \end{bmatrix},$$

is $P > 0$? What is $\|P\|_2$? What is $\|P^{-1}\|_2$? If

$$P = \begin{bmatrix} 10 & \epsilon \\ \epsilon & 20 \end{bmatrix},$$

then what range of values can ϵ take on and still ensure that $P > 0$?

Exercise 2.3 (Norm Properties) Fill in the details of the proof of Example 2.5 by showing that (2.21) holds.

Exercise 2.4 (Signal Norms) Prove that $e^{-5t} \in \mathcal{L}_2$, $\cos(t) \notin \mathcal{L}_2$, $2 - 2e^{-t} \in \mathcal{L}_\infty$, $a^k \in \ell_2$ (for $0 \le a < 1$), and $e^{-2t} \in \mathcal{S}_2(0)$.

Exercise 2.5 (Signals, Systems, Norms) Let

$$\hat{G}(s) = \frac{2}{s+3}$$

be a system with an input $u(t) = e^{-2t}$. Suppose that the output of the system is $y(t)$. What is $\|\hat{G}\|_\infty$ and $\|y\|_2$?

Exercise 2.6 (Continuity) Prove that

$$f(t) = 2 - e^{-t}$$

is uniformly continuous on $D = \mathbb{R}^+$. Prove that

$$f(t) = e^{-2t}$$

is Lipschitz continuous on $D = \mathbb{R}^+$. What is the value of the Lipschitz constant in this case?

Exercise 2.7 (Convergence, Barbalat's Lemma) Suppose in Example 2.7 that $f(x) = -2x^3$. Show $x(t) \to 0$ as $t \to \infty$.

Exercise 2.8 (Stability, Asymptotic Stability) Suppose you are given the scalar differential equation

$$\dot{x} = -ax - bx^3,$$

where $a, b > 0$. Find an equilibrium point. Is it isolated? Is the equilibrium uniformly stable? If so, prove it. Is the equilibrium uniformly asymptotically stable in the large? If so, prove it.

Exercise 2.9 (Exponential Stability) Suppose you are given the scalar differential equation

$$\dot{x} = -a\left(b + e^{-t}\right)x,$$

where $a, b > 0$. Find an equilibrium point. Is it isolated? Is the equilibrium exponentially stable in the large? If so, prove it.

Exercise 2.10 (Lyapunov Stability) Study the trajectory of $V = x_1^2 + x_2^2 + x_3^2$ to prove that the system

$$\begin{aligned} \dot{x}_1 &= -x_1 + x_2 \\ \dot{x}_2 &= -2x_2 + x_3 \\ \dot{x}_3 &= -x_3 - x_3^3 \end{aligned}$$

is exponentially stable.

Exercise 2.11 (Instability) Study the trajectory of $V = x_1^2 + x_2^2$ to prove that the system

$$\begin{aligned} \dot{x}_1 &= x_1 + x_2 \\ \dot{x}_2 &= -x_1 + 2x_2 \end{aligned}$$

is unstable.

Exercise 2.12 (Another Class of Signal Norms) Another useful metric to quantify the size of a signal may be defined by using

$$\mathcal{S}_p(c) = \left\{ x(t) \in \mathsf{R}^n : \left(\lim_{t\to\infty} \frac{1}{t} \int_0^t |x(\tau)|^p d\tau \right)^{1/p} \leq c, \quad c \geq 0 \right\}.$$

We say that a signal $x(t)$ is small in the root mean squared sense if $x \in \mathcal{S}_2(c)$ for some finite $c \geq 0$. We say a function $x(t)$ is small on average if $x \in \mathcal{S}_1(c)$. As an example, consider the signal defined by $x(t) = e^{-2t} + 0.1\sin(t)$. This signal is not in $\mathcal{L}_1$, but $x \in \mathcal{S}_1(0.1)$. Also, $\mathcal{S}_p(a) \subseteq \mathcal{S}_p(b)$ for any $a, b \in \mathsf{R}$ such that $0 \leq a \leq b$, and if $x \in \mathcal{L}_p$ then $x \in \mathcal{S}_p(0)$. As an example, $e^{-t} \in \mathcal{L}_p$, so $e^{-t} \in \mathcal{S}_p(0)$.

Chapter 3

Neural Networks and Fuzzy Systems

3.1 Overview

Few technologies have been used for such a vast variety of applications as neural networks and fuzzy systems. They have been found to be truely interdisciplinary tools appearing in the fields of economics, business, science, psychology, biology, and engineering to name a few.

Based upon the structure of a biological nervous system, artificial neural networks use a number of interconnected simple processing elements ("neurons") to accomplish complicated classification and function approximation tasks. The ability to adjust the network parameters (weights and biases) makes it possible to "learn" information about a process from data, whether it is describing stock trends or the relation between an actuator input and some sensor data. Neural networks typically have the desirable feature that little knowledge about a process is required to sucessfully apply a network to the problem at hand (although if some domain-specific knowledge is known then it can be beneficial to use it). In other words, they are typically regarded as a "black box" technique. This approach often leads to engineering solutions in a relatively short amount of time since expensive system models required by many conventional approaches are not needed. Of course, however, sufficient data is typically needed for effective solutions.

Fuzzy systems are intended to model higher level cognitive functions in a human. They are normally broken into (1) a rule-base that holds a human's knowledge about a specific application domain, (2) an inference mechanism that specifies how to reason over the rule-base, (3) fuzzification which transforms incoming information into a form that can be used by the fuzzy system, and (4) defuzzification which puts the conclusions from the inference mechanism into an appropriate form for the application at hand. Often, fuzzy systems are constructed to model how a human performs a task. They are either constructed manually (i.e., using heuristic domain-specific knowledge in a manner similar to how an expert system is constructed) or in a similar manner to how a neural network is constructed via training with data. While in the past fuzzy systems were exclusively constructed with heuristic approaches we take the view here that they are simply an alternative approximator structure with tunable parameters (e.g., input and output membership function parameters) and hence they can be viewed as a "black box approach" in the same way as neural networks can. Fuzzy systems do, however, sometimes offer the additional beneficial feature of a way to incorporate heuristic information; you simply specify some rules via heuristics and tune the others using data (or even fine tune the ones specified heuristically via the data). In other words, it is sometimes easier to specify a good initial guess for the fuzzy system.

In this chapter we define some basic fuzzy systems and neural networks, in fact, ones that are most commonly used in practice. We do not spend time discussing the heuristic construction of fuzzy systems since this is treated in detail elsewhere. In the next chapter we will provide a variety of optimization methods that may be used to help specify the parameters used to define neural networks and fuzzy systems.

3.2 Neural Networks

The brain is made up of a huge number of different types of neurons interconnected through a complex network. A typical neuron is composed of an input region, which contains numerous small branches called dendrites. The neuron body contains the nucleus and other cell components, while the axon is used to transmit impulses to other cells (see Figure 3.1).

When an impulse is received at the voltage-sensitive dendrite, the cell membrane becomes depolarized. If this potential reaches the cell threshold potential, via pulses received at possibly many dendrites, an action potential which lasts only a millisecond or two is triggered and an impulse is sent out to other neurons via the axon. The magnitude of the impulse which is sent is not dependent upon the magnitude of the voltage potential which triggered the action.

In our model of an artificial neuron, as is typical, we will preserve the underlying structure, but will make convenient simplifications in the actual functional description of its action. We will, for example, typically assume that the magnitude of the output is dependent upon the magnitude of the inputs. Also, we will not assume the inputs and outputs are impulses. Instead, a smoothly varying input will cause a smoothly varying output.

The brain consists of a network of various neurons through which impulses are transmitted, from the axon of one neuron to the dendrites of another. The impulses may be fed back to previous cells within the network. Artificial neural networks in which information may be fed back to

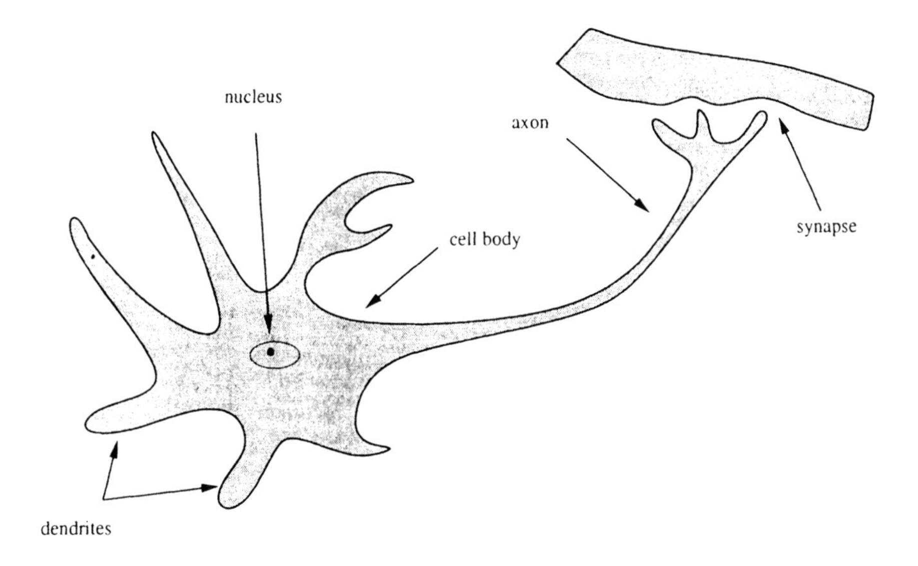

Figure 3.1. Simple representation of a neuron.

previous neurons are called recurrent neural networks, whereas networks in which information is allowed to proceed in a single direction are called feedforward neural networks. Examples of recurrent and feedforward neural networks are shown in Figure 3.2 where each circle represents a neuron, and the lines represent the transmission of signals along axons. To simplify analysis, we will focus on the use of feedforward neural networks for the estimation and control schemes in the chapters to follow.

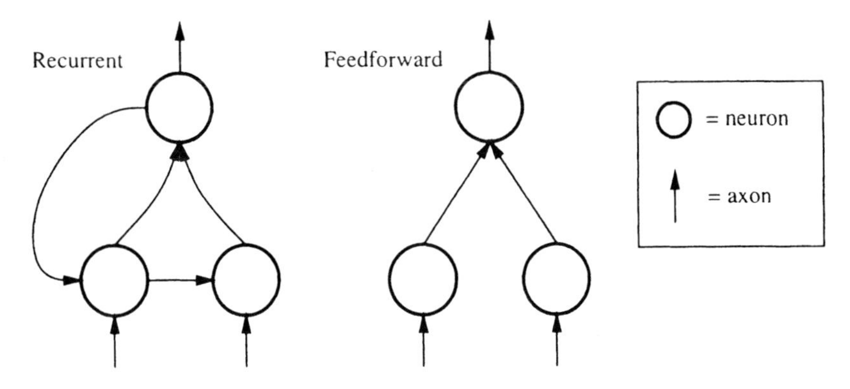

Figure 3.2. Examples of recurent and feedforward neural networks.

The input vector to the neuron is $x = [x_1, \ldots, x_n]^\top$, where x_i, $1 \leq i \leq n$, is the ith input to the neuron. Though in a biological system x_i represents a voltage caused by an electrochemical reaction, in this artificial framework,

x_i may be a variable representing, for example, the value for a temperature or pressure sensor.

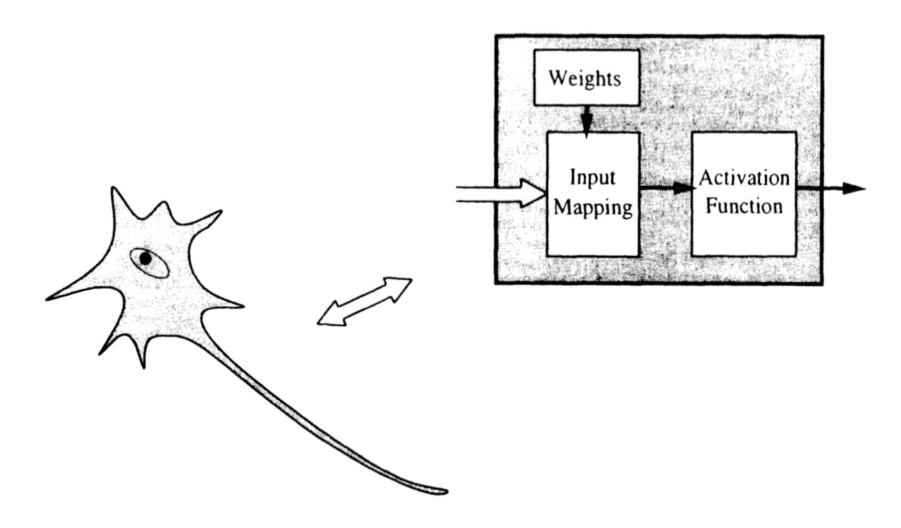

Figure 3.3. Schematic of a neuron and artificial mathematical representation.

Figure 3.3 shows a biological neuron along with our mathematical representation. We will consider two distinct mathematical operations within our model. First, the neuron inputs will undergo an input mapping in which each of the dentrite inputs is combined. After the input mapping, the signal is passed through an activation function to produce the output.

3.2.1 Neuron Input Mappings

The neuron input mapping takes a vector of inputs, $x \in \mathsf{R}^n$, and transforms these into a scalar, denoted by s. The input mapping is dependent upon a vector of weights $w = [w_1, \ldots, w_n]^\top$, which are selected according to some past knowledge and may be allowed to change over time based upon new neural stimuli. Using both the weights and inputs, a mapping is performed to quantify the relation between w and x. This relationship, for example, may describe the "colinearity" of w and x (discussed below). We will denote the input mapping by

$$s = w \odot x.$$

A number of input mappings have been used in the neural network literature, the most popular being the inner product and Euclidean input mappings. Another useful but not as widely used input mapping is the weighted average.

Inner Product: The inner product input mapping (also commonly refered to as the dot product), may be considered to be a measure of the similarity between the orientation of the vectors w and x, defined by

$$s = w \odot x = w^{\top} x. \tag{3.1}$$

This may also be expressed as $s = |w||x| \cos\theta$ where θ is the angle between w and x. Thus, if w and x are orthogonal, then $s(x) = 0$. Similarily, the inner product increases as x and w become colinear (the angle θ decreases and at $\theta = 0$, x and w are colinear).

We will find that for each of the techniques in this book, it is required that we know how a change in the weights will affect the output of the neural network. To determine this, the gradient of each neuron output with respect to the weights will be calculated. For the standard inner product input mapping, we find that

$$\frac{\partial s}{\partial w} = x^{\top}. \tag{3.2}$$

Notice that the gradient for the inner product input mapping with respect to the weights is simply the value of the inputs themselves.

Weighted Average: An input mapping that is closely related to the inner product is the weighted average defined by

$$s = w \odot x = \frac{w^{\top} x}{\sum_{j=1}^{n} x_j}. \tag{3.3}$$

Geometrically speaking, the weighted average again determines to what degree the two vectors are colinear. To ensure that the input mapping is well defined for all x (even $x = 0$), we may choose to rewrite (3.3) as

$$w \odot x = \frac{w^{\top} x}{\gamma + \sum_{j=1}^{n} |x_j|}, \tag{3.4}$$

where $\gamma > 0$. For this weighted average, the partial derivative of the neuron ouptut with respect to the weights is expressed as

$$\frac{\partial s}{\partial w} = \frac{x^{\top}}{\gamma + \sum_{j=1}^{n} |x_j|}. \tag{3.5}$$

This is again similar to the inner product, with the addition of the normalizing term in the denominator.

Euclidean: The Euclidean input mapping is defined by

$$s = w \odot x = |w - x| = \sqrt{(w - x)^{\top}(w - x)}. \tag{3.6}$$

This mapping will result in $s \geq 0$ since it is the Euclidean norm of the vector $w - x$ (recall that norms are non-negative). The Euclidean input mapping has the following gradient with respect to the weights:

$$\frac{\partial s}{\partial w} = \frac{x - w}{|w - x|}. \tag{3.7}$$

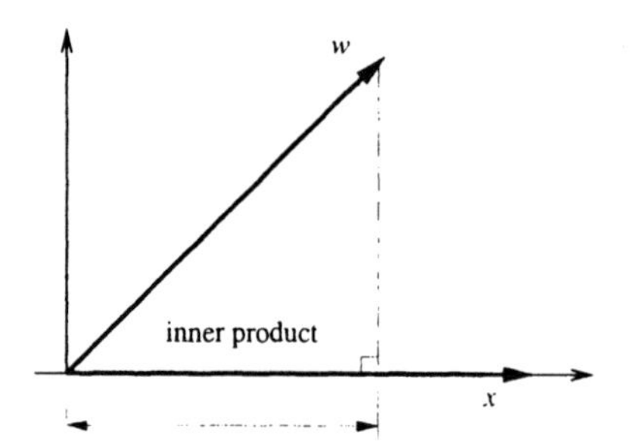

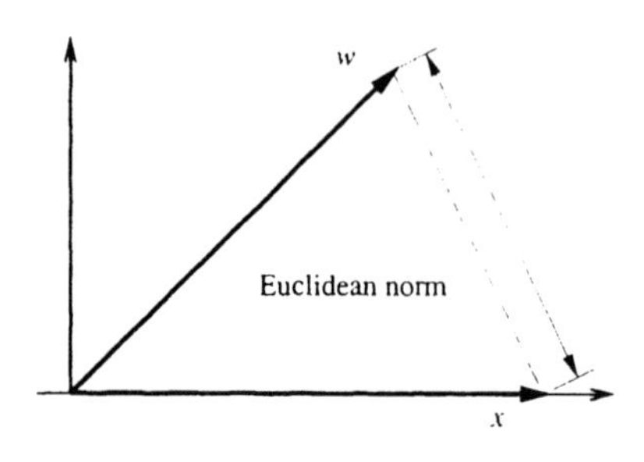

Figure 3.4. Graphical representation of the inner product and Euclidean input mappings.

A geometrical intrepretation for the inner product and Euclidean input mappings is shown in Figure 3.4. Notice that as w and x become orthogonal, the inner product vanishes while the Euclidean norm increases.

Adding a Bias Term: Along with the input vector x, an artificial neuron is often also provided a constant bias as an input so that now $x = [1, x_1, \ldots, x_n]^\top$ with $x \in \mathsf{R}^{n+1}$. We have chosen the bias to be 1 since the input weight associated with the bias may be chosen to arbitrarily scale this value. That is, the weight vector is now changed to

$$w = [w_0, w_1, \ldots, w_n]^\top$$

with w_0 corresponding to the bias input. The same neuron input mappings as described above apply to the case where a bias term is used.

3.2.2 Neuron Activation Functions

According to Figure 3.3, after the input mapping, the neuron produces an output using an activation function. This activation function transforms the value produced by the input mapping to a value which is suitable for another neuron, or possibly a value which may be understood by an external system (e.g., as an input to an actuator).

Definition 3.1: A function $\psi : \mathsf{R} \to \mathsf{R}$ which maps an input mapping to R is said to be an **activation function** if it is piecewise continuous.

Definition 3.2: A function $\psi : \mathsf{R} \to \mathsf{R}$ is said to be a **squashing function** if it is an activation function, $\lim_{x\to\infty} \psi(x) = 1$, and $\lim_{x\to-\infty} \psi(x) = 0$.

Activation functions for artificial neurons may be either **bounded** or **unbounded**. A bounded activation function is one for which $|\psi(s)| \le k$, where $k \in \mathsf{R}$ is a finite constant, and thus it has the property that even if $s \to \infty$, one is still ensured that $\psi(s) \in \mathcal{L}_\infty$. For an unbounded activation function, on the other hand, $\psi(s) \in \mathcal{L}_\infty$ does not hold for all s.

Definition 3.3: A function $\psi : D \to \mathsf{R}$ is said to be **unipolar** on D if $\psi(x) \ge 0$ for all $x \in D$ (**positive unipolar**) or if $\psi(x) \le 0$ (**negative unipolar**) for all $x \in D$.

With $\psi(\cdot)$ continuously differentiable, the change in the neuron output with respect to the neuron weights may be obtained from the chain rule as

$$\frac{\partial \psi(s)}{\partial w} = \frac{\partial \psi}{\partial s}\frac{\partial s}{\partial w}. \tag{3.8}$$

This formula will become useful when determining how to adjust the neuron weights so that the output of the squashing function changes in some desired manner. A few of the more commonly used activation functions will now be defined.

Threshold: The threshold function (or Heavyside function) is one of the original activation functions studied by McCulloch and Pitts. It is defined by

$$\psi(s) = \begin{cases} 1 & s > 0 \\ 0 & \text{otherwise.} \end{cases} \tag{3.9}$$

Since $\psi(s)$ is not continuous for the threshold function, $d\psi/ds$ is not well defined for all $s \in \mathsf{R}$. Though we will not use the threshold activation function in our adaptive schemes (because its derivative is not well defined), it will prove valuable in the function approximation proofs in Chapter 5.

Linear: The simplest of the activation functions is a linear mapping from input to output defined by

$$\psi(s) = s. \tag{3.10}$$

This is a monotonic, unbounded activation function. The gradient of the linear activation function is simply $\frac{\partial \psi(s)}{\partial s} = 1$. We will see that linear activation functions are often used to generate the outputs of multi-layered neural networks.

Saturated Linear: A variant of the linear activation function is the saturated linear activation function, defined as

$$\psi(s) = \text{sat}(s), \tag{3.11}$$

where

$$\text{sat}(x) = \begin{cases} 1 & \text{if } x > 1, \\ x & \text{if } -1 \leq x \leq 1, \\ -1 & \text{otherwise.} \end{cases} \quad . \tag{3.12}$$

This is a monotonic bounded activation function. Since the saturated linear function is not continuously differentiable (why?), often a continuously differentiable approximation is used instead. One such example is the hyperbolic tangent.

Hyperbolic Tangent: The hyperbolic tangent activation function is defined as

$$\psi(s) = \tanh(s) = \frac{1 - \exp(-2s)}{1 + \exp(-2s)}. \tag{3.13}$$

This is a monotonic, bipolar activation function whose derivative is

$$\partial\psi(s)/\partial s = 1 - \psi^2(s).$$

Sigmoid: A frequently used monotonic, unipolar squashing function is the sigmoid. In general a sigmoid is any such "s-shaped" function, and is often specified as

$$\psi(s) = \frac{1}{1 + \exp(-2s)}. \tag{3.14}$$

The gradient of this function is

$$\partial\psi(s)/\partial s = 2\psi(s)(1 - \psi(s)).$$

Note that the gradient is defined in terms of $\psi(\cdot)$ itself.

Radial Basis: One of the most popular non-monotonic activation functions is the radial basis function, commonly defined by

$$\psi(s) = \exp(-s^2/\gamma^2), \tag{3.15}$$

where $\gamma \in \mathsf{R}$. Neural networks composed entirely of radial basis functions have been used in a number of applications owing to their mathematical properties. Their function approximation properties will be discussed in Chapter 5.

Others: There are numerous other activation functions which may be used. They tend to be limited only by one's imagination and practical implementations. Even though we can define and use rather sophisticated activation functions, the ones presented above tend to be sufficient. We will discuss this point in more detail in Chapter 5 during a discussion on universal approximation capabilities of neural networks. To give you an idea of other possibilities for activation functions consider

$$\psi(s) = \frac{s}{1 + |s|} \tag{3.16}$$

or

$$\psi(s) = \frac{s^n}{1 + |s|^n} \text{sgn}^{n-1}(s), \tag{3.17}$$

where $n > 1$. Additional activation functions will be considered in the homework exercises.

Example 3.1 Consider a neuron defined by $\psi(s) = \tanh(s)$, with an inner product input mapping and with $x = [1, 2]^\top$. What set of weights, $w = [w_1, w_2]^\top$, will cause $\psi(s) = 0$? This is equivalent to finding w_1 and w_2 such that $s = w_1 + 2w_2 = 0$. Any point along the line $w_1 = -2w_2$ will satisfy this. Notice that with two adjustable weights, an infinite number of choices for w_1 and w_2 exist which cause $\psi(s) = 0$. △

3.2.3 The Mulitlayer Perceptron

The most common neural network found in applications today is the **feedforward multilayer perceptron** (MLP), also known simply as the multilayer feedforward neural network. It is a collection of artificial neurons in which the output of one neuron is passed to the input of another. The neurons (or nodes) are typically arranged in collections called layers, such that the output of all the nodes in a given layer are passed to the inputs of the nodes in the next layer.

An input layer is the input vector x, while the output layer is the connection between the neural network and the rest of the world. A hidden layer is a collection of nodes which lie between the input and output layers as shown in Figure 3.5. Without a hidden layer, the MLP reduces to a collection of n neurons operating in parallel.

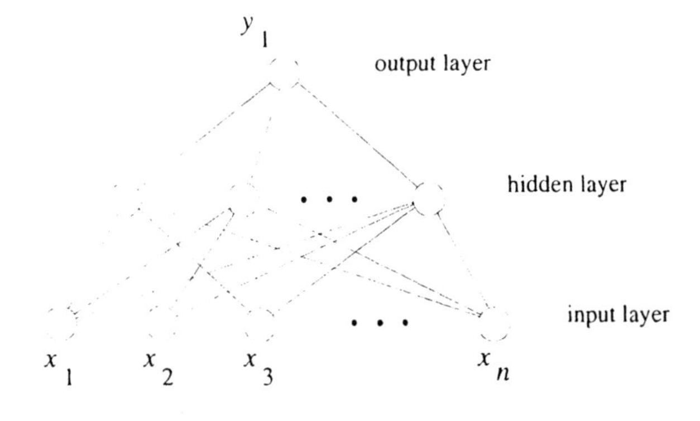

Figure 3.5. Schematic of a mulitlayer perceptron.

In a fully connected MLP, each neuron output is fed to each neuron input within the next layer. If we let θ be a vector of all the adjustable parameters in the network (weights and biases, and sometimes the parameters of the activation functions), then we denote the input-ouptut mapping of the MLP by $\mathcal{F}(x,\theta)$. To better understand the functional form of this mapping, consider the following example:

Example 3.2 Consider the MLP with one hidden layer consisting of nodes with activation functions defined by $\psi_j(s)$ for the jth hidden node and a single output node defined using a linear activation function. The input-output mapping for this MLP is defined by

$$\mathcal{F}(x,\theta) = \sum_{j=1}^{q} c_j \psi_j \left(\sum_{i=1}^{n} w_{ij} x_i + b_j \right) + d, \tag{3.18}$$

where $\theta = [d, c_1, \ldots, c_q, b_1, \ldots, b_q, w_{11}, \ldots, w_{nq}]^\top$ is a vector of adjustable parameters. Notice that each b_j and d are biases which were included with each node. There are q neurons in the hidden layer, n inputs, and one output in this neural network. $\triangle$

Within a multilayer perceptron, if there are many layers and many nodes in each layer, there will be a large number of adjustable parameters (e.g., weights and biases). MLP's with several hundred or thousands of adjustable weights are common in complex real-world applications.

3.2.4 Radial Basis Neural Network

A **radial basis neural network** (RBNN) is typically comprised of a layer of radial basis activation functions with an associated Euclidean input mapping (but there are many ways to define this class of neural networks). The output is then taken as a linear activation function with an inner product or weighted average input mapping. A RBNN with two inputs and 4 nodes is shown in Figure 3.6.

The input-output relationship in a RBNN with $x = [x_1, \ldots, x_n]^\top$ as an input is given by

$$\mathcal{F}(x,\theta) = \sum_{i=1}^{m} w_i \exp(-|x - c_i|^2/\gamma_i^2), \tag{3.19}$$

where $\theta = [w_1, \ldots, w_m]^\top$ when an inner product mapping is used within the output node. Typically, the values of the vectors c_i, $i = 1, \ldots, m$ and the scalar γ are held fixed, while the values of θ are adjusted so that the mapping produced by the RBNN matches some desired mapping. Because the adjustable weights appear linearly, we may express (3.19) as

$$\mathcal{F}(x,\theta) = \theta^\top \zeta(x), \tag{3.20}$$

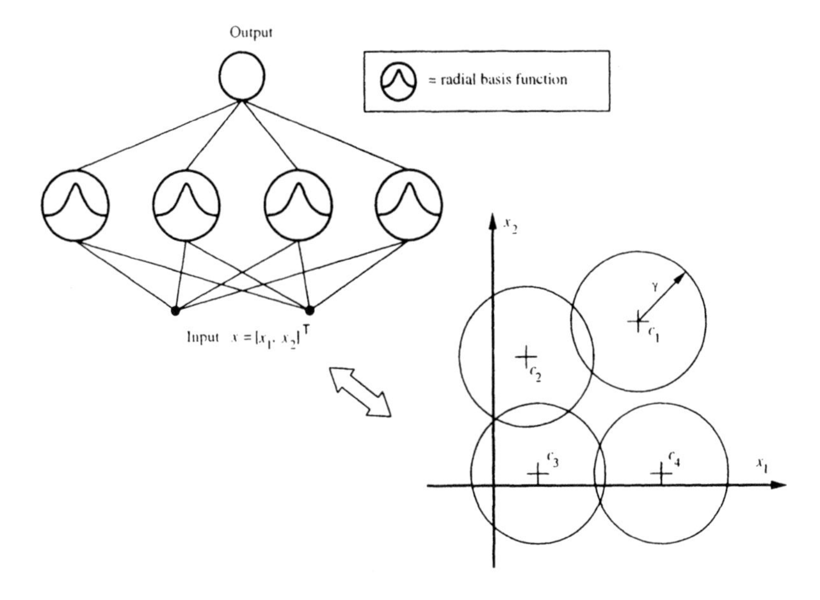

Figure 3.6. Radial basis neural network with 4 nodes.

where $\zeta_i = \exp(-|x - c_i|^2/\gamma^2)$. When a weighted average mapping is used in the output node, the RBNN becomes

$$\mathcal{F}(x, \theta) = \frac{\sum_{i=1}^{m} w_i \exp(-|x - c_i|^2/\gamma^2)}{\sum_{i=1}^{m} \exp(-|x - c_i|^2/\gamma^2)}, \tag{3.21}$$

which may again be expressed as $\mathcal{F}(x, \theta) = \theta^\top \zeta(x)$, now with $\theta = [w_1, \ldots, w_m]^\top$ and

$$\zeta_i = \frac{\exp(-|x - c_i|^2/\gamma^2)}{\sum_{i=1}^{m} \exp(-|x - c_i|^2/\gamma^2)}. \tag{3.22}$$

The Gaussian form of the activation functions lets one view the RBNN as a weighted average when using (3.21), where the value of w_i is weighted heavier when x is close to c_i. Thus the input space is broken into overlapping regions with centers corresponding to the c_i's as shown in Figure 3.6.

3.2.5 Tapped Delay Neural Network

If past input values are also processed by a neural network we obtain what is often referred to as a **tapped delay neural network**. An example using a single input and associated delayed values is shown in Figure 3.7 (here $1/z$ denotes a delay of T).

Here, u is the input and we can, for instance, let $x = [u(k), \ldots, u(k-n)]^\top$ so that the output of the neural network is $y = \mathcal{F}(x, \theta)$ (θ is a vector holding

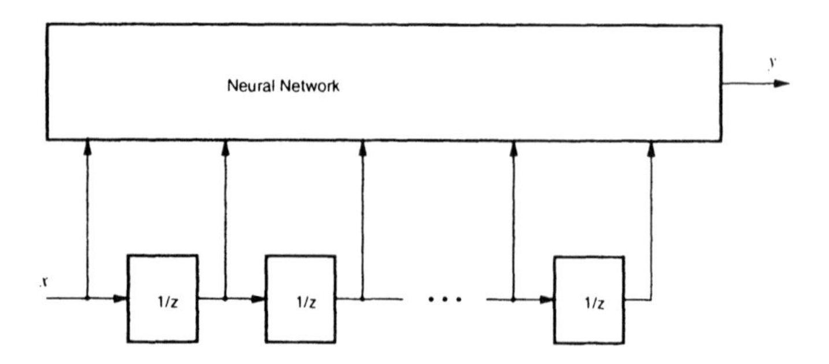

Figure 3.7. Tapped delay neural network.

the parameters of the neural network). In estimation and control applications it is common to let the input to the neural network be a sequence of past inputs and outputs of the process.

3.3 Fuzzy Systems

Traditionally, fuzzy systems have been constructed in a heuristic fashion using application-specific knowledge. For instance, when a fuzzy system is used as a controller for an automobile we capture the heuristic ideas a human has about how to do speed regulation and load these into "rules" that we store in a "rule-base" in the "fuzzy controller." One rule might be "if the current speed is 50 miles per hour and the desired speed is 55 miles per hour then press down on the accelerator a bit more." Other rules may incorporate information about the rate at which the speed is approaching, or departing from, the desired speed. The fuzzy controller uses fuzzy sets and fuzzy logic to implement a set of rules about how to control the vehicle speed. During operation, it determines which control rules apply to the current situation, and applies these in an analogous way to how a human would if he or she were physically controlling the system. In this way, it is said that the fuzzy controller emulates the human cognitive decision-making process (or, in other words, it conducts "inference").

A fuzzy system is shown in Figure 3.8. Here, we show the rule-base that holds the set of rules about, for example, how to control a process. Also, we see explicit inclusion of the "inference mechanism" which is the part of the fuzzy system that decides which rules should be used, and applies them. Not shown here, but discussed below, are the processes that transform information into a form that can be used by the inference mechanism ("fuzzification") and transform the actions of the inference mechanism into a form that can be used in practical applications ("defuzzification").

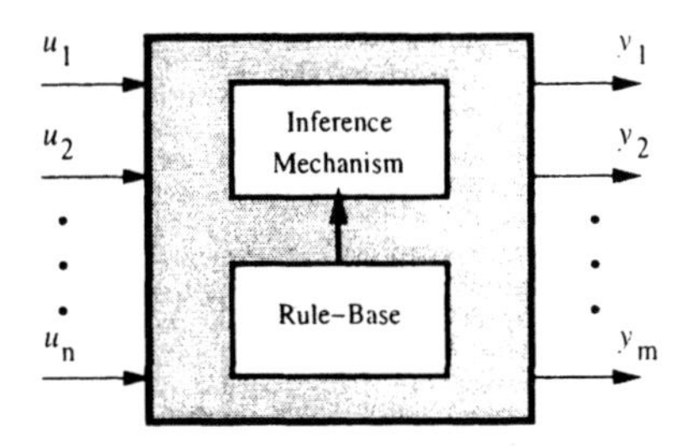

Figure 3.8. An n-input, m-output fuzzy system.

3.3.1 Rule-Base and Fuzzification

A multiple-input single-output (MISO) fuzzy system is a nonlinear mapping from an input vector $x = [x_1, x_2, \ldots, x_n]^\top \in \mathsf{R}^n$ to an output $y \in \mathsf{R}$. To define a MIMO fuzzy system with m outputs simply define m MISO fuzzy systems. The fuzzy system is characterized by a set of p **If – Then** rules, stored in a **rule-base**, expressed as

$$\begin{array}{lllll}
R_1: & \textbf{If} & (\tilde{x}_1 \text{ is } \tilde{F}_1^{k_1} \textbf{ and } \cdots \textbf{ and } \tilde{x}_n \text{ is } \tilde{F}_n^{l_1}) & \textbf{Then} & (\tilde{y} \text{ is } \tilde{G}^{a_1}) \\
\vdots & & \vdots & & \vdots \\
R_p: & \textbf{If} & (\tilde{x}_1 \text{ is } \tilde{F}_1^{k_p} \textbf{ and } \cdots \textbf{ and } \tilde{x}_n \text{ is } \tilde{F}_n^{l_p}) & \textbf{Then} & (\tilde{y} \text{ is } \tilde{G}^{a_p}).
\end{array}$$

Here, $\tilde{F}_b^a$ is the ath linguistic value associated with the linguistic variable $\tilde{x}_b$ that describes input x_b. Similarly, $\tilde{G}^a$ is the ath linguistic value associated with the linguistic variable $\tilde{y}$ that describes the output y. Linguistic variables are simply word descriptions of, for example, numeric variables (e.g., "speed" might be the linguistic variable for the velocity of the vehicle that we denote with $v(t)$). Linguistic variables change over time and hence take on specific linguistic values (typically adjectives). For instance, "speed" is "small" or "speed" is "large." The "linguistic rules" listed above are those gathered from a human expert.

Example 3.3 As an example of a set of linguistic rules, for a cruise control example suppose that $p = 3$, $n = 1$, and x_1 is the error between the desired speed $v_d(t)$ and the actual (sensed) speed $v(t)$ (i.e., $x_1(t) = v_d(t) - v(t)$). In this case, a word description of the fuzzy controller input variable $x_1(t)$ could be "speed-error" so that the linguistic variable is $\tilde{x}_1 =$"speed-error." Suppose that there are three linguistic values for the speed-error linguistic variable and that these are "positive," "zero," and "negative." Suppose that the output of the fuzzy controller is the change in throttle angle that we denote as $y(t)$ and use a linguistic variable $\tilde{y} =$"change-in-throttle." Suppose it has linguistic variables "increase," "stay-the-same," and "decrease."

The three rules, in the rule-base, of the form listed above would be

$$\begin{array}{lllll} R_1: & \textbf{If} & (\tilde{x}_1 \text{ is } \tilde{F}_1^{k_1}) & \textbf{Then} & (\tilde{y} \text{ is } \tilde{G}^{a_1}) \\ R_2: & \textbf{If} & (\tilde{x}_1 \text{ is } \tilde{F}_1^{k_2}) & \textbf{Then} & (\tilde{y} \text{ is } \tilde{G}^{a_2}) \\ R_3: & \textbf{If} & (\tilde{x}_1 \text{ is } \tilde{F}_1^{k_3}) & \textbf{Then} & (\tilde{y} \text{ is } \tilde{G}^{a_3}) \end{array}$$

or, more specifically,

$$\begin{array}{lllll} R_1: & \textbf{If} & (\tilde{x}_1 \text{ is "positive"}) & \textbf{Then} & (\tilde{y} \text{ is "increase"}) \\ R_2: & \textbf{If} & (\tilde{x}_1 \text{ is "negative"}) & \textbf{Then} & (\tilde{y} \text{ is "decrease"}) \\ R_3: & \textbf{If} & (\tilde{x}_1 \text{ is "zero"}) & \textbf{Then} & (\tilde{y} \text{ is "stay-the-same"}). \end{array}$$

Rule R_1 says that if the vehicle is traveling at a speed less than the desired speed, then increase the amount of throttle (i.e., make $y(t)$ positive). Rule R_2 says that if the vehicle is traveling at a speed greater than the desired speed, then decrease the amount of throttle (i.e., make $y(t)$ negative). Rule R_3 says that if the actual speed is close to the desired speed then do not move the throttle angle (i.e., let $y(t) = 0$). $\triangle$

To apply the knowledge represented in the linguistic rules we further quantify the meaning of the rules using fuzzy sets and fuzzy logic. In particular, using fuzzy set theory, the rule-base is expressed as a set of **fuzzy implications**

$$\begin{array}{lllll} R_1: & \textbf{If} & (F_1^{k_1} \textbf{ and } \cdots \textbf{ and } F_n^{l_1}) & \textbf{Then} & G^{a_1} \\ \vdots & & \vdots & & \vdots \\ R_p: & \textbf{If} & (F_1^{k_p} \textbf{ and } \cdots \textbf{ and } F_n^{l_p}) & \textbf{Then} & G^{a_p}, \end{array}$$

where F_b^a and G^a are fuzzy sets defined by

$$F_b^a = \left\{ (x_b, \mu_{F_b^a}(x_b)) : x_b \in \mathsf{R} \right\} \tag{3.23}$$

$$G^a = \left\{ (y, \mu_{G^a}(y)) : y \in \mathsf{R} \right\}. \tag{3.24}$$

The membership functions, $\mu_{F_b^a}, \mu_{G^a} \in [0, 1]$ describe how sure one is of a particular linguistic statement. For example, $\mu_{F_b^a}$ quantifies how well the linguistic variable $\tilde{x}_b$, that represents x_b, is described by the linguistic value $\tilde{F}_b^a$. There are many ways to define membership functions [170]. For instance, Tables 3.1 specifies triangular membership functions with center c and width w, and it specifies Gaussian membership functions with center c and width σ. It is good practice to sketch these six functions, labeling all aspects of the plots.

Example 3.4 Continuing with the above cruise control example we could quantify the meaning of each of the rule premises and consequents

	Triangular
Left	$\mu^L(x) = \begin{cases} 1 & \text{if } x \le c \\ \max\left(0, 1 + \frac{c-x}{0.5w}\right) & \text{otherwise} \end{cases}$
Centers	$\mu(x) = \begin{cases} \max\left(0, 1 + \frac{x-c}{0.5w}\right) & \text{if } x \le c \\ \max\left(0, 1 + \frac{c-x}{0.5w}\right) & \text{otherwise} \end{cases}$
Right	$\mu^R(x) = \begin{cases} \max\left(0, 1 + \frac{x-c}{0.5w}\right) & \text{if } x \le c \\ 1 & \text{otherwise} \end{cases}$

	Gaussian
Left	$\mu^L(x) = \begin{cases} 1 & \text{if } x \le c \\ \exp\left(-\frac{1}{2}\left(\frac{x-c}{\sigma}\right)^2\right) & \text{otherwise} \end{cases}$
Centers	$\mu(x) = \exp\left(-\frac{1}{2}\left(\frac{x-c}{\sigma}\right)^2\right)$
Right	$\mu^R(x) = \begin{cases} \exp\left(-\frac{1}{2}\left(\frac{x-c}{\sigma}\right)^2\right) & \text{if } x \le c \\ 1 & \text{otherwise} \end{cases}$

Table 3.1. Some standard membership functions.

with membership functions and hence, fuzzy sets. For instance, the following triangular membership functions could be used to describe the "speed-error" linguistic values:

1. "positive": μ^R with $c = 5$ and $w = 10$.
2. "negative": μ^L with $c = -5$ and $w = 10$.
3. "zero": μ with $c = 0$ and $w = 5$.

For practice, plot all three of these membership functions on the same axis vs. x_1. Also, define reasonable membership functions to represent the three linguistic values for the change-in-throttle output.

Note that the choice of what membership functions to use is made using insights from the problem at hand and normally not via a specific systematic procedure; however, some membership functions (e.g., triangular ones) result in computationally simpler algorithms and this may be one factor that enters into deciding which function to use. $\triangle$

This completes the description of the rule-base. Fuzzification is simply the process of obtaining values for the inputs x_i and computing $\mu(x_i)$ for each of the input membership functions μ.

3.3.2 Inference and Defuzzification

The premise fuzzy set, which we denote as,

$$F_1 \times F_2 \times \cdots \times F_n$$

(it is a "fuzzy Cartesian product"), of each rule has a membership function $\mu_{F_1^{k_i} \times \cdots \times F_n^{l_i}}(x)$ that is obtained using the t-norm [170]

$$\mu_{F_1^{k_i} \times \cdots \times F_n^{l_i}}(x) = \mu_{F_1^{k_i}}(x_1) \star \cdots \star \mu_{F_n^{l_i}}(x_n),$$

which may be defined by the min-operator (i.e., $\star$ defined as minimum)

$$\mu_{F_1^{k_i} \times \cdots \times F_n^{l_i}}(x) = \min\left\{\mu_{F_1^{k_i}}(x_1), \ldots, \mu_{F_n^{l_i}}(x_n)\right\}, \tag{3.25}$$

or product-operator (i.e., $\star$ defined as a standard mathematical product)

$$\mu_{F_1^{k_i} \times \cdots \times F_n^{l_i}}(x) = \mu_{F_1^{k_i}}(x_1) \cdot \ldots \cdot \mu_{F_n^{l_i}}(x_n), \tag{3.26}$$

among others. The t-norm simply quantifies the conjunction in the premise of each rule (it is a basic operation in fuzzy logic).

Two approaches to quantify fuzzy implications for the ith rule are

$$\mu_{F_1^{k_i} \times \cdots \times F_n^{l_i} \to G^{a_i}}(x, y) = \min\left\{\mu_{F_1^{k_i} \times \cdots \times F_n^{l_i}}(x), \mu_{G^{a_i}}(y)\right\}, \tag{3.27}$$

and

$$\mu_{F_1^{k_i} \times \cdots \times F_n^{l_i} \to G^{a_i}}(x, y) = \mu_{F_1^{k_i} \times \cdots \times F_n^{l_i}}(x) \cdot \mu_{G^{a_i}}(y). \tag{3.28}$$

You use one approach or the other depending on how you want to quantify the fuzzy implication. For a given x, the membership functions (fuzzy sets) on the left-hand sides of these two equations represent the conclusions reached by the inference mechanism. We will call these "implied fuzzy sets."

Next, we introduce the process of defuzzification that converts the implied fuzzy sets into actual numbers that are then the outputs of the system. Basically, in general, more than one rule will apply at each time instant and hence there will be more than one conclusion (i.e., more than one implied fuzzy set with a membership function that is not zero everywhere) reached by the inference mechanism at each time instant (i.e., the conclusions generated by the inference mechanism are a set of implied fuzzy sets, where for the ith rule the membership function of the implied fuzzy set is given by $\mu_{F_1^{k_i} \times \cdots \times F_n^{l_i} \to G^{a_i}}(x, y)$ with x specified). Defuzzification combines the conclusions from all the rules and provides a single number that represents the conclusions.

Using "center-average defuzzification," the output of the fuzzy system, $y = \mathcal{F}(x)$, may be expressed as

$$y = \mathcal{F}(x) = \frac{\sum_{i=1}^{p} c_i \mu_{F_1^{k_i} \times \cdots \times F_n^{l_i}}(x)}{\sum_{i=1}^{p} \mu_{F_1^{k_i} \times \cdots \times F_n^{l_i}}(x)}, \tag{3.29}$$

where each c_i is the point at which $\mu_{G^{a_i}}(y)$ reaches its maximum (if $\mu_{G^{a_i}}(y)$ is symmetric about the point it reaches its maximum, which is the typical case).

For "center-of-gravity defuzzification," another option for combining the conclusions reached by the inference mechanism, the output of the fuzzy system is

$$y = \mathcal{F}(x) = \frac{\sum_{i=1}^{p} c_i \int_{\tau} \mu_{F_1^{k_i} \times \cdots \times F_n^{l_i} \to G^{a_i}}(x, \tau) d\tau}{\sum_{i=1}^{p} \int_{\tau} \mu_{F_1^{k_i} \times \cdots \times F_n^{l_i} \to G^{a_i}}(x, \tau) d\tau}, \tag{3.30}$$

where here c_i is the center of area of $\mu_{G^{a_i}}(y)$ for the ith rule. Again, typically $\mu_{G^{a_i}}(y)$ is chosen to be symmetrical about its maximum so that c_i is the center of $\mu_{G^{a_i}}(y)$. It is assumed that the fuzzy system is defined so that for all $x \in \mathbb{R}^n$, we have $\mu_{F_1^{k_i} \times \cdots \times F_n^{l_i}}(x) > 0$ for at least one rule i, so that (3.29) and (3.30) are well defined.

This completes the definition of a fuzzy system; equations (3.29) and (3.30) provide mathematical charaterizations of all the operations of a fuzzy controller (when we simulate a fuzzy system it is equations of this type that we code).

Because it will be important later, we note that a change in the output of the fuzzy system with respect to the change in c_i is given by $\zeta = [\zeta_1, \ldots, \zeta_p]^\top = \partial \mathcal{F}(x)/\partial \theta$ where $\theta = [c_1, \ldots, c_p]^\top$. For fuzzy systems defined using center-average defuzzification,

$$\zeta_i = \frac{\mu_{F_1^{k_i} \times \cdots \times F_n^{l_i}}(x)}{\sum_{i=1}^{p} \mu_{F_1^{k_i} \times \cdots \times F_n^{l_i}}(x)}, \tag{3.31}$$

and for fuzzy systems using center-of-gravity defuzzification,

$$\zeta_i = \frac{\int_{\tau} \mu_{F_1^{k_i} \times \cdots \times F_n^{l_i} \to G^{a_i}}(x, \tau) d\tau}{\sum_{i=1}^{p} \int_{\tau} \mu_{F_1^{k_i} \times \cdots \times F_n^{l_i} \to G^{a_i}}(x, \tau) d\tau}, \tag{3.32}$$

so that $\mathcal{F}(x) = \theta^\top \zeta$, where $\zeta = [\zeta_1, \ldots, \zeta_p]^\top$. When we want to emphasize the dependence of the fuzzy system on its parameters we will use the notation $y = \mathcal{F}(x, \theta)$.

Example 3.5 Consider defining a fuzzy control system for the system defined by

$$m\ddot{x} = u, \tag{3.33}$$

where m is the mass of an object, x is the position, and u is the input force. If we would like to position the mass at $x = r$ and both the signals x and $\dot{x}$ are available for use in a control law, then one might consider using a simple PD-type control system. That is, we might define $u = k_p e + k_d \dot{e}$ where $e = r - x$ and $\dot{e} = -\dot{x}$ with $k_p, k_d > 0$.

This type of control law has been used in numerous applications due to its simplicity. It also makes intuitive sense in terms of how it works. For example, if we have a large negative error and the derivative indicates that it is becoming more negative, then a large negative force should be applied. On the other hand, if both the error and derivative are near zero, then a small force (if any) should be applied. Using this logic, we may create a small rule base to control the mass as follows:

1. If "e is Negative" and "$\dot{e}$ is Negative" then "u is Negative", or
2. If "e is Zero" and "$\dot{e}$ is Negative" then "u is Negative", or
3. If "e is Posotive" and "$\dot{e}$ is Negative" then "u is Zero", or
4. If "e is Negative" and "$\dot{e}$ is Positive" then "u is Zero", or
5. If "e is Zero" and "$\dot{e}$ is Positive" then "u is Positive", or
6. If "e is Positive" and "$\dot{e}$ is Positive" then "u is Positive".

Consider using triangular membership functions for the inputs and outputs as shown at the top of Figure 3.9. Here, the input e takes on a relatively large negative value and $\dot{e}$ is a small positive value as denoted by the vertical dashed lines. Each rule is then shown in terms of the associated input and output membership functions. To determine the degree of membership of the output of each rule, the minimum value of the input membership functions is used.

After the output membership function value for each rule is determined, the contribution from each rule is combined to form the output of the fuzzy system. This is shown in the lower right-hand side of Figure 3.9. If, for example, center of gravity is used in the defuzzification, we notice that the output of the fuzzy system would be a small negative number (why?). △

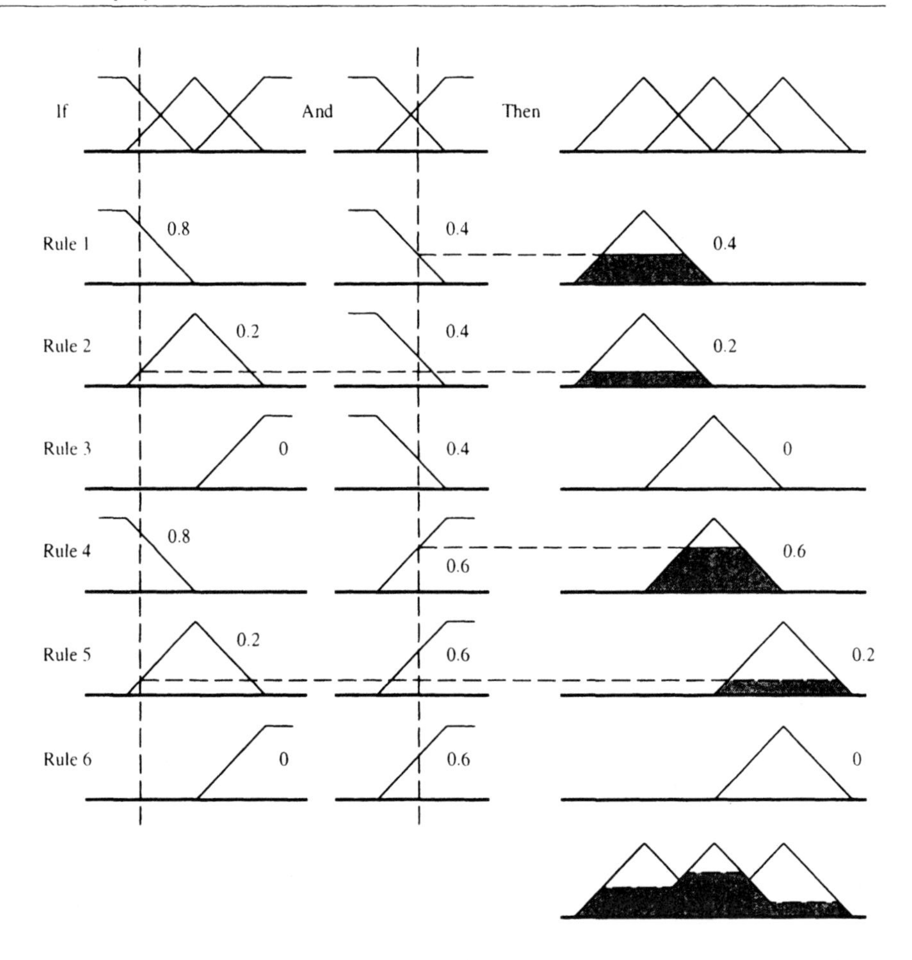

Figure 3.9. Defining the output of the fuzzy system in Example 3.5.

3.3.3 Takagi-Sugeno Fuzzy Systems

A Takagi-Sugeno fuzzy system uses rules of the form:

$$\begin{array}{llll} R_1: & \textbf{If} \ (\tilde{x}_1 \text{ is } \tilde{F}_1^{k_1} \textbf{ and } \cdots \textbf{ and } \tilde{x}_n \text{ is } \tilde{F}_n^{l_1}) & \textbf{Then} & c_1 = g_1(x) \\ \vdots & \vdots & & \vdots \\ R_p: & \textbf{If} \ (\tilde{x}_1 \text{ is } \tilde{F}_1^{k_p} \textbf{ and } \cdots \textbf{ and } \tilde{x}_n \text{ is } \tilde{F}_n^{l_p}) & \textbf{Then} & c_p = g_p(x). \end{array}$$

Here, the premises are exactly the same as for the standard fuzzy systems defined above. The consequents, however, are not fuzzy sets; they are functions. In particular,

$$c_q = g_q(x)$$

is the consequence of the qth rule and the function $g_q : \Re^n \to \Re$.

To compute the output of the Takagi-Sugeno fuzzy system with center-average defuzzification we write

$$y = \frac{\sum_{i=1}^{p} c_i \mu_i}{\sum_{i=1}^{p} \mu_i}, \tag{3.34}$$

where $\mu_i := \mu_{F_1 \times \cdots \times F_n}(x_1, \ldots, x_n)$ is the value that the membership function (defined via (3.25) or (3.26)) for the antecedent of the ith rule takes on at $x = [x_1, \ldots, x_n]^\top$ (notice that this fuzzy system can be viewed as an interpolator between the p output functions). It is assumed that the fuzzy system is defined so that for all $x \in \Re^n$, we have $\sum_{i=1}^{p} \mu_i \neq 0$.

Sometimes we will let the output consequences for each rule be a linear combination of a set of Lipschitz continuous functions, $\gamma_k(x) \in \Re$, $k = 1, 2, \ldots, m-1$, so that

$$c_i = g_i(x) := a_{i,0} + a_{i,1}\gamma_1(x) + \cdots + a_{i,m-2}\gamma_{m-2}(x) + a_{i,m-1}\gamma_{m-1}(x) \tag{3.35}$$

$i = 1, \ldots, p$. In the special case where $m - 1 = n$ and $\gamma_i(x) = x_i$ for all i, the functions on the right-hand side of this equation are linear (actually, affine) and we have the traditional definition of the Takagi-Sugeno system.

Next, define

$$z = \begin{bmatrix} 1 \\ \gamma_1(x) \\ \vdots \\ \gamma_{m-1}(x) \end{bmatrix} \in \Re^m \tag{3.36}$$

and

$$\theta^\top = \begin{bmatrix} \theta_{1,0} & \theta_{1,1} & \cdots & \theta_{1,m-1} \\ \theta_{2,0} & \theta_{2,1} & \cdots & \theta_{2,m-1} \\ \vdots & \vdots & \ddots & \vdots \\ \theta_{p,0} & \theta_{p,1} & \cdots & \theta_{p,m-1} \end{bmatrix}. \tag{3.37}$$

The consequence vector associated with the fuzzy rules is now given by $c = \theta^\top z$, so that the output of the fuzzy system may now be expressed as

$$y = z^\top \theta \zeta = \mathcal{F}(x, \theta), \tag{3.38}$$

where $\zeta^\top = [\mu_1, \cdots, \mu_p] \,/\, [\sum_{i=1}^{p} \mu_i]$. Notice that this definition has θ as a matrix; next we show how to parameterize the fuzzy system so that θ is a vector.

Notice that if $m - 1 = n$ we can let

$$\theta = [\theta_{1,0}, \ldots, \theta_{p,0}, \theta_{1,1}, \ldots, \theta_{p,1}, \ldots, \theta_{1,n}, \ldots, \theta_{p,n}]^\top$$

and

$$\tilde{\zeta} = [\zeta_1, \ldots, \zeta_p, \gamma_1(x)\zeta_1, \ldots, \gamma_1(x)\zeta_p, \ldots, \gamma_n(x)\zeta_1, \ldots, \gamma_n(x)\zeta_p]^\top$$

so that

$$y = \theta^\top \tilde{\zeta},$$

which is a parameterization similar to all the others that we have specified for the neural networks and fuzzy systems (and hence gradients can be found in a similar way).

Finally, we would like to note that while standard fuzzy systems have traditionally been defined in a heuristic fashion using insights gathered from the system to determine the rules, the Takagi-Sugeno fuzzy system has often been specified with both heuristic methods and data (heuristic ideas can be used to come up with the premise membership functions, and then data can be used to train the consequent terms, especially if the consequent functions are linear or affine). Of course, there is no reason that data and training algorithms (e.g., gradient and least squares) cannot be used to train all the parameters in a fuzzy system, in a similar manner to how parameters are trained in a neural network. This is the main topic of the next chapter.

3.4 Summary

Upon completion of this chapter, the reader should have an understanding of

- Neurons, including weights, input mappings, and activation functions.
- Feedforward neural networks, including nodes and layers.
- Radial basis neural networks.
- How to calculate the neural network and fuzzy system gradients with respect to the adjustable parameters.
- Standard fuzzy systems.
- Takagi-Sugeno fuzzy systems.

In summary, this chapter introduces the two main approximator structures that we use for the adaptive schemes presented later in this book.

3.5 Exercises and Design Problems

Exercise 3.1 (Mathematical Representations of Neural Networks and Fuzzy Systems)

(a) Write out the full mathematical equations for a multilayer perceptron with two hidden layers. Be sure to utilize distinct notation for the weights, biases, and activation functions for each layer.

(b) Write out the full mathematical equations for a fuzzy system that uses Gaussian input membership functions, triangular output membership functions, product to represent the premise and implication, and center-average defuzzification.

Exercise 3.2 (Neural Network Implementation) Use your favorite computer language (C, MATLAB, etc.) to implement the MLP and radial basis neural networks. Allow enough flexibility so that the size of the neural networks may easily be changed an so that the weights and biases may be adjusted by another function.

Exercise 3.3 (Fuzzy System Implementation) Use your favorite computer language to implement a fuzzy system. Allow enough flexibility in your implementation so that it is possible to change the number of rules and the type of membership functions used. Also allow for the possibility that the parameters used to define the membership functions may be modified by another function.

Exercise 3.4 (An MLP Update Law) Consider the MLP whose output is given by $\mathcal{F}(x, \theta)$, where θ is a vector of weights and biases. Assume that for a given x we want to find some θ such that $y = r$ where r is the desired constant neural network output. If $e = y - r$ is the neural network output error and $V = e^2$ is a Lyapunov candidate, then describe the stability of $e = 0$ when we define

$$\dot{\theta} = -\gamma \left[\frac{\partial \mathcal{F}}{\partial \theta} \right]^\top e,$$

where $\gamma > 0$. What the the implications if we define the neural network such that $|\partial \mathcal{F}/\partial \theta| > 0$ for all x (rather than $|\partial \mathcal{F}/\partial \theta| \geq 0$).

Exercise 3.5 (Fuzzy Cruise Control) Consider the cruise control problem where the vehicle speed is governed by

$$m\dot{v} = -Av^2 + u, \tag{3.39}$$

where $m = 1200kg$ is the mass of the vehicle, $A = 0.4Nm^2/s^2$ is the aerodynamic drag, v is the vehicle speed in m/s, and u is the input force. If v is measurable and r is the desired vehicle speed, then define a rule base which could be used to regulate the vehicle speed so that $v = r$. Simulate the closed-loop system. Try to adjust the output membership functions to improve the system performmce.

Exercise 3.6 (Fuzzy Control for an Inverted Pendulum) Consider the simple problem of balancing an inverted pendulum on a cart. Let y denote the angle that the pendulum makes with the vertical (in radians), l be the half-pendulum length (in meters), and u be the force input that moves the cart (in Newtons). Use r to denote the desired angular position of the pendulum. The goal is to balance the pendulum in the upright position (i.e., $r = 0$) when it initially starts with some nonzero angle off the vertical (i.e., $y \neq 0$). One model for the inverted pendulum shown is given by

$$\begin{aligned} \ddot{y} &= \frac{9.8\sin(y) + \cos(y)\left[\frac{-\bar{u}-0.25\dot{y}^2\sin(y)}{1.5}\right]}{0.5\left[\frac{4}{3} - \frac{1}{3}\cos^2(y)\right]} \\ \dot{\bar{u}} &= -100\bar{u} + 100u. \end{aligned} \tag{3.40}$$

The first-order filter on u to produce $\bar{u}$ represents an actuator. In your simulations let the initial condition be $y(0) = 0.1$ radians, $\dot{y}(0) = 0$, and $\ddot{y}(0) = 0$.

(a) Develop a fuzzy controller with $e = r - y$ and $\dot{e}$ as inputs, the minimum operator to represent both the "and" in the premise and the implication, and COG defuzzification. Simulate the closed-loop system and plot the output y and input u to demonstrate that your fuzzy controller can balance the pendulum.

(b) Repeat (a) for the case where you use product to represent the premise and implication and center-average defuzzification.

Chapter 4

Optimization for Training Approximators

4.1 Overview

As humans, we are intuitively familiar with the process of optimization because of our constant exposure to it. For instance, in business investments we seek to maximize our profits; in recreational games we seek to maximize our own score or minimize that of our opponent. It is not surprising that optimization plays a key role in engineering and many other fields. In circuit design we may want to maximize power transfer, in motor design we may want to design for the highest possible torque delivery for a given amount of current, or in communication system design we may want to minimize the probability of error in signal transmission. Indeed, in the design of control systems we have the field of "optimal control," where one objective might be to minimize tracking error and control effort (energy) while stabilizing a system.

Here, as in many adaptive control methods, the adaptive schemes are designed to search for a parameter set which minimizes a cost function, while maintaining, or seeking to achieve, certain closed-loop properties (e.g., stability) of the adaptive system. For instance, we may seek to adjust the parameters of a neural network or fuzzy system (which we treat as "approximators") so that the neural network or fuzzy system approximator nonlinearity matches that of the plant, and then this synthesized nonlinearity is used to specify a controller that reduces the tracking error. Optimization then forms a fundamental foundation on which all the approaches rest. It is for this reason that we provide an introduction to optimization here. The reader who is already familiar with optimization methods can skip (or skim) this chapter and go to the next one.

4.2 Problem Formulation

Consider the minimization of the **cost function** $J(\theta) > 0$ where $\theta \in \mathcal{S} \subseteq \mathsf{R}^p$ is a vector of p adjustable parameters. In other words, we wish to find some $\theta^* \in \mathcal{S}$ such that

$$\theta^* = \arg\min_{\theta \in \mathcal{S}} J(\theta). \tag{4.1}$$

This type of optimization problem is referred to as "constrained optimization" since we require that $\theta \in \mathcal{S}$. When $\mathcal{S} = \mathsf{R}^p$, the minimization problem becomes "unconstrainted."

If we wish to find a parameter set θ that shapes a function $\mathcal{F}(x, \theta)$ (that represents a neural network or fuzzy system with tunable parameters θ) so that $\mathcal{F}(x, \theta)$ and $f(x)$ match at $x = \bar{x}$, then one might try to minimize the cost function

$$J(\theta) = |f(\bar{x}) - \mathcal{F}(\bar{x}, \theta)|^2 \tag{4.2}$$

by adjusting θ. If we wish to cause $\mathcal{F}(x, \theta)$ to match $f(x)$ on the region $x \in \mathcal{S}_x$, then minimization of

$$J(\theta) = \sup_{x \in \mathcal{S}_x} |f(x) - \mathcal{F}(x, \theta)|^2 \tag{4.3}$$

would be one possible cost function to consider. Minimizing the difference between a known parameterized function (an "approximator") $\mathcal{F}(x, \theta)$ and another function $f(x)$ which is in general only partially known is referred to as "function approximation." This special optimization problem will be of particular interest to us throughout the study of adaptive systems using fuzzy systems and neural networks.

Practically speaking, however, in our adaptive estimation and control problems we are either only given a finite amount of information in the form of input-output pairs about the unknown function $f(x)$ or we are given such input-output pairs one at a time in a sequence. Suppose that there are n input variables so $x = [x_1, \ldots, x_n]^\top \in \mathsf{R}^n$. Suppose we present the function $f(x)$ with a variety of input data (specific values of the variable x) and collect the outputs of the function. Let the ith input vector of data be denoted by

$$x^i = [x_1^i, \ldots, x_n^i]^\top,$$

where $x^i \in \mathcal{S}_x$ and denote the output of the function by

$$y^i = f(x^i).$$

Furthermore, let the "training data set" be denoted by

$$G = \{(x^i, y^i) : i = 1, 2, \ldots, M\}.$$

Given this, a practical cost function to minimize is given by

$$J(\theta) = \sum_{i=1}^{M} |f(x^i) - \mathcal{F}(x^i, \theta)|^2. \tag{4.4}$$

We will study several ways to minimize this cost function, keeping in mind that we would like to be minimizing a function like the one given in (4.3) at the same time (i.e., even though we can only minimize (4.4) we want to obtain accurate function approximation over a whole continuous range of variables $x \in \mathcal{S}_x$). Clearly, if G does not contain a sufficient number of samples within $\mathcal{S}_x$, it will not be possible to do a good job at lowering the value of (4.3). For instance, you often need some type of "coverage" of the input space $\mathcal{S}_x$ by the x^i data (e.g., uniform on a grid with a small distance between the points). The problem is, however, that in practice you often do not have a choice of how to distribute the data over $\mathcal{S}_x$; often you are forced to use a given G directly as it is given to you (and you cannot change it to improve approximation accuracy). We see that due to issues with how G is constructed, the problem of function approximation, specifically the minimization of the magnitude of the "approximation error"

$$e(x) = f(x) - \mathcal{F}(x, \theta)$$

is indeed a difficult problem.

What do function approximation and optimzation have to do with adaptive control? As a simple example, suppose we wish to drive the output of the system defined by

$$\dot{x} = f(x) + u \tag{4.5}$$

to zero, where $f(x)$ is a smooth but unknown function, x is the scalar system output, and u is the input. Consider the function $\mathcal{F}(x, \theta^*)$, which approximates the unknown function $f(x)$, where θ^* is a vector of ideal parameters for the cost function (4.3). Then the controller defined by

$$u = -\mathcal{F}(x, \theta^*) - kx, \tag{4.6}$$

where $k > 0$ would drive $|x| \to 0$ *assuming* that $|\mathcal{F}(x, \theta^*) - f(x)| = 0$. This happens because the closed-loop dynamics become $\dot{x} = -kx$ which is an exponentially stable system.

In general, it will be our task to find $\theta = \theta^*$ so that the approximator $\mathcal{F}(x, \theta) = \mathcal{F}(x, \theta^*) \approx f(x)$. Notice that even in this simple problem, some key issues with trying to find $\theta = \theta^*$ are present when the cost function (4.4) is used in place of (4.3). For instance, to generate the training data set G we need to assume that we know (can measure) x. However, even though we know u, we do not necessarily know $f(x)$ unless we can also assume that we know $\dot{x}$ (which can be difficult to measure due to noise). If we do know $\dot{x}$ then we can let

$$f(x) = \dot{x} - u.$$

Even though under these assumptions we can gather data pairs $(x^i, f(x^i))$ to form G for training the approximator, we cannot in general pick the x^i unless we can repeatedly initialize the differential equation in (4.5) with every possible $x^i \in \mathcal{S}_x$. Since in many practical situations there is only one initial condition (or a finite number of them) that we can pick, the only data we can gather are constrained by the solution of (4.5). This presents us with the following problem: How can we pick the $u(t)$ so that the data we can collect to put in G will ensure good function approximation? Note that the "controllability" of the *uncertain* nonlinear system in (4.5) will impact our ability to "steer" the state to regions in $\mathcal{S}_x$ where we need to improve approximation accuracy. Also, the connection between G and approximation accuracy depends critically on what optimization algorithm is used to construct θ, as well as on the approximator's structural potential to match the unknown function $f(x)$.

We see that even for our simple scalar problem, a guarantee of approximation accuracy is difficult to provide. Often, the central focus is on showing that even if perfect approximation is not achieved, we still get a stable closed-loop system. In fact, for our example even if $\mathcal{F}(x, \theta)$ does not exactly match $f(x)$, the resulting closed loop system dynamics are stable and converge to a ball around zero assuming that an θ can be found such that $|f(x) - \mathcal{F}(x, \theta)| \leq D$ for all x, where $D > 0$ (see Homework problem (4.7)).

The main focus of this chapter is to provide optimization algorithms for constructing θ so that $\mathcal{F}(x, \theta)$ approximates $f(x)$. As the above example illustrates, the end approximation accuracy will not be paramount. We simply need to show that if we use the optimization methods shown in this chapter to adjust the approximator, then the resulting closed-loop system will be stable. The size of the approximator error, however, will typically affect the performance of the closed-loop system.

4.3 Linear Least Squares

We will first concentrate on solving the least squares problem for the case where

$$J(\theta) = \sum_{i=1}^{M} w_i \left| f(x^i) - \mathcal{F}(x^i, \theta) \right|^2, \tag{4.7}$$

where $w_i > 0$ are some scalars, $f(x)$ is an unknown function, and $\mathcal{F}(x,\theta)$ is an approximator defined by

$$\mathcal{F}(x,\theta) = \frac{\partial \mathcal{F}}{\partial \theta}\theta \tag{4.8}$$

so that θ appears linearly (i.e., a "linear in the parameters" approximator which we will sometimes call a linear approximator). In this chapter, we will use the short hand $\zeta^\top = \partial\mathcal{F}/\partial\theta$. We will later study techniques which consider some $\mathcal{F}(x,\theta)$ such that θ does not necessarily appear linearly in the output of the approximator. Next, we will introduce batch and recursive least squares methods to find $\theta = \theta^*$ which minimizes the cost function (4.7) for input-output data in G assuming the approximator has the form of (4.8).

4.3.1 Batch Least Squares

We will introduce the batch least squares method to train linear approximators by first discussing the solution of the linear system identification problem. Let f denote the physical system that we wish to identify. The training set G is defined by the experimental input-output data that is generated from this system. In linear system identification, we can use a model

$$y(k) = \sum_{i=1}^{q} \theta_{a_i} y(k-i) + \sum_{i=0}^{p} \theta_{b_i} u(k-i),$$

where $u(k)$ and $y(k)$ are the system input and output at time k. This form of a system model is often referred to as an ARMA (AutoRegressive Moving Average) model. In this case the approximator $y(k) = \mathcal{F}(x,\theta)$ is defined with

$$\zeta^\top(k) = \frac{\partial\mathcal{F}}{\partial\theta} = [y(k-1),\cdots,y(k-q),u(k),\cdots,u(k-p)],$$

and

$$\theta = [\theta_{a_1},\cdots,\theta_{a_q},\theta_{b_0},\cdots,\theta_{b_p}]^\top.$$

We have $n = q+p+1$ so that $\zeta(k)$ and θ are $n \times 1$ vectors, and often $\zeta(k)$ is called the "regression vector." System identification amounts to adjusting θ so that (4.7) is minimized. Often, for system identification we choose $\zeta^i = \zeta(i)$, $y^i = y(i)$, and let $G = \{(\zeta^i, y^i) : i = 1,2,\ldots,M\}$.

In the batch least squares method we define

$$Y = [y^1, y^2, \ldots, y^M]^\top$$

to be an $M \times 1$ vector of output data where the y^i, $i = 1, 2, \ldots, M$ come from G (i.e., y^i such that $(\zeta^i, y^i) \in G$). We let

$$\Phi = \begin{bmatrix} (\zeta^1)^\top \\ (\zeta^2)^\top \\ \vdots \\ (\zeta^M)^\top \end{bmatrix}$$

be an $M \times n$ matrix that consists of the ζ^i data vectors stacked into a matrix (i.e., the ζ^i such that $(\zeta^i, y^i) \in G$). Let

$$\epsilon^i = y^i - (\zeta^i)^\top \theta$$

be the error in approximating the data pair $(\zeta^i, y^i) \in G$ using θ. Define

$$E = \left[\epsilon^1, \epsilon^2, \ldots, \epsilon^M\right]^\top$$

so that

$$E = Y - \Phi\theta.$$

Now choose

$$J(\theta) = \frac{1}{2} E^\top E$$

to be a measure of how good the approximation is for all the data, for a given θ, which is (4.7) with $w_i = 1$ for $i = 1, \ldots M$. We want to pick θ to minimize $J(\theta)$. Notice that $J(\theta)$ is convex in θ so that a local minimum is a global minimum in this case.

Using basic ideas from calculus, if we take the partial derivative of J with respect to θ and set it equal to zero, we get an equation for the best estimate (in the least squares sense) of the unknown θ. Another approach to deriving this result is to notice that

$$2J = E^\top E = Y^\top Y - Y^\top \Phi\theta - \theta^\top \Phi^\top Y + \theta^\top \Phi^\top \Phi\theta.$$

Then, we "complete the square" by assuming that $\Phi^\top \Phi$ is invertible and letting

$$\begin{aligned} 2J &= Y^\top Y - Y^\top \Phi\theta - \theta^\top \Phi^\top Y + \theta^\top \Phi^\top \Phi\theta \\ &\quad + Y^\top \Phi(\Phi^\top \Phi)^{-1} \Phi^\top Y - Y^\top \Phi(\Phi^\top \Phi)^{-1} \Phi^\top Y \end{aligned}$$

(where we are simply adding and subtracting the same terms at the end of the equation). Hence,

$$\begin{aligned} 2J &= Y^\top (I - \Phi(\Phi^\top \Phi)^{-1} \Phi^\top) Y \\ &\quad + (\theta - (\Phi^\top \Phi)^{-1} \Phi^\top Y)^\top \Phi^\top \Phi(\theta - (\Phi^\top \Phi)^{-1} \Phi^\top Y). \end{aligned}$$

Since the first term in this equation is independent of θ, we cannot reduce J via this term, so it can be ignored. Thus, to get the smallest value of J, we choose θ so that the second term is equal to zero since its contribution is a non-negative value. We will let θ^* denote the value of θ that achieves the minimization of J, and we notice that

$$\theta^* = (\Phi^\top \Phi)^{-1} \Phi^\top Y, \tag{4.9}$$

since the smallest we can make the last term in the above equation is zero. This is the equation for batch least squares that shows we can directly compute the least squares estimate θ^* from the "batch" of data that is loaded into Φ and Y. If we pick the inputs to the system so that it is "sufficiently excited" [135], then we will be guaranteed that $\Phi^\top \Phi$ is invertible (rank$(\Phi) = n$); furthermore, if the data come from a linear plant with known q and p, then for sufficiently large M we will achieve perfect estimation of the plant parameters.

In "weighted" batch least squares we use

$$J(\theta) = \frac{1}{2} E^\top W E, \tag{4.10}$$

where, for example, W is an $M \times M$ diagonal matrix with its diagonal elements $w_i > 0$ for $i = 1, 2, \ldots, M$. These w_i can be used to weight the importance of certain elements of G more than others. For example, we may choose to have it put less emphasis on older data by choosing $w_1 < w_2 < \cdots < w_M$ when x^2 is collected after x^1, x^3 is collected after x^2, and so on. The resulting parameter estimates can be shown to be given by

$$\theta^* = (\Phi^\top W \Phi)^{-1} \Phi^\top W Y. \tag{4.11}$$

To show this, simply use (4.10) and proceed with the derivation in the same manner as above.

Example 4.1 As a very simple example of how batch least squares can be used, suppose that we would like to identify the coefficients for the system

$$y(k) = \theta_a y(k-1) + \theta_b u(k), \tag{4.12}$$

where $\zeta(k) = [y(k-1), u(k)]^\top$. Suppose that the data that we would like to fit the system to is given by

$$G = \left\{ \left(\begin{bmatrix} 1 \\ 1 \end{bmatrix}, 2 \right), \left(\begin{bmatrix} 2 \\ 1 \end{bmatrix}, 3 \right), \left(\begin{bmatrix} 3 \\ 1 \end{bmatrix}, 4 \right) \right\},$$

so that $M - 3$. We will use (4.9) to compute the parameters for the solution that best fits the data (in the sense that it will minimize the sum of the squared distances between the identified system and the data). To do this we let

$$\Phi = \begin{bmatrix} 1 & 1 \\ 2 & 1 \\ 3 & 1 \end{bmatrix}$$

and

$$Y = \begin{bmatrix} 2 \\ 3 \\ 4 \end{bmatrix}.$$

Hence,

$$\theta^* = (\Phi^\top \Phi)^{-1} \Phi^\top Y = \left(\begin{bmatrix} 14 & 6 \\ 6 & 3 \end{bmatrix} \right)^{-1} \begin{bmatrix} 20 \\ 9 \end{bmatrix} = \begin{bmatrix} 1 \\ 1 \end{bmatrix},$$

and the system

$$y(k) = y(k-1) + u(k)$$

best fits the data in the least squares sense. The same general approach works for larger data sets. △

4.3.2 Recursive Least Squares

While the batch least squares approach has proven to be very successful for a variety of applications, the fact that by its very nature it is a "batch" method (i.e., all the data are gathered, then processing is done) may present computational problems. For small M we could clearly repeat the batch calculation for increasingly more data as they are gathered, but as M becomes larger the computations become prohibitive due to the fact that the dimensions of Φ and Y depend on M. Next, we derive a recursive version of the batch least squares method that will allow us to update our estimate of θ^* each time we get a new data pair, without using all the old data in the computation and without having to compute the inverse of $\Phi^\top \Phi$.

Since we will be successively increasing the size of G, and since we will assume that we increase the size by one each time step, we let a time index $k = M$ and i be such that $0 \le i \le k$. Let the $n \times n$ matrix

$$P(k) = (\Phi^\top \Phi)^{-1} = \left(\sum_{i=1}^{k} \zeta^i (\zeta^i)^\top \right)^{-1} \tag{4.13}$$

and let $\theta(k-1)$ denote the least squares estimate based on $k-1$ data pairs ($P(k)$ is called the "covariance matrix"). Assume that $\Phi^\top \Phi$ is nonsingular for all k. We have $P^{-1}(k) = \Phi^\top \Phi = \sum_{i=1}^{k} \zeta^i (\zeta^i)^\top$ so we can pull the last term from the summation to get

$$P^{-1}(k) = \sum_{i=1}^{k-1} \zeta^i (\zeta^i)^\top + \zeta^k (\zeta^k)^\top$$

and hence

$$P^{-1}(k) = P^{-1}(k-1) + \zeta^k (\zeta^k)^\top. \tag{4.14}$$

Now, using (4.9) we have

$$\begin{aligned}
\theta(k) &= (\Phi^\top \Phi)^{-1} \Phi^\top Y \\
&= \left(\sum_{i=1}^{k} \zeta^i (\zeta^i)^\top \right)^{-1} \left(\sum_{i=1}^{k} \zeta^i y^i \right) \\
&= P(k) \left(\sum_{i=1}^{k} \zeta^i y^i \right) = P(k) \left(\sum_{i=1}^{k-1} \zeta^i y^i + \zeta^k y^k \right). \quad (4.15)
\end{aligned}$$

Hence,

$$\theta(k-1) = P(k-1) \sum_{i=1}^{k-1} \zeta^i y^i$$

so that

$$P^{-1}(k-1)\theta(k-1) = \sum_{i=1}^{k-1} \zeta^i y^i.$$

Now, replacing $P^{-1}(k-1)$ in this equation with the result in (4.14), we get

$$(P^{-1}(k) - \zeta^k (\zeta^k)^\top)\theta(k-1) = \sum_{i=1}^{k-1} \zeta^i y^i.$$

Using the result from (4.15), this gives us

$$\begin{aligned}
\theta(k) &= P(k)(P^{-1}(k) - \zeta^k (\zeta^k)^\top)\theta(k-1) + P(k)\zeta^k y^k \\
&= \theta(k-1) - P(k)\zeta^k (\zeta^k)^\top \theta(k-1) + P(k)\zeta^k y^k \\
&= \theta(k-1) + P(k)\zeta^k (y^k - (\zeta^k)^\top \theta(k-1)). \quad (4.16)
\end{aligned}$$

This provides a method to compute an estimate of the parameters $\theta(k)$ at each time step k from the past estimate $\theta(k-1)$ and the latest data pair that we received, (ζ^k, y^k). Notice that $(y^k - (\zeta^k)^\top \theta(k-1))$ is the error in predicting y^k using $\theta(k-1)$.

To update θ in (4.16) we need $P(k)$, so we could use

$$P^{-1}(k) = P^{-1}(k-1) + \zeta^k (\zeta^k)^\top. \tag{4.17}$$

But then we will have to compute an inverse of a matrix at each time step (i.e., each time we get another input output data pair). Clearly, this computation is not desirable for real-time implementation, so we would like to avoid it. To do so we will use the following **matrix inversion lemma**:

Lemma 4.1: *If $\theta \in \mathrm{R}^{n \times n}, B \in \mathrm{R}^{n \times m}, C \in \mathrm{R}^{m \times n}$, then*

$$(A + BCD)^{-1} = A^{-1} - A^{-1}B(C^{-1} + DA^{-1}B)^{-1}DA^{-1} \tag{4.18}$$

provided A, C, and $(C^{-1} + DA^{-1}B)$ are nonsingular square matrices.

We will use the matrix inversion lemma to remove the need to compute the inverse of $P^{-1}(k)$ that comes from (4.17) so that it can be used in (4.16) to update θ. Notice that

$$\begin{aligned} P(k) &= (\Phi^\top(k)\Phi(k))^{-1} \\ &= (\Phi^\top(k-1)\Phi(k-1) + \zeta^k(\zeta^k)^\top)^{-1} \\ &= (P^{-1}(k-1) + \zeta^k(\zeta^k)^\top)^{-1} \end{aligned}$$

and that if we use the matrix inversion lemma with $A = P^{-1}(k-1)$, $B = \zeta^k$, $C = I$, and $D = (\zeta^k)^\top$, we get

$$P(k) = P(k-1) - P(k-1)\zeta^k(I + (\zeta^k)^\top P(k-1)\zeta^k)^{-1}(\zeta^k)^\top P(k-1), \tag{4.19}$$

which, together with

$$\theta(k) = \theta(k-1) + P(k)\zeta^k(y^k - (\zeta^k)^\top\theta(k-1)) \tag{4.20}$$

(which was derived in (4.16)), is called the "recursive least squares (RLS) algorithm." Basically, the matrix inversion lemma turns a matrix inversion into the inversion of a scalar (i.e., the term $(I + (\zeta^k)^\top P(k-1)\zeta^k)^{-1}$ is a scalar).

We need to initialize the RLS algorithm (i.e., choose $\theta(0)$ and $P(0)$). One approach to do this is to use $\theta(0) = 0$ and $P(0) = P_0$ where $P_0 = \alpha I$ for some large $\alpha > 0$. This is the choice that is often used in practice. Other times, you may pick $P(0) = P_0$ but choose $\theta(0)$ to be the best guess of the true parameters.

There is a "weighted recursive least squares" (WRLS) algorithm also. Suppose that the parameters of the physical system θ vary slowly. In this case it may be advantageous to choose

$$J(\theta, k) = \frac{1}{2}\sum_{i=1}^{k} \lambda^{k-i}(y^i - (\zeta^i)^\top\theta)^2,$$

where $0 < \lambda \leq 1$ is called a "forgetting factor" since it gives the more recent data higher weight in the optimization. Using a similar approach to the above, you can show that the equations for WRLS are given by

$$\begin{aligned} P(k) &= \frac{1}{\lambda}\left(I - P(k-1)\zeta^k(\lambda I + (\zeta^k)^\top P(k-1)\zeta^k)^{-1}(\zeta^k)^\top\right)P(k-1) \\ \theta(k) &= \theta(k-1) + P(k)\zeta^k(y^k - (\zeta^k)^\top\theta(k-1)), \end{aligned} \tag{4.21}$$

where when $\lambda = 1$ we get standard RLS.

It is important to point out that for the RLS (weighted or regular) we must have a "sufficiently rich input signal" $u(k)$ to ensure that the algorithm works properly. Sufficient richness in a signal simply ensures that the input will excite the system enough so that the input-output data that is loaded in the sequence of regression vectors will give us enough information to determine the parameters of the system. Specifically, recall that for the above derivation we had to assume that $\Phi^\top\Phi$ is nonsingular for all k. The question is what happens if this term becomes singular. For instance, suppose that $\zeta^k = 0$, $k \geq 0$ (which is certainly not a sufficiently rich signal) and notice that in this case (4.21) becomes

$$\begin{aligned} P(k) &= \frac{1}{\lambda}P(k-1) \\ \theta(k) &= \theta(k-1). \end{aligned} \tag{4.22}$$

If we had chosen $\theta(0) = 0$ and $P(0) = \alpha I$ for some large $\alpha > 0$ (much bigger than one) then $\theta(k) = 0$ for all $k \geq 0$ and as $k \to \infty$ the diagonal elements of $P(k)$ all approach infinity.

Example 4.2 Consider the discrete-time system defined by

$$y(k) = ay(k-1) + b(k)u(k-1), \tag{4.23}$$

where $a = 0.9$ and $b(k) = 1 + 0.2\sin(0.02\pi k)$. Since $b(k)$ is a time-varying coefficient, the recursive least squares routine with a forgetting factor may be used to estimate a and $b(k)$ using $\zeta(k) = [y(k-1), u(k-1)]^\top$. As the value of λ is decreased, the RLS routine will tend to "forget" older input-output samples more quickly.

Figure 4.1 shows the true value of $b(k)$ along with RLS estimates when $\lambda = 0.2, 0.8, 0.99$ where the input is defined by $u(k) = 0.5\sin(0.2\pi k) + \cos(0.1\pi k)$. As $\lambda \to 1$, the RLS with forgetting factor converges to the batch least squares routine, where a constant which minimizes the sum of the squared errors is estimated for $b(k)$. Even though using $\lambda = 0.2$ in this case caused the RLS estimate to accurately track the true value of $b(k)$, using small values for λ will tend to make errors in the parameter estimates more sensitive to noise in the measurements of y_i and ζ_i. $\triangle$

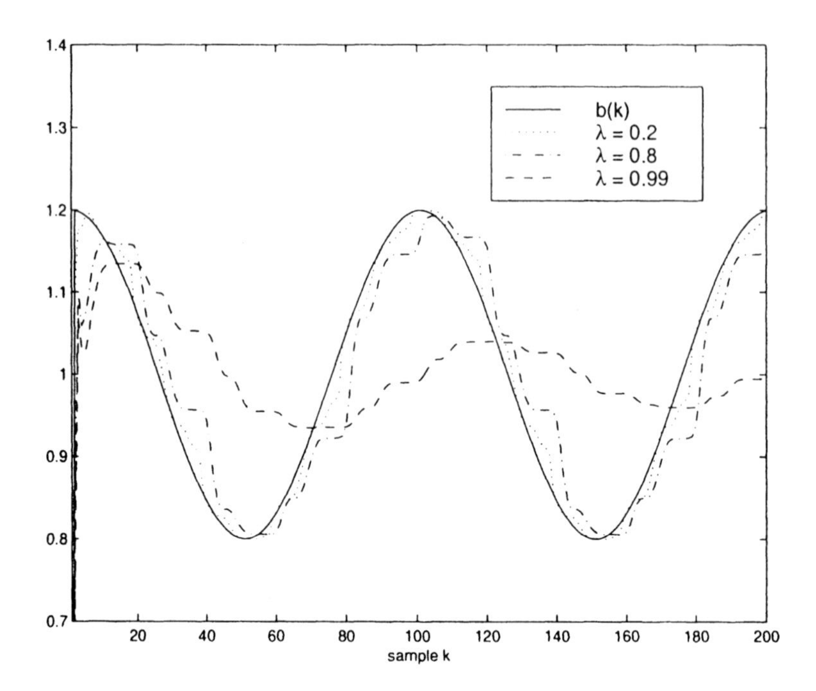

Figure 4.1. Recursive least squares estimate of $b(k)$ using various forgetting factors.

4.4 Nonlinear Least Squares

While in the last section we studied the use of linear in the parameter approximators $\mathcal{F}(x,\theta) = \frac{\partial \mathcal{F}}{\partial \theta}\theta$ to minimize (4.7), here we will also seek to minimize (4.7) (with the $w_i = 1$), but we will consider the adjustment of θ for the general nonlinear in the parameter problem for $\mathcal{F}(x,\theta)$, for the remainder of this chapter. First, we explain how to use gradient methods to adjust θ for only a single training data pair. Next, we generalize the development to the case of multiple (and sequential) training data, and the discrete time case (throughout we will discuss the important issue of convergence). We discuss the constrained optimization problem and close the chapter with a brief treatment of line search and high order techniques for function approximation. This last section will be particularly useful to those who are only concerned with the off-line training of approximators (e.g., for estimators), and in cases when you want to perform off-line tuning of an approximator before adjusting it on-line.

4.4.1 Gradient Optimization: Single Training Data Pair

Consider the situation in which it is desired to cause an approximator $\mathcal{F}(x,\theta)$ to match the function $f(x)$ at only a single point x^1 where $y^1 = f(x^1)$. Given an input x^1 one would like to adjust θ so that the difference between the desired output and approximator output

$$e = y^1 - \mathcal{F}(x^1, \theta)$$

is reduced, where y may be either vector or scalar valued. In terms of an optimization problem, we want to minimize the cost function

$$J(\theta) = \frac{1}{2} e^\top e.$$

Taking infinitesimal steps along the negative gradient of $J(\theta)$ with respect to θ will ensure that $J(\theta)$ is nonincreasing. That is, choose

$$\dot{\theta} = -\eta \left. \frac{\partial J(z)}{\partial z} \right|_{z=\theta}^\top, \tag{4.24}$$

where $\eta > 0$ is a constant and if $\theta = [\theta_1, \ldots, \theta_p]^\top$, then

$$\frac{\partial J(\theta)}{\partial \theta} = \left[\frac{\partial J(\theta)}{\partial \theta_1}, \ldots, \frac{\partial J(\theta)}{\partial \theta_p} \right].$$

To see that J is nonincreasing when θ is adjusted according to (4.24), notice that

$$\frac{dJ(\theta(t))}{dt} = \left. \frac{\partial J(z)}{\partial z} \right|_{z=\theta} \dot{\theta} = -\eta \left| \frac{\partial J}{\partial \theta} \right|^2 \tag{4.25}$$

so J is nonincreasing.

Using the definition for $J(\theta)$ we get

$$\dot{\theta} = -\frac{\eta}{2} \frac{\partial e^\top e}{\partial \theta}$$

or

$$\dot{\theta} = -\frac{\eta}{2} \frac{\partial}{\partial \theta} (y^1 - \mathcal{F}(x^1, \theta))^\top (y^1 - \mathcal{F}(x^1, \theta)),$$

so that

$$\dot{\theta} = -\frac{\eta}{2} \frac{\partial}{\partial \theta} ({y^1}^\top y^1 - 2\mathcal{F}(x^1, \theta)^\top y^1 + \mathcal{F}(x^1, \theta)^\top \mathcal{F}(x^1, \theta)).$$

Now, taking the partial derivative, we get

$$\begin{aligned} \dot{\theta} &= -\eta \left(-\frac{\partial \mathcal{F}(x^1, \theta)^\top}{\partial \theta} y^1 + \frac{\partial \mathcal{F}(x^1, \theta)^\top}{\partial \theta} \mathcal{F}(x^1, \theta) \right) \\ &= \eta \left. \frac{\partial \mathcal{F}(x^1, z)}{\partial z} \right|_{z=\theta}^\top (y^1 - \mathcal{F}(x^1, \theta)), \end{aligned}$$

so

$$\dot{\theta} = \eta\zeta(x^1, \theta)e, \tag{4.26}$$

where $\eta > 0$, and

$$\zeta(x^1, \theta) = \left.\frac{\partial \mathcal{F}(x^1, z)}{\partial z}\right|_{z=\theta}^{\top} \tag{4.27}$$

is the gradient of $\mathcal{F}$ with respect to θ. Since the parameters are updated along the gradient, this is often referred to as a "gradient update law." We will find that each of the update laws presented later in the development of adaptive controllers will use information about the gradient. The above results are summarized in the following theorem, the proof of which is obvious since J is nonincreasing:

Theorem 4.1: *Given the approximator $\mathcal{F}(x^1, \theta)$ such that $\zeta(x^1, \theta)$ in (4.27) is well defined with adaptation law (4.26) will ensure that $e \in \mathcal{L}_\infty$.*

Example 4.3 Consider learning the input-output relationship of a function evaluated at $y^1 = f(x^1)$, with an approximator (neural network) defined by

$$\mathcal{F}(x^1, \theta) = \alpha_1 + \tanh(\alpha_2 x^1),$$

where $\theta = [\alpha_1, \alpha_2]^\top$.

Choosing $V = e^2$ and $\eta > 0$ we find

$$\dot{V} = -2\eta e\zeta^\top \zeta e \tag{4.28}$$

when using the update law (4.26). From the above definition of the approximator, we know that if $\theta = [\alpha_1, \alpha_2]^\top$, then

$$\begin{aligned} \zeta_1 &= 1 \\ \zeta_2 &= \frac{\partial \tanh(s)}{\partial s} x^1, \end{aligned}$$

where $s = \alpha_2 x^1$. Since $\zeta^\top \zeta = 1 + (\frac{\partial \tanh(s)}{\partial s} x^1)^2 \geq 1$, we find

$$\dot{V} \leq -2\eta e^2 = -2\eta V, \tag{4.29}$$

so that $e = 0$ is an exponentially stable equilibrium point. This shows that the gradient law will update the parameter vector θ so that the approximator will exactly match the function at x^1. △

4.4.2 Gradient Optimization: Multiple Training Data Pairs

So far we have seen how to adjust the weights of a nonlinear approximator when a single data pair (x^1, y^1) is to be matched. Now consider the problem when M input-output pairs, or patterns, (x^i, y^i) where $y^i = f(x^i)$ are to be matched for $i = 1, \ldots, M$. In this case, we let

$$e^i = y^i - \mathcal{F}(x^i, \theta),$$

and let the cost function be

$$J(\theta) = \sum_{j=1}^{M} {e^i}^{\mathsf{T}} e^i. \tag{4.30}$$

Using an approach similar to the single input-output pair case, you can show that the gradient update law is defined by

$$\dot{\theta} = \eta \sum_{j=1}^{M} \zeta^i e^i, \tag{4.31}$$

where

$$\zeta^i = \left. \frac{\partial \mathcal{F}(x^i, z)}{\partial z} \right|_{z=\theta}^{\mathsf{T}}. \tag{4.32}$$

This update law will adjust the approximator parameters such that $J(\theta)$ does not increase over time, as stated in the following theorem:

Theorem 4.2: *Given the approximator $\mathcal{F}(x, \theta)$ such that each ζ^i in (4.32) is well defined with adaptation law (4.31) will ensure that $e^i \in \mathcal{L}_\infty$ for $i = 1, \ldots, M$.*

Proof: Let $V = J(\theta)$. Taking the time derivative yields

$$\dot{V} = -2 \sum_{j=1}^{M} {e^i}^{\mathsf{T}} {\zeta^i}^{\mathsf{T}} \dot{\theta} \tag{4.33}$$

$$= -2 \left[\sum_{j=1}^{M} {e^i}^{\mathsf{T}} {\zeta^i}^{\mathsf{T}} \right] \dot{\theta}. \tag{4.34}$$

Choosing

$$\dot{\theta} = \eta \sum_{j=1}^{M} \zeta^i e^i, \tag{4.35}$$

will ensure that $\dot{V} \leq 0$ so that V is a positive nonincreasing variable which implies that the output error for each data pair is bounded for all time. ■

The update law defined by (4.31) is a continuous version of the batch "backpropagation algorithm" which is common in the neural network literature. Once each ζ_i is computed, it is rather easy to apply. To see how to calculate each ζ_i consider the following example:

Example 4.4 Consider the approximator (multilayer perceptron) defined by

$$\mathcal{F}(x,\theta) = \sum_{j=1}^{q} c_j \psi_j \left(\sum_{i=1}^{p} w_{ij} x_i + b_j \right) + d, \tag{4.36}$$

where $\theta = [d, c_1, \ldots, c_q, b_1, \ldots, b_q, w_{11}, \ldots, w_{pq}]^\top$. The gradients are defined as follows:

$$\begin{aligned}
\frac{\partial \mathcal{F}(x,\theta)}{\partial d} &= 1 \\
\frac{\partial \mathcal{F}(x,\theta)}{\partial c_j} &= \psi_j \left(\sum_{i=1}^{p} w_{ij} x_i + b_j \right) \\
\frac{\partial \mathcal{F}(x,\theta)}{\partial b_j} &= \frac{\partial \mathcal{F}(x,\theta)}{\partial s_j} \frac{\partial s_j}{\partial b_j} \\
\frac{\partial \mathcal{F}(x,\theta)}{\partial w_{ij}} &= \frac{\partial \mathcal{F}(x,\theta)}{\partial s_j} \frac{\partial s_j}{\partial w_{ij}},
\end{aligned}$$

where $s_j = \sum_{i=1}^{p} w_{ij} x_i + b_j$. Since $\partial \mathcal{F}(x,\theta)/\partial s_j = c_j \partial \psi_j(s_j)/\partial s_j$, we find

$$\begin{aligned}
\frac{\partial \mathcal{F}(x,\theta)}{\partial b_j} &= c_j \frac{\partial \psi_j(s_j)}{\partial s_j} \\
\frac{\partial \mathcal{F}(x,\theta)}{\partial w_{ij}} &= c_j \frac{\partial \psi_j(s_j)}{\partial s_j} x_i.
\end{aligned}$$

$\triangle$

The cost function for multiple data pairs in (4.30) is typically more useful in off-line training than the one for a single data pair since it may be used to cause an approximator $\mathcal{F}(x,\theta)$ to approximate the continuous function $f(x)$ over $x \in D \subset \mathbb{R}^n$. In other words, it may be possible to choose some θ such that

$$\sup_{x \in D} |f(x) - \mathcal{F}(x,\theta)| < \epsilon, \tag{4.37}$$

where $\epsilon > 0$ is some small constant. It is our hope that if we choose the set $\{x^1, \ldots, x^M\}$ with $x^i \in D$ such that the x^i's are uniformly distributed throughout D, and we choose the approximation structure properly, then

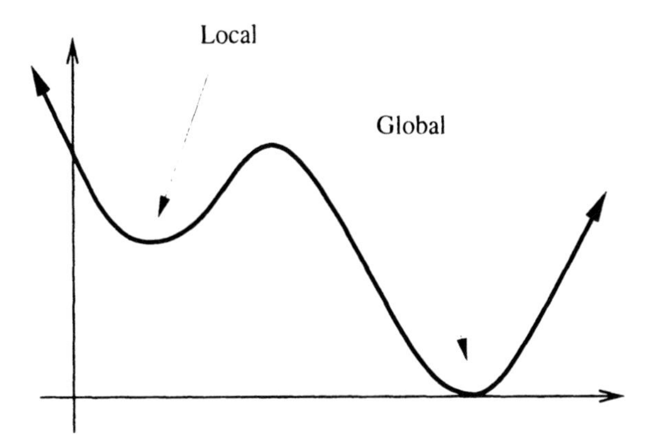

Figure 4.2. Local and global minima in an error surface.

finding some θ such that $|J| \to 0$ will imply that $|f(x) - \mathcal{F}(x,\theta)| \to 0$ on D provided that M is large enough.

Notice that the theorems for the update laws only guarantee that the magnitude of the error will not increase; they do not ensure that $|e^i| \to 0$ for $i = 1, 2, \ldots, M$. The update laws are gradient-based algorithms which modify θ in the direction which decreases $J(\theta)$. Depending upon the approximator structure, situations may exist such that both local and global minima exist as shown in Figure 4.2. A global minimum is the set of approximator parameters such that

$$\theta^* = \arg\min_{\theta \in \mathcal{S}} J(\theta). \tag{4.38}$$

That is, θ^* is the set of approximator parameters which minimizes $J(\theta)$ over all $\mathcal{S}$. A local minimum is found when an arbitrarily small change in $\theta \neq \theta^*$ in any direction will not decrease J. Since $\partial J/\partial\theta = 0$ at a local minimum, $\dot{\theta} = 0$ so that the approximator parameters stop adjusting. Clearly, we would like to find a global minimum, but for multiple data pairs this may be difficult.

Example 4.5 In this example, we will use (4.35) to update the parameters of an approximator (which may be considered to be either a radial basis function neural network or a type of fuzzy system) defined by

$$\mathcal{F}(x,\theta) = \frac{\sum_{i=1}^{p} a_i e^{-((x-c_i)/\sigma)^2}}{\sum_{i=1}^{p} e^{-((x-c_i)/\sigma)^2}}, \tag{4.39}$$

where $p = 20$, $\sigma = 0.5$, and the centers c_i are evenly distributed in

$[0, 2\pi]$. Notice that $\mathcal{F}(x, \theta) = \theta^\top \zeta(x)$, where

$$\zeta_i(x) = \frac{e^{-((x-c_i)/\sigma)^2}}{\sum_{i=1}^{p} e^{-((x-c_i)/\sigma)^2}}.$$

A total of $M = 50$ data points were randomly chosen in $[0, 2\pi]$ for training purposes and are shown in Figure 4.3 as o's along with each $\zeta_i(x)$.

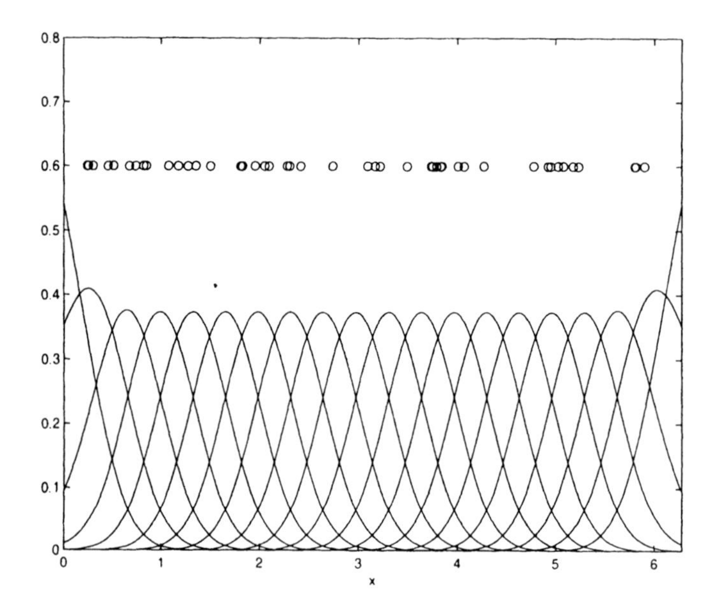

Figure 4.3. The basis functions ζ_i are weighted Gaussian functions with $p = 20$. Each "o" represents an x_i, where $M = 50$.

Using $f(x) = \sin(x)$, the approximator parameter vector θ was updated according to (4.35) for $1sec$ with $\eta = 10$ and $\theta(0) = 0$. Figure 4.4 shows the value of $\sin(x)$ with a solid line and $\mathcal{F}(x, \theta)$ with a dashed line. If now $f(x) = \cos^2(x)$, it is possible to once again update θ so that $\mathcal{F}(x, \theta)$ approximates $\cos^2(x)$ as shown in Figure 4.5.

Notice that in each case, $\mathcal{F}(x, \theta)$ does not approximate the given $f(x)$ near $x = 2\pi$ as well as is does in other regions. Referring back to Figure 4.3, we notice that there are relatively few data points near $x = 2\pi$. This lack of information causes a degradation in the approximation. If there had been no x^i in the region $[\pi, 2\pi]$, then we could not expect the approximator to represent $f(x)$ over that entire region. Thus this approach to function approximation not only requires that a stable update law be defined for θ, but one must also

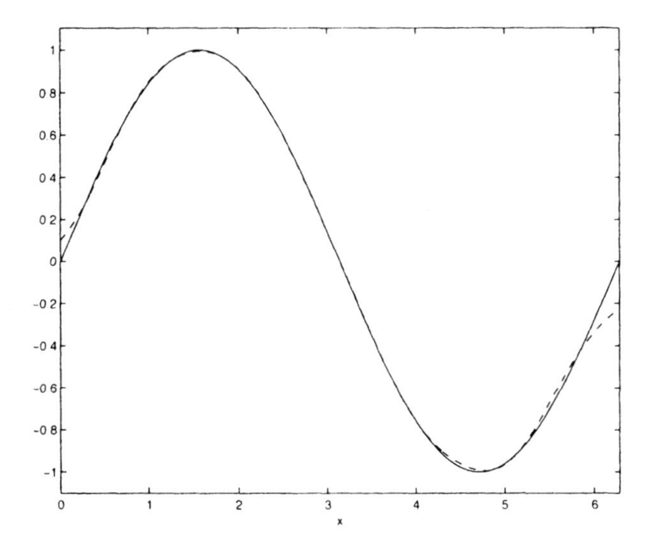

Figure 4.4. The solid line is $\sin(x)$, while the dashed line is its approximation.

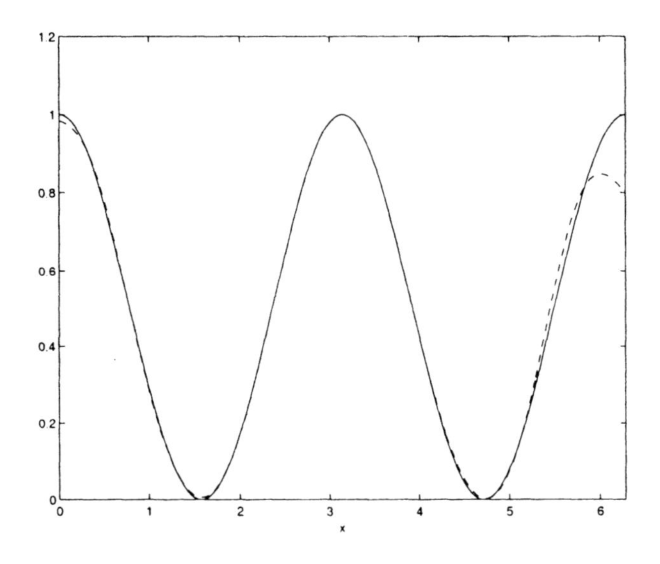

Figure 4.5. The solid line is $\cos^2(x)$, while the dashed line is its approximation.

provide sufficient information before a function may be accurately approximated over all regions. As we will find later, another factor which influences how well an approximator is able to match a function $f(x)$ is the shape and number of ζ_i (we will in particular be interested in the case where each ζ_i is defined using a neural network or fuzzy system). This issue will be addressed with more detail in Chapter 5. △

Here, we saw that a single approximator structure may be used to describe multiple functions simply by changing the value of the parameter vector θ. We will use this idea in the adaptive control of systems where the dynamics are not necessarily known. If $\dot{x} = f(x) + u$, where $f(x)$ is an unknown function, then it may be possible to represent $f(x)$ by $\mathcal{F}(x, \theta^*)$ where θ^* is chosen so that $f(x) = \mathcal{F}(x, \theta^*)$. Letting $u = -\mathcal{F}(x, \theta^*) - kx$ will cause $x = 0$ to be an exponentially stable equilibrium point. Since appropriately choosing θ^* will allow $\mathcal{F}(x, \theta^*)$ to approximate an entire class of functions (as shown in the previous example), the controller $u = -\mathcal{F}(x, \theta^*) - kx$ may be used to control an entire class of systems provided that θ^* is appropriately chosen for each $f(x)$.

4.4.3 Discrete Time Gradient Updates

Next, we consider transforming the continuous gradient descent algorithms into their discrete counterparts. If the input-output pairs (x^i, y^i), $i = 1, \ldots, M$, for which $y^i = f(x^i)$ are to be learned, then we may either adjust the approximator parameters on a single pair (x^i, y^i) at a time (series updating) or based upon the entire collection of data pairs (batch updating).

Series updating is accomplished by selecting a pair (x^i, y^i), where $i \in \{1, \ldots, M\}$ is a random integer chosen at each iteration, and then using Euler's first order approximation of (4.26) so that the parameter update is defined by

$$\theta(k+1) = \theta(k) + \eta {\zeta^i}^\top(k) e(k), \tag{4.40}$$

where k is the iteration step, $e(k) = y^i - \mathcal{F}(x^i, \theta(k))$, and

$$\zeta^i(k) = \left. \frac{\partial \mathcal{F}(x^i, z)}{\partial z} \right|_{z=\theta(k)}^\top .$$

We have absorbed the length of the sampling interval into the adaptation gain η. Since a random presentation of the data pairs is used, the value of θ tends towards a value with minimizes $\sum_{i=1}^{M} (e^i)^2$ on average. A second approach is to use a discretized version of (4.31) so that all the data pairs are considered at once. An Euler approximation gives the update law

$$\theta(k+1) = \theta(k) + \eta \sum_{i=1}^{M} \zeta^i(k) e^i(k), \tag{4.41}$$

where $\eta > 0$. This is often referred to as a gradient update law or **batch back propagation** in the neural network community.

In the derivation of the continuous gradient-based update laws, the learning rate, η, was allowed to be any positive number. Using a discrete update law, however, if η is made to be too large, then the parameter error may converge slowly, oscillate about a fixed value, or diverge, all of which one would like to avoid when updating parameters. The following example shows that if $\eta > 0$ is chosen too large, then the discrete gradient-based algorithms may no longer provide appropriate values for θ.

Example 4.6 Consider the case where the desired approximator output is $y = \mathcal{F}(x, \theta^*)$ where θ^* is a set of ideal parameters which cause the output of the approximator to be y^1 whenever the input x^1 is presented (considering the case where only a single data pair is to be learned). The output error is

$$e(k) = y^1 - \mathcal{F}(x^1, \theta(k)) \tag{4.42}$$

$$= \mathcal{F}(x^1, \theta^*) - \mathcal{F}(x^1, \theta(k)), \tag{4.43}$$

where $\theta(k)$ is the current estimate of θ^*. Defining the parameter error as $\tilde{\theta}(k) = \theta(k) - \theta^*$, a linear representation may be expressed as

$$e(k) = -\zeta^\top \tilde{\theta}(k) + \delta(x^1, \theta, \theta^*). \tag{4.44}$$

Here $|\delta(x^1, \theta, \theta^*)| \leq L|\tilde{\theta}^2|$, with $L > 0$ a finite constant, is the error in representing $e(k)$ by a linear expression (more details on the derivation of δ will be provided in Chapter 5). Here we will assume that we initialize θ such that $|\tilde{\theta}(0)|$ is small and thus $|\delta(x^1, \theta, \theta^*)| \approx 0$. To show that the learning rate, η, needs to be limited for the discrete case, consider the parameter error metric $V(k) = \tilde{\theta}^\top(k)\tilde{\theta}(k)$ (if $\tilde{\theta} \to 0$, then $e(k) \to 0$).

The change in $V(k)$ during an update of the weights is

$$V(k+1) - V(k) = \tilde{\theta}^\top(k+1)\tilde{\theta}(k+1) - \tilde{\theta}(k)^\top \tilde{\theta}(k).$$

Substituting in the update law (4.41),

$$V(k+1) - V(k) = 2\eta\tilde{\theta}^\top(k)\zeta(k)e(k) + \eta^2 e^\top(k)\zeta^\top(k)\zeta(k)e(k),$$

where we have used $\theta(k+1) = \tilde{\theta}(k+1)$. Since $\delta \approx 0$, we have

$$V(k+1) - V(k) \approx -\eta e^\top(k)\left[2I - \eta\zeta^\top(k)\zeta(k)\right]e(k). \tag{4.45}$$

Thus if $0 < \lambda_{\min}(2I - \eta\zeta^\top(k)\zeta(k))$ for all k, then $V(k+1) - V(k) \leq 0$. As η becomes large, however, the boundedness of the approximator output error is no longer guaranteed (that is, the algorithm can become unstable because it will not be the case that $0 < \lambda_{\min}(2I - \eta\zeta^\top(k)\zeta(k))$ so that $V(k)$ increases with k). △

This example shows that one must be careful not to ask for too large of an adaptation rate when dealing with discrete updates. We will see later that this is also true for discrete-time adaptive control systems.

4.4.4 Constrained Optimization

So far, we have not placed any restrictions upon the possible values of θ^*. In some situations, however, we may have a priori knowledge about the feasible values for θ^*. In this case, a constrained optimzation approach may be taken as we discuss next.

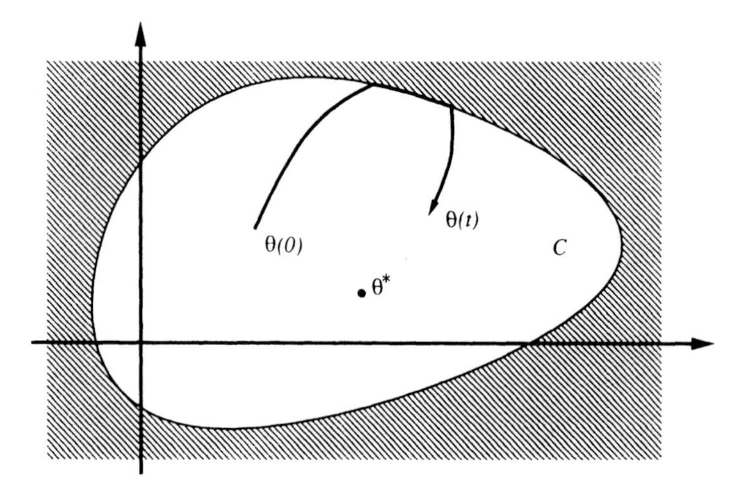

Figure 4.6. Constrained optimization using a projection algorithm.

If it is known that the ideal parameters θ^* belong to a convex set C, then it is possible to modify the above adaptation routines to ensure that they remain within C. We will, in particular, consider the use of a "projection algorithm." Figure 4.6 shows how the projection algorithm works. If the parameters are within C, then the trajectory defined by $\dot{\theta}$ is not changed. If θ reaches the boundary of C (denoted by $\mathcal{B}$), however, then $\dot{\theta}$ must be modified such that θ will not leave C and in particular so that it stays on $\mathcal{B}$ until it moves toward the interior of C.

If we are using, for example, an update law defined by $\dot{\theta} = \eta v(t)$ where $\eta > 0$, then this may be redefined to incorporate the projection as

$$\dot{\theta} = \begin{cases} \Pr(\eta v) & \text{if } \theta \in \mathcal{B} \text{ and } v^\top b_\perp > 0 \\ \eta v & \text{otherwise,} \end{cases} \tag{4.46}$$

where $\Pr(x)$ is the projection of x onto the the hyperplane tangent to $\mathcal{B}$ at θ and $b_\perp$ is the unit vector perpendicular to the hyperplane pointing outward at θ. In this way, only the component of the update which does not move θ outside of $\mathcal{C}$ is used in the update. If a cost function (or Lyapunov function in the study of adaptive systems) defined by $V = \tilde{\theta}^\top \tilde{\theta}$ is used in the stability analysis of the update algorithm with $\tilde{\theta} = \theta - \theta^*$, then the stability results are unaffected by the projection. This is because if we modify $\dot{\theta}$ so that θ does not move outside $\mathcal{C}$, then $|\theta - \theta^*|$ is smaller because of the projection since $\theta^* \in \mathcal{C}$.

When $V = \tilde{\theta}^\top P \tilde{\theta}$ where $P > 0$, the projection algorithm must be changed slightly to ensure that V decreases at least as fast as the case when no projection is used. First we let

$$P = P_1 P_1^\top,$$

so that $V = \tilde{\theta}^\top P_1 P_1^\top \tilde{\theta}$. Using the change of coordinates $\bar{\theta} = P_1^\top \tilde{\theta}$, notice that $V = \bar{\theta}^\top \bar{\theta}$. A standard projection algorithm may now be used for $\bar{\theta}$ by also transforming $\mathcal{C}$ to $\bar{\mathcal{C}}$ and $\mathcal{B}$ to $\bar{\mathcal{B}}$. Since the transformation $\bar{\theta} = P_1^\top \tilde{\theta}$ is linear, $\bar{\mathcal{C}}$ will still be convex.

Example 4.7 Suppose that we wish to design a projection algorithm for the case where $\theta \in \mathcal{C}$, with

$$\mathcal{C} = \left\{ \theta = [\theta_1, \ldots \theta_p]^\top \in \mathrm{R}^p : b_i \leq a_i \leq c_i, \text{for } i = 1, \ldots, p \right\}, \quad (4.47)$$

so that each element of θ is confined to an interval. To do this, as we update the θ_i, if $b_i \leq \theta_i \leq c_i$ then you use the update generated by the gradient method. However, if the update law tells you that the parameter θ_i should go outside the interval, then you place its value on the interval edge. Moreover, if the value of θ_i lies on either edge of the interval and the update law says the next value of the parameter should be in the interval then the update law is allowed to place it there. Clearly such a projection law works for both continuous and discrete time gradient update laws and it is very easy to implement in code. △

4.4.5 Line Search and the Conjugate Gradient Method

Control algorithms that use an approximator with parameters that are modified in real time are referred to as adaptive or on-line approximation techniques. The adaptive estimation and control methods presented later in this book use the least squares and gradient methods presented earlier in this chapter, and these will be shown to provide stable operation for on-line estimation and control methods. In this section, we will depart from the main focus of the book to focus on off-line training (optimization) techniques. These methods can be useful for constructing (nonadaptive) estimators, and for a priori training of approximators that will later be used in an on-line fashion (e.g., in indirect adaptive control). For the off-line training of the approximator $\mathcal{F}(x, \theta)$ to match some unknown nonlinear function $f(x)$ we will not be concerned here with how $\mathcal{F}(x, \theta)$ will cause some system dynamics to behave; here we are only concerned with adjusting θ to make $\mathcal{F}(x, \theta)$ match $f(x)$ as closely as possible.

For example, the Levenberg-Marquardt and Conjugate Gradient optimization methods are popular approaches for neural network and fuzzy system training. Here, we will discuss a line search method and the Conjugate Gradient method.

Line Search

When the optimization problem is reduced to a single dimension, a number of techniques may be used to efficiently find a minimum along the search dimension. Each of these typically requires that a minimum be bracketed such that given points $a < b < c$, we have $J(b) < J(a)$ and $J(b) < J(c)$ so that one or more minimum exists between a and c. Once the minimum has been bracketed, a routine such as the **golden section** search, which is outlined below, may be used to iteratively find its location:

1. Choose values $a < b < c$ such that $J(b) < J(a)$ and $J(b) < J(c)$. Let $R = 0.38197$ (the "golden ratio").
2. If $|c - b| > |b - a|$, then let $t_1 = b$ and $t_2 = b + R(c - b)$, otherwise let $t_1 = b - R(b - a)$ and $t_2 = b$.
3. If $J(t_2) < J(t_1)$, then let $a = t_1$, $t_1 = t_2$, and $t_2 = t_2 + R(c - t_2)$, otherwise $c = t_2$, $t_2 = t_1$, and $t_1 = t_1 - R(t_1 - a)$.
4. If $|c - a| > tol$, go to step 3.
5. If $J(t_1) < J(t_2)$, then return t_1, otherwise return t_2.

There exists a number of other line minimization routines such as Brent's algorithm, which may provide improved convergence. See [181] for further discussion.

Example 4.8 Consider the minimization of the function

$$y = (x - 1)^2 + 1, \tag{4.48}$$

which is minimized with $x = 1$. The golden section search may be used in the minization given some initial bracketing values a, b, and c. Choosing $a = 0$, $b = 0.1$, and $c = 10$, the golden section search is able to minimize (4.48). Figure 4.7 shows the progression of the bracketing values a, t_1, t_2, and c. Notice that the algorithm converges to $x = 1$. △

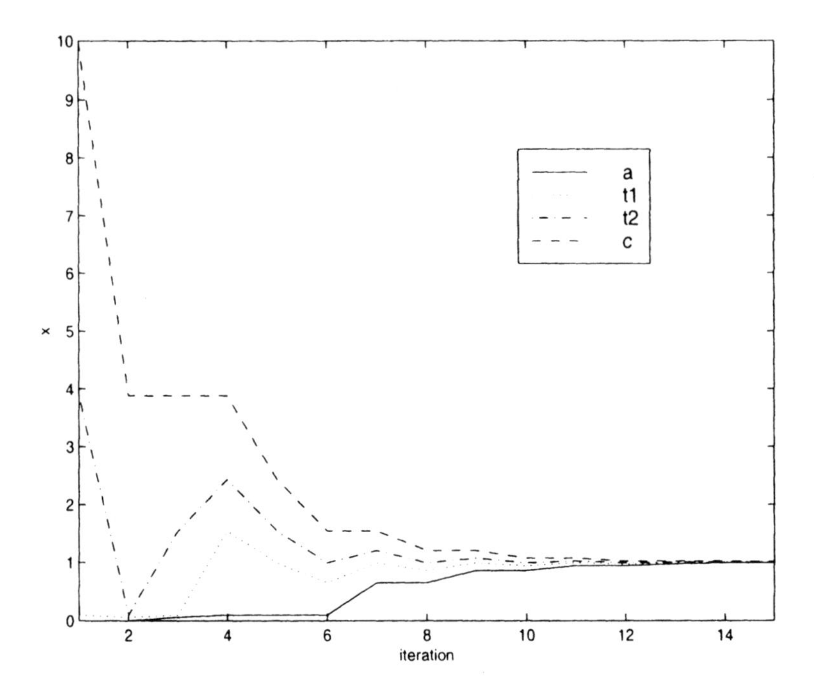

Figure 4.7. Bracketing in the golden section search.

The Conjugate Gradient Method

Consider the general minimization of

$$J = \sum_{j=1}^{M} (e^j)^\top e^j. \tag{4.49}$$

If $e^j = (y^j) - \mathcal{F}(x^j, \theta)$, then

$$\frac{\partial J}{\partial \theta} = -2 \sum_{j=1}^{M} (e^j)^\top \frac{\partial \mathcal{F}(x^j, \theta)}{\partial \theta}. \tag{4.50}$$

If $\theta(k)$ is a guess of θ^* (the value which minimizes J), then let

$$d(k) = \left. \frac{\partial J}{\partial \theta} \right|_{\theta = \theta(k)}^{\top} \tag{4.51}$$

be the "search direction." Since $d(k)$ is along the negative gradient, J will decrease as we move along $\theta(k) + \eta d(k)$ where $\eta > 0$ is the search length with $\theta(k)$ and $d(k)$ held constant. In fact, J will continue to decrease until the gradient in the search direction becomes zero (the minimum occurs when the gradient goes to zero). That is, until

$$d(k)^{\top} \left. \frac{\partial J}{\partial \theta} \right|_{\theta=\theta(k)+\eta d} = 0. \tag{4.52}$$

Once the minimum along d is found, the procedure is repeated with $\theta(k+1) = \theta(k) + \eta d(k)$ until J converges to some desired value or no longer decreases. This is called the method of "steepest descent."

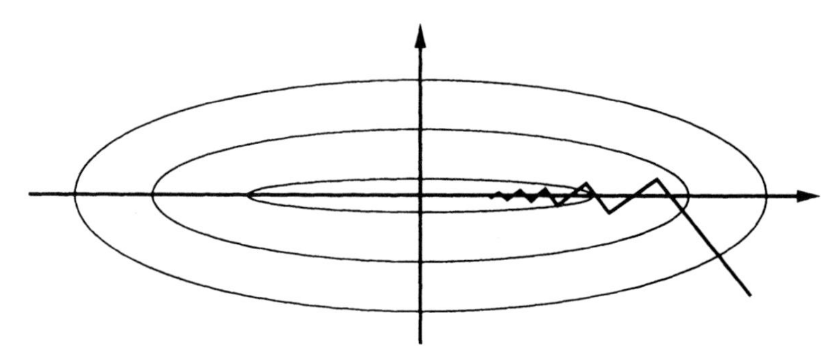

Figure 4.8. Staircase updating of the steepest descent optimization routine.

If a new $d(k)$ is chosen in the negative gradient direction, we see that each search direction is orthogonal to the previous direction since the change in the gradient along the previous direction was exactly zero when we stopped the line search. The weights are thus modified in a staircase fashion toward a minimum of J as shown in Figure 4.8. If $J(\theta)$ consists of long narrow valleys, then the steepest descent algorithm causes the minimization to proceed by taking a number of steps, often repeating directions as we move toward a minimum. Rather than taking orthogonal steps each time, which are not independent of one another, it is desirable to move in new directions which do not redo the minimization which has already been completed. This concept is known as moving in "conjugate directions."

To see how this is accomplished, consider the Taylor expansion of our cost function $J(\theta)$ given by

$$J(\theta) = J(\theta_0) + (\theta - \theta_0)^{\top}\zeta + \frac{1}{2}(\theta - \theta_0)^{\top} H_0(\theta - \theta_0) + h.o.t., \tag{4.53}$$

where $\zeta = \partial J/\partial\theta|_{\theta=\theta_0}$ and $H_0 = H|_{\theta=\theta_0}$ with $H = [h_{ij}]$ and $h_{ij} = \partial^2 J/\partial\theta_i\partial\theta_j$ is the "Hessian matrix." If $J(\theta)$ is quadratic, then it has a global minimum at $\partial J/\partial\theta|_{\theta=\theta^*} = 0$. Ignoring the higher order terms ("h.o.t.") in (4.53), note that

$$\frac{\partial J}{\partial \theta}^{\top} = \zeta + H_0(\theta - \theta_0). \tag{4.54}$$

At the minimum we have $\zeta + H_0(\theta^* - \theta_0) = 0$, or solving we find

$$\theta^* = \theta_0 - H_0^{-1}\zeta. \tag{4.55}$$

Since $J(\theta)$ is not an exact quadratic function in general, we must iterate to find θ^*. Setting θ_0 equal to the current parameter set, an iterative form is found to be

$$\theta(k+1) = \theta(k) - H(k)^{-1}\zeta(k), \tag{4.56}$$

where

$$\zeta(k) = \left.\frac{\partial J}{\partial \theta}\right|_{\theta=\theta(k)} \tag{4.57}$$

$$H(k) = H|_{\theta=\theta(k)}. \tag{4.58}$$

In general, it is very time consuming to calculate H and thus it is not typically used in practice, but this does help explain the method of conjugate gradients. Consider how the gradient changes as we move along some direction, say $\delta\theta$,

$$\delta\left(\frac{\partial J}{\partial \theta}\right) = H_0(\delta\theta).$$

If we have just moved in the direction d and now want to move in the direction u, we desire that the direction be "conjugate" so that

$$0 = d^\top \delta\left(\frac{\partial J}{\partial \theta}\right) = d^\top H_0 u.$$

If all the search directions for a set of vectors are conjugate, then it is said to be a conjugate set. The conjugate gradient method finds successively conjugate search directions without needing to calculate the Hessian. In particular, the Fletcher-Reeves-Polak-Ribiere conjugate gradient method is given as follows:

1. Calculate $\zeta(k)$. Set the search direction equal to $d(k) = -\zeta(k)$.

2. Find $\theta(k+1)$ which minimizes $J(\theta)$ along $d(k)$ (this is achieved via line minimization).

3. Calculate $\zeta(k+1)$.

4. If $|\theta(k+1) - \theta(k)| < tol$ then return $\theta(k+1)$.

5. Set $d(k+1) = -\zeta(k+1) + \eta d(k)$, where

$$\eta = \frac{(\zeta(k+1) - \zeta(k))^\top \zeta(k+1)}{\zeta(k)^\top \zeta(k)}.$$

6. Set $k = k + 1$ and goto 2.

Though a number of alternative optimization methods exists, the above algorithm is suggested for general purpose off-line learning of the approximator parameters when the gradients exist.

Example 4.9 Here, we will apply the method of conjugate gradients to find a parameter set to minimize the cost function (4.49). Consider learning the function

$$y = \cos\left(\frac{\pi}{0.5 + |x|}\right), \tag{4.59}$$

using the approximator (a radial basis function neural network)

$$\mathcal{F}(x, \theta) = \sum_{i=1}^{p} a_i e^{-((x-c_i)/\sigma)^2}, \tag{4.60}$$

where a_i are adjustable parameters and c_i and σ are assumed to be fixed in this example. Here, $M = 100$ data points were taken from a normal distribution about $x = 0$. The Gaussian centers c_i were picked to be evenly spaced between -2 and 2, while $\theta(0) = 0$.

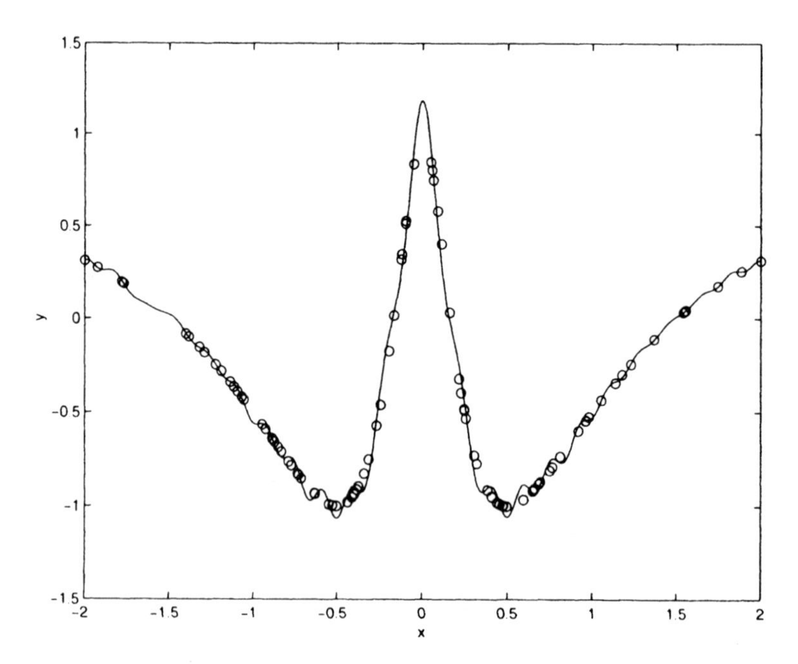

Figure 4.9. The output of the approximator (–) and training points (o's).

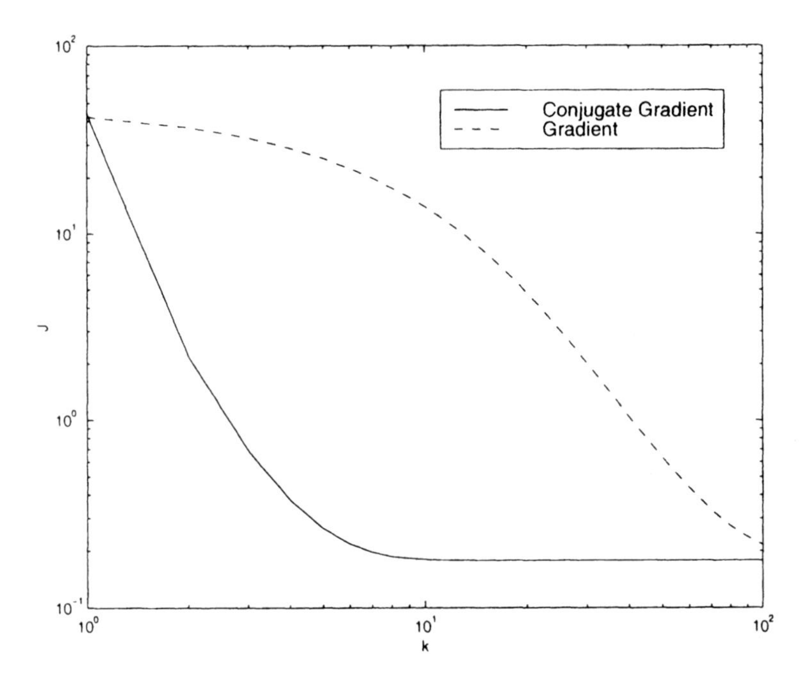

Figure 4.10. Value of the cost function during learning when using the conjugate gradient approach (–) and gradient descent (- -).

The golden section search algorithm was used for the line minimization in the conjugate gradient routine. Figure 4.9 shows the (x^i, y^i) data pairs used for training along with $\mathcal{F}(x, \theta)$ after the conjugate gradient training. In Figure 4.10 we notice that the conjugate gradient algorithm is able to reduce the cost function much more quickly than the gradient routine defined by (4.41) with $\eta = 0.01$. $\triangle$

4.5 Summary

Upon completion of this chapter, the reader should have an understanding of

- Linear least squares techniques (batch and recursive).
- Nonlinear least squares techniques (gradient methods, discrete time and constrained cases).
- Line search and the conjugate gradient method.

The tools provided in this chapter will prove useful when defining the parameters of a fuzzy system or neural network so that some function is approximated. When this approximated function represents some nonlinearity describing the dynamics of a system we wish to control, it may be possible to incorporate the fuzzy system or neural network in the control law to improve closed-loop performance.

It was also shown that it may be possible to use the same approximator structure to represent multiple functions by simply changing the value of some parameters (see Example 4.5). This is an important property of fuzzy systems and neural networks. In general, we will find that fuzzy systems and neural networks are able to approximate any continuous function if enough adjustable parameters are included in their definition. This will be the focus of the next chapter.

4.6 Exercises and Design Problems

Exercise 4.1 (Batch and Recursive Least Squares Derivation) In this problem you will derive several of the least squares methods that were developed in this chapter. First, using basic ideas from calculus, take the partial of J in (4.7) with respect to θ and set it equal to zero. From this derive an equation for how to pick θ^*. Compare it to (4.9). (*Hint:* If m and b are two $n \times 1$ vectors and θ is an $n \times n$ symmetric matrix (i.e., $\theta = \theta^\top$), then $\frac{d}{dm} b^\top m = b$, $\frac{d}{dm} m^\top b = b$, and $\frac{d}{dm} m^\top \theta m = 2\theta m$.) Repeat for the weighted batch least squares approach. Finally, derive (4.21) for the weighted recursive least squares approach.

Exercise 4.2 (Batch Least Squares) Suppose that for Example 4.1 we use the three training data pairs in the training data set G, but add one more. In particular, add the pair

$$\left(\begin{bmatrix} 1.5 \\ 1 \end{bmatrix}, 2.2 \right)$$

to G (to get $M = 4$). Find θ^* using the (nonweighted) batch least squares approach. Plot on a three-dimensional plot (with the z axis as $y(k)$, and the other two axes $y(k-1)$ and $u(k)$) the training data for Example 4.1 (the three points) and the resulting least squares fit (it is a plane). Repeat this for the case considered above where we add one data point. Plot the new plane and data points on the same plot and compare. Does the change in slope of the plane from the $M = 3$ to $M = 4$ case make sense? In what sense?

Exercise 4.3 (Recursive Least Squares) Suppose that for Example 4.2 we use $b(k) = 1 + 0.2\sin(0.01\pi k)$ (i.e., we halve the frequency of the time-varying parameter). Repeat the example, and in particular focus on finding the highest value for λ that will cause the estimate to achieve as good of tracking of $b(k)$ as in Example 4.2 (i.e., as good as the case for $\lambda = 0.2$). Compare the value you find to the one found in the example. Is it bigger or smaller? Why?

Exercise 4.4 (Flat Gradients in Neural Networks) Show that if all the weights of a feedforward neural network are initialized to zeros, then $\zeta = 0$. This is undesirable since if ζ is used within a parameter update routine, then the weights will never change. This is why neural network weights are typically initialized to small random values.

Exercise 4.5 (Gradient Tuning of Neural Networks)

(a) Repeat Example 4.5, but for the case where $f(x) = \sin^2(x)$, and where you use a multilayer perceptron with a single hidden layer as the approximator. Tune the gradient algorithm as necessary to get good approximation. Provide a plot of the function $f(x)$ and the approximator on the same graph to compare the estimation accuracy.

(b) Repeat (a) but use a radial basis function neural network.

Exercise 4.6 (Gradient Training of Fuzzy Systems)

(a) Consider a single-input, single-output (standard) fuzzy system with a total of 20 triangular input and output membership functions. If the rule-base is defined such that input membership function i is associated with the ith output membership function, then a total of 20 rules are formed. Assume the input membership functions are held constant while the centers of the output membership functions are allowed to vary. Use the gradient descent routine to minimize the error between the output of the fuzzy system and the function $y = \sin(x)$ over $x \in [-\pi, \pi]$ using a total of 50 random test points selected from $[-\pi, \pi]$.

(b) Repeat (a) but use a Takagi-Sugeno fuzzy system with output functions that are affine.

Exercise 4.7 (Controller Design) Consider the system defined by

$$\dot{x} = f(x) + u. \tag{4.61}$$

If an approximation to $f(x)$ exists such that $\sup |f(x) - \mathcal{F}(x,\theta)| \leq W$ exists, then show that using the controller $u = -kx - \mathcal{F}(x,\theta)$ will ensure that

$$\lim_{t\to\infty} |x| \leq \frac{W}{\sqrt{k}}. \tag{4.62}$$

Hint: Use the Lyapunov candidate $V = x^2$ to show that $\dot{V} \leq -kV + W^2/k$.

Exercise 4.8 (Line Search) Use the golden section search to find an x that minimizes the following functions:

- $f(x) = x^2 + x + 1$
- $f(x) = \left(x^2 + 0.1\right)\exp(-x^2)$.

Plot each $f(x)$ and comment on the ability of the golden section search to find a global minimum.

Exercise 4.9 (Conjugate Gradient Optimization) Use the conjugate gradient routine to adjust the weights of a MLP so that it reasonably matches the following functions over $x \in [-\pi, \pi]$:

- $f(x) = \sin(x)$
- $f(x) = \cos(x)$
- $f(x) = 1 + \sin^2(x)$.

Try the above for various numbers of nodes in the network.

Chapter 5

Function Approximation

5.1 Overview

The use of function approximation actually has a long history in control systems. For instance, we use function approximation ideas in the development of models for control design and analysis, and conventional adaptive control generally involves the on-line tuning of linear functions (linear approximators) to match unknown linear functions (e.g., tuning a linear model to match a linear plant with constant but unknown parameters) as we discussed in Chapter 1. The adaptive routines we will study in this book may be described as on-line function approximation techniques where we adjust approximators to match unknown nonlinearities (e.g., plant nonlinearities).

In Chapter 4, we discussed the tuning of several candidate approximator structures, and especially focused on neural networks and fuzzy systems. In this chapter, we will show that fuzzy systems or neural networks with a given structure possess the ability to approximate large classes of functions simply by changing their parameters; hence, they can represent, for example, a large class of plant nonlinearities. This is important since it provides a theoretical foundation on which the later techniques are built. For instance, it will guarantee that a certain ("ideal") level of approximation accuracy is possible, and whether or not our optimization algorithms succeed in achieving it, this is what the stability and performance of our adaptive systems typically depends on. It is for this reason that neural network or fuzzy system approximators are preferred over linear approximators (like those studied in adaptive control for linear systems). Linear approximator structures cannot represent as wide of a class of functions, and for many nonlinear functions the parameters of a neural network or fuzzy system may be adjusted to get a lower approximation error than if a linear approximator were used. The theory in the later chapters will allow us to translate this improved potential for approximation accuracy into improved performance guarantees for control systems.

5.2 Function Approximation

In the material to follow, we will denote an approximator by $\mathcal{F}(x)$, showing an obvious connection to the notation used in the two previous chapters. When a particular parameterization of the approximator is of importance, we may write the approximator as $\mathcal{F}(x, \theta)$, where $\theta \in \mathsf{R}^p$ is a vector of parameters which are used in the definition of the approximator mapping. Suppose that $\Omega^p \subset \mathsf{R}^p$ denotes the set of all values that the parameters of an approximator may take on (e.g., we may restrict the size of certain parameters due to implementation constraints). Let

$$\mathcal{G} = \{\mathcal{F}(x, \theta) : \theta \in \Omega^p, p \geq 0\}$$

be the "class" of functions of the form $\mathcal{F}(x, \theta)$, $\theta \subset \Omega^p$, for any $p \geq 0$. For example, $\mathcal{G}$ may be the set of all fuzzy systems with Gaussian input membership functions and center-average defuzzification (no matter how many rules and membership functions this fuzzy system uses). In this case, note that p generally increases as we add more rules or membership functions to the fuzzy system, as p describes the number of adjustable parameters of the fuzzy system (similar comments hold for neural networks, with weights and biases as parameters). In this case, when we say "functions of class $\mathcal{G}$" we are not saying how large p is.

Uniform approximation is defined as follows:

Definition 5.1: A function $f : D \to \mathsf{R}$ may be **uniformly approximated** on $D \subseteq \mathsf{R}^n$ by functions of class $\mathcal{G}$ if for each $\epsilon > 0$, there exists some $\mathcal{F} \in \mathcal{G}$ such that $\sup_{x \in D} |\mathcal{F}(x) - f(x)| < \epsilon$.

It is important to highlight a few issues. First, in this definition the choice of an appropriate $\mathcal{F}(x)$ can depend on ϵ; hence, if you pick some $\epsilon > 0$, certain $\mathcal{F}(x) \in \mathcal{G}$ may result in $\sup_{x \in D} |\mathcal{F}(x) - f(x)| < \epsilon$, while others may not. Second, when we say $\mathcal{F}(x) \in \mathcal{G}$ in the above definition we are not specifying the value of $p \geq 0$, that is the number of parameters defining $\mathcal{F}(x)$ needed to achieve a particular $\epsilon > 0$ level of accuracy in function approximation. Generally, however, we need larger and larger values of p (i.e., more parameters) to ensure that we get smaller and smaller values of ϵ (however, for some classes of functions f, it may be that we can bound p).

Next, a universal approximator is defined as follows:

Definition 5.2: A mathematical structure defining a class of functions $\mathcal{G}_1$ is said to be a **universal approximator** for functions of class $\mathcal{G}_2$ if each $f \in \mathcal{G}_2$ may be uniformly approximated by $\mathcal{G}_1$.

We may, for example, say that "radial basis neural networks are a universal approximator for continuous functions" (which will be proven later in this chapter). Stating the class of functions for which a structure is a universal approximator helps qualify the statement. It may be the case that a particular neural network or fuzzy system structure is a universal approximator for continuous functions, for instance, and at the same time that structure may not able to uniformly approximate discontinuous functions. Thus we must be careful when making statements such as "neural networks (fuzzy systems) are universal approximators," since each type of neural network is a universal approximator for only a class of functions $\mathcal{G}$, where $\mathcal{G}$ is unique to the type of neural network or fuzzy system under investigation.

Additionally, when one chooses an implementation strategy for a fuzzy system or neural network, certain desirable approximation properties may no longer hold. Let $\mathcal{G}_1$ be the class of all radial basis neural networks. Within this class is, for example, the class of radial basis networks with 100 or fewer nodes $\mathcal{G}_2 \subset \mathcal{G}_1$. Just because continuous functions may be uniformly approximated by $\mathcal{G}_1$ does not necessarily imply that they may also be uniformly approximated by $\mathcal{G}_2$. Strictly speaking, a universal approximator is rarely (if ever) implemented for a meaningful class of functions. As we will see, to uniformly approximate the class of continuous functions with an arbitrary degree of accuracy, an infinitely large fuzzy system or neural network may be necessary. Fortunately, the adaptive techniques presented later will not require the ability to approximate a function with arbitrary accuracy; rather we will require that a function may be approximated over a bounded subspace with some finite error.

In the remainder of this section we will introduce certain classes of functions that can serve as uniform or universal approximators for other classes of functions. It should be kept in mind that the proofs to follow will establish conditions so that, given an approximator with a *sufficient* number of tuned parameters, the approximator will match some function $f(x)$ with arbitrary accuracy. The proofs, however, do not place bounds on the minimum number of adjustable parameters required. This issue is discussed later.

5.2.1 Step Approximation

Our first approximation theorem will use a step function to uniformly approximate a continuous function in one dimension. A step function may be defined as follows:

Definition 5.3: The function $\mathcal{F}(x) : D \to \mathsf{R}$ for $D \subset \mathsf{R}$ is said to be a **step function** if it takes on only a finite number of distinct values, with each value assigned over one or more disjoint intervals. The parameters describing a step function characterize the values the step function takes and the intervals over which these values hold. Let $\mathcal{G}_s$ denote the class of all step functions.

Notice that we require distinct values when defining the step function. If all the values were the same, then a "step" would not occur. The following example helps clarify this definition:

Example 5.1 Consider the step function defined by

$$f(x) = \begin{cases} 1.5 & -1 < x \leq 1 \\ -1 & 1 < x \leq 2 \\ 4 & 2 < x \leq 6 \\ 3 & 6 < x \leq 8 \end{cases} \tag{5.1}$$

A plot of this step function is shown in Figure 5.1.

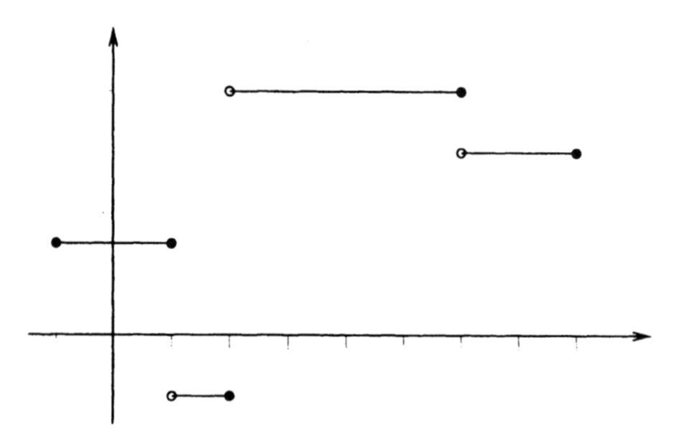

Figure 5.1. Plot of the step function defined by (5.1).

△

Let $\mathcal{G}_{cb}(n, D)$ be the set of all scalar-valued continuous functions defined on a bounded subset $D \subset \mathsf{R}^n$. The first of our uniform approximation theorems is given as follows:

Theorem 5.1: *Step functions defining the class $\mathcal{G}_s$ are universal approximators for $f \in \mathcal{G}_{cb}(1, D)$, $D = [a, b]$.*

Proof: Since f is continuous on D and D is a compact set, f is uniformly continuous on D (the "uniform continuity theorem" [14] may be used to show that f is uniformly continuous since D is a closed, bounded interval), so for any given $\epsilon > 0$ there exists some $\delta(\epsilon) > 0$ such that if $x, y \in D$ and $|x - y| < \delta(\epsilon)$, then $|f(x) - f(y)| < \epsilon$. Divide the interval $D = [a, b]$ into m nonintersecting intervals of equal length $h = (b - a)/m$, with the intervals defined by

$$I_1 = [a, a + h]$$

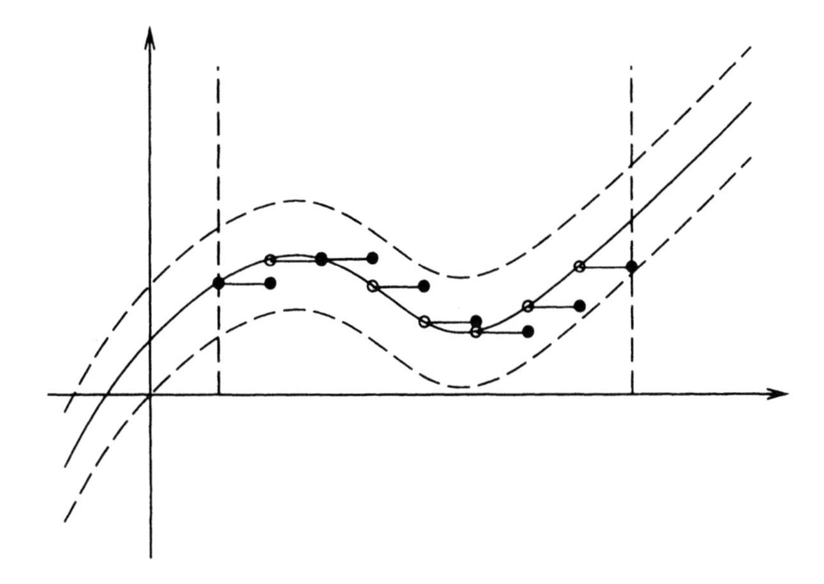

Figure 5.2. Approximating a continuous function with a step function.

$$\begin{aligned} I_2 &= (a+h, a+2h] \\ &\vdots \\ I_k &= (a+(k-1)h, a+kh] \\ &\vdots \\ I_m &= (b-h, b]. \end{aligned} \tag{5.2}$$

Choose m sufficiently large such that $h < \delta(\epsilon)$ for the given ϵ, so that the magnitude of the difference between any two values of f in I_k is less than ϵ. Define the step function as $\mathcal{F}(x) = f(a + (k-1)h)$ on $x \in I_k$. Since the value of the step function on an interval is simply the value of f at the left end point of the interval, we find $|f(x) - \mathcal{F}(x)| < \epsilon$. ■

In the above proof notice that the continuity of f and the restriction that $D = [a, b]$ for some $a, b \in \mathsf{R}$ play key roles in the ability of $\mathcal{F}$ to be a universal approximator. These restrictions ensure that for a given ϵ there will exist some $f \in \mathcal{G}_s$ that will result in an ϵ-accurate approximation. Notice that for smaller values of $\epsilon > 0$, for functions f that have higher slopes and that are defined on larger intervals (i.e., with larger $|b - a|$) we will generally need a larger value of m and hence more parameters in the step function to achieve ϵ-accuracy in function approximation.

Next, note that while we restrict f to be a scalar function defined on $[a, b]$ it should be clear that the above result will generalize to the class of all functions $f \in \mathcal{G}_{cb}(n, D)$ for any n, where $D \subset \mathsf{R}^n$ such that D is a compact set (i.e., it is closed and bounded). Of course, in this case we would have to use appropriately defined multidimensional step functions as the approximator structure.

We may now use the above result to prove that a class of neural networks are universal approximators for continuous functions. Recall that the threshold function, which is used to define "McCulloch-Pitts" nodes, is defined by

$$H(x) = \begin{cases} 0 & x \leq 0 \\ 1 & x > 0 \end{cases}. \tag{5.3}$$

Using the McCulloch-Pitts nodes, we may establish the following:

Theorem 5.2: *Two layer neural networks with threshold-based hidden nodes and a linear output node are universal approximators for* $f \in \mathcal{G}_{cb}(1, D)$, $D = [a, b]$.

Proof: Assume we use the proof of Theorem 5.1 to define the I_k, $k = 1, 2, \ldots, m$ for a given $\epsilon > 0$. If m intervals were required in the proof of Theorem 5.1, then define the McCulloch-Pitts neural network by

$$\mathcal{F}(x, \theta) = c_1 + \sum_{i=2}^{m} c_i H(x - s_i), \tag{5.4}$$

where $\theta = [c_1, \ldots, c_m, s_2, \ldots, s_m]^\top$ which is clearly a function of class $\mathcal{G}_s$. Notice that c_1 is a bias term in the neural network. Here, we will explain how to pick the parameter vector θ to achieve a specified $\epsilon > 0$ accuracy in function approximation. First, define each s_k as the left endpoint of the interval I_k in Theorem 5.1. That is, $s_k = a + (k-1)h$, where $h = (b-a)/m$. From the definition of the Heavyside function, for $0 < \delta \leq h$

$$\begin{aligned} \mathcal{F}(s_k + \delta, \theta) &= c_1 + \sum_{i=2}^{m} c_i H(s_k + \delta - s_i) \\ &= c_1 + \sum_{i=2}^{k} c_i. \end{aligned} \tag{5.5}$$

From the proof of Theorem 5.1, we desire $\mathcal{F}(s_k + \delta, \theta) = f(s_k)$ so we use

$$\begin{bmatrix} 1 & 0 & \cdots & 0 \\ 1 & 1 & \ddots & \vdots \\ \vdots & & \ddots & 0 \\ 1 & \cdots & 1 & 1 \end{bmatrix} \begin{bmatrix} c_1 \\ c_2 \\ \vdots \\ c_{m-1} \\ c_m \end{bmatrix} = \begin{bmatrix} f(s_1) \\ f(s_2) \\ \vdots \\ f(s_{m-1}) \\ f(s_m) \end{bmatrix} \tag{5.6}$$

to solve for each c_i. Notice that the summation of (5.5) is accomplished through the matrix multiplication on the left-hand side of (5.6). Since the matrix on the left hand side of (5.6) is lower triangular, it is invertible so there is a unique solution for each c_i. ■

This shows that there exists a class of neural networks which are universal approximators. We will later want to be able to take the gradient of the neural network with respect to the adjustable parameters to define parameter update laws. Since McCulloch-Pitts activation functions are discontinuous, the gradient is not well defined. Fortunately, nodes with arbitrary "sigmoid functions" may also be used to create neural networks which are universal approximators as described in the following theorem.

Theorem 5.3: *Two layer neural networks with hidden nodes defined by a sigmoid function $\psi : \mathsf{R} \to [0,1]$ and a linear output node are universal approximators for $f \in \mathcal{G}_{cb}(1, D)$, $D = [a, b]$.*

Proof: To complete this proof, we will first show that the sigmoid function $\psi : \mathsf{R} \to [0,1]$ may uniformly approximate the Heavyside function on $\mathsf{R} - \{0\}$. By the definition of a sigmoid function, for each $\delta' > 0$, we have $\lim_{a\to\infty} \psi(a\delta') = 1$ and $\lim_{a\to\infty} \psi(-a\delta') = 0$. This ensures that for any $x \neq 0$ and $\epsilon > 0$ there exists some $a > 0$ such that $|H(x) - \psi(ax)| < \epsilon$ and $|H(-x) - \psi(-ax)| < \epsilon$. These two inequalities thus ensure that for any $\epsilon > 0$ there exists some $a > 0$ such that $|H(x) - \psi(ax)| < \epsilon$ where $x \in \mathsf{R} - \{0\}$. This is shown graphically in Figure 5.3.

Define the neural network by

$$\mathcal{F}(x, \theta) = c_1 + \sum_{i=2}^{m} c_i \psi(a(x - \theta_i)), \tag{5.7}$$

where c_i and θ_i are as defined in Theorem 5.2 for step functions. Then

$$\begin{aligned} |f(x) - \mathcal{F}(x, \theta)| &= \left| f(x) - c_1 - \sum_{i=2}^{m} c_i \psi(a(x - \theta_i)) \right| \\ &\leq \left| f(x) - c_1 - \sum_{i=2}^{m} c_i H(x - \theta_i) \right| \\ &\quad + \left| \sum_{i=2}^{m} c_i \left[\psi(a(x - \theta_i)) - H(x - \theta_i) \right] \right|. \end{aligned}$$

From Theorem 5.2, for any $\epsilon > 0$, we may choose m such that

$$|f(x) - \mathcal{F}(x, \theta)| \leq \epsilon/3 + \left| \sum_{i=2}^{m} c_i \left[\psi(a(x - \theta_i)) - H(x - \theta_i) \right] \right|. \tag{5.8}$$

That is, we define sufficiently many step functions so that the magnitude of the difference between $f(x)$ and the collection of step functions defined by

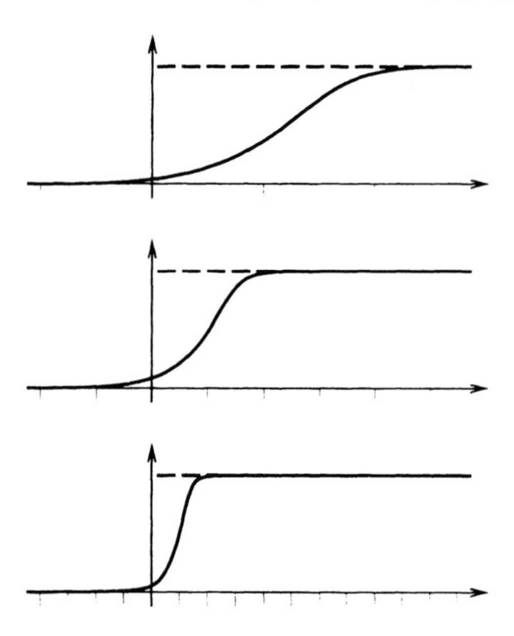

Figure 5.3. Approximating a Heavyside function with a sigmoid function. Notice that as the axis is scaled, the sigmoid function looks more like a Heavyside function which is shown by the dashed line.

(5.4) is no greater than $\epsilon/3$. Notice that this also requires that $|c_k| \leq \epsilon/3$ for $k > 1$ since the step function is held constant on the interval between steps and the magnitude of change in (5.4) is $|c_k|$ when moving from I_k to I_{k+1}. Assume that $x \in I_k$ so that

$$\begin{aligned} |f(x) - \mathcal{F}(x,\theta)| &\leq \epsilon/3 + \left| \sum_{i=2, i\neq k}^{m} c_i \left[\psi(a(x-\theta_i)) - H(x-\theta_i) \right] \right| \\ &\quad + |c_k| \left| \psi(a(x-\theta_k)) - H(x-\theta_k) \right| . \end{aligned}$$

Each c_k describes the magnitude of the step required when moving from the I_{k-1} to I_k intervals, thus $|c_k| \leq \epsilon/3$. Choose $a > 0$ such that $1 - \psi(ah) < 1/(m-2)$ and $\psi(-ah) < 1/(m-2)$. Thus

$$\begin{aligned} |f(x) - \mathcal{F}(x,\theta)| &\leq \epsilon/3 + \left| \sum_{i=2, i\neq k}^{m} c_i \left[\psi(a(x-\theta_i)) - H(x-\theta_i) \right] \right| + \epsilon/3 \\ &\leq \epsilon/3 + \left| \sum_{i=2, i\neq k}^{m} \frac{\epsilon}{3} \frac{1}{m-2} \right| + \epsilon/3 = \epsilon, \end{aligned}$$

which completes the proof. ∎

5.2.2 Piecewise Linear Approximation

Another intuitive approach to approximate functions is the use of piecewise linear functions.

Definition 5.4: The function $f : D \to \mathsf{R}$ for $D \subseteq \mathsf{R}$ is said to be **piecewise linear** on D if D may be broken into a finite number of nonintersecting intervals, denoted $I_1, \ldots I_m$, such that f is linear on each I_k, $k = 1, \ldots, m$.

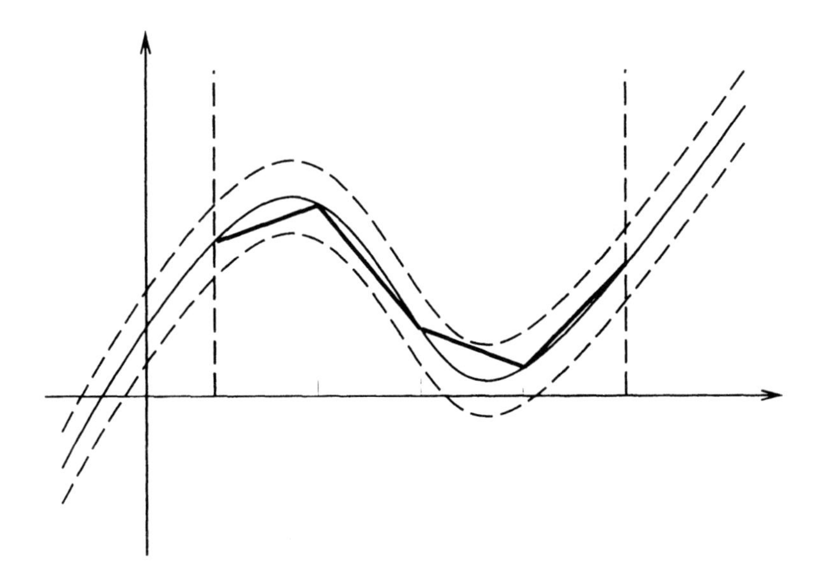

Figure 5.4. Approximating a continuous function with a piecewise linear function.

Theorem 5.4: *A continuous function* $f : D \to \mathsf{R}$ *may be uniformly approximated on* $D = [a, b]$ *by a piecewise linear function* $\mathcal{F} : D \to \mathsf{R}$.

Proof: Since f is uniformly continuous on D, for any given $\epsilon > 0$ there exists some $\delta(\epsilon) > 0$ such that if $x, y \in D$ and $|x - y| < \delta(\epsilon)$, then $|f(x) - f(y)| < \epsilon$. As was done for the step approximation proof, divide the interval $D = [a, b]$ into m nonintersecting intervals of equal length $h = (b - a)/m$, with the intervals defined in (5.2).

Choose m sufficiently large such that $h < \delta(\epsilon')$ for $\epsilon' = \epsilon/2$, so that the difference between any two values of f in I_k is less than $\epsilon/2$. Define the piecewise linear function $\mathcal{F}$ such that it takes on the value of f at the interval endpoints (see Figure 5.4). If s_k is the value of $\mathcal{F}$ at the left endpoint of I_k, then $\mathcal{F}(x) = s_k + z_k(x)$ on I_k where $z_k(x)$ is a ramp with $z_k = 0$ at the left endpoint of I_k. By the definition of m, we know that $|z_k(x)| < \epsilon/2$ on I_k since z_k ramps to the difference between the right and left endpoint values of f in I_k. Thus

$$\begin{aligned} |\mathcal{F}(x) - f(x)| &= |s_k - f(x) + z_k(x)| \\ &\leq |s_k - f(x)| + |z_k(x)| \\ &< \epsilon/2 + \epsilon/2 = \epsilon. \end{aligned}$$

■

In the proof of Theorem 5.4, we actually showed that a continuous function may be uniformly approximated by a *continuous* piecewise linear function. Since the set of continuous piecewise linear functions is a subset of the set of all piecewise linear functions, Theorem 5.4 holds. This fact, however, leads us to the following important theorem.

Theorem 5.5: *Fuzzy systems with triangular input membership functions and center average defuzzification are universal approximators for $f \in \mathcal{G}_{cb}(1, D)$ with $D = [a, b]$.*

Proof: By construction, it is possible to show that any given continuous piecewise linear function may be described exactly by a fuzzy system with triangular input membership functions and center average defuzzification on an interval $D = [a, b]$. To show this, consider the example in Figure 5.5 where $g(x)$ is a given piecewise linear function which is to be represented by a fuzzy system. The fuzzy system may be expressed as

$$\mathcal{F}(x, \theta) = \frac{\sum_{i=1}^{m} c_i \mu_i(x)}{\sum_{i=1}^{m} \mu_i(x)}, \tag{5.9}$$

where θ is a vector of parameters that include the c_i (output membership function centers) and parameters of the input membership functions. Let $I_k = (\underline{\sigma}_k, \sigma_k^c]$ and $I_{k+1} = (\sigma_k^c, \bar{\sigma}_k]$ for $k = 1, 2, \ldots, m$ be defined so that $g(x)$ is a line in any I_k. For $k \neq 1$ and $k \neq m$ choose $\mu_k(x)$ to be a triangular membership function such that $\mu_k(\underline{\sigma}_k) = \mu_k(\bar{\sigma}_k) = 0$ and $\mu(\sigma_k^c) = 1$. See Figure 5.5. For $k = 1$ choose $\mu_1(x) = 1$ for $x \leq \underline{\sigma}_1$ and let $\mu_1(x)$, $\underline{\sigma}_1 < x < \sigma_1^c$, be a line from the pair $(\underline{\sigma}_1, 1)$ to $(\sigma_1^c, 0)$ and $\mu_1(x) = 0$ for $x > \sigma_1^c$. For $k = m$, construct μ_m in a similar manner but so that it saturates at unity on the right rather than the left. For $i = 1$ let $c_1 = g(\underline{\sigma}_1)$ and for $i = m$ let $c_m = g(\bar{\sigma}_m)$. For $i \neq 1$ and $i \neq m$ let $c_i = g(\sigma_i^c)$ and we leave it to the reader to show that in this case that $\mathcal{F}(x, \theta) = g(x)$ for $x \in D$: To do this simply show that the fuzzy system exactly implements the lines on the intervals defined by g.

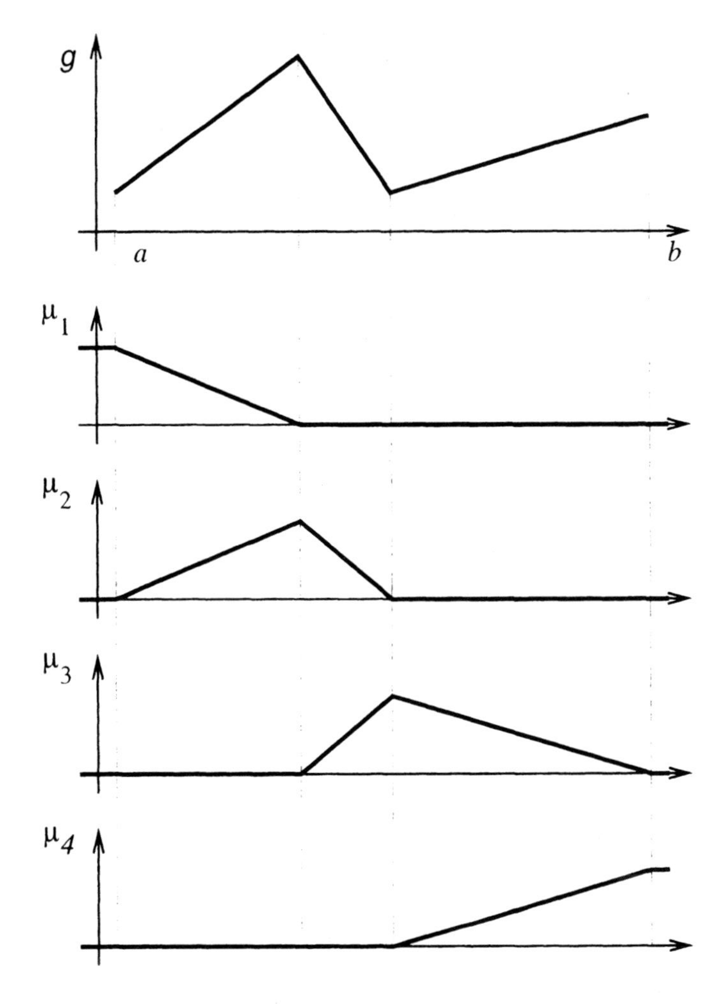

Figure 5.5. Approximating a continuous piecewise linear function with a fuzzy system on an interval.

Since continuous piecewise linear functions (which are universal approximators for continuous functions $f : D \to \mathbb{R}$) may be exactly represented by fuzzy systems with center average defuzzification, we establish the result. ∎

5.2.3 Stone-Weierstrass Approximation

In this section we introduce the Weierstrass theorem, Stone-Weierstrass theorem, and ideas on how to use them in the study of approximator structures. The methods of this section provide very general and useful ways to determine if approximators are universal approximators for the class of functions that are continuous and defined on a compact set.

To begin, we need to define some approximator structures that are based on polynomials.

Definition 5.5: The function $g : D \to \mathsf{R}$ for $D \subseteq \mathsf{R}$ is said to be a **polynomial function** if it is in the class of functions defined by

$$\mathcal{G}_{pf} = \left\{ g(x) = \sum_{i=0}^{p} a_i x^i : a_0, \ldots, a_p \in \mathsf{R}, p \geq 0 \right\}.$$

Let $x = [x_1, x_2, \ldots, x_n]^\top$ be the input. Similarly, define $\mathcal{G}_{mpf}$ to be the class of multivariable polynominal functions, $g : D \to \mathsf{R}$, $D \subset \mathsf{R}^n$, where

$$\mathcal{G}_{mpf} = \left\{ g(x) = \sum_{i,j,\ldots,k} a_{i,j,\ldots,k} x_1^i x_2^j \cdots x_n^k : a_{i,j,\ldots,k} \in \mathsf{R} \right\}.$$

Theorem 5.6: *(Weierstrass) Polynomials are universal approximators for $f \in \mathcal{G}_{cb}(n, D)$, $D \subset \mathsf{R}^n$.*

We will wait to prove this result until we have the tools provided below. What the Weierstrass theorem tells us is that there exists a polynomial which may approximate a continuous function over an interval with arbitrary accuracy. Readers familiar with Taylor series expansion may not be too surprised with this result . A Taylor series expansion, however, is performed about a point rather than across an interval whose size is only confined to an interval of the real numbers.

Theorem 5.7: *(Stone-Weierstrass) A continuous function $f : D \to \mathsf{R}$ may be uniformly approximated on $D \subseteq \mathsf{R}^n$ by functions of class $\mathcal{G}$ if*

(1) *The constant function $g(x) = 1, x \in D$ belongs to $\mathcal{G}$,*

(2) *If g_1, g_2 belong to $\mathcal{G}$, then $ag_1 + bg_2$ belongs to $\mathcal{G}$ for all $a, b \in \mathsf{R}$,*

(3) *If g_1, g_2 belong to $\mathcal{G}$, then $g_1 g_2$ belongs to $\mathcal{G}$, and*

(4) *If $x_1 \neq x_2$ are two distinct points in D, then there exists a function in $g \in \mathcal{G}$ such that $g(x_1) \neq g(x_2)$.*

The proof of the Stone-Weierstrass theorem is beyond the scope of this book, but the interested reader is encouraged to see [14] for more details. We may, however, use the results of the Stone-Weierstrass theorem to prove the Weierstrass approximation theorem. The following example shows how to do this for the $n = 1$ case (the case for general n is similar):

Example 5.2 Here, we will prove the Weierstrass approximation theorem using the Stone-Weierstrass approximation theorem. To do so, we must show that items (1)-(4) hold for the class of polynomial functions $\mathcal{G}_{pf}$.

Using the definition of polynomial functions with $a_0 = 1$ and $a_k = 0$ for $k \neq 0$, (1) is established. If $g_1 = \sum_{i=0}^{n} \alpha_i x^i$ and $g_2 = \sum_{i=0}^{n} \beta_i x_i$, then

$$ag_1 + bg_2 = \sum_{i=0}^{n} (a\alpha_i + b\beta_i)x^i.$$

Since $ag_1 + bg_2$ is a polynomial function, (2) is established. Notice that we may choose g_1 and g_2 to both be defined with $n+1$ coefficients without loss of generality since it is possible to set coefficients to zero such that the proper polynomial order is obtained (e.g., if $g_1 = 1 + 2x$ and $g_2 = x + x^2$, then may let $g_1 = \alpha_0 + \alpha_1 x + \alpha_2 x^2$ and $g_2 = \beta_0 + \beta_1 x + \beta_2 x^2$ where $\alpha_2 = \beta_0 = 0$).

Similarly, multiplying two polynomial functions results in another polynomial function, establishing (3). If we let $g(x) = x$, which is a member of polynomial functions, then $g(x_1) \neq g(x_2)$ for all $x_1 \neq x_2$, establishing (4). $\triangle$

Directly applying the Stone-Weierstrass approximation theorem, we obtain the following result:

Theorem 5.8: *Fuzzy systems with Gaussian membership functions and COG defuzzification are universal approximators for $f \in \mathcal{G}_{cb}(n, D)$ with $D \subseteq \mathsf{R}^n$.*

Proof: The proof is left as an exercise but all that you need to do is show that items (1)-(4) of the Stone-Weierstrass theorem hold and you do this by working directly with the mathematical definition of the fuzzy system. ∎

This implies that if a sufficient number of rules are defined within the fuzzy system, then it is possible to choose the parameters of the fuzzy system such that the mapping produced will approximate a continuous function with arbitrary accuracy.

To show that multi-input neural networks are universal approximators, we first prove the following:

Theorem 5.9: *The function that is used to define the class*

$$\mathcal{G}_{cos} = \left\{ g(x) = \sum_{i=1}^{m} a_i \cos(b_i^\top x + c_i) : a_i, c_i \in \mathsf{R}, b_i \in \mathsf{R}^n \right\} \tag{5.10}$$

is a universal approximator for $f \in \mathcal{G}_{cb}(n, D)$ for $D \subset \mathsf{R}^n$.

Proof: Part (1) follows by letting $a_i = 1$ and $c_i, b_i = 0$. Part (2) follows from the form of $g(x)$. We may show Part (3) using the following trigonometric identity

$$\cos(a)\cos(b) = \frac{1}{2}\left[\cos(a+b) + \cos(a-b)\right]. \tag{5.11}$$

Part (4) may be shown using a proof by contradiction argument. ■

Using the above result, it is now possible to show the following:

Theorem 5.10: *Two layer neural networks with hidden nodes each defined by a sigmoid function, and a linear output node, are universal approximators for $f \in \mathcal{G}_{cb}(n, D)$ for $D \subset \mathrm{R}^n$.*

Proof: By Theorem 5.9, given some $\epsilon > 0$, there exists a function

$$g(x) = \sum_{i=1}^{m} a_i \cos(b_i^\top x + c_i),$$

such that $|f(x) - g(x)| \leq \epsilon/2$. Define $z_i = b_i^\top x + c_i$ and $I_i = \{z_i \in \mathrm{R} : z_i = b_i^\top x + c_i, x \in D\}$. Since $b_i^\top x$ is linear in x, we know that I_i is an interval on the real line if D is a compact region. Since $h_i(z_i) = a_i \cos(z_i)$ is continuous on I_i, we find each $h_i(z_i)$ may be continuously approximated by a two layer neural network as defined in Theorem 5.3, such that

$$\left| h_i(z_i) - d_{i,1} - \sum_{j=2}^{p_i} d_{i,j}\psi(\alpha_{i,j}(z_i - \theta_{i,j})) \right| \leq \frac{\epsilon}{2m}. \tag{5.12}$$

Thus

$$\begin{aligned} |f(x) - \mathcal{F}(x,\theta)| &\leq |f(x) - g(x)| + |g(x) - \mathcal{F}(x,\theta)| \\ &\leq \epsilon/2 + |g(x) - \mathcal{F}(x,\theta)|. \end{aligned} \tag{5.13}$$

But we also know that

$$\begin{aligned} |g(x) - \mathcal{F}(x,\theta)| &\leq \left| g(x) - \sum_{i=1}^{m}\left(d_{i,1} + \sum_{j=2}^{p_1} d_{i,j}\psi(\alpha_j(z_i - \theta_{i,j}))\right)\right| \\ &\leq \left| \sum_{i=1}^{m}\left(h_z(z_i) - d_{i,1} - \sum_{j=2}^{p_1} d_{i,j}\psi(\alpha_{i,j}(z_i - \theta_{i,j}))\right)\right| \\ &\leq \sum_{i=1}^{m} \frac{\epsilon}{2m}. \end{aligned}$$

Thus $|f(x) - \mathcal{F}(x,\theta)| \leq \epsilon$, which completes the proof. ■

Thus the proof of the single-input case proof based upon step functions is now extended to the general multi-input case. In the derivation, a very large number of adjustable parameters were defined. It should be noted, however, that the proof provided existence of a sufficiently large neural network to uniformly approximate a nonlinear function. In practice, much smaller neural networks may typically be used to sufficiently approximate a given nonlinear function for a specific application. Some of these issues will be addressed in more detail in the remainder of this chapter.

5.3 Bounds on Approximator Size

So far, we have shown that a number of approximation schemes may be used to uniformly approximate functions. After choosing an approach to approximate a function, however, one must determine how large an approximator must be to achieve a particular level of approximation accuracy. If a neural network is going to be used, then we must determine how many layers and nodes are required. Or, if we use a fuzzy system, how many rules and membership functions do we need? The bounds presented in this section deal with linear in the parameter approximators. We will later see that less conservative results may be possible if more general nonlinear parameterization is used.

5.3.1 Step Approximation

An upper bound on the required size of the approximators based upon step functions will now be investigated.

Theorem 5.11: *A step function $\mathcal{F}$ defined with $m \geq \frac{(b-a)L}{\epsilon}$ intervals may approximate a continuously differentiable function $f : D \to \mathsf{R}$ with an error bounded by $|f(x) - \mathcal{F}(x, \theta)| \leq \epsilon$, where $|df/dx| \leq L$ on $D = [a, b]$.*

Proof: Assume that $x \in I_k$ with $I_k = (c, d]$, so that $\mathcal{F}(x) = f(c)$ by the definition of the step function approximator of Theorem 5.1. Since $|df/dx| \leq L$, we find $|f(d) - f(c)| \leq hL$, where $h = (b-a)/m$. We thus require that $\epsilon \geq hL$, which is satisfied when $m \geq (b-a)L/\epsilon$. ■

This theorem may now be directly applied to single input neural networks with sigmoid functions. Following the steps of Theorem 5.3, we find that $m \geq 3(b-a)L/\epsilon$ sigmoid functions would be required to approximate a continuously differentiable function with an error of $|f(x) - \mathcal{F}(x, \theta)| \leq \epsilon$. This shows why you generally need more nodes in a neural network to achieve a higher approximation accuracy over a larger domain $[a, b]$.

Also L increases with the spatial frequency of $f(x)$. To see this, consider the case where $f(x) = \sin(\omega x)$, where $\omega > 0$ is the spatial frequency of $f(x)$. Since $\partial f / \partial x = \omega \cos(\omega x)$, we may choose $L = \omega$. Thus the number of adjustable parameters required to approximate a function with a given level of accuracy increases with the spatial frequency of the function to be approximated.

Similar derivations could be made for the n-dimension approximation where the approximator $\mathcal{F}(x)$ is to be defined over $x_i \in [a_i, b_i]$ for $i = 1, \ldots, n$. In this case, one could define an n-dimensional grid on which the approximator is defined. If $||\partial f/\partial x|| \leq L$ for all $x_i \in [a_i, b_i]$, where $\partial f/\partial x \in \mathrm{R}^{n \times n}$ and $L \in \mathrm{R}$, one may require that the ith dimension have $m_i \geq \frac{(b_i - a_i)L}{\epsilon}$ intervals. This will require a total of

$$\prod_{i=1}^{n} \frac{(b_i - a_i)L}{\epsilon} = \left(\frac{L}{\epsilon}\right)^n \prod_{i=1}^{n} (b_i - a_i) \tag{5.14}$$

grid points in the approximation. Thus the number of points increases exponentially with the dimension of the system. This explosion in the number of grid points is often referred to as the curse of dimensionality.

5.3.2 Piecewise Linear Approximation

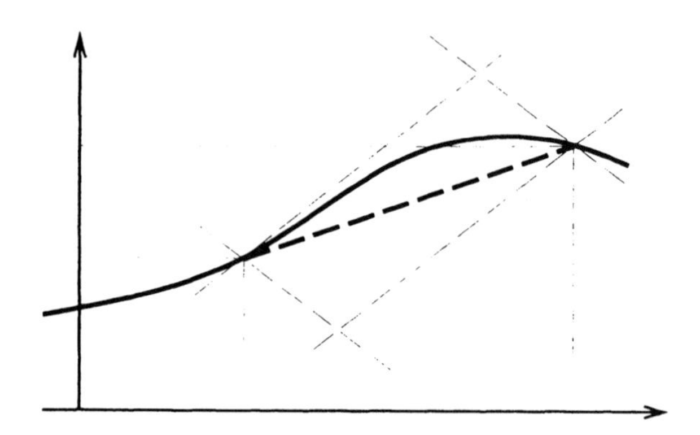

Figure 5.6. Bound for piecewise linear approximation.

A similar argument for the required number of grid points in a piecewise linear approximation may be made as follows:

Theorem 5.12: *A piecewise continuous function $\mathcal{F}$ with $m \geq \frac{(b-a)L}{2\epsilon}$ intervals may approximate a continuously differentiable function $f : D \to \mathrm{R}$ with an error bounded by $|f(x) - \mathcal{F}(x, \theta)| \leq \epsilon$, where $|df/dx| \leq L$ on $D = [a, b]$.*

Proof: Assume that $x \in I_k$ with $I_k = (c, d]$ some interval as we define in the proof of Theorem 5.5. Define the fuzzy system as in Theorem 5.5 so that $\mathcal{F}(c) = f(c)$ and $\mathcal{F}(d) = f(d)$. Since $|df/dx| \leq L$, we have

$$f(c) - (x - c)L \leq f(x) \leq f(c) + (x - c)L \tag{5.15}$$

and

$$f(d) - (d - x)L \leq f(x) \leq f(d) + (d - x)L \tag{5.16}$$

for $x \in I_k$. These inequalities are shown in Figure 5.6. The point $x = x_q$ is at the intersection of $g_a(x) = f(c) - (x - c)L$ and $g_b(x) = f(d) - (d - x)L$ and the maximum inaccuracy is given by

$$q = max_{x \in I_k} \{\min \{\mathcal{F}(x) - g_a(x), \mathcal{F}(x) - g_b(x)\}\}. \tag{5.17}$$

From the definition of g_a and g_b we have

$$2x_q L = f(c) - f(d) + (c + d)L \tag{5.18}$$

so $q = \mathcal{F}(x_q) - [f(c) - (x_q - c)L]$. Or after some manipulation (letting $h = \frac{b-a}{m}$),

$$q = \frac{hL}{2} - \frac{(f(c) - f(d))^2}{2hL}. \tag{5.19}$$

The worst case is when $f(c) = f(d)$, so $q \leq \frac{hL}{2}$. We wish for $q \leq \epsilon$, which is satisfied when $m \geq \frac{(b-a)L}{2\epsilon}$. ■

This shows why you may want many rules in the rule-base of a fuzzy system to achieve higher accuracy in approximation. We have seen that it is possible to determine an upper bound on the required size of a neural network or fuzzy system so that given some Lipschitz function $f : D \to \mathsf{R}$ and $\epsilon > 0$, it is possible to find a fuzzy system or neural network such that $|f(x) - \mathcal{F}(x, \theta)| \leq \epsilon$ on D. Since we have only required that a Lipschitz constant be known for f, the results tend to be rather conservative. If, for example, we find that a fuzzy system with 100 rules is sufficient to approximate f on D with an error no greater than ϵ by the above theorems, that does not necessarily exclude the possibility that there exists a fuzzy system with only 50 rules which will also approximate f with error ϵ.

Though the above theorems (both the step approximation and piecewise linear approximation) deal with nonlinear functions, the parameterization may be viewed as linear. This is because for a specified grid spacing (based on the size of L), the output of the approximator is assumed to be linear with respect to the value at the grid points. It should be noted that it has been found that nonlinear parameterizations are often more effective approximators since fewer adjustable parameters may be required for the same level of approximation. This will be discussed in more detail later in this chapter.

5.4 Ideal Parameter Set and Representation Error

We have seen that by making a fuzzy system or neural network large enough, it is possible to approximate a function $f : D \to \mathsf{R}$ arbitrarily well on D. In practice, however, we are typically limited to using approximators of only moderate size due to computer hardware limitations.

If an approximator $\mathcal{F} : D \times \Omega^p \to \mathsf{R}$ with inputs $x \in D$ and adjustable parameters $\theta \in \Omega^p$ is to approximate a continuous function $f : D \to \mathsf{R}$, then there exists an ideal parameter set defined by

$$\theta^* = \left\{ \theta \in \Omega^p : \theta = \arg\min_{\theta \in \Omega^p} \left(\sup_{x \in D} |f(x) - \mathcal{F}(x, \theta)| \right) \right\} \tag{5.20}$$

(we assume that the operator "arg" simply picks one of the $\theta^* \in \Omega^p$). Thus θ^* is the parameter set which causes $\mathcal{F}(x, \theta^*)$ to best approximate $f(x)$ on D in the sense measured by $|\cdot|$. Any $\mathcal{F}(x, \theta^*)$ where $\theta^* \in \theta^*$ is called an **ideal representation** for $f(x)$. We have defined θ^* as the parameter set which causes a fuzzy system or neural network with a given structure to best represent a continuous function on D. In general θ^* may contain more than a single element, so that given some $\theta_1^* \in \theta^*$, there may exist another $\theta_2^* \in \theta^*$ with $\theta_1^* \neq \theta_2^*$. This may be seen by the following example.

Example 5.3 Consider the neural network defined by

$$\mathcal{F}(x, \theta) = \tanh(w_1 x) + \tanh(w_2 x) \tag{5.21}$$

with $x, w_1, w_2 \in \mathsf{R}$ and $\theta = [w_1, w_2]^\top$. If given some $f(x)$ we find $\theta = \theta_1^* = [w_1^*, w_2^*]^\top$ minimizes $|f(x) - \mathcal{F}(x, \theta)|$ on D, then θ_1^* is an ideal parameter set. Notice that

$$\tanh(w_1^* x) + \tanh(w_2^* x) = \tanh(w_2^* x) + \tanh(w_1^* x). \tag{5.22}$$

If $w_1^* \neq w_2^*$, then if we let $\theta_2^* = [w_2^*, w_1^*]^\top$ such that $\theta_1^* \neq \theta_2^*$, we find that θ_2^* is also in the ideal parameter set. $\triangle$

Definition 5.6: The approximator $\mathcal{F} : D \times \Omega^p \to \mathsf{R}$ for $f : D \to \mathsf{R}$ has an **ideal representation error** $w(x) = f(x) - \mathcal{F}(x, \theta^*)$ on D given some $\theta^* \in \theta^*$ with θ^* defined by (5.20).

It is important to keep in mind that the ideal representation error for an approximator is defined using some θ^* in the ideal parameter set. If we have some parameter vector $\theta(t)$ which is a time-varying estimate of θ^*, then the ideal representation error for $\mathcal{F}(x, \theta(t))$ is still defined by $|f(x) - \mathcal{F}(x, \theta^*)|$. That is, the ideal representation error is defined in terms of how well an approximator $\mathcal{F}(x, \theta(t))$ with a given structure may represent some $f(x)$

when $\theta(t) = \theta^*$, rather than how well it is approximating $f(x)$ at some time t with an arbitrary $\theta(t)$. Additionally, the ideal representation error is dependent upon which θ^* is chosen, even though its bound on D is independent of this choice.

There will be times when we know that D is a closed and bounded (compact) set, but we may or may not know an explicit bound on it. In this case we know that there exists some $W > 0$ such that $w(x) \leq W$ for all $x \in D$. We will often use W in this book and will call it the bound on the approximation error. Notice that by using the results of the previous sections, if f is continuous and defined on a compact set we can always find an approximator structure that allows us to choose W very small; however, reducing W will in general require that we increase the size of the approximator.

5.5 Linear and Nonlinear Approximator Structures

In this section we explain how approximators can either be linear or nonlinear in their parameters, discuss properties of both linear and nonlinear in the parameter approximators, then show how to use linearization to begin with a nonlinear in the parameter approximator and obtain a linear in the parameter approximator.

5.5.1 Linear and Nonlinear Parameterizations

Most classes of neural networks and fuzzy systems can be linearly parameterized, at least in terms of some of their parameters (e.g., many fuzzy systems are linear in their output membership function centers). In this case we may write $\mathcal{F}(x, \theta) = \theta^\top \zeta(x)$, so that $\zeta(x)$ is not a function of θ and the parameters θ enter linearly. Note also that in this case $\partial \mathcal{F}(x,\theta)/\partial\theta = \zeta(x)^\top$.

When the parameters in the vector θ in $\mathcal{F}(x,\theta)$ include, for example, parameters of the activation functions of a multi-layer neural network, or input membership function parameters of a fuzzy system, then $\mathcal{F}(x,\theta)$ is a nonlinear function of the parameters θ, so that $\partial \mathcal{F}(x,\theta)/\partial\theta = \zeta(x,\theta)^\top$. The following examples demonstrate how to specify $\zeta(x,\theta)$ when the set of adjustable parameters does not appear linearly.

Example 5.4 Consider the simple feedforward neural network defined as

$$\mathcal{F}(x,\theta) = w_0 + w_1 \tanh(w_2 x),$$

where $x \in \mathbb{R}$ and $\theta = [w_0, w_1, w_2]^\top$. Then

$$\zeta(x,\theta) = \left[\frac{\partial \mathcal{F}(x,\theta)}{\partial \theta}\right]^\top = \begin{bmatrix} 1 \\ \tanh(w_2 x) \\ w_1 x(1 - w_2^2) \end{bmatrix}$$

If you pick $\theta = [w_0, w_1]^\top$, then the neural network is linear in the parameters, for this definition of parameters. $\triangle$

Example 5.5 Suppose that for a fuzzy system we only use input membership functions of the "center" Gaussian form shown in Table 3.1. For the ith rule, suppose that the input membership function is

$$\exp\left(-\frac{1}{2}\left(\frac{x_j - c_j^i}{\sigma_j^i}\right)^2\right)$$

for the jth input universe of discourse. Here, for $i = 1, 2, \ldots, R$ and $j = 1, 2, \ldots, n$, c_j^i (σ_j^i) is the center (spread) of the input membership function on the jth universe of discourse for the ith rule. Let b_i, $i = 1, 2, \ldots, R$, denote the center of the output membership function for the ith rule, use center-average defuzzification, and product to represent the conjunctions in the premise. Then,

$$\mathcal{F}(x, \theta) = \frac{\sum_{i=1}^{R} b_i \prod_{j=1}^{n} \exp\left(-\frac{1}{2}\left(\frac{u_j - c_j^i}{\sigma_j^i}\right)^2\right)}{\sum_{i=1}^{R} \prod_{j=1}^{n} \exp\left(-\frac{1}{2}\left(\frac{u_j - c_j^i}{\sigma_j^i}\right)^2\right)} \tag{5.23}$$

is an explicit representation of the fuzzy system. Notice that if we fix the parameters of the input membership functions and choose

$$\theta = [b_1, b_2, \ldots, b_R]^\top,$$

then the fuzzy system is linear in its parameters. If, on the other hand, we choose

$$\begin{aligned} \theta \;=\; & [b_1, b_2, \ldots, b_R, c_1^1, c_1^2, \ldots, c_1^R, \ldots, c_n^1, c_n^2, \ldots, c_n^R, \\ & \sigma_1^1, \sigma_1^2, \ldots, \sigma_1^R, \ldots, \sigma_n^1, \sigma_n^2, \ldots, \sigma_n^R]^\top, \end{aligned} \tag{5.24}$$

then the fuzzy system $\mathcal{F}(x, \theta)$ is nonlinear in its parameters. In this case we can compute $\zeta(x, \theta)$ as we did in the above example using simple rules from calculus. $\triangle$

5.5.2 Capabilities of Linear vs. Nonlinear Approximators

Recall that W is the bound on the representation error of the unknown function $f(x)$ with the approximator $\mathcal{F}(x, \theta)$. For a *given* approximator structure $\mathcal{F}(x, \theta)$ all we know is that the bound on the approximation error

$W > 0$ exists; however, we may not know how small it is. The universal approximation property simply says that we may increase the size of the approximator structure and properly define the parameters of the approximator to achieve any desired accuracy (i.e., to make W as small as we want); it does not say how big the approximator must be, or if you fix the structure $\mathcal{F}(x, \theta)$ how small W is.

Do we want to use linear or nonlinear in the parameter approximators? Barron's work in approximation theory [13] gives us some clues as to how to answer this question:

- He shows that for a nonlinear in the parameter approximator (like a single layer nonlinear network with sigmoids for the activation functions), for a certain wide class of functions (with certain smoothness properties) that we would like to approximate, if we tune the parameters of the approximator properly (including the ones that enter in a nonlinear fashion), then the integral squared error over the approximation domain is less than
$$\frac{C}{N},$$
where N is the number of nodes (squashing functions). The value of C depends on the size of the domain over which the approximation takes place (it increases for larger domains), and how oscillatory the function is that we are trying to approximate (with more oscillations C increases). For certain general classes of functions, C can increase exponentially as the dimension n increases, but for a fixed n, and a fixed domain size, Barron's results show that by adding more sigmoidal functions, *if we tune the parameters properly,* we will get a definite decrease in the approximation error (and with only a linear increase in approximator size, that is, a linear increase in the number of parameters). Certain types of approximators with translates of Gaussian functions also hold this property [61].

- For linear in the parameter approximators, for the same type of functions to be approximated as in the nonlinear case discussed above, Barron shows that there is *no way* to tune the parameters that enter linearly (given a fixed number of "basis functions") so that the approximation error is better than
$$\frac{C_L}{N^{2/n}}.$$
Here, C_L has similar dependencies as C had for the nonlinear in the parameter case. Note, however that there is a dependence on the dimension n in the bound, so that for high-dimensional function approximation, a nonlinear in the parameter approximator may better avoid the curse of dimensionality. Also, one should be careful with

the choice of the nonlinear part of the network, in order not to add more approximator structure while not gaining any more ability to reduce the approximation error.

To summarize, it may be desirable to use approximators that are nonlinear in their parameters, since a nonlinear in the parameters approximator can be simpler than a linear in the parameters one (in terms of the size of its structure and hence number of parameters) yet achieve the same approximation accuracy (i.e., the same W above). Since good design of the approximator structure is important to reduce W we would like to be able to tune nonlinear in the parameter approximators; this is the main subject we discuss below.

The general problem is that, on the one hand, we know how to tune linear in the parameter approximators, but in certain cases they may not be able to reduce the approximation error sufficiently; on the other hand, we do not know too much about how to effectively tune nonlinear in the parameter approximators, but we know that if we can tune them properly we may be able to reduce the approximation error more than what a linear in the parameter approximator with the same complexity could. We emphasize, however, that finding the best approximator structure is a difficult and as yet unsolved problem. Determination of the best nonlinear in the parameter structure or advantages and disadvantages of nonlinear versus linear in the parameter approximator structures is a current research topic.

5.5.3 Linearizing an Approximator

The last section showed that it may be desirable to use approximators that are nonlinear in their parameters, since a nonlinear in the parameters approximator can be simpler than a linear in the parameters one (in terms of the size of its structure and hence number of parameters) yet achieve the same approximation accuracy (i.e., the same W above). In this section, we show how to linearize a nonlinear in the parameter approximator, which will later enable us to tune its parameters when used in adaptive controllers.

Consider the class of approximators which are Lipschitz continuous in the adjustable parameters (which may enter in a linear or nonlinear fashion), and are such that the parameters $\theta \in \Omega$, where Ω is a convex set. Define, for a given $\theta^* \in \Omega$,

$$E(x,\theta) = \mathcal{F}(x,\theta^*) - \mathcal{F}(x,\theta),$$

where $\theta \in \Omega$, as the difference between the ideal representation of $f(x)$ and its current representation. Here, we say that $\mathcal{F}(x,\theta^*)$ is an ideal representation of $f(x)$ if

$$\theta^* = \arg\min_{\theta\in\Omega} \left[\sup_{x\in D} |\mathcal{F}(x,\theta) - f(x)| \right], \tag{5.25}$$

where we assume D is a compact set. Thus, we may write

$$f(x) = \mathcal{F}(x, \theta^*) + w(x),$$

where $w(x)$ is the representation error, and from the universal approximation property we know that $w(x) \leq W$ for some $W > 0$. That is, for a *given* approximation structure our representation error W is finite but generally unknown. However, as discussed above, simply by properly increasing the size of the approximator structure we can reduce W to be arbitrarily small so that if we pick any $W > 0$ a priori there exists an approximator structure that can achieve that representation accuracy. Also, note that D is a compact set. Normally, to reduce W by choosing the approximator structure we have to make sure that the structure's parameters result in good "coverage" of D so that appropriate parameters θ in $\mathcal{F}(x, \theta)$ can be tuned to reduce the representation error over the entire set D. Next, we will study some properties of our Lipschitz continuous approximators that will later allow us to tune parameters when dealing with adaptive systems.

Using the mean value theorem one obtains

$$E(x, \theta) = E(x, \theta^*) + \frac{\partial E(x, z)}{\partial z}\tilde{\theta},$$

where the parameter error is $\tilde{\theta} = \theta - \theta^*$, and z is some point on the line segment $z \in \bar{L}(\theta, \theta^*)$ (i.e., $z = \theta^* + \bar{\alpha}(\theta - \theta^*)$ for some $\bar{\alpha} \in [0, 1]$). Note that $|z - \theta| \leq |\tilde{\theta}|$ for any $z \in \bar{L}(\theta, \theta^*)$. Since $E(x, \theta^*) = 0$, we have

$$\begin{aligned} E(x, \theta) &= \frac{\partial E(x, \theta)}{\partial \theta}\tilde{\theta} + \left[\frac{\partial E(x, z)}{\partial z} - \frac{\partial E(x, \theta)}{\partial \theta}\right]\tilde{\theta} \\ &= -\frac{\partial \mathcal{F}(x, \theta)}{\partial \theta}\tilde{\theta} + \delta(x, \theta, \theta^*), \end{aligned} \tag{5.26}$$

where $\delta(x, \theta, \theta^*) = \left[\frac{\partial E(x,z)}{\partial z} - \frac{\partial E(x,\theta)}{\partial \theta}\right]\tilde{\theta}$. Using Cauchy's inequality,

$$|\delta(x, \theta, \theta^*)| \leq \left|\frac{\partial E(x, z)}{\partial z} - \frac{\partial E(x, \theta)}{\partial \theta}\right| |\tilde{\theta}|. \tag{5.27}$$

If $\partial E(x, z)/\partial z$ is Lipschitz continuous on $z \in \bar{L}(\theta, \theta^*)$, then since $|z - \theta| \leq |\tilde{\theta}|$, we have

$$\left|\frac{\partial E(x, z)}{\partial z} - \frac{\partial E(x, \theta)}{\partial \theta}\right| \leq L|\tilde{\theta}|, \tag{5.28}$$

where L is a Lipschitz constant (which we can determine since it is in terms of the known approximator structure). Thus

$$|\delta(x, \theta, \theta^*)| \leq L|\theta|^2,$$

so if we are able to find a way to adjust θ so that we reduce $|\tilde{\theta}|^2$, then θ will tend toward θ^* so that $\mathcal{F}(x,\theta)$ will tend toward $\mathcal{F}(x,\theta^*)$.

It is interesting to note that since $E(x,\theta)$ is continuously differentiable, a first order Taylor expansion is possible and hence

$$E(x,\theta-\tilde{\theta}) = E(x,\theta) - \frac{\partial E(x,\theta)}{\partial\theta}\tilde{\theta} + o(|\tilde{\theta}|) \tag{5.29}$$

where $o(|\tilde{\theta}|)$ is a function of $\tilde{\theta}$ such that

$$\lim_{|\tilde{\theta}|\to 0} \frac{o(|\tilde{\theta}|)}{|\tilde{\theta}|} \to 0. \tag{5.30}$$

Rearranging (5.29) and simplifying, we get

$$\begin{aligned} E(x,\theta) &= \frac{\partial E(x,\theta)}{\partial\theta}\tilde{\theta} - o(|\tilde{\theta}|) \\ &= -\frac{\partial \mathcal{F}(x,\theta)}{\partial\theta}\tilde{\theta} + \delta(x,\theta,\theta^*), \end{aligned} \tag{5.31}$$

where $\delta(x,\theta,\theta^*) = -o(|\tilde{\theta}|)$. Hence, $\delta(x,\theta,\theta^*)$ is the contribution from higher order terms in a Taylor series expansion, and from the use of the mean value theorem, comparing (5.26) and (5.31), we see that all the higher order terms are bounded by $L|\tilde{\theta}|^2$.

Using (5.26), and the fact that $f(x) = \mathcal{F}(x,\theta^*) + w(x)$ we will later express $f(x) - \mathcal{F}(x,\theta)$ as

$$\begin{aligned} f(x) - \mathcal{F}(x,\theta) &= \mathcal{F}(x,\theta^*) + w(x) - \mathcal{F}(x,\theta) && (5.32) \\ &= -\tilde{\theta}^\top\zeta(x,\theta) + \delta(x,\theta,\theta^*) + w(x), && (5.33) \end{aligned}$$

where $\zeta(x,\theta) = (\partial\mathcal{F}(x,\theta)/\partial\theta)^\top$. Letting $\bar{w} = w(x) + \delta(x,\theta,\theta^*)$, we find

$$f(x) - \mathcal{F}(x,\theta) = -\tilde{\theta}^\top\zeta(x,\theta) + \bar{w}(x). \tag{5.34}$$

If $|\tilde{\theta}|^2$ is bounded, then there exists some $\bar{W} \geq 0$ such that $\bar{w}(x) \leq \bar{W}$ for all $x \in D$. Hence, at times we will be concerned with showing that $|\tilde{\theta}|$ is bounded. Notice that schemes which are nonlinear in the parameters introduce the $\delta(x,\theta,\theta^*)$ term which, in general, increases the representation error. In some instances, however, using the nonlinear in the parameter schemes may allow for smaller representation errors with fewer adjustable parameters than the linear in the parameter schemes, thus justifying their use in some applications.

5.6 Discussion: Choosing the Best Approximator

As we have seen, there are several methods to uniformly approximate continuous functions. Along with fuzzy systems and neural networks, there exists a number of other techniques such as polynomials, splines, and trigonometric series, to name a few. Before making a decision on which technique to use, it is important to consider the following issues:

- *Ease of implementation:* Depending on the application and on the particular function to be approximated, certain approximation techniques may be easier to implement than others, and yet perform adequately. For example, if we happen to know that the function is an nth order polynomial, then a simple polynomial approximator may be preferable to a more general fuzzy system or neural network.

- *Choice of structure and number of adjustable parameters:* The more knowledge we have about the function to be approximated, the better are the chances to make a good choice for the approximator structure. For instance, when using Takagi-Sugeno fuzzy systems, the right-hand side (non)linear functions may be chosen to match known parts of the unknown function. The choice of the number of adjustable parameters is also very application dependent, but one has to be careful not to overparameterize (think of the problem of interpolating between a sample of points with a polynomial; if the order of the polynomial is increased too much, the approximator will match the samples well, but it may significantly deviate from the real function between samples). Note also that, as a rule of thumb, one may expect a nonlinear in the parameters approximator to require less parameters than a linear in the parameters one for the same accuracy, with the disadvantage that the nonlinear in the parameters approximator may be more difficult to tune effectively.

- *Realize that you may need to try several structures:* In general, you may want to try several approximator structures to see what the trade-off is between performance and approximator complexity. In particular, it may be wise to consider the physics of the problem and perhaps some plots of the data to try to determine the type of nonlinearities that are present. Then, try to pick the approximator structure so that it contains nonlinearities similar to those seen (e.g., if it seems like the mapping has two linear regions, with a nonlinear transition between them, you may want to try a Takagi-Sugeno fuzzy system with two rules, one for each linear region). Next, you should try a simple approximator structure to see how it works. Then, you should increase its complexity (e.g., size p) until you get the performance you want. Realize, however, that with your implementation constraints, and choice of structure, you may not be able to achieve the performance you need so you may need to invest in more computational power, and switch to a different structure.

5.7 Summary

In this chapter we have studied approximation properties of conventional, neural, and fuzzy system approximator structures. In particular, we have covered the following topics:

- Uniform and universal approximation neural networks and fuzzy systems.
- Bounds on approximation error.
- Linear and nonlinear in the parameter approximation structures.

This chapter is used in the remaining chapters in several ways. First, the theory is used at several points in the proofs. Second, it provides insights into choices of approximator structures for practical applications.

A few trends to keep in mind when trying to approximate a nonlinearity with a fuzzy system or neural network are as follows:

- As the spatial frequency of the nonlinearity increases, more parameters should be included in the fuzzy system or neural network.
- As the dimension of the approximator increases, the number of required adjustable parameters will increase.
- As the desired fidelity of the approximation improves, the number of required adjustable parameters will increase.
- As the size of the region over which the approximation is to hold increases, more parameters should be included in the fuzzy system or neural network.

We have also seen that linear in the parameter approximators may require more adjustable parameters than their nonlinear counterparts.

5.8 Exercises and Design Problems

Exercise 5.1 (Uniform Approximation) Show that the class of functions

$$\mathcal{G}_1 = \left\{g(x) = a\sin^2(x) + b\sin(x)\cos(x) + c\cos^2(x) : a, b, c \in \mathsf{R}\right\}$$

may be uniformly approximated by

$$\mathcal{G}_2 = \{g(x) = p\sin(2x) + q\cos(2x) + r : p, q, r \in \mathsf{R}\} \tag{5.35}$$

on $x \in \mathsf{R}$.

Exercise 5.2 (Radial Basis Function Neural Networks) Let

$$\mathcal{G}_{rbnn} = \left\{ g(x) = \sum_{i=1}^{p} a_i \exp\left(-\gamma_i \left|x - c_i\right|^2\right) \right\} \tag{5.36}$$

be the class of radial basis neural network functions with $a_i, \gamma_i \in \mathsf{R}$ and $c_i \in \mathsf{R}^n$ for $i = 1, \ldots, n$. Use the Stone-Weierstrass theorem to show that $\mathcal{G}_{rbnn}$ is a universal approximator for $f \in \mathcal{G}_{cb}(n, D)$.

Exercise 5.3 (Fuzzy Systems) Let

$$\mathcal{G}_{fs} = \left\{ g(x) = \frac{\sum_{i=1}^{p} a_i \exp\left(-\gamma_i \left|x - c_i\right|^2\right)}{\sum_{i=1}^{p} \exp\left(-\gamma_i \left|x - c_i\right|^2\right)} \right\} \tag{5.37}$$

be the class of fuzzy systems defined with Gaussian input membership functions where $a_i, \gamma_i \in \mathsf{R}$ and $c_i \in \mathsf{R}^n$ for $i = 1, \ldots, n$. Use the Stone-Weierstrass theorem to show that $\mathcal{G}_{fs}$ is a universal approximator for $f \in \mathcal{G}_{cb}(n, D)$ (proving Theorem 5.5).

Exercise 5.4 (Approximator Size) Use Theorem 5.12 to find a sufficient number of intervals to approximate

- $f(x) = 1 + x^2$
- $f(x) = \sin^2(x) + 2\cos^2(4x)$
- $f(x) = x \sin(10/x)$

using a piecewise linear approximation over $x \in [-1, 1]$.

Exercise 5.5 (Taylor Expansion) The Taylor expansion of the continuous function $f : \mathsf{R} \to \mathsf{R}$ about $x = 0$ is given by

$$f(x) = f(0) + \frac{df}{dx}(0)x + \frac{1}{2!}\frac{d^2 f}{dx^2}(0)x^2 + \cdots = \sum_{i=0} \frac{f^{(i)}(0)}{i!} x^i. \tag{5.38}$$

Given some small $\delta, \epsilon > 0$, how many terms of the Taylor expansion are needed so that

$$\sup_{x \in D} \left| f(x) - \sum_{i=0}^{n} \frac{f^{(i)}}{i!} \right| < \epsilon, \tag{5.39}$$

where $D = \{x \in \mathsf{R} : |x| \leq \delta\}$.

Part II

State-Feedback Control

Chapter 6

Control of Nonlinear Systems

6.1 Overview

The purpose of this chapter is to summarize a collection of standard control design techniques for certain classes of nonlinear systems. Later we will use these control techniques to develop adaptive control approaches that are suitable for use when there is additional uncertainty in the plant dynamics. Since the linear concept of phase does not carry over to the nonlinear world, we will not consider many of the traditional control design techniques such as using Bode and Nyquist plots. Instead, we will use Lyapunov-based design techniques where a controller is chosen to help decrease a measure of the system error.

Let

$$\begin{aligned} \dot{x} &= f(x,u) \\ y &= h(x,u) \end{aligned} \tag{6.1}$$

define the dynamics of a system with state $x \in R^n$, input $u \in R^m$, and output $y \in R^p$. Given a control law $u = \nu(t,x)$, it is assumed that $f(t,x)$ is locally Lipschitz in x and piece-wise continuous in t so that given the initial state $x(0)$, there exists a unique trajectory satisfying (6.1). Throughout this book we will use the notation $u = \nu(z)$ to define a control law where $z(t)$ is a vector of appropriate signals for the particular application. The vector z may contain, for example, reference signals, states, dynamic signals, or combinations of any of these. We will only consider controllers where the components of z are measurable signals. The purpose of the controller is typically to force $y \to r(t)$ where $r \in R^p$ is a reference signal. When r is time-varying, defining a control law $u = \nu(z)$ to force $y \to r(t)$ is called the tracking problem. If r is a constant, the problem is commonly referred to as set-point regulation.

To help develop general control techniques, we will study certain canonical forms of the system dynamics. If the original system is defined by

$$\dot{\xi} = f_\xi(\xi, u), \tag{6.2}$$

then a *diffeomorphism* may be used to create the state representation $x = T(\xi)$. Here T is a diffeomorphism (a diffeomorphism is a continuously differentiable mapping with a continuously differentiable inverse) which is used to form the new system representation

$$\begin{aligned} \dot{x} &= \tfrac{\partial T}{\partial \xi} f_\xi(\xi, u) = f(x, u) \\ y &= h_\xi(\xi, u) = h(x, u), \end{aligned} \tag{6.3}$$

where $f(x, u)$ and $h(x, u)$ may take on a special form when dealing with canonical representations. Thus in the stability analysis throughout the remainder of this book, we will typically consider the x, rather than the ξ representation. It is important to keep in mind that the change of coordinates only changes the representation of the dynamics and not the input-to-output characteristics of the system.

When deriving a control law, we will first define an error system, $e = \chi(t, x)$ with $e \in \mathbb{R}^q$, which provides a quantitative (usually instantaneous) measure of the closed-loop system performance. The system dynamics are then used with the definition of the error system to define the error dynamics, $\dot{e} = \alpha(t, x, u)$. A Lyapunov candidate, $V(e)$ with $V : \mathbb{R}^q \to \mathbb{R}$, is then used to provide a scalar measurement of the error system in a similar fashion that a cost function is used in traditional optimization. The purpose of the controller is then to reduce V along the solutions of the error dynamics.

The initial control design techniques presented here will assume that the plant dynamics are known for all $x \in \mathbb{R}^n$. Once we understand some of the basic tools of nonlinear control design for ideal systems, we will study the control of systems which posses certain types of uncertainty. In particular, it will be shown how nonlinear damping and dynamic normalization may be used to stabilize possibly unbounded system uncertainties.

The control design techniques will assume that any uncertainty in the plant model may be bounded by known functions (with possible multiplicative uncertainty). If in reality the models and/or bounds are only valid when $x \in S_x$, where S_x is a compact set, then there may be cases when the stability analysis is invalid. This is often seen when a controller is designed based on the linearization of a nonlinear plant. If the state travels too far away from the nominal operating point (i.e., the point about which the linearization was performed), it is possible for the plant nonlinearities to drive the system unstable. In this chapter, we will derive bounds on the state trajectory using the properties of the Lyapunov function to ensure that x never leaves the space over which the plant dynamics are understood. Since we will place bounds on x, additional properties of the plant dynamics, such as Lipschitz continuity, only need to hold on S_x. Throughout the remainder of this book, we will use the notation S_y to represent the space over which the signal $y \in R^m$ may travel.

6.2 The Error System and Lyapunov Candidate

Before we present any control techniques, the concepts of an error system and Lyapunov candidate must be understood. We will see that the choice of the error system and Lyapunov candidate will actually be used in the definition of the controller much in the same way a cost function will influence the form of an optimization routine.

6.2.1 Error Systems

For any control system there is a collection of signals that one wants to ensure is bounded or possibly converge to a desired value. The size of the mismatch between the current signal values and the space of desired values is measured by an error variable $e \in R^q$. The definition of the system error is typically driven by both the desired system performance specification and the structure of the system dynamics. Consider the system dynamics

$$\dot{x} = f(x, u) \tag{6.4}$$

with output $y = h(x)$, where $x, u \in R$ are scalar signals. If one wants to drive $y \to r(t)$, where r is a reference signal, then choosing the error system $e = \chi(t, x) = y - r(t)$ would provide a measure of the closed-loop tracking performance, typically referred to as the tracking error.

In the more general case when $x \in R^n$, $u \in R^m$, and $y \in R^p$, choosing $e = y - r(t)$ may not provide a satisfactory measure of the tracking performance, in that the trajectory $e(t)$ may not provide sufficient information about the internal dynamics of the closed-loop system.

Example 6.1 Consider the following simple linear system

$$\begin{aligned} \dot{x}_1 &= -x_1 + u \\ \dot{x}_2 &= x_2 - u \end{aligned} \tag{6.5}$$

with $y = x_1$, and assume we want y to follow the reference trajectory $r(t) = \sin t$. The choice of error system

$$e = \chi(t, x) = y - r(t) = x_1 - \sin t$$

yields

$$\dot{e} = -x_1 + u - \cos t. \tag{6.6}$$

The error dynamics are easily stabilized by setting $u = \cos t + \sin t$ so that

$$\dot{e} = -e \tag{6.7}$$

and $e(t)$ decreases to zero exponentially fast. However, this choice of u yields an unstable closed-loop system because the x_2-dynamics become

$$\dot{x}_2 = x_2 + \cos t + \sin t, \tag{6.8}$$

which defines a linear unstable system with a nonzero input. Here the problem is generated by the wrong choice of the error system. A better choice is given by

$$e = \chi(t, x) = \begin{bmatrix} x_1 - \sin t \\ x_2 - \cos t \end{bmatrix},$$

yielding the error dynamics

$$\begin{aligned} \dot{e}_1 &= -x_1 - \cos t + u \\ \dot{e}_2 &= x_2 - u + \sin t. \end{aligned} \tag{6.9}$$

The choice $u = \sin t + \cos t - e_1 + e_2$ yields

$$\begin{aligned} \dot{e}_1 &= -2e_1 + e_2 \\ \dot{e}_2 &= -e_1, \end{aligned} \tag{6.10}$$

which defines an asymptotically stable linear system with eigenvalues at -1. The stability of the new error dynamics also implies that the system states $x_1(t)$ and $x_2(t)$ are bounded functions of time since $x_1(t) = e_1(t) + \sin t$ and $x_2(t) = e_2(t) + \cos t$. △

From the example above, it should be clear that the choice of the error system is crucial to the solution of the tracking problem and that the error system should possess two basic features

1. $e = 0$ should imply $y(t) = r(t)$ or $y(t) \to r(t)$.

2. The boundedness of the error system trajectory $e(t)$ should imply the boundedness of the system state $x(t)$.

These two requirements are summarized in the following assumption.

Assumption 6.1: Assume the error system $e = \chi(t, x)$ is such that $e = 0$ implies $y(t) \to r(t)$ and that the function χ satisfies $|x| \le \psi_x(t, |e|)$ for all t, where $\psi_x : \mathsf{R}^+ \times \mathsf{R}^+ \to \mathsf{R}$ is bounded for any bounded e. Additionally, $\psi_x(t, s)$ is nondecreasing with respect to $s \in \mathsf{R}^+$ for each fixed t.

If Assumption 6.1 is satisfied, then boundedness of the error system will imply boundedness of the state trajectories of (6.3). In addition, if there exists some signal $\eta(t) \geq |e|$ for all t, then $|x| \leq \psi_x(t, \eta)$ since $\psi_x(t, \eta)$ is nondecreasing with respect to $\eta \in \mathbb{R}^+$. Because of Assumption 6.1, we will require not only that the error system provides a measure of the closed-loop system performance, but also that it places bounds on the system states.

Given a general dynamical system (6.1), an error system satisfying Assumption 6.1 can be found by defining the *stable inverse* of the system.

Definition 6.1: Given a bounded reference trajectory $r(t)$, a pair of functions $(x^r(t), c^r(t))$ is said to a **stable inverse** of (6.4) if, for all $t \geq 0$, $x^r(t)$ and $c^r(t)$ are bounded, $x^r(t)$ is differentiable and

$$\begin{aligned} \dot{x}^r(t) &= f(x^r, c^r) \\ r(t) &= h(x^r(t)). \end{aligned} \tag{6.11}$$

Once a stable inverse has been found, the error system can be chosen as

$$e = \chi(t, x) = x - x^r(t). \tag{6.12}$$

It is easy to see that this error system satisfies the two requirements in Assumption 6.1:

- When $e(t) = 0$, we have $x(t) = x^r(t)$ and thus $y(t) = r(t)$.
- If $e(t)$ is bounded for all $t \geq 0$ then $x(t) = e(t) + x^r(t)$ is also bounded because $x^r(t)$ is bounded. In particular, $|x| \leq |e| + |x^r(t)| = \psi_x(t, |e|)$ for all t, where clearly ψ is nondecreasing with respect to its second argument, for each fixed t.

Example 6.2 We now return to Example 6.1 and find the stable inverse of the plant for the reference trajectory $r(t) = \sin t$. To this end, according to Definition 6.1, we seek to find two bounded functions of time $x^r(t)$ and $c^r(t)$ satisfying

$$\begin{aligned} \dot{x}^r{}_1 &= -x^r{}_1 + c^r \\ \dot{x}^r{}_2 &= x^r{}_2 - c^r \\ \sin t &= x^r{}_1(t). \end{aligned} \tag{6.13}$$

In this case the stable inverse is easily found to be $(x^r(t), c^r(t)) = \left([\sin t, \cos t]^\top, \sin t + \cos t\right)$. Return now to the second error system defined in Example 6.1 and note that $\chi(t, x)$ is precisely defined as

$$e = \chi(t, x) = x - x^r(t).$$

$\triangle$

Notice that the error system (6.12) has dimension n. Sometimes one may be able to find lower dimensional error systems satisfying Assumption 6.1. Once an error system has been chosen, the system dynamics may be used to calculate the error dynamics. Given the system dynamics governed by

$$\dot{x} = f(x) + g(x)u, \tag{6.14}$$

the error dynamics become

$$\begin{aligned} \dot{e} &= \frac{\partial \chi}{\partial t} + \frac{\partial \chi}{\partial x}\dot{x} && (6.15) \\ &= \alpha(t,x) + \beta(x)u, && (6.16) \end{aligned}$$

where

$$\alpha(t,x) = \tfrac{\partial \chi}{\partial t} + \tfrac{\partial \chi}{\partial x} f(x) \qquad \beta(x) = \tfrac{\partial \chi}{\partial x} g(x).$$

We will refer to (6.15) as the error dynamics. Since the plant dynamics were affine in the input, the error dynamics defined by (6.16) are also affine in the input. We will later use the error dynamics (6.16) in the development of adaptive controllers.

6.2.2 Lyapunov Candidates

One could directly study the trajectory of the error dynamics (6.15) under feedback control, $u = \nu(z)$, to analyze the closed-loop system performance. Since the error dynamics are nonlinear in general, however, closed-form solutions exist only for a limited number of simple systems. To greatly simplify the analysis, a scalar Lyapunov candidate $V : \mathsf{R}^q \to \mathsf{R}$ is used. The Lyapunov candidate, $V(e)$, for the error system, $e = \chi(t,x)$, is chosen to be positive definite with $V(e) = 0$ if and only if $e = 0$. Thus if a controller may be defined such that V is decreasing along the solutions of (6.15), then the "energy" associated with the error system must be decreasing. If in addition it can be shown that $V \to 0$, then $e \to 0$.

A common choice for a Lyapunov candidate is to use

$$V = e^{\top} P e, \tag{6.17}$$

where $P \in \mathsf{R}^{q \times q}$ is positive definite. Assume that some control law $u = \nu(z)$ is chosen so that $\dot{V} \leq -k_1 V + k_2$ where $k_1 > 0$ and $k_2 \geq 0$ are bounded constants. According to Lemma 2.1, we find

$$V(t) \leq \frac{k_2}{k_1} + \left(V(0) - \frac{k_2}{k_1}\right) e^{-k_1 t}.$$

Then using the Rayleigh-Ritz inequality defined in (2.23), we obtain

$$\begin{aligned} |e|^2 &\leq \frac{V}{\lambda_{\min}(P)} \\ &\leq \frac{k_2}{k_1 \lambda_{\min}(P)} + \left(\frac{V(0)}{\lambda_{\min}(P)} - \frac{k_2}{k_1 \lambda_{\min}(P)}\right) e^{-k_1 t}. && (6.18) \end{aligned}$$

Thus we see that studying the trajectory of V directly places bounds on $|e|$. Using this concept, we will define controllers so that for a given positive definite $V(e)$ we achieve $\dot{V} \leq -k_1 V + k_2$ (or a similar relationship) implying boundedness of V to ensure $|e|$ is bounded. Assumption 6.1 will then be used to find bounds for $|x|$.

6.3 Canonical System Representations

To develop general procedures for control design, we will consider special canonical representations of the plant dynamics for the design model. If the dynamics are not originally in a canonical form, we will use the diffeomorphism $x = T(\xi)$ to obtain a canonical representation. Once the dynamics have been placed in a canonical form, we will find that an appropriate error system and Lyapunov candidate may be generated.

We will find that the design of a controller for a nonlinear system will generally use the following steps:

1. Place the system dynamics into some canonical representation.
2. Choose an error system satisfying Assumption 6.1 and Lyapuonv candidate.
3. Find a control law $u = \nu(z)$ such that $\dot{V} \leq -k_1 V + k_2$, where $k_1 > 0$ and $k_2 \geq 0$.

As we will see, placing the system dynamics into a canonical form often allows for an easy choice of an error system and Lyapunov candidate for which an appropriate control law may be defined. We will find that the particular choice of the error system, and thus the Lyapunov candidate, will generally influence the specification of the control law used to force $\dot{V} \leq -k_1 V + k_2$. Since the goal of the control law is to specify the way in which the Lyapunov function decreases, this approach to control is referred to as Lyapunov-based design.

6.3.1 State-Feedback Linearizable Systems

A system is said to be state-feedback linearizable if there exists a diffeomorphism $x = T(\xi)$, with $T(0) = 0$, such that

$$\dot{x} = Ax + B\left(f(x) + g(x)u\right), \tag{6.19}$$

where $x \in R^n$, $u \in R^m$, and (A, B) form a controllable pair. The functions $f : R^n \to R^m$ and $g : R^n \to R^{m \times m}$ are assumed to be Lipschitz and $g(x)$ is invertible. For the state feedback problem all the states are measurable, and thus we may say the plant output is $y = x$. We will now see how to choose an error system, Lyapunov candidate, and controller for systems satisfying (6.19) for both the set-point regulation and tracking problems.

The Set-Point Regulation Problem

Consider the state regulation problem for a system defined by (6.19) where we wish to drive $x \to r$ where $r \in \mathbb{R}^n$ is the desired state vector. The regulation problem naturally suggests the error system

$$e = x - r, \tag{6.20}$$

which is a measure of the difference between the current and desired state values. This way if $e \to 0$, the control objectives have been met. As long as $|r|$ is bounded, Assumption 6.1 is met since $|x| \leq |e| + |r|$. Since r is a constant value, the error dynamics become

$$\begin{aligned} \dot{e} &= Ax + B(f(x) + g(x)u) \\ &= Ae + B(f(x) + g(x)u) + Ar \end{aligned} \tag{6.21}$$

according to (6.19). We will now consider the Lyapunov candidate $V = e^\top Pe$, where P is a symmetric positive definite matrix, to help establish stability of the closed-loop system.

Consider the control law $u = \nu(z)$ (with $z \equiv x$) defined by

$$\nu(z) = g^{-1}(x)\left(-f(x) + Ke\right), \tag{6.22}$$

where the feedback gain matrix K is chosen such that $A_k = A + BK$ is Hurwitz. With this choice of $\nu(z)$ we see that the plant nonlinearities are cancelled so $\dot{e} = A_k e + Ar$. If $\dot{e} = 0$ when $e = 0$ (it is an equilibrium point) we will require that $Ar = 0$. Thus $r = 0$ is always a valid set-point. Depending upon the structure of A, however, other choices may be available (this will be demonstrated shortly in Example 6.3).

The rate of change of the Lyapunov candidate now becomes

$$\dot{V} = e^\top (PA_k + A_k^\top P)e = -e^\top Qe,$$

where $PA_k + A_k^\top P = -Q$ is a Lyapunov matrix equation with Q positive definite.

It is now possible to find explicit bounds on $|e|$. By using the Rayleigh-Ritz inequality defined in (2.23) we find

$$\begin{aligned} \dot{V} &\leq -\lambda_{\min}(Q)|e|^2 \\ &\leq -\frac{\lambda_{\min}(Q)}{\lambda_{\max}(P)} V. \end{aligned} \tag{6.23}$$

Using Lemma 2.1 it is then possible to show that

$$|e| \leq \sqrt{\frac{V(0)}{\lambda_{\min}(P)}} e^{-ct/2},$$

where $c = \lambda_{\min}(Q)/\lambda_{\max}(P)$.

The above shows that if the functions $f(x)$ and $g(x)$ are known, then it is possible to design a controller which ensures an exponentially stable closed-loop system. We also saw that the error dynamics may be expressed in the form (6.16) with $\alpha = Ae + Bf(x)$ and $\beta = Bg(x)$. We will continue to see this form for the error dynamics throughout the remainder of this chapter. In fact, we will later take advantage of this form when designing adaptive controllers.

Example 6.3 Here we will design a satellite orbit control algorithm which is used to ensure that the proper satellite altitude and orbital rate are maintained. The satellite is assumed to have mass m and potential energy k_g/r. The satellite dynamics may be expressed as

$$\begin{aligned} \dot{z} &= v \\ \dot{v} &= z\omega^2 - \frac{k_g}{mz^2} + \frac{u_1}{m} \\ \dot{\omega} &= -\frac{2v\omega}{z} + \frac{u_2}{mz}, \end{aligned}$$

where r and v represent the radial position and velocity, respectively, while ω is the angular velocity of the satellite [149]. Assume we wish to define a controller so that $z \to z_d$ and $\omega \to \omega_d$ where z_d and ω_d are constants.

The first step in developing a controller for this problem using the above feedback linearization approach is to place the system dynamics into the form (6.19). Letting $x = [z, v, \omega]^\top$ we find

$$\dot{x} = Ax + B\left(f(x) + g(x)u\right), \tag{6.24}$$

with

$$A = \begin{bmatrix} 0 & 1 & 0 \\ 0 & 0 & 0 \\ 0 & 0 & 0 \end{bmatrix} \qquad B = \begin{bmatrix} 0 & 0 \\ 1 & 0 \\ 0 & 1 \end{bmatrix}$$

and

$$f(x) = \begin{bmatrix} z\omega^2 - \frac{k_g}{mz^2} \\ -\frac{2v\omega}{z} \end{bmatrix} \qquad g(x) = \begin{bmatrix} \frac{1}{m} & 0 \\ 0 & \frac{1}{mz} \end{bmatrix}.$$

The error signal then becomes $e = x - r$, where $r = [z_d, 0, \omega_d]^\top$. Since $Ar = 0$, we may choose the control law defined by (6.22). All that is left to do is choose some K such that $A_k = A + BK$ is Hurwitz. Choosing

$$K = \begin{bmatrix} -\lambda_r^2 & -2\lambda_r & 0 \\ 0 & 0 & -\lambda_\omega \end{bmatrix},$$

we find

$$A_k = \begin{bmatrix} 0 & 1 & 0 \\ -\lambda_r^2 & -2\lambda_r & 0 \\ 0 & 0 & -\lambda_\omega \end{bmatrix},$$

which is Hurwitz for any $\lambda_r, \lambda_\omega > 0$. The values of λ_r and λ_ω may be used to set the eigenvalues of the radial and angular channels, respectively. △

The Tracking Problem

For the tracking problem, we will define a diffeomorphism $x = T(\xi)$ so that

$$\dot{x} = A_c x + b_c \left(f(x) + g(x)u\right), \tag{6.25}$$

where (A_c, b_c) fits a controllable canonical form and $y = x$. To simplify the analysis for the tracking problem, we will assume single-input single-output systems. For single-input systems, we have

$$A_c = \begin{bmatrix} 0 & I \\ 0 & 0 \end{bmatrix},$$

and $b_c = [0, \ldots, 0, 1]^\top$, so (6.25) is equivalent to

$$\begin{aligned} \dot{x}_1 &= x_2 \\ &\vdots \\ \dot{x}_{n-1} &= x_n \\ \dot{x}_n &= f(x) + g(x)u. \end{aligned} \tag{6.26}$$

We will now define conditions which may be used to help define a diffeomorphism, $x = T(\xi)$, so that we may place a system into the form defined by (6.25). First assume that the system dynamics are affine in the control so that $\dot{\xi} = f_\xi(\xi) + g_\xi(\xi)u$. Since $x = T(\xi)$ we find

$$\begin{aligned} \dot{x} &= \frac{\partial T}{\partial \xi}\left(f_\xi(\xi) + g_\xi(\xi)u\right) && (6.27) \\ &= A_c x + b_c(f(x) + g(x)u). && (6.28) \end{aligned}$$

Using the definitions for A_c and b_c for a single-input system,

$$\begin{aligned} \frac{\partial T}{\partial \xi} f_\xi &= A_c T(\xi) + b_c f(T(\xi)) && (6.29) \\ &= \begin{bmatrix} T_2 \\ \vdots \\ T_n \\ f(T(\xi)) \end{bmatrix} && (6.30) \end{aligned}$$

when $u = 0$, where $T(\xi) = [T_1(\xi), \ldots, T_n(\xi)]^\top$. Matching the remaining terms we obtain

$$\frac{\partial T}{\partial \xi} g_\xi = b_c g(x) \tag{6.31}$$

$$= \begin{bmatrix} 0 \\ \vdots \\ 0 \\ g(T(\xi)) \end{bmatrix}. \tag{6.32}$$

If we find some diffeomorphism $x = T(\xi)$ which satisfies

$$\begin{array}{c} \frac{\partial T_i}{\partial \xi} g_\xi = 0 \quad \text{for } i = 1, \ldots, n-1 \\ \frac{\partial T_n}{\partial \xi} g_\xi \neq 0 \end{array}, \tag{6.33}$$

where $T_{i+1} = \frac{\partial T_i}{\partial \xi} f_\xi$ for $i = 1, \ldots, n-1$, then

$$f = \frac{\partial T_n}{\partial \xi} f_\xi \tag{6.34}$$

and

$$g = \frac{\partial T_n}{\partial \xi} g_\xi. \tag{6.35}$$

The following example demonstrates how a diffeomorphism may be defined so that (6.33) is satisfied.

Example 6.4 Consider the system defined by

$$\begin{aligned} \dot{\xi}_1 &= \xi_1^2 + \xi_2 \\ \dot{\xi}_2 &= \xi_1 \xi_2 + u \end{aligned} \tag{6.36}$$

so that

$$f_\xi = \begin{bmatrix} \xi_1^2 + \xi_2 \\ \xi_1 \xi_2 \end{bmatrix} \qquad g_\xi = \begin{bmatrix} 0 \\ 1 \end{bmatrix}.$$

We will want to choose $T(\xi) = [T_1(\xi), T_2(\xi)]^\top$ such that

$$\frac{\partial T_1}{\partial \xi} g_\xi = 0 \qquad \frac{\partial T_2}{\partial \xi} g_\xi \neq 0,$$

with $T_2 = \frac{\partial T_1}{\partial \xi} f_\xi$. Since

$$\frac{\partial T_1}{\partial \xi} g_\xi = \frac{\partial T_1}{\partial \xi_2} = 0,$$

we choose T_1 to be independent of ξ_2. Thus

$$T_2 = \begin{bmatrix} \frac{\partial T_1}{\partial \xi_1} & 0 \end{bmatrix} \begin{bmatrix} \xi_1^2 + \xi_2 \\ \xi_1 \xi_2 \end{bmatrix} = \frac{\partial T_1}{\partial \xi_1}(\xi_1^2 + \xi_2). \tag{6.37}$$

The choice $T_1(\xi_1) = \xi_1$ satisfies (6.33) since then $T_2(\xi) = \xi_1^2 + \xi_2$ and

$$\frac{\partial T_2}{\partial \xi} g_\xi = \begin{bmatrix} 2\xi_1 & 1 \end{bmatrix} \begin{bmatrix} 0 \\ 1 \end{bmatrix} = 1. \tag{6.38}$$

Thus the diffeomorphism

$$x = T(\xi) = \begin{bmatrix} \xi_1 \\ \xi_1^2 + \xi_2 \end{bmatrix} \tag{6.39}$$

transforms the system into

$$\dot{x} = \begin{bmatrix} \xi_1^2 + \xi_2 \\ 2\xi_1(\xi_1^2 + \xi_2) + \xi_1 \xi_2 \end{bmatrix} + \begin{bmatrix} 0 \\ 1 \end{bmatrix} u. \tag{6.40}$$

Since

$$\xi = T^{-1}(x) = \begin{bmatrix} x_1 \\ x_2 - x_1^2 \end{bmatrix}, \tag{6.41}$$

we find $\dot{x} = A_c + b_c(f(x) + g(x)u)$, where

$$f(x) = 2x_1 x_2 + x_1(x_2 - x_1^2),$$

and

$$g(x) = 1.$$

△

The reader interested in proving the existence of a diffeomorphism which places the state $x = T(\xi)$ into the form (6.19) is urged to see the work in [149].

We will now study some examples in which control laws are defined for systems satisfying (6.25). We will see that different choices for the error system and Lyapunov candidate may produce rather different control laws. The first approach will use an error system similar to that used in the state regulation problem, while the second approach will define the error system based on a stable manifold.

Example 6.5 In this example we will design a controller for (6.25) such that $x_1 \to r(t)$. Since x_{i+1} is the derivate of x_i, we will define the error system

$$e = x - \begin{bmatrix} r \\ \dot{r} \\ \vdots \\ r^{(n-1)} \end{bmatrix}. \tag{6.42}$$

If $\left|\left[r, \dot{r}, \cdots, r^{(n-1)}\right]^\top\right| \leq \bar{r}$, then $|x| \leq |e| + \bar{r}$ so Assumption 6.1 is satisfied. The error dynamics may now be expressed as

$$\begin{aligned} \dot{e} &= \dot{x} - \dot{r} \\ &= A_c e + b_c \left(-r^{(n)} + f(x) + g(x)u\right), \end{aligned} \tag{6.43}$$

where we have used the structure of A_c and b_c.

Now consider the Lyapunov candidate $V = e^\top P e$, where P is a positive definite symmetric matrix to be chosen shortly. We now choose the control law $u = \nu(z)$ as

$$\nu = \frac{r^{(n)} - f(x) - k^\top e}{g(x)}, \tag{6.44}$$

where

$$A_k = \begin{bmatrix} 0 & & \\ \vdots & & I \\ 0 & & \\ -k_1 & \cdots & -k_n \end{bmatrix} \tag{6.45}$$

is Hurwitz with $k = [k_1, \ldots, k_n]^\top$. With this control law, $\dot{e} = A_k e$ so

$$\dot{V} = e^\top (P A_k + A_k^\top P) e. \tag{6.46}$$

Choosing P to satisfy the Lyapunov matrix equation $P A_k + A_k^\top P = -Q$, where Q is positive definite, one obtains

$$\dot{V} = -e^\top Q e,$$

so that the closed-loop system is stable. △

As an alternative to the "traditional" approach (shown above) to the tracking problem for state-feedback linearizable systems, one may define the error system using a stable manifold as shown in the following example:

Example 6.6 Here we will consider the tracking problem for the single-input state feedback linearizable system (6.25) where we wish to drive $x \to r(t)$. The error system may be defined using a stable manifold so that $e = \chi(t, x)$ where

$$\chi(t, x) = k_1(x_1 - r) + \cdots + k_{n-1}(x_{n-1} - r^{(n-2)}) + x_n - r^{(n-1)}, \tag{6.47}$$

and

$$L = \begin{bmatrix} 0 & & \\ \vdots & & I \\ 0 & & \\ -k_1 & \cdots & -k_{n-1} \end{bmatrix}$$

is Hurwitz (we also say that the polynomial $s^{n-1}+k_{n-1}s^{n-2}+\cdots+k_1$ is Hurwitz). For now we will simply assume that this error system satisfies Assumption 6.1 and continue the control design so that $e \to 0$. We will then show that defining the error system using a stable manifold does indeed satisfy Assumption 6.1.

Taking the derivative of e we find

$$\dot{e} = k_1(x_2 - \dot{r}) + \cdots + k_{n-1}(x_n - r^{(n-1)}) - r^{(n)} + f(x) + g(x)u$$

so $\dot{e} = \alpha(t,x) + \beta(x)u$ with

$$\alpha(t,x) = -r^{(n)} + k_1(x_2 - \dot{r}) + \cdots + k_{n-1}(x_n - r^{(n-1)}) + f(x),$$

and $\beta(x) = g(x)$. Now consider the Lyapunov candidate $V = \frac{1}{2}e^2$ so that

$$\dot{V} = e\left(\alpha(t,x) + \beta(x)u\right). \tag{6.48}$$

Using the control law $u = \nu(z)$ with

$$\nu(z) = \frac{-\alpha(t,x) - \kappa e}{\beta(x)}, \tag{6.49}$$

and $\kappa > 0$ we find $\dot{V} = -\kappa e^2 = -2\kappa V$ so $e = 0$ is an exponentially stable equilibrium point (the trajectory converges to the manifold $S_x(t) = \{x \in \mathrm{R}^n : \chi(t,x) = 0\}$).

We will now show that bounding $|e|$ implies that $|x|$ is bounded. It is reasonable that bounding $|e|$ should bound the plant states since

$$(y-r)(s) = \frac{1}{s^{n-1} + k_{n-1}s^{n-2} + \cdots + k_1} e(s),$$

with the denominator poles in the left half plane. To show that Assumption 6.1 is satisfied when using the error system defined by (6.47), first let

$$\mu = \begin{bmatrix} x_1 - r \\ \vdots \\ x_{n-1} - r^{(n-2)} \end{bmatrix},$$

so $\dot{\mu} = L\mu + be$, where $b = [0,\ldots,0,1]^\top \in \mathrm{R}^{n-1}$. Since L is Hurwitz, we may use (2.91) and conclude that $\mu(t) \le \psi_\mu(t,|e|)$, where

$$\psi_\mu = c_1 e^{-c_2 t}|e(0)| + \int_0^t c_1 e^{-c_2(t-\tau)}|e(\tau)|d\tau, \tag{6.50}$$

and c_1, c_2 are chosen to satisfy $|e^{Lt}| \le c_1 e^{-c_2 t}$ (one may use the results of Example 2.16 to calculate values for c_1 and c_2). Using the definition of the error system, we find

$$\begin{bmatrix} x_1 \\ \vdots \\ x_n \end{bmatrix} = \begin{bmatrix} r \\ \vdots \\ r^{(n-1)} \end{bmatrix} + \begin{bmatrix} \mu \\ \zeta \end{bmatrix},$$

where $\zeta = e - k_1(x_1 - r) - \cdots - k_{n-1}(x_{n-1} - r^{(n-2)})$ so

$$|x| \leq (1 + |k_1| + \cdots + |k_{n-1}|)\psi_\mu(t, |e|) + |e| + \bar{r}.$$

Thus we can set

$$\psi_x(t, e) = (1 + |k_1| + \cdots + |k_{n-1}|)\psi_\mu(t, |e|) + |e| + \bar{r},$$

which satisfies Assumption 6.1. △

From the previous example, we see that when a stable manifold is used to define an error system, it may not be possible to define a static bounding relationship between the error system and system states. This occurs since it is possible for $e = 0$ as the state trajectory slides along the surface S_x even though $y \neq r(t)$ due to the system initial conditions. Notice that $e \in \mathbb{R}$ in the above example, while $e \in \mathbb{R}^n$ in Example 6.5. Either choice forms a valid error system for the tracking problem.

It should also be noted that the controllers defined by (6.44) and (6.49) in the previous examples are identical, up to the choice of the gains on the various error terms. This is because the feedback terms are linear in both cases, and thus each approach simply suggests different coefficients be used. We will later see that it may be possible to add nonlinear feedback terms based on the definition of the error system and Lyapunov candidate to improve closed-loop robustness. In these cases the resulting control law may not be as similar.

6.3.2 Input-Output Feedback Linearizable Systems

Consider the system

$$\begin{aligned} \dot{\xi} &= f_\xi(\xi) + g_\xi(\xi)u \\ y &= h(\xi), \end{aligned} \tag{6.51}$$

where $\xi \in \mathbb{R}^{n+d}$ with $d \geq 0$. The system is said to have strong relative degree n if

$$L_{g_\xi} h(\xi) = L_{g_\xi} L_{f_\xi} h(\xi) = \cdots = L_{g_\xi} L_{f_\xi}^{n-2} h(\xi) = 0$$

and $L_{g_\xi} L_{f_\xi}^{n-1} h(\xi)$ is bounded away from zero. Here $L_{g_\xi} h$ is the Lie derivative defined as $L_{g_\xi} h \equiv \frac{\partial h}{\partial \xi} g_\xi$ and $L_{g_\xi}^2 h \equiv L_{g_\xi}\left(L_{g_\xi} h\right)$. Consider the change of coordinates $[q^\top x^\top]^\top = T(\xi)$ with $q \in \mathrm{R}^d$ and $x \in \mathrm{R}^n$. We will consider the choice of the diffeomorphism $T = [T_1, \ldots, T_{n+d}]^\top$ such that $T(0) = 0$ and

$$\frac{\partial T_i(\xi)}{\partial \xi} g_\xi(\xi) = 0$$

for $i = 1, \ldots, d$ and

$$\begin{aligned} T_{d+1} &= h(\xi) \\ T_{d+2} &= L_{f_\xi} h(\xi) \\ &\vdots \\ T_{n+d} &= L_{f_\xi}^{n-1} h(\xi). \end{aligned}$$

With this change of coordinates, the system dynamics become

$$\begin{aligned} \dot{q} &= \phi(q, x) \\ \dot{x}_1 &= x_2 \\ &\vdots \\ \dot{x}_{n-1} &= x_n \\ \dot{x}_n &= f(q, x) + g(q, x) u \end{aligned} \tag{6.52}$$

with the output $y = x_1$ which is said to be in input-output feedback linearizable form. Notice that $f(q, x) = L_{f_\xi}^n h(\xi)$ and $g(q, x) = L_{g_\xi} L_{f_\xi}^{n-1} h(\xi)$, where $\xi = T^{-1}[q^\top, x^\top]^\top$. If $d = 0$, then the system is said to be simultaneously state-feedback linearizable and input-output linearizable (which is in the form (6.25)). It is assumed that both the q and x state vectors are measurable and that the functions ϕ, f, and g are Lipschitz. The dynamics of $\dot{q} = \phi(q, 0)$ are referred to as zero dynamics of the system. The next example provides a motivation for this definition.

Example 6.7 Consider the case where the states q, x define a linear system with transfer function from plant input-to-output given by

$$P(s) = k \frac{s^d + b_{d-1} s^{d-1} + \ldots + b_0}{s^{n+d} + a_{n+d-1} s^{n+d-1} + \ldots + a_0}. \tag{6.53}$$

Then a minimal realization may be defined with

$$A = \begin{bmatrix} 0 & & \\ \vdots & & I \\ 0 & & \\ -a_0 & \cdots & -a_{n+d-1} \end{bmatrix} \quad B = \begin{bmatrix} 0 \\ \vdots \\ 0 \\ k \end{bmatrix}, \tag{6.54}$$

and $C = [b_0, \ldots, b_{d-1}, 1, 0, \ldots 0]$.

Since $y = x_1$, we see that

$$\begin{aligned} x_1 = C\xi &= b_0\xi_1 + \ldots + b_{d-1}\xi_d + \xi_{d+1} \\ &\vdots \\ x_n = CA^{n-1}\xi &= b_0\xi_n + \ldots + b_{d-1}\xi_{n+d-1} + \xi_{n+d}. \end{aligned} \tag{6.55}$$

We are now free to define the remaining d states so that $[q^\top, x^\top]^\top = T\xi$ where $T \in \mathbb{R}^{(n+d)\times(n+d)}$ is invertible. Here T is the transformation matrix from the ξ to x, q states. We thus choose

$$\begin{aligned} q_1 &= \xi_1 \\ &\vdots \\ q_d &= \xi_d \end{aligned}$$

so that T is a lower triangular matrix (and thus has full rank). It is easy to show that T is lower triangular since

$$\left[\begin{array}{c} q \\ \hline x \end{array}\right] = \left[\begin{array}{cc} I & 0 \\ \hline \multicolumn{2}{c}{C} \\ \multicolumn{2}{c}{\vdots} \\ \multicolumn{2}{c}{CA^{n-1}} \end{array}\right] \xi.$$

Notice that

$$\begin{aligned} \dot{q}_1 &= q_2 \\ &\vdots \\ \dot{q}_d &= q_{d+1} = \xi_{d+1} = x_1 - b_0\xi_1 - \ldots - b_{d-1}\xi_d, \end{aligned} \tag{6.56}$$

where we have used the definition of x_1 in (6.55). Thus

$$\dot{q} = \begin{bmatrix} 0 & & \\ \vdots & I & \\ 0 & & \\ -b_0 & \cdots & -b_{d-1} \end{bmatrix} q + \begin{bmatrix} 0 & & & \\ \vdots & & \bigcirc & \\ 0 & & & \\ 1 & 0 & \cdots & 0 \end{bmatrix} x. \tag{6.57}$$

We see that with $\dot{q} = A_q q + B_q x$ (where A_q and B_q are defined by (6.57)) the eigenvalues of A_q are equivalent to the zeros of $P(s)$. It is for this reason that $\dot{q} = \phi(q, 0)$ are often referred to as the zero dynamics of the system. △

From (6.57) we see that even if a controller is defined so that the x states are bounded, we still have $q \to \infty$ if $\dot{q} = A_q q$ is not a stable system. Because of this, it will be necessary to require additional conditions upon the dynamics governing the q states to ensure stability. In particular we will assume the q-subsystem, with x as an input, is input-to-state stable, so that there exists some positive definite V_q such that

$$\gamma_{q1}(|q|) \le V_q(q) \le \gamma_{q2}(|q|) \tag{6.58}$$

$$\dot{V}_q \le -\gamma_{y3}(|y|), \quad \forall |y| \ge \psi(|x|), \tag{6.59}$$

where γ_{q1} and γ_{q2} are class $\mathcal{K}_\infty$, and γ_{q3}, ψ are class-$\mathcal{K}$. When $x = 0$ the input-to-state stability assumption (6.59) becomes $\dot{V}_q \le -\gamma_{y3}(|y|)$, for all y, thus implying global asymptotic stability of the origin of the zero dynamics $\dot{q} = \phi(q, 0)$, which are then said to be minimum phase. With these assumptions, it is possible to design a controller by ignoring the trajectory of the zero dynamics using the procedure for state-feedback linearizable systems. The following theorem guarantees that if a controller is designed to stabilize the x dynamics, then the q dynamics are also stable if they satisfy (6.58)-(6.59).

Theorem 6.1: *Let $e = \chi(t, x)$ be an error system satisfying Assumption 6.1. Assume there exists a controller $u = \nu(z)$ and Lyapunov function $V(e)$ such that $\gamma_e(|e|) \le V(e)$ with γ_e class-$\mathcal{K}_\infty$ and $\dot{V} \le 0$ along the trajectories of (6.52) when $V \ge V_r$ assuming ν is well defined for all $q \in \mathbb{R}^d$, $x \in \mathbb{R}^n$ and $t \in \mathbb{R}^+$. If the q-subsystem, with x as an input, is input-to-state stable, that is, there exists some positive definite V_q such that (6.58)-(6.59) hold, then the controller $u = \nu(z)$ ensures that x and q are uniformly bounded.*

Theorem 6.1 may be proven by first showing that the error system is stable (independent of q), and then showing that the q dynamics are therefore bounded. Notice that the above theorem only ensures uniform boundedness of the trajectories. The properties of the e states may be used to prove stronger stability results. If for example, we are ensured that $\dot{V} \le -kV$, then $e = 0$ is an exponentially stable equilibrium point (though $|q|$ may still only be bounded). The following example demonstrates how a controller may be defined for a system with nonlinear zero dynamics using feedback linearization.

Example 6.8 A nonlinear system is defined by

$$\dot{q}_1 = 1 - q_1 - q_1^3 + q_1 x_1^2 \tag{6.60}$$

$$\dot{x}_1 = q_1^2 + u, \tag{6.61}$$

with $y = x_1$. If we wish to drive $x_1 \to 0$, then consider the error system $e = x_1$ and the Lyapunov candidate $V = \frac{1}{2}e^2$. Using the concepts of state-feedback linearization, we choose $u = \nu(z)$ with

$$\nu(q, x) = -q_1^2 - \kappa e \tag{6.62}$$

so that $\dot{V} = -\kappa e^2$ (here, $z = [x_1, q_1]^\top$). If it can be shown that q_1 is bounded so that ν is well defined, then $e = 0$ is an exponentially stable equilibrium point.

From Theorem 6.1 we now simply need to show that the q dynamics satisfy (6.58)-(6.59). Let $V_q = \frac{1}{2}q_1^2$ which satisfies (6.58) with $\gamma_{q1} = \gamma_{q2} = q_1^2/2$. Then

$$\begin{aligned} \dot{V}_q &= q_1\left(1 - q_1 - q_1^3 + q_1 x_1^2\right) && (6.63)\\ &\leq -q_1^2 + |q_1| - q_1^4 + q_1^2|x_1|^2. && (6.64) \end{aligned}$$

Since $-2x^2 \pm 2xy \leq -x^2 + y^2$, we find $-q_1^2 + |q_1| \leq -q_1^2/2 + 1/2$ and $-q_1^4 + q_1^2|x_1|^2 \leq x_1^4/4$. Thus

$$\dot{V}_q \leq -\frac{q_1^2}{2} + \frac{1}{2} + \frac{|x_1|^4}{4}, \tag{6.65}$$

so $\dot{V}_q \leq -V_q + \psi(x)$ where $\psi(x) = (2 + x_1^4)/4$. Using Theorem 6.1, we are ensured that q_1 is uniformly bounded so $e = 0$ is an exponentially stable equilibrium point. △

The systems considered thus far are in a form so that the plant nonlinearities may be easily cancelled by the input. That is, we are able to use feedback to force the closed-loop system to act as a linear system with arbitrary eigenvalues using traditional pole placement techniques. We will now consider a special class of feedback linearizable systems in which, to guarantee global stability, it is not necessary to cancel the plant nonlinearities and assign the closed-loop poles using the input. The particular structure of the system is exploited to achieve robustness in the presence of uncertainties, as we will see in what follows.

6.3.3 Strict-Feedback Systems

A single-input system is said to be in pure-feedback form if there exists a diffeomorphism, $x = T(\xi)$, which renders the system dynamics as

$$\begin{aligned} \dot{x}_1 &= f_1(x_1, x_2)\\ &\vdots\\ \dot{x}_{n-1} &= f_{n-1}(x)\\ \dot{x}_n &= f_n(x, u), \end{aligned} \tag{6.66}$$

where each f_i is Lipschitz. Since $\dot{x}_i$ only depends upon the signal vector $[x_1, \ldots, x_{i+1}]^\top$ for $i = 1, \ldots, n-1$, this system has a triangular structure. A special class of pure-feedback systems, called strict-feedback systems, is found when each successive state and control input is affine so that

$$\begin{aligned} \dot{x}_1 &= f_1(x_1) + g_1(x_1)x_2 \\ &\vdots \\ \dot{x}_{n-1} &= f_{n-1}(x_1, \ldots, x_{n-1}) + g_{n-1}(x_1, \ldots, x_{n-1})x_n \\ \dot{x}_n &= f_n(x) + g_n(x)u, \end{aligned} \tag{6.67}$$

where each g_i is bounded away from zero. Before defining a controller suitable for (6.67), we will first present the concept of integrator backstepping summarized in the following theorem:

Theorem 6.2: **(Integrator Backstepping)** *Let $e = \chi(t, x)$ be an error system satisfying Assumption 6.1 with error dynamics*

$$\dot{e} = \alpha(t, x) + \beta(x)u, \tag{6.68}$$

where $u \in \mathrm{R}$. Let $u = \nu(z)$ be a continuously differentiable globally stabilizing controller such that the radially unbounded Lyapunov function $V(e)$ satisfies

$$\dot{V} \leq -k_1 V + k_2$$

along the solutions of (6.68) when $u = \nu(z)$, where k_1, k_2 are positive costants. Then there exists a stabilizing controller $v = \nu_c(z, q)$ for the composite system

$$\begin{aligned} \dot{e} &= \alpha(t, x) + \beta(x)q \\ \dot{q} &= v, \end{aligned}$$

where $q \in \mathrm{R}$ and v is a new input.

Proof: We will introduce a new error term $e_q = q - \nu(z)$ and Lyapunov candidate

$$V_c(e_c) = V(e) + \frac{1}{2}e_q^2. \tag{6.69}$$

for the composite system with $e_c = [e^\top, e_q]^\top$. Taking the derivative, we find

$$\dot{V}_c = \frac{\partial V}{\partial e}(\alpha(t, x) + \beta(x)(q - \nu(z) + \nu(z))) + e_q(\dot{q} - \dot{\nu}). \tag{6.70}$$

Since $q = e_q + \nu$, we find

$$\dot{V}_c \leq -k_1 V + k_2 + \frac{\partial V}{\partial e}\beta(x)e_q + e_q(\nu_c - \dot{\nu}), \tag{6.71}$$

where $v = \nu_c(z, q)$. Choosing

$$\nu_c = \dot{\nu} - \frac{\partial V}{\partial e}\beta(x) - \frac{k_1}{2}e_q \tag{6.72}$$

results in

$$\dot{V}_c \le -k_1 V + k_2 - \frac{k_1}{2}e_q^2 = -k_1 V_c + k_2. \tag{6.73}$$

This ensures that V_c is bounded so the new error system e_c is also bounded. ■

In the derivation of the proof of Theorem 6.2 we found a stabilizing controller satisfying the theorem. It should be emphasized, however, that (6.72) is just one of many controllers which satisfy Theorem 6.2. In the following example, we will now use the techniques presented in the proof of Theorem 6.2 to create an error system and stabilizing controller for the system defined by (6.67).

Example 6.9 Consider the strict-feedback system defined by (6.67). We will see that a procedure may be used to construct an appropriate error system and stabilizing controller using the techniques employed to prove Theorem 6.2.

x_1 subsystem To create a stabilizing controller, we will begin by considering the subsystem defined by

$$\dot{x}_1 = f_1(x_1) + g_1(x_1)v, \tag{6.74}$$

where v is a virtual input. If we wish to force x_1 to track the reference signal $r(t)$, then we will define the first component of the error system as $e_1 = x_1 - r$. Using (6.74) we find the error dynamics to be

$$\dot{e}_1 = -\dot{r} + f_1(x_1) + g_1(x_1)v.$$

A Lyapunov candidate for the subsystem may then be defined as $V_1 = \frac{1}{2}e_1^2$ so that the feedback linearizing control law

$$v = \nu_1(z_1) = \frac{\dot{r} - f_1(x_1) - \kappa e_1}{g_1(x_1)}$$

ensures $\dot{V}_1 = -\kappa e_1^2 = -2\kappa V_1$. Here $z_1 = [x_1, r, \dot{r}]^\top$. Notice that $\dot{e}_1 = -\kappa e_1$ when $v = \nu_1(z_1)$.

x_2 subsystem We will now use the procedure used to prove Theorem 6.2 to design a controller for the new subsystem

$$\begin{aligned}\dot{x}_1 &= f_1(x_1) + g_1(x_1)x_2\\ \dot{x}_2 &= v,\end{aligned}$$

where v is a new virtual input. Now consider the new error variable $e_2 = x_2 - \nu_1(z_1)$. Using the definition of the x_2 subsystem, the error dynamics for e_1 become $-\dot{r} + \dot{e}_1 = f_1 + g_1(x_2 - \nu_1 + \nu_1)$. The overall error dynamics are therefore

$$\dot{e} = \begin{bmatrix} -\kappa e_1 + g_1 e_2 \\ v - \dot{\nu}_1 \end{bmatrix}. \tag{6.75}$$

A new Lyapunov candidate is now defined as $V_2 = V_1 + \frac{1}{2}e_2^2$ so

$$\begin{aligned} \dot{V}_2 &= e_1[\dot{e}_1] + e_2[\dot{e}_2] \\ &= -\kappa e_1^2 + e_1 g_1 e_2 + e_2[v - \dot{\nu}_1]. \end{aligned} \tag{6.76}$$

Therefore, we let $v = \bar{\nu}_2(z_2)$ with $z_2 = [x_1, x_2, r, \dot{r}, \ddot{r}]^\top$ and

$$\bar{\nu}_2 = \dot{\nu}_1 - e_1 g_1 - \kappa e_2 \tag{6.77}$$

so that $\dot{V}_2 = -\kappa(e_1^2 + e_2^2) = -2\kappa V_2$.

When the subsystem is defined by

$$\begin{aligned} \dot{x}_1 &= f_1(x_1) + g_1(x_1)x_2 \\ \dot{x}_2 &= f_2(x_1, x_2) + g_1(x_1, x_2)u, \end{aligned}$$

the controller $\nu_2 = (-f_2 + \bar{\nu}_2)/g_2$ ensures that $\dot{V}_2 = -2\kappa V_2$.

x_n system This procedure may be repeated until a controller for the system (6.67) is designed. Let $e_i = x_i - \nu_{i-1}(z_{i-1})$, where

$$\nu_i = \frac{-f_i + \dot{\nu}_{i-1} - e_{i-1}g_{i-1} - \kappa e_i}{g_i}$$

for $i = 1, \ldots, n-1$. With this definition, the error dynamics for the system become

$$\dot{e} = \begin{bmatrix} -\kappa e_1 + g_1 e_2 \\ \vdots \\ g_{n-2}e_{n-2} - \kappa e_{n-1} + g_{n-1}e_n \\ -\dot{\nu}_{n-1} + f_n(x) \end{bmatrix} + \begin{bmatrix} 0 \\ \vdots \\ 0 \\ g_n(x) \end{bmatrix} u. \tag{6.78}$$

Notice that the error dynamics again fit the form $\dot{e} = \alpha(t, x) + \beta(x)u$. Defining the controller

$$u = \nu(z) = \frac{-f_n(x) + \dot{\nu}_{n-1} - e_{n-1}g_{n-1} - \kappa e_n}{g_n(x)} \tag{6.79}$$

renders $\dot{V}_n = -2\kappa V_n$ so $e = 0$ is an exponentially stable equilibrium point. △

The above example showed that the error system can be driven to zero. We have not shown, however, that bounding $|e|$ will place bounds on $|x|$ as required by Assumption 6.1. The following lemma is often helpful in proving the existence of bounds on $|x|$.

Lemma 6.1: *If $f : \mathbb{R}^n \to \mathbb{R}$ is locally Lipschitz with constant L when $x \in D$, then there exists some smooth class-$\mathcal{K}$ function γ and $d \in \mathbb{R}$ such that $|f(x)| \leq d + \gamma(|x|)$ when $x \in D$.*

Proof: Since $|f(x)| - |f(0)| \leq |f(x) - f(0)|$ for all x, we find $|f(x)| - |f(0)| \leq L|x|$, where L is a Lipschitz constant. Since $2xy \leq x^2 + y^2$ we find $|f(x)| - |f(0)| \leq |x|^2/2 + L^2/2$. Letting $d = |f(0)| + L^2/2$ and $\gamma(|x|) = |x|^2/2$ proves the lemma. ∎

The following example shows how Lemma 6.1 may be used with other error systems defined using integrator backstepping.

Example 6.10 Consider the error system defined by

$$\begin{aligned} e_1 &= x_1 \\ e_2 &= x_2 + g_2(\bar{x}_2) \\ &\vdots \\ e_n &= x_n + g_n(x), \end{aligned} \tag{6.80}$$

where $\bar{x}_i = [x_1, \ldots x_i]^\top$ and each $g_i(\bar{x}_{i-1})$ is locally Lipschitz. Since $|x_1| = |e_1|$, there exists some k_1 and class-$\mathcal{K}$ function γ_1 such that $|x_1| \leq k_1 + \gamma_1(|e_1|)$. Assume that there exists k_{i-1} and γ_{i-1} such that $|x_{i-1}| \leq k_{i-1} + \gamma_{i-1}(|\bar{e}_{i-1}|)$ where $\bar{e}_i = [e_1, \ldots, e_i]^\top$. Then since g_i is Lipschitz, there exist some constant d_i and class-$\mathcal{K}$ function ψ_i such that

$$\begin{aligned} |x_i| &\leq |e_i| + |g_i(\bar{x}_i)| \\ &\leq |e_i| + d_i + \psi_i(|\bar{x}_{i-1}|) \\ &\leq |e_i| + d_i + \psi_i(k_{i-1} + \gamma_{i-1}(|\bar{e}_{i-1}|)) = \zeta_i(\bar{e}_i) \end{aligned}$$

when $x \in S_x$. Since ζ_i is Lipschitz, we find there exist some constant k_i and class-$\mathcal{K}$ function γ_i such that $|x_i| \leq k_i + \gamma_i(|\bar{e}_i|)$ for all $i = 1, \ldots, n$ by recursion. Also

$$|x| \leq \sum_{i=1}^{n} |x_i| \leq \sum_{i=1}^{n} (k_i + \gamma_i(|\bar{e}_i|)) = \eta_x(|e|). \tag{6.81}$$

Since η_x is Lipschitz, there exists some constant k_x and class-$\mathcal{K}$ function γ_k such that $|x| \leq k_x + \gamma_x(|e|)$ when $x \in S_x$. Assumption 6.1 thus holds for this choice of error system. △

Steps similar to the ones in the previous example may be taken to include the reference trajectory. The example demonstrates that bounding the error does naturally bound the state when an error system is developed using the integrator backstepping approach. To find more explicit bounds on the state, one must consider the particular form of the error system. Bounds for each of the states may then be achieved by first considering the bound on x_1. Once the bound on x_1 has been found, bounds may be established for x_2, and so on.

A multi-input strict-feedback system may be expressed as

$$\begin{aligned}
\dot{x}_{1,1} &= f_{1,1}(\bar{x}_{1,1}) + g_{1,1}(\bar{x}_{1,1})x_{1,2} \\
&\vdots \\
\dot{x}_{1,q_1-1} &= f_{1,q_1-1}(\bar{x}_{1,q_1-1}) + g_{1,q_1-1}(\bar{x}_{1,q_1-1})x_{1,q_1} \\
\dot{x}_{1,q_1} &= f_{1,q_1}(x) + \sum_{i=1}^{m} g^i_{1,q_1}(x)u_i \\
&\vdots \\
\dot{x}_{m,1} &= f_{m,1}(\bar{x}_{m,1}) + g_{m,1}(\bar{x}_{m,1})x_{m,2} \\
&\vdots \\
\dot{x}_{m,q_m-1} &= f_{m,q_m-1}(\bar{x}_{m,q_m-1}) + g_{m,q_m-1}(\bar{x}_{m,q_m-1})x_{m,q_m} \\
\dot{x}_{m,q_m} &= f_{m,q_m}(x) + \sum_{i=1}^{m} g^i_{m,q_m}(x)u_i
\end{aligned} \tag{6.82}$$

with $u = [u_1, \ldots, u_m]^\top \in \mathbb{R}^m$. It is assumed that

$$G(x) = \begin{bmatrix} g^1_{1,q_1}(x) & \cdots & g^m_{1,q_1}(x) \\ \vdots & & \vdots \\ g^1_{m,q_m}(x) & \cdots & g^m_{m,q_m}(x) \end{bmatrix}$$

is nonsingular for all $x \in \mathbb{R}^n$ with $n = \sum_i q_i$ the number of states in the system. A controller may be designed for (6.82) such that $x_{1,i} \to r_i(t)$ using integrator backstepping starting with each $x_{1,i}$ for $i = 1, \ldots, m$. Let $e_{i,j} = x_{i,j} - \nu_{i,j-1}$ for $i = 1, \ldots, m$ and $j = 1, \ldots, q_i$ with

$$\nu_{i,j} = \frac{-f_{i,j} + \dot{\nu}_{i,j-1} - e_{i,j-1} - \kappa e_{i,j}}{g_{i,j}}.$$

If $V = \sum_{i=1}^{m} \sum_{j=1}^{q_i} e^2_{i,j}$, then

$$\dot{V} = \frac{1}{2} \sum_{i=1}^{m} \Big[g_{i,q_i-1} e_{i,q_i-1} e_{i,q_i} \tag{6.83}$$

$$+ e_{i,q_i}\left(-\dot{\nu}_{i,q_i-1} + f_{i,q_i} + \left[g^1_{i,q_i}(x), \ldots, g^m_{i,q_i}(x)\right] u\right) - \sum_{j=1}^{q_i-1} \kappa e^2_{i,j}\Bigg] .$$

Letting $u = \nu(z)$ with

$$\nu(z) = G^{-1}(x)\begin{bmatrix} -f_{1,q_1} + \dot{\nu}_{1,q_1-1} - e_{1,q_1-1} - \kappa e_{1,q_1} \\ \vdots \\ -f_{m,q_m} + \dot{\nu}_{m,q_m-1} - e_{m,q_m-1} - \kappa e_{m,q_m} \end{bmatrix} \tag{6.84}$$

renders $\dot{V} = -2\kappa V$, so $e = 0$ is an exponentially stable equilibrium point.

So far each control design technique has assumed that the equations used to define the system dynamics are known. The remainder of this chapter will be devoted to techniques which may be used to enhance the closed-loop system robustness to allow the control of systems which possess various degrees of uncertainty.

6.4 Coping with Uncertainties: Nonlinear Damping

As shown in the previous section, it is possible to define meaningful error systems for each of the canonical representations so that the error dynamics become

$$\dot{e} = \alpha(t, x) + \beta(x)u, \tag{6.85}$$

and a stabilizing controller $u = \nu(z)$ exists. If the control law $u = \nu(z)$ is defined such that $\dot{V} \leq -k_1 V + k_2$, for a positive definite Lyapunov function, $V(e)$, then it is possible to include an additional stabilizing component to increase the robustness of the closed-loop system to compensate for additional uncertainty. Consider the control law $u = \nu(z) + \nu_d(z)$. Then

$$\dot{V} = \frac{\partial V}{\partial e}\left[\alpha(t, x) + \beta(x)\left(\nu + \nu_d\right)\right]. \tag{6.86}$$

Since $\dot{V} \leq -k_1 V + k_2$ when $u = \nu$, we find

$$\dot{V} \leq -k_1 V + k_2 + \frac{\partial V}{\partial e}\beta(x)\nu_d.$$

Thus as long as $\nu_d \in \mathsf{R}^m$ is chosen so that

$$\frac{\partial V}{\partial e}\beta(x)\nu_d \leq 0,$$

the value of V decreases more quickly. This type of control action is often referred to as nonlinear damping due to its ability to "remove energy" from the error system.

6.4.1 Bounded Uncertainties

The choice of ν_d is often governed by the form of possible destabilizing uncertainties in the system and the desired type of control action (e.g., do we require a smooth control law?). Consider designing a controller for a multi-input system with error dynamics

$$\dot{e} = \alpha(t,x) + \beta(x)(u + \Delta(t,x)), \tag{6.87}$$

where $|\Delta(t,x)| \leq \rho$ represents some bounded uncertainty with $\rho > 0$ a known constant.

If $u = \nu + \nu_d$ where ν is a control law defined for the case with $\Delta \equiv 0$, then the derivative of the Lyapunov candidate becomes

$$\dot{V} \leq -k_1 V + k_2 + \frac{\partial V}{\partial e}\beta(x)(\nu_d + \Delta(t,x)). \tag{6.88}$$

Recall that here we are assuming that there exists some Lyapunov function V and control law $u = \nu(z)$ such that $\dot{V} \leq -k_1 V + k_2$ when $\Delta \equiv 0$. Let $\mu = \left(\frac{\partial V}{\partial e}\beta(x)\right)^{\top}$. Consider the nonlinear damping term $\nu_d = -\rho \text{sgn}(\mu)$ where $y = \text{sgn}(x)$ is defined element-wise with

$$y_i = \begin{cases} 1 & x_i > 0 \\ -1 & x_i < 0 \end{cases}.$$

With this definition for the nonlinear damping term, we find

$$\begin{aligned} \dot{V} &\leq -k_1 V + k_2 - \rho \sum_{i=1}^{m} |\mu_i| + \frac{\partial V}{\partial e}\beta(x)\Delta(t,x) && (6.89) \\ &\leq -k_1 V + k_2. && (6.90) \end{aligned}$$

The addition of the nonlinear damping term ν_d preserves the stability properties of the closed-loop system. Since ν_d is discontinuous, however, one may no longer be guaranteed existence and uniqueness of solutions. This issue has been addressed in a number of papers, largely in the variable structure control literature [204, 227]. To avoid the discontinuity issues, it is possible to choose the "smoothed" version of the nonlinear damping term

$$\nu_d = -\rho \frac{\left(\frac{\partial V}{\partial e}\beta(x)\right)^{\top}}{\left|\frac{\partial V}{\partial e}\beta(x)\right| + c}, \tag{6.91}$$

where $c > 0$ is a small constant. Using (6.91), we find

$$\dot{V} = -k_1 V + k_2 + \frac{\partial V}{\partial e}\beta(x)\left(-\rho \frac{\left(\frac{\partial V}{\partial e}\beta(x)\right)^{\top}}{\left|\frac{\partial V}{\partial e}\beta(x)\right| + c} + \Delta(t,x)\frac{\left|\frac{\partial V}{\partial e}\beta(x)\right| + c}{\left|\frac{\partial V}{\partial e}\beta(x)\right| + c}\right).$$

Since $|\Delta(t,x)| \leq \rho$ for all t, notice that

$$\begin{aligned} \dot{V} &\leq -k_1 V + k_2 + c\rho \frac{\left|\frac{\partial V}{\partial e}\beta(x)\right|}{\left|\frac{\partial V}{\partial e}\beta(x)\right| + c} \\ &\leq -k_1 V + k_2 + c\rho. \end{aligned} \tag{6.92}$$

When $V > (k_2 + c\rho)/k_1$, we find $\dot{V} < 0$, thus V is bounded. Making c small helps reduce the size of the space to which V will converge. On the other hand, reducing c will increase the feedback gain when $\frac{\partial V}{\partial e}\beta(x)$ is near zero, possibly resulting in "chattering" when implemented in, for instance, a discrete-time system.

6.4.2 Unbounded Uncertainties

So far we have designed nonlinear damping terms for the case when the uncertainty may be overbounded by a constant. Now consider the multi-input system

$$\dot{e} = \alpha(t,x) + \beta(x)(u + \Delta(t,x)), \tag{6.93}$$

where $|\Delta(t,x)| \leq \rho\psi(x)$ with ρ an unknown constant and $\psi : \mathsf{R}^p \to \mathsf{R}$ is a known nonnegative function. It is assumed that ψ is bounded for any bounded $x \in \mathsf{R}^n$. In this case we may let $u = \nu + \nu_d$ to find

$$\dot{V} \leq -k_1 V + k_2 + \frac{\partial V}{\partial e}\beta(x)(\nu_d + \Delta(t,x)), \tag{6.94}$$

where Δ may grow without bound if $x \to \infty$. Now consider the stabilizing feedback term

$$\nu_d = -\eta \left(\frac{\partial V}{\partial e}\beta(x)\right)^\top \psi^2(x), \tag{6.95}$$

with $\eta > 0$. The time derivative of the Lyapunov candidate becomes

$$\dot{V} \leq -k_1 V + k_2 - \eta \left|\frac{\partial V}{\partial e}\beta(x)\right|^2 \psi^2(x) + \frac{\partial V}{\partial e}\beta(x)\Delta. \tag{6.96}$$

Since $|\Delta(t,x)| \leq \rho\psi(x)$, we find

$$\dot{V} \leq -k_1 V + k_2 - \eta \left|\frac{\partial V}{\partial e}\beta(x)\right|^2 \psi^2(x) + \rho \left|\frac{\partial V}{\partial e}\beta(x)\right| \psi(x). \tag{6.97}$$

Using $-x^\top x \pm 2x^\top y \leq y^\top y$, notice

$$\dot{V} \leq -k_1 V + k_2 + \frac{\rho^2}{4\eta}. \tag{6.98}$$

If $V > \left(k_2 + \frac{\rho^2}{4\eta}\right)/k_1$, then $\dot{V} < 0$. This guarantees that V is bounded and the closed-loop system is stable.

6.4.3 What if the Matching Condition Is Not Satisfied?

Up to this point, we have presented a number of techniques which ensure that *if an uncertainty appears at the same location as the control input*, then it is possible to define a stabilizing controller using nonlinear damping. Through the use of backstepping, it is also possible to stabilize systems even if such "matching conditions" are not satisfied. This is demonstrated in the following example:

Example 6.11 Consider the system defined by

$$\begin{aligned}\dot{x}_1 &= x_2 + \Delta(t, x_1)\\ \dot{x}_2 &= u,\end{aligned}$$

where $\Delta(t, x_1) \leq \rho\psi(x_1)$ with ρ an unknown bounded constant and ψ is a known smooth nonnegative function. Assume that we want to drive $x_1 \to r(t)$, so define the first component of the error system as $e_1 = x_1 - r(t)$. If the x_1 subsystem were defined by

$$\dot{x}_1 = \Delta(t, x) + v,$$

where v is a virtual input, then we could use nonlinear damping to stabilize the system. In particular, we may choose the Lyapunov candidate $V_1 = \frac{1}{2}e_1^2$ for the x_1 subsystem and the control law $v = \nu_1(z_1)$ with

$$\nu_1(z_1) = \dot{r} - \kappa e_1 - \eta\psi^2(x_1)e_1$$

so that

$$\begin{aligned}\dot{V}_1 &= e_1\left[-\kappa e_1 - \eta\psi^2 e_1 + \Delta(t, x)\right] &\quad (6.99)\\ &\leq -\kappa e_1^2 + \frac{\rho^2}{4\eta}. &\quad (6.100)\end{aligned}$$

Using the backstepping method of Theorem 6.2, we now choose the second component of the error system as $e_2 = x_2 - \nu_1$ and a new Lyapunov candidate $V = V_1 + \frac{1}{2}e_2^2$. Taking the derivative we find

$$\begin{aligned}\dot{V} &= e_1[\dot{r} + \Delta(t, x) + (x_2 - \nu_1 + \nu_1)] + e_2[u - \dot{\nu}_1]\\ &\leq -\kappa e_1^2 + \frac{\rho^2}{4\eta} + e_1 e_2 + e_2[u - \dot{\nu}_1]. &\quad (6.101)\end{aligned}$$

If $\dot{\nu}_1 = \frac{\partial \nu_1}{\partial t} + \frac{\partial \nu_1}{\partial x_1}(x_2 + \Delta(t, x_1))$ were known, then one could directly define the control law

$$\bar{\nu} = \frac{\partial \nu_1}{\partial t} + \frac{\partial \nu_1}{\partial x_1}(x_2 + \Delta(t, x_1)) - e_1 - \kappa e_2,$$

which stabilizes the system. Since Δ is unknown, we will instead use a control law which only cancels the known components of $\dot{\nu}_1$. In particular choose $u = \nu(z_2)$ with

$$\nu(z_2) = \frac{\partial \nu_1}{\partial t} + \frac{\partial \nu_1}{\partial x_1} x_2 - e_1 - \kappa e_2 - w(z_2) e_2, \tag{6.102}$$

where $w(z_2) \geq 0$ is an additional nonlinear damping term to be defined next.

Notice that with our choice for the control we obtain

$$\dot{V} \leq -\kappa e_1^2 - \kappa e_2^2 + \frac{\rho^2}{4\eta} + e_2 \left[-w(z_2) e_2 - \frac{\partial \nu_1}{\partial x_1} \Delta(t, x_1) \right],$$

so choosing $w = \eta \left(\frac{\partial \nu_1}{\partial x_1} \psi(x_1) \right)^2$ we obtain

$$\dot{V} \leq -2\kappa V + \frac{\rho^2}{2\eta}. \tag{6.103}$$

This ensures that e is uniformly ultimately bounded. $\triangle$

6.5 Coping with Partial Information: Dynamic Normalization

So far we have developed controllers for plants where all the states are assumed to be measurable. Here we will study a special class of systems where some of the states are not measurable, but where a dynamic normalizing signal may be created to compensate for their effects. Output feedback techniques will be presented later in the book in which provide an alternative solution.

Consider the system defined by

$$\begin{aligned} \dot{q} &= \phi(q, x) \\ \dot{x} &= f(x) + g(x)(\Delta(t, q, x) + u), \end{aligned}$$

where $q \in \mathsf{R}^d$, $x \in \mathsf{R}^n$ and $u \in \mathsf{R}^m$. The uncertainty $\Delta \in \mathsf{R}^m$ is assumed to satisfy

$$|\Delta(t, q, x)| \leq \rho \psi(|q|, x) \tag{6.104}$$

with $\rho \in \mathsf{R}$ an unknown constant and $\psi : \mathsf{R}^+ \times \mathsf{R}^n \to \mathsf{R}^+$ a known nonnegative function which is nondecreasing with respect to $|q|$. If q were measurable, then it would be possible to use nonlinear damping to stabilize the x dynamics. Here, we will consider the case where q is not assumed to be measurable. To stabilize the system, we will use the properties of the q

dynamics to define a dynamic signal which may be used to dominate the effects of q.

If the subsystem $\dot{q} = \phi(q, x)$, with x as an input, is input-to-state practically stable, then, there exists some positive definite $V_q(q)$ such that

$$\gamma_{q1}(|q|) \leq V_q(q) \leq \gamma_{q2}(|q|) \tag{6.105}$$

$$\frac{\partial V_q}{\partial q}\phi(q, x) \leq -cV_q + \gamma(|x|) + d, \tag{6.106}$$

where $c > 0$, $d \geq 0$, and $\gamma_{q1}, \gamma_{q2}, \gamma$ are class-$\mathcal{K}_\infty$. Using (6.106) it is possible to define a scalar dominating signal

$$\dot{\eta} = -c\eta + \gamma(|x|) + d \tag{6.107}$$

such that $\eta \geq V_q$ assuming that $\eta(0) \geq V_q(0)$.

Assume that the error system $e = \chi(t, x)$ has dynamics described by

$$\dot{e} = \alpha(t, x) + \beta(x)\left(\Delta(t, q, x) + u\right). \tag{6.108}$$

It is then possible to use concepts from nonlinear damping and the dynamic normalizing signal $\eta(t)$ to define a stable controller assuming there is a known control law ν a controller which stabilizes the error dynamics when $\Delta \equiv 0$.

Theorem 6.3: *Let $e = \chi(t, x)$ be an error system satisfying Assumption 6.1 with dynamics described by (6.108). Let $u = \nu(z_1)$ be a stabilizing controller with a radially unbounded Lyapunov function $V(e)$ such that*

$$\dot{V} \leq -k_1 V + k_2 \tag{6.109}$$

along the solutions of (6.108) when $\Delta \equiv 0$. If $\Delta(t, q, x)$ satisfies (6.104) and the q-subsystem is input-to-state practically stable, then there exists a stabilizing controller $u = \nu_c(z)$ with $z = [z_1^\top, \eta]^\top$ where η (defined by (6.107)) is a dynamic dominating signal.

Proof: Letting $\nu_c(z) = \nu + \nu_d$, the derivative of V becomes

$$\dot{V} \leq -k_1 V + k_2 + \frac{\partial V}{\partial e}\beta(x)\left(\Delta(t, q, x) + \nu_d\right). \tag{6.110}$$

From (6.105) we see that $|q| \leq \gamma_{q1}^{-1}(V_q) \leq \gamma_{q1}^{-1}(\eta)$. If

$$\nu_d = -\gamma\left(\frac{\partial V}{\partial e}\beta(x)\right)^\top \psi^2(\gamma_{q1}^{-1}(\eta), x) \tag{6.111}$$

with $\gamma > 0$, we obtain

$$\begin{aligned} \dot{V} &\leq -k_1 V + k_2 + \frac{\partial V}{\partial e}\beta(x)\nu_d + \rho\left|\frac{\partial V}{\partial e}\beta(x)\right|\psi(\gamma_{q1}^{-1}(\eta), x) && (6.112) \\ &\leq -k_1 V + k_2 + \frac{\rho^2}{4\gamma}, && (6.113) \end{aligned}$$

which is independent of η. Thus (6.105)-(6.106), together with (6.113), imply that e and q are uniformly ultimately bounded. ■

6.6 Using Approximators in Controllers

So far we have not incorporated a fuzzy system or neural network into the design of the control law. Rather we have only considered more classical approaches to the controller design. When incorporating approximators, one often needs to consider

1. The approximator error, and
2. The space over which the approximation is valid.

As we will see, the nonlinear damping technique is usually capable of compensating for approximation errors. Thus far, however, we have not had to consider cases where the error (or possibly the plant state) must remain within some predefined space for all t. We will find that by proper choice of the system initial conditions and controller parameters we are able to confine the state trajectory so that the inputs to an approximator remain within some valid subspace.

6.6.1 Using Known Approximations of System Dynamics

Often the plant dynamics are approximated using experimental data or first principles. If $f(x)$ is a function used to define some component of the system dynamics, then assume that $\mathcal{F}(x, \theta)$ is an approximation of $f(x)$ available a priori for control design. The parameter vector $\theta \in \mathrm{R}^p$ is chosen such that the approximation error $w(x) = \mathcal{F}(x, \theta) - f(x)$ is bounded with $|w(x)| \leq W$ for all $x \in \mathrm{R}^n$. If such an approximation is made, it is possible to design a controller assuming that $\mathcal{F}(x, \theta) = f(x)$, and then include a nonlinear damping term to compensate for the effects of the approximation error as shown in the next example. Later, we will relax the global boundedness of the approximation error by only requiring it to be bounded on a suitable compact set.

Example 6.12 Consider the single-input feedback linearizable system

$$
\begin{aligned}
\dot{x}_1 &= x_2 \\
&\vdots \\
\dot{x}_{n-1} &= x_n \\
\dot{x}_n &= f(x) + u,
\end{aligned}
\tag{6.114}
$$

where $y = x_1$ is the output which we wish to drive to $y \to r(t)$. Assume that the function $f(x)$ is not known, but an approximation of $f(x)$ was obtained using experimental data. We will also assume that the approximator $\mathcal{F}(x,\theta)$ is defined such that the approximation error $w(x) = \mathcal{F}(x,\theta) - f(x)$ is bounded by $|w(x)| \leq W$ for all x.

To define a controller using the static approximator, $\mathcal{F}(x,\theta)$, we will first define an error system using a stable manifold. Let $e = \chi(t,x)$, where

$$\chi(t,x) = k_1(x_1 - r) + \cdots + k_{n-1}(x_{n-1} - r^{(n-2)}) + x_n - r^{(n-1)} \quad (6.115)$$

and

$$L = \begin{bmatrix} 0 & & \\ \vdots & & I \\ 0 & & \\ -k_1 & \cdots & -k_{n-1} \end{bmatrix}$$

is chosen to be Hurwitz. The error dynamics then become

$$\dot{e} = k_1(x_2 - \dot{r}) + \cdots + k_{n-1}(x_n - r^{(n-1)}) + f(x) - r^{(n)} + u.$$

We will now consider the Lyapunov candidate $V = \frac{1}{2}e^2$ so that

$$\dot{V} = \frac{\partial V}{\partial e}\left(k_1(x_2 - \dot{r}) + \cdots + k_{n-1}(x_n - r^{(n-1)}) + f(x) - r^{(n)} + u\right).$$

Assume for a moment that $f(x)$ is known. Then it is easy to define the control control law $u = \nu_f(z)$ with

$$\begin{aligned} \nu_f(z) &= -k_1(x_2 - \dot{r}) - \cdots - k_{n-1}(x_n - r^{(n-1)}) \\ &\quad -f(x) + r^{(n)} - \kappa\frac{\partial V}{\partial e} \end{aligned}$$

so that $\dot{V} = -2\kappa V$ and $e = 0$ is an exponentially stable equilibrium point.

Since $f(x)$ is not known, we will instead consider the control law $u = \nu(z) = \nu_{\mathcal{F}} - \eta\frac{\partial V}{\partial e}$, where

$$\begin{aligned} \nu_{\mathcal{F}}(z) &= -k_1(x_2 - \dot{r}) - \cdots - k_{n-1}(x_n - r^{(n-1)}) \\ &\quad - \mathcal{F}(x,\theta) + r^{(n)} - \kappa\frac{\partial V}{\partial e}, \end{aligned}$$

so that the functional form of $\nu_{\mathcal{F}}$ is based on ν_f. The nonlinear damping term $-\eta\frac{\partial V}{\partial e} = -\eta e$ is added to compensate for the mismatch between $f(x)$ and $\mathcal{F}(x,\theta)$.

When $\nu(z)$ is used as the control law, we find

$$\begin{aligned} \dot{V} &= \frac{\partial V}{\partial e}\left[\alpha(t,x) + (\nu_f - \nu_f + \nu_{\mathcal{F}} - \eta e)\right] \\ &= -2\kappa V + e\left(\nu_{\mathcal{F}} - \nu_f - \eta e\right). \end{aligned} \quad (6.116)$$

Since $|\nu_{\mathcal{F}} - \nu_f| = |\mathcal{F} - f| = |w(x)| \leq W$ we obtain

$$\begin{aligned} \dot{V} &\leq -2\kappa V - \eta e^2 + W|e| \\ &\leq -2\kappa V + \frac{W^2}{4\eta}. \end{aligned} \tag{6.117}$$

Thus V and e are bounded. Since $\dot{V} < 0$ when $V > \frac{W^2}{8\kappa\eta}$, we find V asymptotically converges to the positively invariant set $V \leq \frac{W^2}{8\kappa\eta}$. By making κ and/or η large, we see that the invariant set may be made arbitrarily small. △

In the previous example, we found that V converges to $\frac{W^2}{8\kappa\eta}$. Recall that W describes the approximator error. Thus a better approximation will cause W to be smaller so that the ultimate bound will decrease.

6.6.2 When the Approximator Is Only Valid on a Region

Up to this point, we have assumed that the dynamics of the plant model are valid for all $x \in \mathbb{R}^n$. We will now use the properties of the Lyapunov function to find bounds on the trajectory of x when the closed-loop system is stable. One may need to complete such an analysis if

1. An approximator is used that is only valid on a subspace, or
2. The system signals must remain within critical limits for performance or safety reasons.

If a control law may be defined such that $x \in S_x$ for all t, then it is not required that the plant model be accurate outside S_x. This is particularly important when an approximator is used in the design of a controller since in general, an approximator is only valid on some region. To place bounds on $|e|$ and thus on $|x|$ we will study the properties of the Lyapunov function.

Consider for a moment the case where one wishes to design a controller for the scalar system

$$\dot{x} = f(z) + u, \tag{6.118}$$

where $f(z)$ is some nonlinearity which may be approximated by $\mathcal{F}(z, \theta)$ when $z \in S_z$. Here z is a vector of measurable signals, θ is a vector of approximator parameters (such as weights in a neural network), and S_z is the space over which the approximator is defined to reasonably represent $f(z)$. We will assume that if it is possible to confine $e \in B_r$ where B_r is the ball defined by $B_r = \{e \in \mathbb{R}^q : |e| \leq r\}$, then $z \in S_z$. The goal of the control system will be to at least ensure that $e \in B_e$, where $B_e \subseteq B_r$. For the system defined by (6.118) one may then choose the control law $u = \nu$, where

$$\nu = \mathcal{F}(z, \theta) - \kappa e,$$

and the error is defined by $e = x$ so $e \to 0$ implies $x \to 0$.

To use an approximator which is only valid on $z \in S_z$, one must first determine some B_r so that $e \in B_r$ implies $z \in S_z$. If S_z is based on the range over which e may travel (i.e., the approximator has e and an input), then it is usually easy to find some B_r such that $e \in B_r$ implies $z \in S_z$ as long as S_z contains the origin. If S_z specifies the range over which the state x may travel (i.e., the approximator has x as an input), then one may use Assumption 6.1 to determine the range over which e may travel and still ensure that the state is confined such that $z \in S_z$. These cases will be further investigated in the examples throughout the remainder of this book.

The following theorem may be used to place bounds on $|e|$ using the properties of the Lyapunov function.

Theorem 6.4: *Let $V : \mathsf{R}^q \to \mathsf{R}$ be a continuously differentiable function such that*

$$\gamma_1(|e|) \leq V(e) \leq \gamma_2(|e|), \tag{6.119}$$

where γ_1 and γ_2 are class-$\mathcal{K}_\infty$. Assume that for a given error system, a control law $u = \nu(z)$ is defined such that $\dot{V} \leq 0$ along the solutions of $\dot{e}(t, x, u)$ when $|e| \geq b$ where $b \geq 0$. Then

$$|e| \leq \gamma_1^{-1} \circ \gamma_2 \left(\max(|e(0)|, b)\right) \tag{6.120}$$

for all $t \geq 0$.

Proof: From (6.119) we find

$$V \geq \gamma_2(b) \Rightarrow |e| \geq b \Rightarrow \dot{V} \leq 0. \tag{6.121}$$

If $V(0) \leq \gamma_2(b)$, then $0 \leq V \leq \gamma_2(b)$ for all t since V is positive definite and can not grow larger than $\gamma_2(b)$ according to (6.121). If $V(0) > \gamma_2(b)$, then $\dot{V} \leq 0$ until $V \leq \gamma_2(b)$, thus $0 \leq V \leq \max(V(0), \gamma_2(b))$ for all t. From (6.119) we know that

$$\gamma_1(|e|) \leq V \leq \max(V(0), \gamma_2(b)) \leq \gamma_2\left(\max(|e(0)|, b)\right),$$

so $|e| \leq \gamma_1^{-1} \circ \gamma_2\left(\max(|e(0)|, b)\right)$ for all t. ■

In the above proof, we find that $e \in B_e$ for all t where

$$B_e = \left\{ e \in \mathsf{R}^q : |e| \leq \gamma_1^{-1} \circ \gamma_2\left(\max(|e(0)|, b)\right)\right\} \tag{6.122}$$

is the ball containing the error trajectory. Unlike an ultimate bound, B_e also includes the effects of system initial conditions. Using Assumption 6.1 it is then possible to then place bounds on $|x|$. Bounds on the reference signal are typically known a priori. It may therefore be possible to find the range of all the input signals used to define z for the control law $u = \nu(z)$.

Since e never leaves the ball B_e it is not required that $\dot{V} \leq 0$ when $e \in \mathsf{R}^q - B_e$. In other words, we do not care that the Lyapunov function does not necessarily decrease outside the space through which the error trajectory is able to travel. This is summarized in the following corollary.

Corollary 6.1: *Let $V : \mathsf{R}^q \to \mathsf{R}$ be a continuously differentiable function such that*

$$\gamma_1(|e|) \leq V(e) \leq \gamma_2(|e|), \tag{6.123}$$

where γ_1 and γ_2 are class-$\mathcal{K}_\infty$. Assume that for a given error system, a control law $u = \nu(z)$ is defined such that $\dot{V} \leq 0$ when $e \in B_e - B_b$, where B_e is defined by (6.122) and $B_b = \{e \in \mathsf{R}^q : |e| \leq b\}$ with $b \geq 0$. Then

$$|e| \leq \gamma_1^{-1} \circ \gamma_2 (\max(|e(0)|, b)) \tag{6.124}$$

for all $t \geq 0$.

We will use this corollary to show that state trajectories are bounded when using approximators to cancel system nonlinearities even though the approximator may not be capable of representing the nonlinearities for all $x \in \mathsf{R}^n$. As long as an approximation is valid when $e \in B_e$ the stability results hold.

A controller which uses an approximator defined only on a region will typically be created using the following steps:

1. Place the system dynamics into some canonical representation.

2. Choose an error system satisfying Assumption 6.1 and Lyapunov candidate.

3. Choose a control law $u = \nu(z, \mathcal{F}(z))$ using an approximator, $\mathcal{F}(z)$, such that $\dot{V} \leq -k_1 V + k_2$, when $z \in S_z$ with $k_1 > 0$ and $k_2 \geq 0$. For now ignore the case when $z \notin S_z$.

4. Given the approximator, determine some B_r such that $e \in B_r$ implies that $z \in S_z$.

5. Choose the parameters of the control law such that $e \in B_e$ where $B_e \subseteq B_r$.

Thus developing a control law which incorporates an approximator is very similar to the case where an approximation holds globally. In the case where the approximator is only accurate over a region, we must further ensure that the control parameters are properly chosen. In particular, we will require that $e \in B_e$ which will then ensure that the inputs to the approximator remain in an appropriate region. The following example helps demonstrate this approach.

Example 6.13 Here we will again consider the feedback linearizable system in Example 6.12 where it is desired that $x_1 \to r$. This time, however, it is assumed that $|w(x)| \leq W$ only when $|x| \leq d$ where $d > 0$ is a known constant. Given that the controller must ensure $|x| \leq d$, we will now find restrictions on $|x(0)|$, r, and the controller parameters.

In Example 6.12 we used $V = \frac{1}{2}e^2$, so $\gamma_1(|e|) \leq V(e) \leq \gamma_2(|e|)$, where $\gamma_1 = \gamma_2 = \frac{1}{2}e^2$. Additionally, with control $u = \nu(z)$ defined by

$$\nu(z) = \nu_{\mathcal{F}} - \eta \frac{\partial V}{\partial e},$$

we found $\dot{V} \leq -\kappa e^2 + \frac{W^2}{4\eta}$. Thus $\dot{V} < 0$ if $|e| > b$ where $b = \sqrt{W^2/(4\kappa\eta)}$. Using Corollary 6.1, we find $e \in B_e$ for all t where

$$B_e = \{e \in \mathsf{R} : |e| \leq \max(|e(0)|, b)\}.$$

We must now place restriction on the system initial conditions and controller parameters to ensure that $e \in B_e$ implies that $|x| \leq d$. By the definition of the error system, we find

$$|x| \leq (1 + |k_1| + \cdots + |k_{n-1}|)\psi_\mu + |e| + \bar{r},$$

where

$$\psi_\mu = c_1 \mathrm{e}^{-c_2 t}|e(0)| + \int_0^t c_1 \mathrm{e}^{-c_2(t-\tau)}|e(\tau)|d\tau, \tag{6.125}$$

and c_1, c_2 are chosen so that $|e^{Lt}| \leq c_1 \mathrm{e}^{-c_2 t}$. Thus c_1 and c_2 are defined by the choice of the error system. Assuming that c_1 and c_2 are fixed, we will find requirements on the controller gains κ and η.

For simplicity, we will assume that $r(0)$ is chosen such that $e(0) = 0$. Then

$$\psi_\mu \leq \int_0^t c_1 \mathrm{e}^{-c_2(t-\tau)} b d\tau \leq \frac{c_1 b}{c_2},$$

so that

$$|x| \leq \left(1 + \frac{c_1(1 + |k_1| + \cdots + |k_{n-1}|)}{c_2}\right) b + \bar{r}. \tag{6.126}$$

But we require that $|x| \leq d$ for the approximation to hold. We must therefore choose the controller parameters used to define b such that

$$\left(1 + \frac{c_1(1 + |k_1| + \cdots + |k_{n-1}|)}{c_2}\right) b \leq d - \bar{r}.$$

That is,

$$b \leq \frac{c_2(d - \bar{r})}{c_2 + c_1(1 + |k_1| + \cdots + |k_{n-1}|)}. \tag{6.127}$$

Since $b = \sqrt{W^2/(4\kappa\eta)}$, choosing

$$\kappa\eta \geq \frac{W^2}{4}\left[\frac{c_2 + c_1(1 + |k_1| + \cdots + |k_{n-1}|)}{c_2(d - \bar{r})}\right]^2 \tag{6.128}$$

will guarantee that $|x| \leq d$ so that the closed-loop system is indeed stable. △

Since we may ensure that $x \in S_x$ for all t, the assumption that $x = T(\xi)$ is a global diffeomorphism may be relaxed. However, unlike the local and global case, there is no constructive procedure available for finding a transformation T defined on a compact set. In fact, existence conditions for such a transformation are not known either.

6.7 Summary

In this chapter, we studied how to define controllers for a variety of different systems. In general, it was shown that a stabilizing controller may be defined for systems in state-feedback, input-output feedback, and strict-feedback canonical forms. The controller was constructed by first defining an error system and Lyapunov candidate. The control law was then constructed to ensure that the Lyapunov candidate decreases over time.

When uncertainty is included in the system dynamics, one may add nonlinear damping terms to help increase the rate at which the Lyapunov function decays. The nonlinear damping term is defined in such a way that the beneficial effect of the nonlinear damping term will dominate the destabilizing effect of the uncertainty (at least as the error grows large). With a proper choice of the control law, we found that we could always force

$$\dot{V} \leq -k_1 V + k_2,$$

along the solution of the error dynamics, where $k_1 > 0$ and $k_2 \geq 0$. We will later use this fact in the design of the adaptive control laws in which additional system uncertainty may be present.

We also found that when approximators defined on a region are used in the definition of the plant dynamics or a control law, one must pay special attention to the specification of both the intial conditions and control parameters. This way it is possible to guarantee that the inputs to the approximator remain within the region for which an accurate approximation is achieved.

6.8 Exercises and Design Problems

Exercise 6.1 (Domain of Attraction) Consider the nonlinear system

$$\dot{x} = x^3 + u.$$

Since $x = 0$ is an equilibrium point of the system with $u = 0$, the linearized approximation is given by $\dot{x} = u$. Based on this consider the control law $u = -x$ and prove that $x = 0$ is a stable equilibrium point. Find the domain of attraction for the nonlinear representation of the plant.

Exercise 6.2 (Backstepping) Design a control law for the system (6.67) using the integrator backstepping approach so that

$$\dot{V} \leq -f(x_1)V, \tag{6.129}$$

where $f(x_1)$ is any smooth positive function.

Exercise 6.3 (Lyapunov Matrix Equation) Given the Lyapunov matrix equation $A^\top P + PA = -Q$ with $Q > 0$ and A Hurwitz, the quantity

$$\alpha = \frac{\lambda_{\min}(Q)}{\lambda_{\max}(P)} \tag{6.130}$$

may be used to describe the rate of decay of the system error (see (6.23)). Show that (6.130) is maximized when Q is chosen as $Q = I$.

Exercise 6.4 (Sontag's Universal Formula) Given the error dynamics

$$\dot{e} = \alpha(t,x) + \beta(x)u$$

and positive definite radially unbounded function $V(e)$, show that the continuous control law $u = \nu$, where

$$\nu = \begin{cases} -\dfrac{\frac{\partial V}{\partial e}\alpha + \sqrt{\left(\frac{\partial V}{\partial e}\alpha\right)^2 + \left(\frac{\partial V}{\partial e}\beta\left(\frac{\partial V}{\partial e}\beta\right)^\top\right)^2}}{\frac{\partial V}{\partial e}\beta\left(\frac{\partial V}{\partial e}\beta\right)^\top} & \text{if } \frac{\partial V}{\partial e}\beta \neq 0 \\ 0 & \text{otherwise} \end{cases} \tag{6.131}$$

globally asmyptotically stabilizes the origin $e = 0$.

Exercise 6.5 (Input Uncertainty) Consider the error dynamics defined by

$$\dot{e} = \alpha(t,x) + \beta(x)[\Delta(t,x) + \Pi(t,x)u], \tag{6.132}$$

where both additive Δ and multiplicative Π uncertainty is present. Assume that $\Delta \leq \rho\psi(x)$ where $\rho > 0$ is a bounded unknown constant and $\psi(x)$ is a known non-negative function. Also assume that $0 <$

$\pi_1 \leq \Pi(t,x) \leq \pi_2$ for all t and x. Assume that there exists some control law $u = \nu(z)$ and Lyapunov function such that $\dot{V} \leq -k_1 V + k_2$ for some $k_1 > 0$ and $k_2 \geq 0$ when $\Delta \equiv 0$ and $\Pi \equiv 1$. Define a stabilizing controller such that $\dot{V} \leq -k_3 V + k_4$ for some $k_3 > 0$ and $k_4 \geq 0$. Hint: Add $\nu(z) - \nu(z)$ to the additive uncertainty.

Exercise 6.6 (Input Gain Dynamics) Assume that there exists a stabilizing controller $\nu_s(z)$ and Lyapunov function V_s associated with the error dynamics

$$\dot{e} = \alpha(t,x) + \begin{bmatrix} 0 \\ \vdots \\ 0 \\ 1 \end{bmatrix} u$$

such that $\dot{V}_s \leq -k_1 V_s + k_2$ when $u = \nu_s$. Use the Lyapunov candidate

$$V_a = \int^{V_s} \frac{1}{\beta(V_s)}$$

to show that $\nu_s(z)$ also stabilizes the error system

$$\dot{e} = \alpha(t,x) + \begin{bmatrix} 0 \\ \vdots \\ 0 \\ \beta(V_s) \end{bmatrix} u,$$

where $\beta(V_s) > 0$.

Exercise 6.7 (Dead Zones) Define a nonlinear damping term similar to (6.91) using the continuous dead zone $y = D_c(x/\epsilon)$ where $\epsilon > 0$ and

$$D_c(x) = \begin{cases} 1 & \text{if } x \geq 1 \\ x & \text{if } |x| < 1 \\ -1 & \text{if } x \leq -1 \end{cases} . \tag{6.133}$$

Exercise 6.8 (Bounds in Triangular Lyapunov Systems) Show that if

$$\begin{aligned} \dot{V}_x &\leq -k_1 V_x + k_2 \\ \dot{V}_y &\leq -k_3 V_y + \psi(|x|) \end{aligned}$$

for some positive definite functions V_x, V_y (with $\gamma(|x|) \leq V_x$) and non-negative function $\psi : \mathsf{R}^+ \to \mathsf{R}^+$, then

$$V_x(t) \leq \frac{k_2}{k_1} + \left(V_x(0) - \frac{k_2}{k_1} \right) e^{\; k_1 t}$$

and

$$V_y(t) \leq \frac{k_4}{k_3} + \left(V_y(0) - \frac{k_4}{k_3} \right) e^{-k_3 t},$$

where $k_4 = \psi \circ \gamma^{-1} \circ \max(V(0), k_2/k_1)$.

Exercise 6.9 (Dynamic Normalization) Consider the system defined by

$$\begin{aligned} \dot{q} &= \phi(q, x) \\ \dot{x}_1 &= x_2 + \Delta(q, x_1) \\ \dot{x}_1 &= u, \end{aligned}$$

where $|\Delta| \leq |q|^2 |x_1|^2$ and the q dynamics satisfy (6.105)-(6.106). Use the backstepping method to define a controller such that the ultimate bound on the error $e_1 = x_1 - r(t)$ may be made arbitrarily small.

Exercise 6.10 (Adding an Integrator) Consider the system defined by

$$\dot{x}_1 = \Delta(x_1) + u,$$

where $\Delta(x)$ is an uncertainty. Define controllers such that $x \to r(t)$ when

- $\Delta(x) = c + x^2$
- $\Delta(x) = c \sin(x) \cos(x)$
- $\Delta(x) = 1 + cx \cos(x)$,

where $c > 0$ is an unknown constant. Now define controllers for the system

$$\begin{aligned} \dot{x}_0 &= x_1 \\ \dot{x}_1 &= \Delta(x_1) + u, \end{aligned}$$

where we wish to drive $x_0 \to \int r(t)$ (i.e., we still drive $x_1 \to r(t)$). Compare $x_1 - r(t)$ for the above two cases when $r(t)$ is a square wave, sinusoid, and when $r(t)$ is a ramp.

Exercise 6.11 (Point Mass) Consider the point mass whose dynamics are described by

$$m\ddot{x} = f(x) + u,$$

where x is the position, m is the mass, and u is a force input. Here $f(x)$ is a position dependent uncertainty. Assume that $f(x)$ may be approximated by $\mathcal{F}(x, \theta)$ on $x \in D$, where θ is a set of appropriate parameters. Define a control law such that the error $e_1 = x - r$ is driven toward zero. Then consider the cases where

- $f(x) = \sin(x)$
- $f(x) = x + x^3$.

For each $f(x)$ define a fuzzy system that approximates it and plot the response of the closed-loop system.

Exercise 6.12 (Ball and Beam) Consider the ball and beam system defined by

$$\begin{aligned}
\dot{x} &= v \\
\dot{v} &= -g\sin\theta + x\omega^2 \\
\dot{\theta} &= \omega \\
\dot{\omega} &= -\frac{2mxv\omega}{J+mx^2} - \frac{mgx\cos\theta}{J+mx^2} + u,
\end{aligned}$$

where x is the ball position relative to the center of the beam, v is the ball velocity, θ is the angular position of the beam, and ω is the beam's angular rate. Show that $[x - r, v, \theta, \omega] = 0$ is an equlibrium point and linearize the system about this point. Use pole placement to design a stable linear control system.

Exercise 6.13 (M-Link Robot) Consider the dynamics of an m-link robot defined by

$$M(q)\ddot{q} + C(q, \dot{q}) = u,$$

where $q \in \mathsf{R}^m$ is a vector of generalized coordinates describing the position of the robot linkages. The generalized mass matrix M is invertible, and $C(q, \dot{q})$ accounts for centrifugal, Coriolis, and gravitational forces. Define a controller $u = \nu(t, q, \dot{q})$ such that $q \to r(t)$.

Exercise 6.14 (Inverted Pendulum) The dynamics of an inverted pendulum are given by

$$\begin{aligned}
(M+m)\ddot{x} + ml\cos\theta\ddot{\theta} - ml\sin\theta\dot{\theta}^2 &= u_1 \\
ml\cos\theta\ddot{x} + ml^2\ddot{\theta} - mgl\sin\theta &= u_2,
\end{aligned}$$

where x is the cart position and θ is the angle of the pendulum (with $\theta = 0$ when the pendulum is perfectly inverted). The parameters of the system are defined as M is the mass of the cart, m is the point mass attached to the end of the pendulum, l is the length of the pendulum, and g is the constant of gravitational acceleration. Define a control law for u_1 and u_2 that ensure that $x \to x_r$ and $\theta \to 0$ assuming that the states are measurable.

Exercise 6.15 (Speed Control) Consider the longitudinal dynamics of a vehicle define by

$$\begin{aligned}\dot{v} &= \frac{1}{m}(F_m - F_b) - \frac{A_\rho v^2}{m} \\ \dot{F}_m &= -\tau_m F_m + u_m \\ \dot{F}_b &= -\tau_b F_b + u_b,\end{aligned}$$

where v is the vehicle speed, F_m is the force applied by the motor and F_b is the force applied by the brakes. Here m is the vehicle mass, A_ρ is the coefficient of aerodynamic drag, τ_m is the motor time constant, and τ_b is the brake time constant. Design a controller such that the error $e_1 = v - v_r$ with $v_r > 0$ is minimized when the inputs u_m and u_b are confined to be positive values. Try to design the controller so that both the brake and motor are not actuated at the same time.

Exercise 6.16 (Induction Motor) Consider the model of an induction motor [149] given by

$$\begin{aligned}\dot{\omega} &= \frac{n_p M}{J L_r}(\psi_a i_b - \psi_b i_a) - \frac{T_L}{J} \\ \dot{\psi}_a &= -\frac{R_r}{L_r}\psi_a - n_p \omega \psi_b + \frac{R_r}{L_r} M i_a \\ \dot{\psi}_b &= -\frac{R_r}{L_r}\psi_b + n_p \omega \psi_a + \frac{R_r}{L_r} M i_b,\end{aligned}$$

where ω is the rotor angular rate and ψ_a, ψ_b are the rotor fluxes. Here M is the mutual inductance, J is the inertia, n_p is the number of pole pairs, L_r is the rotor inductance, and R_r is the rotor resistance. A controller is to be designed for the inputs i_a and i_b so that $\omega = \omega_r$ and $\psi_a^2 + \psi_b^b = \psi_r$ given the load torque T_L.

Exercise 6.17 (Telescope Pointing) A model of the drive system on a telescope is given by

$$\begin{aligned}J\ddot{\theta} &= T \\ \tau\dot{T} &= -T + u,\end{aligned}$$

where θ is the angular position of the telescope, and T is the drive torque applied to the telescope base. Here J is the moment of inertia, and τ is the motor time constant. If it is only known that the moment of inertia satisfies $0 < J_1 \leq J \leq J_2$, then use the backstepping technique to find a stabilizing control law assuming that the other variables are measurable.

Exercise 6.18 (Magnetic Levitation) Consider the magnetically levitated system defined by

$$\begin{aligned} \dot{x} &= v \\ m\dot{v} &= -g + \frac{ku}{(G-x)^2}, \end{aligned}$$

where x is the position, and v is the velocity. Here g is the gravitational acceleration, k is the electromagnet constant, m is the mass, and G is the nominal gap. Define a control law $u = \nu$ such that $x \to 0$. Find the initial conditions that ensure x remains away from the singularity at $x = G$.

Exercise 6.19 (Field-Controlled DC Motor) The model for a field controlled DC motor [202] is given by

$$\begin{aligned} L_a\dot{i}_a &= -R_a i_a - c_1 i_f \omega + v_a \\ L_f\dot{i}_f &= -R_f i_f + v_f \\ J\dot{\omega} &= -B\omega + c_2 i_f i_a, \end{aligned}$$

where i_a is armature current, i_f is the field current, and ω is the angular rate. Here L_a and R_a are the armature circuit inductance and resistance, respectively, and L_f and R_f are the associated field circuit inductance and resistance. J is the moment of inertia, B is the back EMF constant, and c_1, c_2, are motor constants. Assume that v_a is held constant, and design a controller for v_f such that $\omega \to r(t)$.

Exercise 6.20 (Flexible Joints) The dynamics for a single-link manipulator with flexible joints [212] is described by

$$\begin{aligned} I\ddot{q}_1 + MgL\sin q_1 + k(q_1 - q_2) &= 0 \\ J\ddot{q}_2 - k(q_1 - q_2) &= u, \end{aligned}$$

where q_1 and q_2 are angular positions. The parameters for the system are described by I and J are moments of inertia, M is the mass, L is the distance, and u is the torque input. Define a stable feedback controller such that $q_1 \to r(t)$.

Exercise 6.21 (Antenna Pointing) Consider the dynamics for the antenna pointing system

$$J\ddot{\theta} + B\dot{\theta} = \Delta + u,$$

where θ is the antenna angular position, J is the moment of inertia, B is the coefficient for viscous friction, and Δ is an uncertainty. Use nonlinear damping to help design a control law $u = \nu(t, x)$ such that θ tracks $r(t)$ when it is known that

- $\Delta = k_1\theta + k_2\dot{\theta}$
- $\Delta = (k_1\theta + k_2\dot{\theta})^2$,

where k_1, k_2 are unknown constants.

Exercise 6.22 (Simplified Nonlinear Ball and Beam) To simplify the design of a controller for the ball and beam experiment described in Exercise 6.12 when nonlinearities are considered, we will study the control of

$$\begin{aligned}
\dot{x} &= v \\
\dot{v} &= x\Delta(t) - g\theta \\
\dot{\theta} &= \omega \\
\dot{\omega} &= u,
\end{aligned}$$

where it is assumed that $|\Delta| \le \Omega^2$, and $\Omega > 0$ is some known constant. Use the backstepping approach to design a Lyapunov function, V, and controller, $u = \nu$, such that $\dot{V} \le -k_1 V + k_2$, where $k_1, k_2 > 0$ may be set by the design of the controller.

Chapter 7

Direct Adaptive Control

7.1 Overview

In Chapter 6 we found that it is possible to define static (non-adaptive) stabilizing controllers, $u = \nu_s(z)$ with $u \in \mathbb{R}^m$, for a wide variety of nonlinear plants. In addition to being able to define control laws for systems in input-output feedback linearizable and strict-feedback forms, it was shown how nonlinear damping and dynamic normalization may be used to compensate for system uncertainty. In this and subsequent chapters we will consider using the dynamic (adaptive) controller $u = \nu_a(z, \hat{\theta})$ where now $\hat{\theta}(t)$ is allowed to vary with time.

In general, we will consider two different approaches to developing the adaptive control law. The first is a direct adaptive approach in which a set of parameters in the control law is directly modified to form a stable closed-loop system. In an indirect approach, components of a stabilizing control law are first estimated, and then combined to form the overall control law. For example, if for a given scalar error system $\dot{e} = \alpha(x) + \beta(x)u$ one is able to adaptively approximate $\alpha(x)$ and $\beta(x)$ with $\mathcal{F}_\alpha$ and $\mathcal{F}_\beta$, respectively, then the adaptive control law $\nu_a = (-\kappa e - \mathcal{F}_\alpha)/\mathcal{F}_\beta$ might be suggested as a possible stabilizing controller assuming $\mathcal{F}_\alpha \approx \alpha$ and $\mathcal{F}_\beta \approx \beta$. Design tools for the indirect approach will be studied in greater detail in the next chapter.

As for the case of static controller development, it is useful to study the trajectory of an error system, $e = \chi(t, x)$, which quantifies the controller performance. We will be particularly interested in the tracking problem where we wish to drive $y \to r(t)$ and the set-point regulation problem where $y \to r$ with r a constant. Recall that according to Assumption 6.1, the error system is also chosen such that if $|e|$ is bounded, then it is possible to place bounds on $|x|$. In particular, we will require that $\chi : \mathbb{R}^+ \times \mathbb{R}^n \to \mathbb{R}^q$ is defined such that $|x| \le \psi_x(t, |e|)$ for all t where $\psi_x(t, s)$ is nondecreasing with respect to $s \in \mathbb{R}^+$ for each fixed t.

Throughout this chapter we will assume the dynamics for the error system are defined by $\dot{e} = \alpha(t,x) + \beta(x)u$, so that the error dynamics are affine in the input. As seen in the last chapter, it is possible to define meaningful error systems for a wide class of nonlinear control problems such that this holds. Recall that the time dependence in $\alpha(t,x)$ results indirectly from the time varying reference signal $r(t)$. In fact, for the set-point regulation problem (where r is a constant), we have $\dot{e} = \alpha(x) + \beta(x)u$ when the plant is autonomous.

The direct adaptive control approach studied here will first assume that there exists some possibly unknown static controller $u = \nu_s(z)$ which provides desirable closed-loop performance. Since the static control law $\nu_s(z)$ is a function of known variables, it is possible to approximate $\nu_s(z)$ with $\mathcal{F}_\nu(z,\theta)$ over $z \in S_z$. The value of $\theta \in R^p$ is chosen such that the ideal approximation error is bounded by $|\mathcal{F}_\nu - \nu| \leq W$ whenever $z \in S_z$ with $W \geq 0$ assuming the form of $\mathcal{F}_\nu$ is appropriately chosen. When designing a direct adaptive controller, we may choose $u = \mathcal{F}_\nu(z,\hat{\theta})$, where $\hat{\theta}$ is an estimate of θ. It will then be shown how to choose update laws for $\hat{\theta}(t)$ which result in a stable closed-loop system.

7.2 Lyapunov Analysis and Adjustable Approximators

In this chapter we will investigate the use of an adjustable approximator as a controller. That is, we will let $u = \mathcal{F}_\nu(z,\hat{\theta})$ where $\hat{\theta} \in R^p$ is a set of adjustable parameters. If there exists some θ such that $\mathcal{F}_\nu(z,\theta)$ is able to approximate the static stabilizing control law $u = \nu_s(z)$ with some degree of accuracy when $z \in S_z$, then we would expect that one could directly use an approximator as the controller. In this chapter we will consider the case where some θ is not necessarily known. Instead, we will use an update routine for $\hat{\theta}$ so that the controller $u = \mathcal{F}(z,\hat{\theta})$ produces a stable closed-loop system.

If $f(z)$ is a function to be approximated by an adjustable universal approximator $\mathcal{F}(z,\hat{\theta})$, then there exists some parameter vector θ such that $|f(z) - \mathcal{F}(z,\theta)| \leq W$ for all $z \in S_z$. When $\mathcal{F}(z,\hat{\theta})$ is a universal approximator such as a neural network or fuzzy system, then W may be made arbitrarily small by choosing sufficiently many adjustable parameters in the approximator. Since the approximation error is only guaranteed to be valid when $z \in S_z$, we will need to ensure that at no time will the trajectory of z leave S_z. If z were to leave S_z, then the inequality $|f(z) - \mathcal{F}(z,\theta)| \leq W$ may no longer hold.

Placing bounds on the input to the approximator is an important difference from traditional adaptive control. The benefit of ensuring that $z \in S_z$ for all t goes beyond being able to use universal approximators in the control law. It also allows, for example, the use of traditional adaptive feedback linearization when the model of the plant dynamics is only valid over a region. If, for example, the dynamics were obtained from experimental data, one often only obtains plant characteristics for nominal operating conditions. Traditional control techniques, however, often assume that the approximation holds for all x possibly resulting in an unstable closed-loop system. In high-consequence systems, this may be very dangerous since the instability may not be apparent unless the closed-loop system is pushed to its limits. If the system subsequently becomes unstable at these extreme operating conditions (such as at high velocity in a vehicle), the consequence of the instability may be catastrophic. The following example demonstrates how ignoring the range over which the validity of an approximation holds may lead to control design with hidden instabilities.

Example 7.1 Consider the scalar plant defined by

$$\begin{aligned} \dot{x} &= f(x) + u \\ &= \epsilon x^2 + \theta x + u, \end{aligned} \tag{7.1}$$

where $\theta, \epsilon \in \mathsf{R}$ are unknown constants and $\epsilon > 0$ is assumed to be small (notice that $f(x) = \epsilon x^2 + \theta x$). Assume that we wish to define a controller which will force $x \to 0$ even when θ is unknown. If we wish to drive $x \to 0$, then define the error system $e = x$ and Lyapunov candidate $V_s = \frac{1}{2}e^2$. The error dynamics become

$$\dot{e} = \epsilon x^2 + \theta x + u,$$

so

$$\dot{V}_s = e\left(\epsilon x^2 + \theta x + u\right). \tag{7.2}$$

Let $w(x) = f(x) - \theta x$ define the error in representing $f(x)$ by θx. When $x \in [-1, 1]$ we find $|w(x)| \leq \epsilon$. Thus θx may be considered a good approximation of $f(x)$ since ϵ is assumed to be small. The time derivative of the Lyapunov function may now be expressed as

$$\dot{V}_s = e\left(w(x) + \theta x + u\right), \tag{7.3}$$

with $w(x)$ a bounded uncertainty when $x \in [-1, 1]$.

If θ is known, then the static control law $u = \nu_s(x, \theta)$ with

$$\nu_s(x, \theta) = -\kappa e - \epsilon \operatorname{sgn}(e) - \theta x, \tag{7.4}$$

and $\kappa > 0$ renders

$$\begin{aligned} \dot{V}_s &= -\kappa e^2 + w(x)e - \epsilon|e| & (7.5) \\ &\leq -2\kappa V_s & (7.6) \end{aligned}$$

as long as $x \in [-1, 1]$. This ensures that $x \to 0$ if $x(0) \in [-1, 1]$ and $x(t) \in [-1, 1]$, $t \geq 0$.

But θ is not known, so we will consider the use of an adaptive controller. Using the form of the static feedback controller, an adaptive controller is defined by $u = \mathcal{F}(x, \hat{\theta})$ with

$$\mathcal{F}(x, \hat{\theta}) = -\kappa e - \epsilon \text{sgn}(e) - \hat{\theta} x, \tag{7.7}$$

where $\hat{\theta}$ is an adaptive estimate of θ. A new Lyapunov candidate is now chosen to be $V = V_s + \frac{1}{2}\Gamma^{-1}\tilde{\theta}^2$, where $\tilde{\theta} = \hat{\theta} - \theta$ is the parameter estimate error and $\Gamma > 0$. The time derivative of the Lyapunov candidate becomes

$$\begin{aligned} \dot{V} &= \frac{\partial V_s}{\partial e}\dot{e} + \Gamma^{-1}\tilde{\theta}\dot{\hat{\theta}} \\ &= e\left(w(x) + \theta x + \mathcal{F}(x, \hat{\theta})\right) + \Gamma^{-1}\tilde{\theta}\dot{\hat{\theta}} \\ &= e\left(-\kappa e + w(x) - \epsilon \text{sgn}(e) - \tilde{\theta} x\right) + \Gamma^{-1}\tilde{\theta}\dot{\hat{\theta}}. \end{aligned} \tag{7.8}$$

If $x \in [-1, 1]$, then $|w| \leq \epsilon$, so we find

$$\dot{V} \leq -2\kappa V_s - e\tilde{\theta} x + \Gamma^{-1}\tilde{\theta}\dot{\hat{\theta}}. \tag{7.9}$$

Choosing $\dot{\hat{\theta}} = \Gamma x e$, we obtain $\dot{V} \leq -2\kappa V_s$.

We might then be tempted to use the LaSalle-Yoshizawa theorem to conclude that $x \to 0$ if $x(0) \in [-1, 1]$ as was the case for the static feedback controller. Unfortunately, it is not possible to conclude that $x \to 0$ even if $x(0) \in [-1, 1]$. It is possible for V to decrease while x is increasing due to the parameter error term in the definition of the Lyapunov candidate (that is, the e^2 term may increase and $\tilde{\theta}^2$ may decrease such that the sum defined by V is decreasing). If x leaves the set $[-1, 1]$, then V may also start to increase since the approximation $|w| \leq \epsilon$ is no longer valid, which may indicate that the closed-loop system is no longer stable. Figure 7.1 shows the trajectory of $x(t)$ for various values of $\hat{\theta}(0)$ when $\theta = 1$ and $\epsilon = 0.01$. When $\hat{\theta}(0) \in \{0, -2\}$, the trajectory remains stable, however, when $\hat{\theta}(0) \in \{-4, -6\}$ the closed-loop system becomes unstable. Thus the initial conditions of the parameter estimates may influence the stability of an adaptive system when using approximations which only hold over a compact set, such as is often the case when using system models obtained from experimental data. △

The above example demonstrates the need to ensure that x remains in a region in which a good approximation may be obtained. We will need to show that for a given controller and set of initial conditions, that the state trajectory is bounded such that $z \in S_z$ for all t where S_z represents

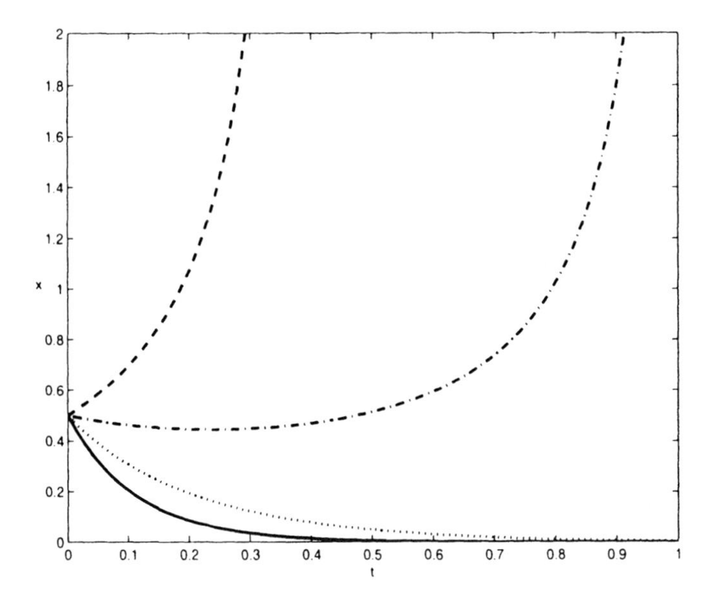

Figure 7.1. State trajectory when $\hat{\theta}(0) = 0$ (–), $\hat{\theta}(0) = -2$ (···), $\hat{\theta}(0) = -4$ (– · –), and $\hat{\theta}(0) = -6$ (- -).

the region over which a good approximation is achievable. In the case of adaptive control, we will only be concerned with the region where an approximation is "achievable." We use the word achievable since there may be no guarantee that given a current set of approximator parameters, a good approximation takes place. However, we will require that some ideal parameter set does exist even if we never use it. This point will become more apparent later when looking at the stability analysis of the direct adaptive controller.

To help guarantee that the state trajectories do not leave the region $z \in S_z$ over which a reasonable approximation may be established, we will use the following theorem.

Theorem 7.1: *Let $V : \mathsf{R}^q \times \mathsf{R}^p \to \mathsf{R}$ be a continuously differentiable function such that*

$$\gamma_{e1}(|e|) + \gamma_{\tilde{\theta}1}(|\tilde{\theta}|) \leq V(e, \tilde{\theta}) \leq \gamma_{e2}(|e|) + \gamma_{\tilde{\theta}2}(|\tilde{\theta}|), \tag{7.10}$$

where $\gamma_{e1}, \gamma_{e2}, \gamma_{\tilde{\theta}1}, \gamma_{\tilde{\theta}2}$ are class-$\mathcal{K}_\infty$. Assume that for a given error system, a control law $u = \nu$ is defined such that both $|e| \geq b_e$ implies $\dot{V} \leq 0$ and $|\tilde{\theta}| \geq b_{\tilde{\theta}}$ implies $\dot{V} \leq 0$. Then $e \in B_e$ for all t with

$$B_e = \left\{ e \in \mathsf{R}^p : |e| \leq \gamma_{e1}^{-1}\left(\max(V(0), V_r)\right) \right\}, \tag{7.11}$$

where $V_r = \gamma_{e2}(b_e) + \gamma_{\tilde{\theta}2}(b_{\tilde{\theta}})$.

Proof: If $V > V_r$, then either $|e| > b_e$ or $|\tilde{\theta}| > b_{\tilde{\theta}}$ (or both). Thus $V > V_r$ implies $\dot{V} \leq 0$. If $V(0) \leq V_r$, then $0 \leq V(t) \leq V_r$ for all t since V is positive definite and cannot grow larger than V_r (an invariant set). If $V(0) > V_r$, then $\dot{V} \leq 0$ until $V \leq V_r$, thus $0 \leq V(t) \leq \max(V(0), V_r)$ for all t. From (7.10) we know that

$$\gamma_e(|e|) \leq V(t) \leq \max(V(0), V_r)$$

so $|e| \leq \gamma_{e1}^{-1}(\max(V(0), V_r))$ for all t. ■

The above theorem will be used to study the range over which e (and x) may travel when an adaptive controller is used. From Assumption 6.1, we know that $|x| \leq \psi_x(t, |e|)$ where ψ_x is nondecreasing with respect to $|e|$. Thus $x \in B_x$ where

$$B_x = \left\{ x \in \mathbb{R}^n : |x| \leq \sup_t \psi_x\left(t, \gamma_e^{-1}(\max(V(0), V_r))\right) \right\} \tag{7.12}$$

for all t. Since $e \in B_e$ for all t, the above theorem may trivially be modified as follows:

Corollary 7.1: *Let $V : \mathbb{R}^q \times \mathbb{R}^p \to \mathbb{R}$ be a continuously differentiable function such that*

$$\gamma_{e1}(|e|) + \gamma_{\tilde{\theta}1}(|\tilde{\theta}|) \leq V(e, \tilde{\theta}) \leq \gamma_{e2}(|e|) + \gamma_{\tilde{\theta}2}(|\tilde{\theta}|), \tag{7.13}$$

where $\gamma_{e1}, \gamma_{e2}, \gamma_{\tilde{\theta}1}, \gamma_{\tilde{\theta}2}$ are class-$\mathcal{K}_\infty$. Assume that for a given error system, a control law $u = \nu$ is defined such that both $e \in B_e - B_b$ implies $\dot{V} \leq 0$ and $|\tilde{\theta}| \geq b_{\tilde{\theta}}$ implies $\dot{V} \leq 0$, where $B_b = \{e \in \mathbb{R}^q : |e| \leq b_e\}$ and B_e is defined by (7.11). Then $e \in B_e$ for all t.

We will find that Corollary 7.1 is useful in the study of adaptive systems using approximators that are defined only over a region. Since we require that $\dot{V} < 0$ for $|e| > b_e$ only when $e \in B_e$, the closed-loop system does not necessarily need to be stable for $e \notin B_e$. This will then place bounds on the range of the approximator input variables used in the control law $u = \mathcal{F}(z, \hat{\theta})$. If a fuzzy system, for example, is used in an adaptive controller, it may not be necessary for the input membership functions to cover all possible control inputs. Instead, the fuzzy system only needs to be defined such that $e \in B_e$ implies all the inputs to the fuzzy system remain in valid region.

7.3 The Adaptive Controller

The goal of the adaptive controller is to provide stable control of systems with significant uncertainty. As seen in the previous chapter, control laws may be defined for many uncertain nonlinear systems using techniques such as nonlinear damping and dynamic normalization. Intuitively, these techniques tend to increase robustness of the closed-loop system by including high gain terms which dominate the effects of the uncertainty. High feedback gain is often undesirable in implementation since it may lead to actuator saturation or may possibly excite other unmodeled dynamics which may lead to instability. Additionally we are often not guaranteed that $e \to 0$ when the nonlinear damping technique is used (especially when the feedback gain is reduced). These are just a few of the reasons that an adaptive control approach may be used in place of a static control law, even with the added complexity associated with the adaptive control laws.

In addition to these performance issues, an adaptive control approach may allow the designer to develop a controller which is "more robust" than its static equivalent. We will see how to use universal approximators, for example, so that systems with wide classes of uncertainties may be controlled even if the exact *functional form* of the uncertainty is unknown. It may also be possible for the adaptive controller to compensate for system faults in which the plant dynamics change due to some component failure or degradation.

For a given control problem, the designer must define an error system $e = \chi(t, x)$ which quantifies the closed-loop system performance and at the same time may be used to place bounds on the system states as required by Assumption 6.1. We will additionally assume that the error dynamics are affine in the control input so that

$$\dot{e} = \alpha(t, x) + \beta(x)u, \tag{7.14}$$

where $e \in \mathsf{R}^q$ and $u \in \mathsf{R}^m$. Note that as explained in the previous chapter this includes several classes of nonlinear systems. The remainder of this section will be devoted to defining update laws $\dot{\hat{\theta}} = \phi(t, x, \hat{\theta})$ so that the control law $u = \mathcal{F}(z, \hat{\theta}(t))$ guarantees that the closed-loop system is stable. Specifically, we will try to define an adaptive controller so that $e \to 0$ for (7.14) and x and $\hat{\theta}$ remain bounded.

7.3.1 σ-modification

Our goal here is to design an update law which modifies the adjustable parameter vector $\hat{\theta} \in \mathsf{R}^p$ so that the controller $u = \mathcal{F}(z, \hat{\theta})$ provides closed-loop stability. To ensure that it is possible to define an update law resulting in a stable adaptive controller, we will require that a static stabilizing controller *exists*. In particular, we will require the following assumption:

Assumption 7.1: There exists an error system $e = \chi(t, x)$ satisfying Assumption 6.1 and static control law $u = \nu_s(z)$ with z measurable, such that for a given radially unbounded, decrescent Lyapunov function $V_s(t, e)$, we find $\dot{V}_s \leq -k_1 V_s + k_2$ along the solutions of (7.14) when $u = \nu_s(z)$.

In addition, we must know how each input affects the states of a plant relative to the other inputs. In particular, we will make the following assumption:

Assumption 7.2: Given the error dynamics (7.14), assume that $\beta(x) = \frac{1}{c}\bar{\beta}(x)$, where $c > 0$ is a possibly unknown scalar constant and $\bar{\beta}(x)$ is known.

This requires that we know the functional form of $\beta(x)$, though we do not necessarily need to know the overall gain. Thus the scalar c allows a degree of freedom in terms of knowledge about the system dynamics. The following example shows how this degree of freedom may be used when controlling poorly understood systems.

Example 7.2 As shown in the previous chapter, there are a number of control problems with error dynamics defined by (7.14) where $\beta = [0, \ldots, 0, \pi]^\top$ with $\pi > 0$ a possibly unknown constant. In this case, we may let $c = 1/\pi$. Since $\bar{\beta} = [0, \ldots, 0, 1]^\top$ is known, Assumption 7.2 is satisfied even when the magnitude of the input gain is not known. △

Here, we will consider using the σ-modified update defined by

$$\dot{\hat{\theta}} = -\Gamma \left[\left(\frac{\partial V_s}{\partial e} \bar{\beta}(x) \frac{\partial \mathcal{F}(z, \hat{\theta})}{\partial \theta} \right)^\top + \sigma \left(\hat{\theta} - \theta^0 \right) \right], \tag{7.15}$$

where $\Gamma \in \mathsf{R}^{p \times p}$ is a positive definite, symmetric matrix used to set the rate of adaptation and $\sigma > 0$ is a term used to increase the robustness of the closed-loop system. Here we are using the notation

$$\frac{\partial \mathcal{F}(z, y)}{\partial \theta} \equiv \left. \frac{\partial \mathcal{F}(z, \theta)}{\partial \theta} \right|_{\theta = y}. \tag{7.16}$$

The vector $\theta^0 \in \mathsf{R}^p$ may be used to include a best guess of some $\theta \in \mathsf{R}^p$, where θ is an ideal parameter vector defined in Theorem 7.2.

Theorem 7.2: *Let Assumption 7.1 and Assumption 7.2 hold with $\gamma_{e1}(|e|) \leq V_s(e) \leq \gamma_{e2}(|e|)$, where γ_{e1} and γ_{e2} are class-$\mathcal{K}_\infty$. If for a given linear in the parameter approximator $\mathcal{F}(z, \hat{\theta})$ there exists some θ such that $|\mathcal{F}(z, \theta) - \nu_a(z)| \leq W$ for all $z \in S_z$, where $e \in B_r$ implies $z \in S_z$, and*

$$\nu_a = \nu_s - \eta \left(\frac{\partial V_s}{\partial e} \bar{\beta}(x) \right)^\top, \tag{7.17}$$

with $\eta > 0$, *then the parameter update law (7.15) with adaptive controller* $u = \mathcal{F}(z, \hat{\theta})$ *guarantee that the solutions of (7.14) are bounded given* $B_e \subseteq B_r$, *where* B_e *is defined by (7.25).*

Proof: Consider the Lyapunov candidate

$$V_a = cV_s + \frac{1}{2}\tilde{\theta}^\top \Gamma^{-1} \tilde{\theta}, \tag{7.18}$$

where Γ is positive definite and symmetric, and $c > 0$ such that $\bar{\beta}(x) = c\beta(x)$. Taking the derivative we find

$$\dot{V}_a = c\left[\frac{\partial V_s}{\partial t} + \frac{\partial V_s}{\partial e}\left(\alpha(t,x) + \beta(x)\mathcal{F}(z,\hat{\theta})\right)\right] + \tilde{\theta}^\top \Gamma^{-1} \dot{\tilde{\theta}}.$$

Also

$$\begin{aligned} \mathcal{F}(z,\hat{\theta}) &= \mathcal{F}(z,\hat{\theta}) - \mathcal{F}(z,\theta) + \mathcal{F}(z,\theta) \\ &= \mathcal{F}(z,\hat{\theta}) - \mathcal{F}(z,\theta) + \nu_s(z) - \eta\left(\frac{\partial V_s}{\partial e}\bar{\beta}\right)^\top + w, \end{aligned} \tag{7.19}$$

where $w = \mathcal{F}(z,\theta) - \nu_a$ with $|w| \le W$ for all $z \in S_z$.

Using (7.19) we find

$$\begin{aligned} \dot{V}_a \le\; & -ck_1 V_s + ck_2 + \tilde{\theta}^\top \Gamma^{-1} \dot{\tilde{\theta}} \\ & + \frac{\partial V_s}{\partial e}\bar{\beta}(x)\left(\mathcal{F}(z,\hat{\theta}) - \mathcal{F}(z,\theta) - \eta\left(\frac{\partial V_s}{\partial e}\bar{\beta}\right)^\top + w\right), \end{aligned} \tag{7.20}$$

where we have used the assumption that $\dot{V}_s \le -k_1 V_s + k_2$ when $u = \nu_s(z)$. Since $\mathcal{F}(z,\hat{\theta}) - \mathcal{F}(z,\theta) = \frac{\partial \mathcal{F}}{\partial \theta}\tilde{\theta}$ for a linearly parameterized approximator and $\dot{\tilde{\theta}} = \dot{\hat{\theta}}$, we find

$$\dot{V}_a \le -ck_1 V_s + ck_2 - \eta\left|\frac{\partial V_s}{\partial e}\bar{\beta}(x)\right|^2 + \left|\frac{\partial V_s}{\partial e}\bar{\beta}(x)\right| W - \sigma\tilde{\theta}^\top\left(\hat{\theta} - \theta^0\right)$$

whenever $z \in S_z$. Using the inequality $-x^\top x \pm 2x^\top y \le y^\top y$ we find

$$-\eta\left|\frac{\partial V_s}{\partial e}\bar{\beta}(x)\right|^2 + \left|\frac{\partial V_s}{\partial e}\bar{\beta}(x)\right| W \le \frac{W^2}{4\eta}. \tag{7.21}$$

Also since $-2x^\top x \pm 2x^\top y \le -x^\top x + y^\top y$ we obtain

$$-\tilde{\theta}^\top\left(\hat{\theta} - \theta^0\right) = -\tilde{\theta}^\top(\tilde{\theta} + \theta - \theta^0) \le -\frac{|\tilde{\theta}|^2}{2} + \frac{|\theta - \theta^0|^2}{2}. \tag{7.22}$$

Using (7.21) and (7.22), we find

$$\dot{V}_a \leq -ck_1 V_s + ck_2 + \frac{W^2}{4\eta} - \sigma\frac{|\tilde{\theta}|^2}{2} + \sigma\frac{|\theta - \theta^0|^2}{2}. \tag{7.23}$$

Since $\gamma_{e1}(|e|) \leq V_s(e)$, we are assured that

$$\dot{V}_a \leq -ck_1\gamma_{e1}(|e|) - \frac{\sigma|\tilde{\theta}|^2}{2} + d, \tag{7.24}$$

where $d = ck_2 + \frac{W^2}{4\eta} + \sigma\frac{|\theta-\theta^0|^2}{2}$. If $|e| > b_e$ or $|\tilde{\theta}| > b_{\tilde{\theta}}$ where $b_e = \gamma_{e1}^{-1}\left(\frac{d}{ck_1}\right)$ and $b_{\tilde{\theta}} = \sqrt{\frac{2d}{\sigma}}$, then $\dot{V} < 0$.

Using Corollary 7.1, with $\gamma_{\tilde{\theta}2} = \lambda_{\max}(\Gamma^{-1})|\tilde{\theta}|^2$ and the bounds on V_s given in the statement of the theorem, we see that $e \in B_e$ with

$$B_e = \left\{e \in \mathbb{R}^q : |e| \leq \gamma_{e1}^{-1}\left(\max(V_a(0), V_r)/c\right)\right\}, \tag{7.25}$$

where

$$\begin{aligned} V_r &= c\gamma_{e2}(b_e) + \frac{\lambda_{\max}(\Gamma^{-1})b_{\tilde{\theta}}^2}{2} \\ &= c\gamma_{e2} \circ \gamma_{e1}^{-1}\left(\frac{d}{ck_1}\right) + \frac{d\lambda_{\max}(\Gamma^{-1})}{\sigma}. \end{aligned}$$

By properly choosing the values of k_1, σ, and Γ^{-1} it is possible to make V_r arbitrarily small. Thus if the initial conditions may be chosen such that $V_a(0)$ is sufficiently small, it is then possible to ensure $B_e \subseteq B_r$. ■

The above theorem shows that if there exists a nonadaptive (static) stabilizing controller, $\nu_s(z)$, then there also exists an adaptive stabilizing controller as long as there exists some θ such that the approximator $\mathcal{F}(z, \theta)$ reasonably approximates $\nu_a(z)$. Thus the existence of a stable direct adaptive controller reduces to proving the existence of a stabilizing static controller (which was the topic of the previous chapter) and a suitable approximator structure.

The direct adaptive controller is typically defined using the following steps:

1. Place the plant in a canonical representation so that an error system may be defined.

2. Define an error system and Lyapunov candidate V_s for the static problem.

3. Define a static control law $u = \nu_s$ which ensures that $\dot{V}_s \leq -k_1 V_s + k_2$ (that is, satisfy Assumption 7.1).

4. Choose an approximator $\mathcal{F}(z, \hat{\theta})$ such that there exists some θ guaranteeing $|\mathcal{F}(z, \theta) - \nu_a(z)| \le W$ for all $z \in S_z$, where ν_a is defined by (7.17). Estimate upper bounds for W and $|\theta - \theta^0|$ where θ^0 may be viewed as a "best guess" of θ.

5. Find some B_r such that $e \in B_r$ implies $z \in S_z$.

6. Choose the initial conditions, control parameters, and update law parameters such that $B_e \subseteq B_r$ with B_e defined by (7.25).

It should be emphasized that the static control law ν_s does not need to be implementable. It may be defined, for example, using unknown parameters, unknown functions, or other uncertainties as long as an approximator may be defined such that $|\mathcal{F}(z, \theta) - \nu_a(z)| \le W$ where ν_a is given by (7.17).

The approach to the analysis of the direct adaptive controller using an approximator is somewhat different from traditional stability analysis. In traditional stability analysis, one typically either defines a controller which ensures that $\dot{V} \le -W(x)$ with $W \ge 0$, or tries to find some V for a specific controller so that $\dot{V} \le -W(x)$. Here we simply require that *there exists* a controller such that $\dot{V} \le -W$ for a known V and that there exists some θ such that an approximator $\mathcal{F}(z, \theta)$ is able to match the controller with some accuracy. We do not require that either the stabilizing controller $\nu_s(z)$ or ideal parameter vector θ be known (though there are some initialization requirements for the approximator parameters since B_e and thus the domain of the approximator is dependent upon $\tilde{\theta}(0)$).

In the above theorem, the term $w = \mathcal{F}(z, \theta) - \nu_s(z)$ is often referred to as the ideal representation error since it is the difference between the approximator with ideal parameter vector θ and the true function it is representing. If there exists some θ such that $|w| = 0$ for all $z \in S_z$ (e.g., if you know the structure of the nonlinearity and only have parametric uncertainty such as is typically the case in adaptive control for linear systems), then it is possible to set $\eta = 0$ in the above theorem. The stability analysis may then be carried out as before, but with $W = 0$ so $d = ck_2 + \sigma\frac{|\theta - \theta^0|^2}{2}$ in the definition of V_r.

Theorem 7.2 only tells us that the solutions will remain bounded such that the approximator is well behaved. Using the results of the above proof it is possible to place more explicit bounds upon the RMS tracking error. In particular, notice that from (7.24) we find

$$\dot{V}_a \le -ck_1\gamma_{e1}(|e|) + d. \tag{7.26}$$

Rearranging terms and integrating, we see that

$$\int^t \gamma_{e1}(|e|)d\tau \le \int^t \left(-\frac{\dot{V}_a}{ck_1} + \frac{d}{ck_1} \right) d\tau. \tag{7.27}$$

Since V_a is bounded we find that

$$\lim_{t\to\infty} \frac{1}{t}\int^{t} \gamma_{e1}(|e|)d\tau \leq \frac{d}{ck_1}. \tag{7.28}$$

If, for example, we choose $\gamma_{e1}(s) = \frac{1}{2}s^2$, then

$$\lim_{t\to\infty} \frac{1}{t}\int^{t} \frac{1}{2}|e|^2 d\tau \leq \frac{d}{ck_1},$$

so the RMS error is bounded by

$$\sqrt{\lim_{t\to\infty} \frac{1}{t}\int^{t} |e|^2 d\tau} \leq \sqrt{\frac{2d}{ck_1}}. \tag{7.29}$$

From (7.29) we see that to improve the RMS error, one must either decrease d, or increase k_1. The value of d may be decreased by decreasing k_2, W, or $\sigma|\theta - \theta^0|$. Often the choice of which parameters to adjust is dictated by the control problem.

Since $\dot{V}_a < 0$ when $|e| > b_e$ or $|\tilde{\theta}| > b_{\tilde{\theta}}$, we are ensured that V_a will decrease to $V \leq V_r$ where $V_r = \gamma_{e1}(b_e) + \gamma_{\tilde{\theta}2}(b_{\tilde{\theta}})$ with $\gamma_{\tilde{\theta}2} = \lambda_{\max}(\Gamma^{-1})|\tilde{\theta}|^2$. Thus $|e|$ will decrease such that $|e| \leq \gamma_{e1}^{-1}(V_r)$. Again since the control and update law parameters may be chosen such that b_e and $b_{\tilde{\theta}}$ may be made arbitrarily small, the ultimate bound on $|e|$ may be made arbitrarily small.

We will now see how to apply the direct adaptive controller using the σ-modification to adjust its parameters. We will start by studying the problem in which the approximator used to define the adaptive controller does not have limits on its inputs. Thus $B_r = \mathbb{R}^q$ since the approximator input z may take on any value. We will then study a different problem in which a fuzzy system is used with a finite domain associated with its input membership functions.

Example 7.3 Assume that the system dynamics for a particular plant may be transformed into

$$\begin{aligned} \dot{x}_1 &= \Delta_1(x) + x_2 \\ \dot{x}_2 &= \Delta_2(x) + u, \end{aligned} \tag{7.30}$$

where x_1, x_2 are measurable, and Δ_1, Δ_2 are defined by

$$\begin{aligned} \Delta_1 &= \rho_1 \\ \Delta_2 &= \begin{bmatrix} x_1 & x_2 \end{bmatrix} \begin{bmatrix} \rho_2 & \rho_3 \\ \rho_3 & \rho_4 \end{bmatrix} \begin{bmatrix} x_1 \\ x_2 \end{bmatrix} \end{aligned}$$

with ρ_i unknown for $i = 1, \ldots, 4$. Suppose that we wish to define a direct adaptive controller that drives $x_1 \to r(t)$, where $r(t)$ and its first two derivatives are measurable. To do this, we must define an error system defined by measurable signals and Lyapunov candidate for which there exists a static control law that stabilizes the system.

Using the backstepping procedure, we start by considering the control of the system

$$\dot{x}_1 = \Delta_1(x) + v, \tag{7.31}$$

where v is a virtual input. Define the first error variable as $e_1 = x_1 - r$. If $V_1 = \frac{1}{2}e_1^2$, then the controller $v = \nu_1(z_1)$ with

$$\nu_1 = \dot{r} - \Delta_1(x) - \kappa e_1 \tag{7.32}$$

will ensure that $\dot{V}_1 \leq -2\kappa V_1$, so that the e_1 dynamics are stabilized. As long as r is finite, then the x_1 trajectory will also remain bounded. A second error variable might then be defined using ν_1, such as $e_2 = x_2 - \nu_1$. But Δ_1 is unknown and thus ν_1 may not be formed.

Ignoring Δ_1 for now, consider the error $e_2 = x_2 + 3\kappa e_1/2 - \dot{r}$ (which is measurable and thus may be used in the control law) and Lyapunov candidate $V_s = \frac{1}{2}e_1^2 + \frac{1}{2}e_2^2$. Since

$$\begin{aligned}
\dot{e}_1 &= \Delta_1 - \frac{3\kappa e_1}{2} + e_2 \\
\dot{e}_2 &= -\ddot{r} + \Delta_2 + \frac{3\kappa}{2}(\Delta_1 + e_2 - \frac{3\kappa}{2}e_1) + u,
\end{aligned}$$

the derivative of the Lyapunov candidate becomes

$$\begin{aligned}
\dot{V}_s &= e_1\dot{e}_1 + e_2\dot{e}_2 \\
&= e_1\left(\Delta_1 + e_2 - \frac{3\kappa}{2}e_1\right) \\
&\quad + e_2\left(-\ddot{r} + \Delta_2 + \frac{3\kappa}{2}(\Delta_1 + e_2 - \kappa e_1) + u\right).
\end{aligned} \tag{7.33}$$

Notice that $\beta = [0, 1]^\top$ in (7.14).

Now consider the control law $u = \nu_s(z)$ with

$$\nu_s = \ddot{r}(t) - e_1 - \Delta_2 - \frac{3\kappa}{2}(\Delta_1 + e_2 - \kappa e_1) - \kappa e_2. \tag{7.34}$$

Since $-x^2/2 + xy \leq y^2/2$, we find

$$\begin{aligned}
\dot{V}_s &= -\frac{3\kappa}{2}e_1^2 + e_1\rho_1 - \kappa e_2^2 \\
&\leq -2\kappa V_s + \frac{\rho_1^2}{2\kappa}.
\end{aligned} \tag{7.35}$$

Here $z = [\ddot{r}, e_1, e_2, x_1, x_2]^\top$. Notice that ν_s may not be implemented since it is defined using Δ_1 and Δ_2 which are unknown. The direct adaptive controller, however, may be implemented since Assumption 7.1 simply requires the existence of a stabilizing control law.

Notice that the controller may now be expressed as

$$\begin{aligned}\nu_s(z) &= \ddot{r} - (1-\kappa^2)e_1 - \frac{5}{2}\kappa e_2 - \Delta_2 - \kappa\Delta_1 \\ &= \ddot{r} - (1-\kappa^2)e_1 - \frac{5}{2}\kappa e_2 + \theta^\top \begin{bmatrix} \kappa \\ x_1^2 \\ x_1 x_2 \\ x_2^2 \end{bmatrix}, \end{aligned} \tag{7.36}$$

where

$$\theta = \begin{bmatrix} -\rho_1 \\ -\rho_2 \\ -2\rho_3 \\ -\rho_4 \end{bmatrix}. \tag{7.37}$$

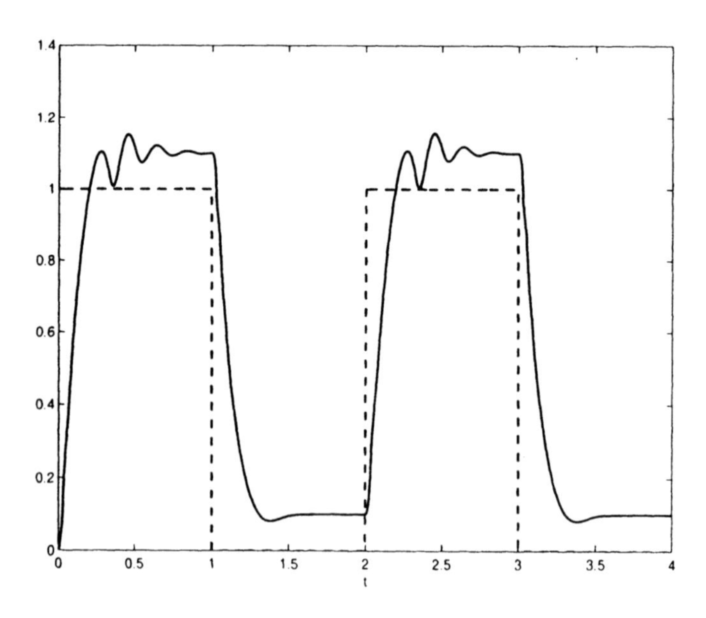

Figure 7.2. Closed-loop performance when x (-) is commanded to track a reference r (- -) defined by a square wave.

To use the direct adaptive controller, we must define a linear in the parameter approximator. Let

$$\mathcal{F}(z,\hat{\theta}) = \ddot{r} - (1-\kappa^2)e_1 - \frac{5}{2}\kappa e_2 + \hat{\theta}^\top \begin{bmatrix} \kappa \\ x_1^2 \\ x_1 x_2 \\ x_2^2 \end{bmatrix}. \tag{7.38}$$

Notice that this choice of the approximator is based on the form of the stabilizing controller defined by (7.36). Since $v_s(z) = \mathcal{F}(z,\theta)$ for all z, it was not necessary to include a term to account for the nonlinear damping term $\eta\left(\frac{\partial V_s}{\partial e}\bar{\beta}(x)\right)^\top$ in (7.17). Using σ-modification, the update law becomes

$$\dot{\hat{\theta}} = -\Gamma\left(\begin{bmatrix} \kappa \\ x_1^2 \\ x_1 x_2 \\ x_2^2 \end{bmatrix} e_2 + \sigma\hat{\theta}\right), \tag{7.39}$$

where Γ is positive definite and symmetric, and $\sigma > 0$. Notice we have chosen $\theta^0 = 0$ in (7.15).

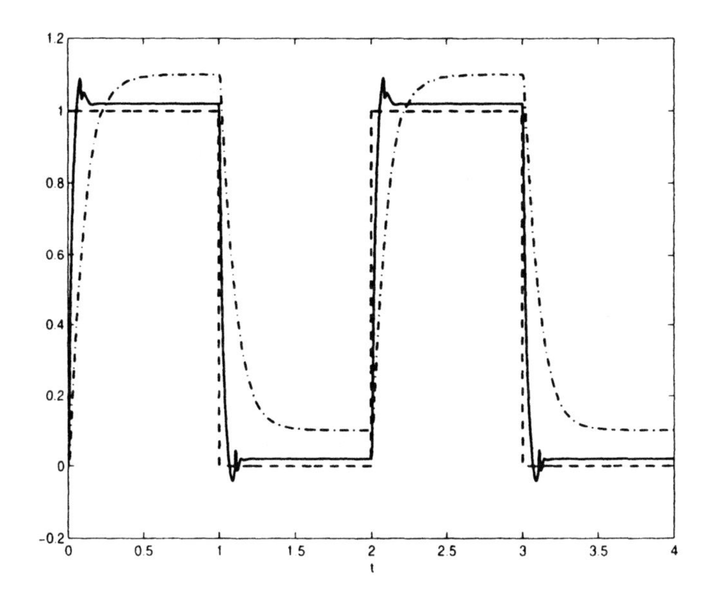

Figure 7.3. Improving the transient performance of the closed-loop system by increasing κ (–) and Γ ($- \cdot -$).

Assume that $\rho_1 = 1$, $\rho_2 = -1$, $\rho_3 = 2$, and $\rho_4 = 1$. Let $\kappa = 10$, $\eta = 1$, $\Gamma = 10I$ (where I is the identity matrix), and $\sigma = 0.1$. Figure 7.2 shows the trajectory of the closed-loop system when $r(t)$ is a square wave. Notice that there is a bit of steady-state error and ringing. The steady-state error may be improved by increasing κ, while the ringing may be reduced by increasing the rate of adaptation. Figure 7.3 shows the case where $\kappa = 50$ and $\Gamma = 50I$. $\triangle$

In the previous example, we designed the approximator based on knowledge of the form of the underlying nonlinearity. Based on this form, it led to a choice for the approximator structure. This approach to developing an approximator may be referred to as a "physically motivated" approach. As an alternative approach, we will also use universal approximators in which the approximator structure is chosen so that it may be applicable to a wider class of underlying nonlinearities. The physically motivated approach often falls under the classification of "nonlinear adaptive control," while the use of universal approximators (such as fuzzy systems or neural networks) in the design of an adaptive controller may be referred to as adaptive fuzzy control, neural control, intelligent control, among others. In this book we hope to remove some of the distinctions between these different fields since the primary difference is the motivation of the choice of the approximator, and not necessarily the stability analysis.

Thus far we have seen how to define an adaptive controller when it is possible for the approximator to accurately represent some desired controller for all z (that is $B_r \in \mathrm{R}^q$). According to Theorem 7.2, we may also define stable adaptive controllers when an approximator is used which is only well defined on the region $z \in S_z$ so long as there is some B_r such that $e \in B_r$ implies $z \in S_z$. The following example shows how to design a stable adaptive controller when using a finite fuzzy system.

Example 7.4 Consider the velocity control of an automobile whose dynamics are defined by

$$m\dot{x} = -\rho x^2 + u, \tag{7.40}$$

where $\rho \in (0, 0.4]$ is the unknown coefficient of aerodynamic drag, m is the known vehicle mass, and x is the vehicle speed. If we wish to drive $x \to r(t)$, then define the error system by

$$e = x - r.$$

Notice that $\dot{e} = -\dot{r} - \rho x^2/m + u/m$. Using the representation defined by (7.14), we find $\dot{e} = \alpha(t,x) + \beta u$, where $\alpha = -\dot{r} - \rho x^2/m$ and $\beta = 1/m$.

Consider the Lyapunov candidate $V_s = \frac{1}{2}e^2$ and the static control law $u = \nu_s(z)$, where

$$\nu_s(z) = m(\dot{r} - \kappa e) + \rho x^2 \tag{7.41}$$

with $\kappa > 0$ and $z = [\dot{r}, e, x]^\top$. This choice of the control law renders $\dot{V}_s = -2\kappa V_s$, so $k_1 = 2\kappa$ and $k_2 = 0$ in Assumption 7.1. Since ρ is unknown, however, this static controller may not be implemented.

Because the aerodynamic drag is unknown, we might want to use a fuzzy system to approximate a control term which compensates for its effects. In particular, consider the controller $u = \mathcal{F}(z, \hat{\theta})$, where

$$\mathcal{F}(z, \hat{\theta}) = m(\dot{r} - \kappa e) - \eta e/m + \frac{\sum_{i=1}^{p} \hat{\theta}_i \mu_i(x)}{\sum_{i=1}^{p} \mu_i(x)}, \tag{7.42}$$

with $\eta > 0$. Each μ_i is an input membership function for the fuzzy system and

$$\frac{\partial \mathcal{F}}{\partial \hat{\theta}} = \left[\frac{\mu_1}{\sum_i \mu_i}, \ldots, \frac{\mu_p}{\sum_i \mu_i} \right].$$

The fuzzy system will be used to cancel the effects of the aerodynamic drag, while the term $\eta e/m$ has been included to account for the additional nonlinear damping term in (7.17). Assume that the fuzzy system is defined using $p = 10$ triangular input membership functions as shown in Figure 7.4 so that speeds up to $40 m/s$ will be considered. Since m (and thus β) is known, we may choose the update law according to

$$\dot{\hat{\theta}} = -\Gamma \left[e \frac{\partial \mathcal{F}(z, \hat{\theta})}{\partial \theta} / m + \sigma \hat{\theta} \right], \tag{7.43}$$

where we have used $\partial V_s / \partial e = e$ and $\bar{\beta} = \beta$ in (7.15).

Before choosing the controller parameters, we will estimate a bound on W (the magnitude of the representation error). An estimate of W will be needed since it will influence our choices of the controller parameters to ensure that $B_e \subseteq B_r$ (where B_r will also be defined shortly). When the parameters of the fuzzy system are fixed, it simply behaves as a linear interpolator. Notice that when x is at one of the input membership function centers, the degree of membership is zero for the other input membership functions. Considering the aerodynamic drag term we choose $\theta_i = \rho c_i^2$, where c_i is the center for the ith membership function so that the approximation will be perfect at the membership function centers.

Define the representation error $w = \nu_a - \mathcal{F}(z, \theta)$, where $\nu_a = \nu_s - \eta e/m$. Notice that

$$w(x) = \rho x^2 - \frac{\sum_{i=1}^{p} \theta_i \mu_i(x)}{\sum_{i=1}^{p} \mu_i(x)}.$$

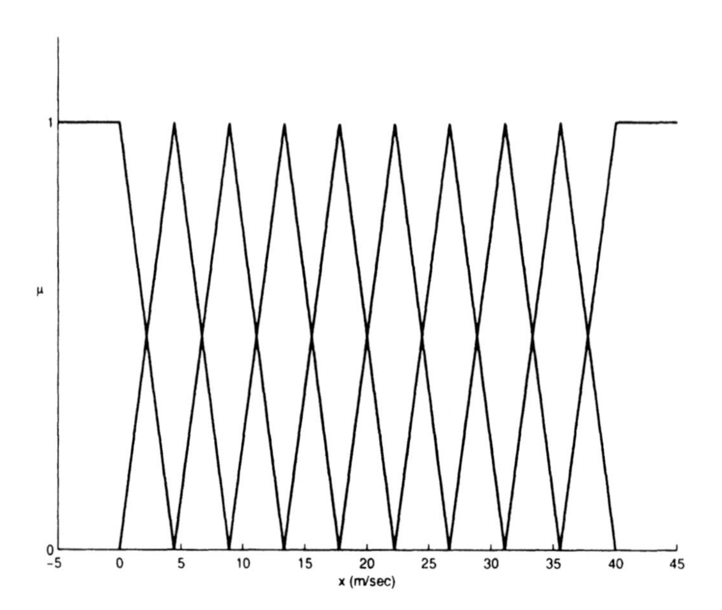

Figure 7.4. Input membership functions for the fuzzy system.

When $\rho = 0.4$, we find $|w| < 2$ as shown in Figure 7.5 for our choice of θ. Note that the peaks in Figure 7.5 show the error in the interpolation. Clearly, if we used more approximator structure (e.g., more input membership functions), then generally we can reduce the amount of approximation error. As ρ decreases, the bound on $|w|$ also decreases, so $|w| < 2$ for all possible ρ when $x \in [0, 40]$. When $\rho = 0.4$, we also find that $|\theta| = 978.4$, so $|\theta| \leq 980$ for any $\rho \in (0, 0.4]$.

We now need to find some B_r such that $e \in B_r$ implies that $x \in [0, 40]$. If the desired velocity is defined to be in the range $r \in [1.7, 40 - 1.7]$, then $|e| \leq 1.7$ implies that $x \in [0, 40]$. We now need to choose the controller parameters such that $e \in B_e$ with $B_e \subseteq B_r$ where $B_r = \{e \in \mathbf{R} : |e| < 1.7\}$.

From Theorem 7.2, we may choose $c = 1$ (where c is defined in Assumption 7.2) so that

$$B_e = \left\{ e \in \mathbf{R} : |e| \leq \sqrt{2 \max(V_a(0), V_r)} \right\}, \tag{7.44}$$

where $V_a(0) = V_s(0) + \frac{1}{2}\tilde{\theta}^\top(0)\Gamma^{-1}\tilde{\theta}(0)$ and

$$V_r = d\left[\frac{1}{2\kappa} + \frac{\lambda_{\max}(\Gamma^{-1})}{\sigma}\right].$$

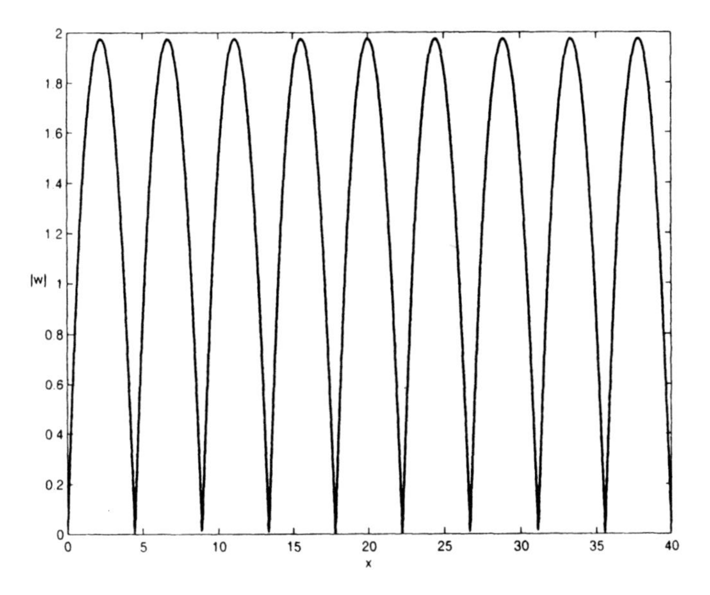

Figure 7.5. The difference between the fuzzy system with ideal parameters and the true aerodynamic drag when $\rho = 0.4$.

From the proof of Theorem 7.2 one finds

$$\begin{aligned} d &= k_2 + \frac{W^2}{4\eta} + \sigma\frac{|\theta|^2}{2} \\ &\leq 0 + \frac{4}{4\eta} + \frac{\sigma 980^2}{2}. \end{aligned} \tag{7.45}$$

Choosing $\eta = 10$ and $\sigma = 2e^{-7}$, we find $d \leq 0.196$.

We will assume that the reference is chosen such that $r(0) = x(0)$. If $\hat{\theta}(0) = 0$, then $V_a(0) = \frac{1}{2}\theta^\top\Gamma^{-1}\theta$. We must now choose Γ and κ such that $\sqrt{2\max(V_a(0), V_r)} \leq 1.7$. This is accomplished with $\kappa = 1/10$ and

$$\Gamma = \begin{bmatrix} 2.5e^6 & & 0 \\ & \ddots & \\ 0 & & 2.5e^6 \end{bmatrix},$$

since $V_r \leq 1.4$ and $V(0) = \frac{1}{2}\theta^\top\Gamma^{-1}\theta \leq 0.2$.

The performance of the adaptive controller using a fuzzy system is shown in Figure 7.6. When $\Gamma = 0$ (adaptation turned off), there is more steady-state error. $\triangle$

In the previous example, we were able to find a bound on the representation error w. In many real applications, however, it may not be possible to

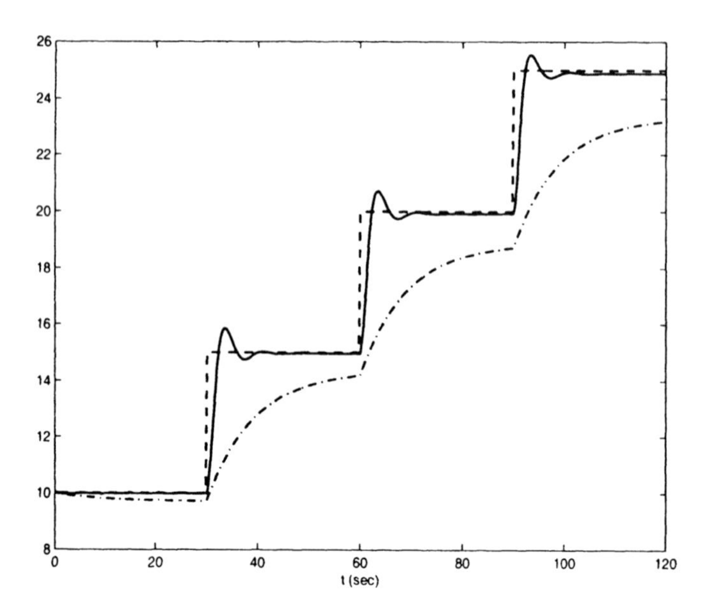

Figure 7.6. Transient performance for the direct adaptive fuzzy controller (–) when tracking $r(t)$ (- -). Compare to the case when $\Gamma = 0$ ($- \cdot -$).

find W, possibly because the plant dynamics are too poorly understood. In these cases, we must use knowledge about the approximator and the function it is to represent to determine a bound $|w| \le W$ using rules of thumb. From Chapter 5, for example, we know that as the spatial frequency of the function we wish to approximate increases, w will, in general, increase. Fortunately, the controller iteself does not directly use W, rather we need knowledge of W to place bounds on B_e. If a conservative value for W is chosen, then we will obtain a conservative estimate of B_e.

7.3.2 ϵ-modification

In the previous section we saw that the σ-modification defined by

$$\dot{\hat{\theta}} = -\Gamma \left[\left(\frac{\partial V_s}{\partial e} \bar{\beta}(x) \frac{\partial \mathcal{F}(z,\hat{\theta})}{\partial \theta} \right)^\top + \sigma \left(\hat{\theta} - \theta^0 \right) \right],$$

is able to adjust the parameters of the direct adaptive controller so that the closed-loop system is stable. The term $\sigma \left(\hat{\theta} - \theta^0 \right)$ in the σ-modification is used to improve the robustness of the adaptive controller since it keeps the parameter estimate, $\hat{\theta}$, from growing without bound when uncertainty exists in the system. Unfortunately, when $\left| \frac{\partial V_s}{\partial e} \bar{\beta}(x) \right|$ becomes small, $\dot{\hat{\theta}}$ is dominated by the $\sigma\left(\hat{\theta}-\theta^0\right)$ term. This causes $\hat{\theta}$ to be driven toward θ^0. If θ^0 is not a good approximation of the ideal parameter vector θ, then $|e|$ may start to increase.

To overcome this problem with the σ-modification, it is possible to modify the update law so that

$$\dot{\hat{\theta}} = -\Gamma\left[\left(\frac{\partial V_s}{\partial e}\beta(x)\frac{\partial \mathcal{F}(z,\hat{\theta})}{\partial \theta}\right)^{\top} + \varepsilon(e)\left(\hat{\theta}-\theta^0\right)\right], \tag{7.46}$$

where Γ is a symmetric positive definite matrix, θ^0 is a best guess of θ, and $\varepsilon(e) > 0$ is a new robust term. A common choice for the ϵ-modification is to use

$$\varepsilon = \sigma|e|, \tag{7.47}$$

with $\sigma > 0$. Notice that with this choice, when $|e|$ is small, then the contribution from the robust term is reduced. The ϵ-modification does require a slightly different set of assumptions from the σ-modification. In particular, we will require the following:

Assumption 7.3: There exists an error system $e = \chi(t,x)$ satisfying Assumption 6.1 and static control law $u = \nu_s(z)$ with z measurable. There exists some known V_s satisfying $k_3|e|^2 \leq V_s(t,e) \leq k_4|e|^2$ and $\left|\frac{\partial V_s}{\partial e}\beta(x)\right| \leq k_5|e|$ such that $\dot{V}_s \leq -k_1 V_s + k_2|e|$ along the solutions of (7.14) when $u = \nu_s(z)$.

With Assumption 7.3 the following theorem holds when using the ϵ-modification.

Theorem 7.3: *Assume Assumption 7.2 and Assumption 7.3 hold. If for a given linear in the parameter approximator $\mathcal{F}(z,\hat{\theta})$ there exists some θ such that $|\mathcal{F}(z,\theta) - \nu_s(z)| \leq W$ for all $z \in S_z$ where $e \in B_r$ implies that $z \in S_z$, then the parameter update law (7.46) with adaptive controller $u = \mathcal{F}(z,\hat{\theta})$ guarantee that the solutions of (7.14) are bounded given $B_e \subseteq B_r$ with B_e defined by (7.50).*

Proof: Define the representation error as $w = \mathcal{F}(z,\theta) - \nu_s$. Following the steps up to (7.20) in the proof of Theorem 7.2, we find

$$\dot{V}_a \leq -ck_1V_s + ck_2|e| + \frac{\partial V_s}{\partial e}\beta(x)\left(\mathcal{F}(z,\hat{\theta}) - \mathcal{F}(z,\theta) + w\right) + \tilde{\theta}^{\top}\Gamma^{-1}\dot{\tilde{\theta}},$$

where we have used Assumption 7.3. Since $\mathcal{F}(z,\hat{\theta}) - \mathcal{F}(z,\theta) = \frac{\partial \mathcal{F}}{\partial \theta}\tilde{\theta}$ for a linearly parameterized approximator, we find

$$\begin{aligned}\dot{V}_a &\leq -ck_1V_s + (ck_2 + k_5W)|e| + \frac{\partial V_s}{\partial e}\beta(x)\frac{\partial \mathcal{F}}{\partial \hat{\theta}}\tilde{\theta} + \tilde{\theta}^{\top}\Gamma^{-1}\dot{\tilde{\theta}} \\ &= -ck_1V_s + (ck_2 + k_5W)|e| - \sigma|e|\tilde{\theta}^{\top}\left(\hat{\theta}-\theta^0\right),\end{aligned}$$

where we have used the definition of the update law (7.46) and the assumption that $|w| \le W$ when $z \in S_z$.

Using $-2x^\top x \pm 2x^\top y \le -x^\top x + y^\top y$ we find that

$$-\tilde{\theta}^\top\left(\hat{\theta} - \theta^0\right) = -\tilde{\theta}^\top(\tilde{\theta} + \theta - \theta^0) \le -\frac{|\tilde{\theta}|^2}{2} + \frac{|\theta - \theta^0|^2}{2},$$

so

$$\dot{V}_a \le -ck_1 V_s + (ck_2 + k_5 W)|e| + \frac{\sigma|e|}{2}\left(-|\tilde{\theta}|^2 + |\theta - \theta^0|^2\right). \quad (7.48)$$

Since $-V_s \le -k_3|e|^2$ we may combine terms to obtain

$$\dot{V}_a \le |e|\left[-ck_1 k_3|e| + (ck_2 + k_5 W) + \frac{\sigma}{2}\left(-|\tilde{\theta}|^2 + |\theta - \theta^0|^2\right)\right]. \quad (7.49)$$

It is now possible to define some $b_e, b_{\tilde{\theta}} > 0$ such that $\dot{V}_s \le 0$ when $|e| \ge b_e$ or $|\tilde{\theta}| \ge b_{\tilde{\theta}}$. In particular, let

$$\begin{aligned} b_e &= \frac{2(ck_2 + k_5 W) + \sigma|\theta - \theta^0|^2}{2ck_1 k_3} \\ b_{\tilde{\theta}} &= \sqrt{\frac{2(ck_2 + k_5 W)}{\sigma} + |\theta - \theta^0|^2}. \end{aligned}$$

We will now use Corollary 7.1 to complete the proof. Letting

$$V_r = ck_4 b_e^2 + \frac{\lambda_{\max}(\Gamma^{-1}) b_{\tilde{\theta}}^2}{2},$$

we find that $e \in B_e$ for all t with

$$B_e = \left\{ e \in R^q : |e| \le \sqrt{\frac{\max(V(0), V_r)}{ck_3}} \right\}. \quad (7.50)$$

Since the controller parameters may be chosen to make b_e and $b_{\tilde{\theta}}$ arbitrarily small, it is always possible to ensure that $B_e \subseteq B_r$ by proper choice of the initial conditions. ■

The above theorem places explicit bounds on the trajectory of e. As for the σ-modification, however, we will also want to know how the controller parameters affect the closed-loop system performance. Let $d = ck_2 + k_5 W + \sigma|\theta - \theta^0|^2/2$. Starting from (7.49) we find

$$\begin{aligned} \dot{V}_a &\le -ck_1 k_3|e|^2 + d|e| \\ &\le -\frac{ck_1 k_3|e|^2}{2} + \frac{d^2}{2ck_1 k_3}. \end{aligned}$$

Rearranging terms and integrating as done for the σ-modification, the RMS error is bounded by

$$\sqrt{\lim_{t\to\infty}\frac{1}{t}\int^{t}|e|^2 d\tau} \leq \frac{d}{ck_1k_3}. \tag{7.51}$$

Notice that the RMS error for the ϵ-modification is adjusted similarly to the case for the σ-modification and may be made arbitrarily small. For example, increasing k_1 improves the RMS error in both cases.

7.4 Inherent Robustness

We have shown that a direct adaptive controller may be defined to stabilize a wide class of nonlinear systems. In this section we will study the robustness of the resulting closed-loop system.

7.4.1 Gain Margins

Since c may be any positive constant in Assumption 7.2, the direct adaptive controller has infinite gain margin. That is, it is insensitive to an overall static feedback gain variation. This in itself may be considered an improvement over static feedback linearization as shown by the following example.

Example 7.5 Given the system

$$\dot{x} = x^2 + u, \tag{7.52}$$

it is possible to use feedback linearization to define a controller which drives $e = x$ to zero. The controller $u = \nu(x)$ designed by feedback linearization becomes

$$\nu(x) = -x^2 - \kappa e,$$

where a Lyapunov function for the nominal system was chosen as $V = \frac{1}{2}e^2$. If the system is truly defined by

$$\dot{x} = x^2 + \pi u, \tag{7.53}$$

where $\pi > 0$, then

$$\dot{V} = -\kappa x^2 + (1-\pi)x^3. \tag{7.54}$$

If $\pi < 1$ and $x > \kappa/(1-\pi)$, then $\dot{V} > 0$ for all t and $x \to \infty$. If $\pi > 1$, and $x < -|\kappa/(1-\pi)|$, then $x \to -\infty$. Thus it is possible for a controller designed by feedback linearization to have no gain margin since it is now possible to find $x(0)$ such that $x \to \infty$ for any $\pi \neq 1$.

7.4.2 Disturbance Rejection

Using a universal approximator as the controller provides enough flexibility to create a closed-loop system which is robust with respect to certain classes of unmodeled unbounded disturbances. This means that if we designed the controller for a system without disturbances, then that controller is robust with respect to disturbances without modification as long as the approximator is capable of also modeling a robust nonadaptive controller. Thus the direct adaptive controller using a universal approximator is inherently robust with respect to disturbances.

Consider the error dynamics defined by

$$\dot{e} = \alpha(t,x) + \beta(x)(\Delta(t,x) + u), \tag{7.55}$$

where $\Delta(t,x)$ is a possibly unbounded disturbance. Assume that there exists some positive definite scalar function $\psi(x)$ such that $|\Delta(t,x)| \le \rho\psi(x)$ with $\rho \ge 0$. If ρ is bounded and $\psi(x) \ge 0$ is well defined for all x, then it is possible to use nonlinear damping to define a static stabilizing controller assuming that there exists a stabilizing controller for the case when $\Delta \equiv 0$.

Theorem 7.4: *Let Assumption 7.1 and Assumption 7.2 hold with $\gamma_{e1}(|e|) \le V_s(e) \le \gamma_{e2}(|e|)$ where γ_{e1} and γ_{e2} are class-$\mathcal{K}_\infty$. If for a given linear in the parameter approximator $\mathcal{F}(z,\hat{\theta})$ there exists some θ such that $|\mathcal{F}(z,\theta) - \nu_a(z)| \le W$ for all $z \in S_z$, where $e \in B_r$ implies $z \in S_z$, and*

$$\nu_a = \nu_s - \eta\left(\frac{\partial V_s}{\partial e}\bar{\beta}(x)\right)^\top - \eta_\Delta\left(\frac{\partial V_s}{\partial e}\bar{\beta}(x)\right)^\top \psi^2(x), \tag{7.56}$$

with $\eta, \eta_\Delta > 0$, then the parameter update law (7.15) with adaptive controller $u = \mathcal{F}(z,\hat{\theta})$ guarantee that the solutions of (7.55) are bounded given $B_e \subseteq B_r$ where B_e is defined by (7.25).

Proof: The proof follows that for Theorem 7.2. When $\Delta \ne 0$, we find $e \in B_e$ where

$$B_e = \left\{e \in \mathsf{R}^q : |e| \le \gamma_{e1}^{-1}\left(\max(V(0), V_r)/c\right)\right\} \tag{7.57}$$

with

$$V_r = c\gamma_{e2} \circ \gamma_{e1}^{-1}\left(\frac{d}{ck_1}\right) + \frac{d\lambda_{\max}(\Gamma^{-1})}{\sigma},$$

and $d = ck_2 + \frac{W^2}{4\eta} + \frac{\rho^2}{4\eta_\Delta} + \sigma\frac{|\theta-\theta^0|^2}{2}$. ■

With an extremely flexible approximator (one that is able to represent a large number of stabilizing controllers), it is possible to obtain a "very robust" closed-loop system using a direct adaptive controller. If a universal approximator such as a fuzzy system or neural network is used to define $\mathcal{F}(z,\hat{\theta})$, then the same controller and update law may be used to compensate for wide classes of disturbances. Assume that there exists parameter vectors θ_1 and θ_2 such that $|\mathcal{F}(z,\theta_1) - \nu_a| \leq W$ when $\Delta \equiv \Delta_1$ and $|\mathcal{F}(z,\theta_2) - \nu_a| \leq W$ when $\Delta \equiv \Delta_2$. It is then possible to use the same controller to compensate for either disturbance without modifying the control structure or update routine assuming the controller parameters are properly chosen to handle either case.

Assume that for a given disturbance $\Delta(t,x)$ there exists some θ that includes a nonlinear damping term to help compensate for the disturbance. Since nonlinear damping helps stabilize the closed-loop system, the same ideal parameter vector θ may be used to analyze a system in which $\Delta = 0$. Thus a single ideal parameter vector θ may be used for multiple systems. Similarly, it is possible to use multiple values for θ to prove stability for a given plant. For example, some θ_1 may be chosen for a nominal system. Since adding a nonlinear damping term will not destabilize the system, it is then possible to choose θ_2 which includes nonlinear damping. Since there may be multiple stabilizing θ for a particular system, we will not be interested in determining if $\hat{\theta} \to \theta$. In fact, if we did somehow force $\hat{\theta}$ to some θ, then we may actually decrease the direct adaptive controller's ability to compensate for wide classes of disturbances and system uncertainties. Because of this, we will not be concerned with issues of persistency of excitation (which has been an important topic in traditional adaptive control techniques to guarantee proper parameter estimation) in our treatment of adaptive control.

7.5 Improving Performance

We have seen that using a flexible approximator to define the direct adaptive controller allows wide classes of systems to be stabilized even in the presence of possibly unbounded disturbances. In this section, we will see how the performance of the direct adaptive controller may be improved.

It was shown that bounds on the RMS value of e may be obtained when using either the σ-modification or the ϵ-modification. It was shown, for example, the RMS error obtained when using the σ-modification may be bounded by

$$\sqrt{\lim_{t\to\infty} \frac{1}{t}\int^{t} |e|^2 d\tau} \leq \sqrt{\frac{2d}{ck_1}},$$

where $d = ck_2 + \frac{W^2}{4\eta} + \sigma\frac{|\theta - \theta^0|^2}{2}$. Thus increasing η or k_1 may improve the RMS error. In addition to properly setting the controller parameters, there are additional ideas which may be used to improve closed-loop system performance. These will be studied next.

7.5.1 Proper Initialization

Notice that for both the σ-modification and the ϵ-modification, the error is bounded by

$$|e| \leq \gamma_{e1}^{-1}\left(\max(V_a(0), V_r)/c\right), \tag{7.58}$$

where $V_r \geq 0$ and $\gamma_{e1} \leq V_s(t,e)$ with γ_{e1} a class-$\mathcal{K}$ function. Thus the bound on $|e|$ is dependent upon $V_a(0)$. Recall that $V_a = cV_s + \frac{1}{2}\tilde{\theta}^\top \gamma^{-1}\tilde{\theta}$ and $V_s \leq \gamma_{e2}(|e|)$. Thus decreasing $|e(0)|$ will also tend to decrease $V_a(0)$. For the tracking problem, it is possible to define a new reference trajectory which will ensure that $V_s(0) = 0$ as shown in the next example.

Example 7.6 Consider the system defined by

$$\begin{aligned} \dot{x}_1 &= f_1(x_1) + x_2 \\ \dot{x}_2 &= f_2(x) + u, \end{aligned}$$

where we wish to drive $x_1 \to r$. Here we will not concern ourselves with finding a stabilizing controller since the point of the example is simply to show how to help improve the initial conditions. Consider the error system defined by

$$\begin{aligned} e_1 &= x_1 - r \\ e_2 &= x_2 - \dot{r} + \kappa e_1 + f_1(x_1). \end{aligned}$$

If $r(t)$ is defined by some external reference generator, then it may not be the case that $e(0) = 0$.

To help reduce $|e(0)|$, we will now consider the error system defined by

$$\begin{aligned} e_1 &= x_1 - q_1 \\ e_2 &= x_2 - q_2 + \kappa e_1 + f_1(x_1), \end{aligned}$$

where

$$\begin{aligned} \dot{q}_1 &= q_2 \\ \dot{q}_2 &= -k_2(q_2 - \dot{r}) - k_1(q_1 - r), \end{aligned} \tag{7.59}$$

with $s^2 + k_2 s + k_1$ a Hurwitz polynomial. Thus q_1 is simply a filtered version of r. Now we may choose $q_1(0) = x_1(0)$ and $q_2(0) = x_2(0) + f_1(x_1(0))$. This will ensure that $|e(0)| = 0$ for our new error system. △

7.5.2 Redefining the Approximator

In the previous chapter we saw that it is possible to make control laws "more robust" with respect to system uncertainty via nonlinear damping. Unfortunately this may lead to terms in the feedback algorithm that may be characterized by high gain and/or high spatial frequency which are often difficult to model accurately by fuzzy systems or neural networks with a reasonable number of adjustable parameters. If one increases the number of adjustable parameters to obtain a better representation of the ideal controller nonlinearities, then the initialization of the approximator becomes more restrictive since $\tilde{\theta}^\top(0)\Gamma^{-1}\tilde{\theta}(0)$ increases with p where $\theta \in \mathrm{R}^p$. Since the bound $|e| \leq \gamma_{e1}^{-1}(\max(V_a(0), V_r)/c)$ is dependent upon $|\tilde{\theta}(0)|$ and $|\theta - \theta^0|$, using a large number of adjustable parameters may increase the bound on e.

Rather than using a universal approximator to represent all the terms of an ideal control law, we may consider splitting the control law into separate parts. It is then possible to explicitly define strong nonlinearities in a static term within the control law and simply let the adjustable portion of the approximator match easily approximated terms. This is demonstrated in the following example.

Example 7.7 Assume that for a given system we wish the direct adaptive controller to match

$$\nu_a = \nu_s - \eta\left(\frac{\partial V_s}{\partial e}\bar{\beta}\right)^\top, \tag{7.60}$$

where

$$\nu_s = -f(x) - \eta_2\psi^2(x)\left(\frac{\partial V_s}{\partial e}\bar{\beta}\right)^\top.$$

If $\nu_a(z)$ is a smooth function, then we could define a multi-input fuzzy system to directly represent ν_a. If $\psi^2(x)$ is a function with high spatial frequency, however, it may be difficult for a fuzzy system to approximate ν_a with a small number of rules.

If $f(x)$ is a smooth function which may be approximated relatively easily by a fuzzy system, then it may be advantageous to use the fuzzy system to only model $f(x)$. Consider the approximator

$$\mathcal{F}(z,\hat{\theta}) = -\eta\left(\frac{\partial V_s}{\partial e}\bar{\beta}\right)^\top - \eta_2\psi^2(x)\left(\frac{\partial V_s}{\partial e}\bar{\beta}\right)^\top - \mathcal{F}_{fs}(x,\hat{\theta}), \tag{7.61}$$

where $\mathcal{F}_{fs}$ is a fuzzy system used to approximate the term $f(x)$. The representation error defined in Theorem 7.2 then becomes

$$\begin{aligned} w &= \mathcal{F}(z,\theta) - \nu_a(z) \\ &= -\mathcal{F}_{fs}(x,\theta) + f(x), \end{aligned}$$

where θ is some ideal parameter vector. Since the nonlinear damping terms are no longer represented by the fuzzy system, it is possible that the bound on the representation error w will decrease (it is easier for the fuzzy system to represent $f(x)$ rather than $f(x)$ plus some other high-frequency terms). $\triangle$

As seen in this section, the closed-loop system performance when using a direct adaptive controller may be improved by properly selecting the controller parameters, good initialization, and by choosing an approximator which suits the particular control application. In addition to these techniques, there may be other ways to improve the performance of the controller. If, for example, steady-state error is an important factor, then it is possible to include the integral error term $\int^t (x - r)d\tau$ when trying to drive $x \to r$.

7.6 Extension to Nonlinear Parameterization

The theorems presented thus far show that stable update laws may be defined for approximators that are linearly parameterized. In this section, we will see how the analysis may be extended to the case of nonlinear parameterization.

The most straight forward approach is to transform the problem into a linear in the parameter form through algebraic manipulation. This typically results in an overparameterization of the problem as shown in the following exmaple.

Example 7.8 Consider the approximator defined by

$$\mathcal{F}(x, \theta) = (x + \theta)^2,$$

where $\theta \in \mathrm{R}$. Multiplying terms we find

$$\mathcal{F}(x, \theta) = x^2 + 2\theta x + \theta^2.$$

It is now possible to define a new parameter vector $\bar{\theta} = [2\theta, \theta^2]$ so that

$$\mathcal{F}(x, \phi) = x^2 + \bar{\theta}_1 x + \bar{\theta}_2. \tag{7.62}$$

Thus by increasing the number of unknown parameters, it is possible to define an approximator that is linear in the new parameter set. $\triangle$

In the case where the transformation to a linear representation is not practical, we must directly consider the effects of the nonlinear parameterization. Recall from Chapter 5 that when using an approximator with a nonlinear parameterization, we have

$$\mathcal{F}(z,\hat{\theta}) - \mathcal{F}(z,\theta) = \frac{\partial \mathcal{F}(z,\hat{\theta})}{\partial \theta}\tilde{\theta} + \delta, \tag{7.63}$$

where $|\delta| \le L|\tilde{\theta}|^2$ when $z \in S_z$ with $L \ge 0$ a Lipschitz constant. In some cases, it may be known that $|\delta|$ is small when the approximator inputs are bounded with $z \in S_z$. If we have $|\delta| \le W_\delta$, when $z \in S_z$, then it is possible to use the σ-modification and ϵ-modification presented before where the bound on the representation error, W, is simply replaced with $W + W_\delta$.

Often, however, it is not possible to place explicit bounds on δ. In this case, we must include an additional stabilizing term to compensate for the effects of the nonlinear parameterization. If the σ-modification is to be used with the direct adaptive controller defined using a nonlinearly parameterized approximator, then we must ensure that there exists some θ that allows for the definition of a controller which is robust enough to compensate for the approximator nonlinearities. The following is an extension of Theorem 7.2 to the nonlinear parameterization case:

Theorem 7.5: *Let Assumption 7.1 and Assumption 7.2 hold with $\gamma_{e1}(|e|) \le V_s(e) \le \gamma_{e2}(|e|)$ where γ_{e1} and γ_{e2} are class-$\mathcal{K}_\infty$. If for a given possibly nonlinear approximator $\mathcal{F}(z,\hat{\theta})$ there exists some θ such that $|\mathcal{F}(z,\theta) - \nu_a(z)| \le W$ for all $z \in S_z$, where $e \in B_r$ implies that $z \in S_z$, and*

$$\nu_a = \nu_s - \eta\left(\frac{\partial V_s}{\partial e}\bar{\beta}(x)\right)^\top - \eta_d\left|\hat{\theta} - \theta^0\right|^4\left(\frac{\partial V_s}{\partial e}\bar{\beta}(x)\right)^\top, \tag{7.64}$$

with $\eta, \eta_d > 0$, then the parameter update law (7.15) with adaptive controller $u = \mathcal{F}(z,\hat{\theta})$ guarantee that the solutions of (7.14) are bounded given $B_e \subseteq B_r$ with B_e defined by (7.68).

Proof: Following the steps in Theorem 7.2 up to (7.19) we find

$$\begin{aligned} \dot{V}_a \le\; & -ck_1 V_s + ck_2 + \tilde{\theta}^\top \Gamma^{-1}\dot{\tilde{\theta}} \\ & + \frac{\partial V_s}{\partial e}\bar{\beta}(x)\left(\mathcal{F}(z,\hat{\theta}) - \mathcal{F}(z,\theta) - \eta\left(\frac{\partial V_s}{\partial e}\bar{\beta}\right)^\top\right. \\ & \left. -\eta_d\left|\hat{\theta} - \theta^0\right|^4\left(\frac{\partial V_s}{\partial e}\bar{\beta}(x)\right)^\top + w\right), \end{aligned} \tag{7.65}$$

where $w = \mathcal{F}(z,\theta) - \nu_a$. Notice that

$$\mathcal{F}(z,\hat{\theta}) - \mathcal{F}(z,\theta) = \frac{\partial \mathcal{F}}{\partial \theta}\tilde{\theta} + \delta, \tag{7.66}$$

where $|\delta| \le L|\tilde{\theta} + \theta^0 - \theta^0|^2 \le 2L|\hat{\theta} - \theta^0|^2 + 2L|\theta - \theta^0|^2$. The last inequality was found using $|x+y|^2 \le |x+y|^2 + |x-y|^2 = 2|x|^2 + 2|y|^2$.

Notice that

$$\left(W + 2L|\theta - \theta^0|^2\right)\left|\frac{\partial V_s}{\partial e}\bar{\beta}(x)\right| - \eta\left|\frac{\partial V_s}{\partial e}\bar{\beta}(x)\right|^2 \le \frac{\left(W + 2L|\theta - \theta^0|^2\right)^2}{4\eta},$$

and

$$2L|\hat{\theta} - \theta^0|^2\left|\frac{\partial V_s}{\partial e}\bar{\beta}(x)\right| - \eta_d|\hat{\theta} - \theta^0|^4\left|\frac{\partial V_s}{\partial e}\bar{\beta}(x)\right|^2 \le \frac{L^2}{\eta_d}.$$

Using these inequalities, we find

$$\dot{V}_a \le -ck_1V_s + ck_2 + \frac{\left(W + 2L|\theta - \theta^0|^2\right)^2}{4\eta} + \frac{L^2}{\eta_d} - \sigma\tilde{\theta}^\top\left(\hat{\theta} - \theta^0\right)$$

when $z \in S_z$. Since $-2\tilde{\theta}^\top(\hat{\theta} - \theta^0) \le -|\tilde{\theta}|^2 + |\theta - \theta^0|^2$, the derivative of the Lyapunov candidate is

$$\dot{V}_a \le -ck_1\gamma_{e1}(|e|) - \frac{\sigma|\tilde{\theta}|^2}{2} + d, \tag{7.67}$$

where

$$d = ck_2 + \frac{\left(W + 2L|\theta - \theta^0|^2\right)^2}{4\eta} + \frac{L^2}{\eta_d} + \frac{\sigma|\theta - \theta^0|^2}{2}.$$

Following the reset of the steps in the proof of Theorem 7.2, we obtain

$$B_e = \left\{e \in \mathbf{R}^q : |e| \le \gamma_{e1}^{-1}\left(\max(V(0), V_r)/c\right)\right\}, \tag{7.68}$$

where

$$V_r = c\gamma_{e2} \circ \gamma_{e1}^{-1}\left(\frac{d}{ck_1}\right) + \frac{d\lambda_{\max}(\Gamma^{-1})}{\sigma}.$$

Since we may choose $k_1, k_2, \eta, \eta_d, \sigma$, and Γ^{-1}, we may make B_e arbitrarily small so it is always possible to choose $B_e \subseteq B_r$. ■

Even though the bounds when using a nonlinear parameterization are more strongly influenced by the magnitude of $|\theta - \theta^0|$, the reduction in the required number of adjustable parameters may provide a greater advantage. Even though the Lipschitz constant, L, is not explicitly used in the definition of the control law, an upper bound on L is required to ensure that $B_e \subseteq B_r$.

7.7 Summary

In this chapter we learned how to define stable direct adaptive controllers for a variety of nonlinear plants. It was shown that if a static stabilizing controller exists, then it may be possible to define a static adaptive controller using either the σ-modification or ϵ-modification to update the controller's adjustable parameters. The direct adaptive controller is defined by the following steps:

1. Place the system in a canonical representation so that an error system may be defined.

2. Define an error system and Lyapunov candidate V_s for the static problem.

3. Define a static control law $u = \nu_s$ that ensures that $\dot{V}_s \leq -k_1 V_s + k_2$ for the σ-modification approach, or $\dot{V}_s \leq -k_1 V_s + k_2|e|$ when using the ϵ-modification.

4. Choose an approximator $\mathcal{F}(z, \hat{\theta})$ such that there exists some θ where $|\mathcal{F}(z, \theta) - \nu_a(z)| \leq W$ for all $z \in S_z$ where ν_a is defined by (7.17) for the σ-modification and $\nu_a = \nu_s$ for the ϵ-modification. Estimate upper bounds for W and $|\theta - \theta^0|$ where θ^0 may be viewed as a "best guess" of θ.

5. Find some B_r such that $e \in B_r$ implies $z \in S_z$.

6. Choose the initial conditions, control parameters, and update law parameters such that $B_e \subseteq B_r$ with B_e the bound on the size of the error trajectory.

As long as one choose the initial conditions, controller parameters, and update law such that $B_e \subseteq B_r$, then the error trajectory will remain bounded, which implies that the states will also be bounded. It was additionally shown that by proper choice of the control parameters, it is possible to guarantee that the errors will converge to an arbitrarily small value (unlike B_e, this value is independent of the initial conditions).

It was then shown that the direct adaptive controller is robust with respect to static gain uncertainty and to various additive system uncertainties. Also, by choosing a fuzzy system or neural network in the definition of the control law that is able to represent wide classes of functions, the resulting closed-loop system is made robust with respect to wide classes of additive uncertainty. This is a great advantage when implementing a control law for a system in which the exact functional form of all possible uncertainties is unknown.

When a nonlinear approximator is used in the definition of the control law, we also found that the direct adaptive controller may still be used so long as an additional nonlinear damping term is added to ensure that the approximator nonlinearity does not destabilize the closed-loop system. Since nonlinear in the parameter approximators may represent a wider class of functions than a linear in the parameter approximator with the same number of adjustable parameters, this extension may prove beneficial in some applications.

7.8 Exercises and Design Problems

Exercise 7.1 (Time-Varying Input Gain) Consider the error system dynamics

$$\dot{e} = \alpha(t, x) + \begin{bmatrix} 0 \\ \vdots \\ 0 \\ \beta(t) \end{bmatrix} u, \tag{7.69}$$

where $\bar{x} = [x_1, \ldots, x_{n-1}]^\top$. The gain is bounded such that $0 < \beta_1 \leq \beta(t) \leq \beta_2$ and it is known that $|\dot{\beta}| \leq B$. Use the Lyapunov candidate

$$V_a = \frac{V_s}{\beta(t)} + \tilde{\theta}^\top \Gamma^{-1} \tilde{\theta}$$

to define a stable adaptive controller, where V_s is a Lyapunov function for the static control law $u = \nu_s$.

Exercise 7.2 (ϵ-Modification Revisited I) Other values may be picked for $\varepsilon(e)$ in the ϵ-modification. Show that the update law (7.46) using

$$\varepsilon(e) = \frac{\sigma |e|}{|e| + \eta} \tag{7.70}$$

with $\eta > 0$ results in a stable closed-loop system. Notice that this modification is similar to the σ-modification when $|e|$ is large.

Exercise 7.3 (ϵ-Modification Revisited II) Show that the update law (7.46) using

$$\varepsilon(e) = \sigma D_c \left(\frac{|e|}{\eta} \right), \tag{7.71}$$

where $\sigma, \eta > 0$ and

$$D_c(x) = \begin{cases} 1 & \text{if } x \geq 1 \\ x & \text{if } |x| < 1 \\ -1 & \text{if } x \leq -1 \end{cases}$$

results in a stable closed-loop system.

Exercise 7.4 (Control of a Linear System) Consider the single-input plant defined by

$$\begin{aligned} \dot{x}_1 &= x_2 \\ \dot{x}_2 &= k_1 x_1 + k_2 x_2 + u, \end{aligned}$$

where k_1, k_2 are unknown constants and we wish to drive $y = x_1 \to r$ with r a constant. Consider the error system defined by $e = [x_1 - r, x_2]^\top$ so that the error dynamics become

$$\begin{aligned} \dot{e}_1 &= e_2 \\ \dot{e}_2 &= k_1 x_1 + k_2 x_2 + u. \end{aligned}$$

Develop a direct adaptive controller using the σ-modification and show the closed-loop system response for various values of k_1 and k_2.

Exercise 7.5 (Time-Varying Disturbance) Consider the plant defined by

$$\dot{x} = \rho x \sin(\omega t + \phi) + u, \tag{7.72}$$

where $\rho, \phi \in \mathrm{R}$ are unknown, but $\omega > 0$ is known. Show that (7.72) may be transformed into

$$\dot{x} = \theta_1 x \sin(\omega t) + \theta_2 x \cos(\omega t) + u$$

using trigonometric identities. Define a direct adaptive controller to drive $x \to 0$.

Exercise 7.6 (Point Mass) Consider the point mass whose dynamics are described by

$$m\ddot{x} = -kx - b\dot{x} + u, \tag{7.73}$$

where k, b and m are unknown constants. Define a direct adaptive controller that drives $x \to r(t)$.

Exercise 7.7 (Sliding-Mode Controller) Consider the system

$$\begin{aligned} \dot{x}_1 &= x_2 \\ \dot{x}_2 &= \theta_1 x_2^2 + \Delta(t) + \theta_2 u, \end{aligned}$$

where θ_1, θ_2 are unknown constants and $|\Delta(t)| \leq \rho$, with $\rho > 0$ a known constant. Design a direct adaptive controller based on a stable manifold to drive $x \to r(t)$.

Exercise 7.8 (Backstepping) Consider the system

$$\begin{aligned} \dot{x}_1 &= \theta_1 \sin(x_1) + x_2 \\ \dot{x}_2 &= \theta_2 x_2^2 + \theta_3 u, \end{aligned}$$

where $\theta_1, \theta_2, \theta_3$ are unknown constants. Design a direct adaptive controller based on the backstepping method to drive $x \to r(t)$.

Exercise 7.9 (Adaptive Fuzzy Control) Consider the system

$$\dot{x} = f(x) + u,$$

where $f(x)$ is to be approximated by the adjustable fuzzy system $\mathcal{F}_{fs}(x, \hat{\theta})$ on $x \in [-1, 1]$. Design a direct adaptive controller so that $x \to r(t)$ when $f(x)$ may be defined by any of the following functions:

- $f(x) = 1$
- $f(x) = x + x^2$
- $f(x) = \sin(x)$.

Exercise 7.10 (Adaptive Neural Control) Repeat Example 7.9 using a multilayer perceptron.

Exercise 7.11 (Surge Tank I) A model for a surge tank is given by

$$A(x)\dot{x} = -c\sqrt{2gx} + u, \tag{7.74}$$

where x is the liquid level, and u is the input flow (assume it can be both positive and negative). Here $A(x)$ is the cross-sectional area of the tank, $g = 9.81 m/s^2$, and c is the unknown cross-sectional area of the output pipe. Design a direct adaptive controller given that $A(x) = ax^2$, where $a > 0$ is an unknown constant.

Exercise 7.12 (Surge Tank II) Consider the surge tank described in Exercise 7.11. Show that the controller $u = \nu_s$ with

$$\nu_s = c\sqrt{2gx} + \frac{1}{A(x)}\left[\dot{r} - \kappa(x - r(t))\right], \tag{7.75}$$

where $\kappa > 0$ stabilizes the system such that $\dot{V}_s \leq -2\kappa V_s$ when $V_s = \frac{1}{2}e^2$ with $e = x - r(t)$. Show how the direct adaptive controller

$$\nu_a = c\sqrt{2gx} + \left[\dot{r} - \kappa(x - r(t))\right]\mathcal{F}(x, \hat{\theta}) - \eta e \tag{7.76}$$

may be used to stabilize the system.

Exercise 7.13 (Three-Phase Motor) The dynamics for a permanent magnet 3-phase motor are described by

$$\begin{aligned}
\dot{\theta} &= \omega \\
J\dot{\omega} &= -B\omega + K\sin(N\theta)i_a + K\sin(N(\theta + 2\pi/3))i_b \\
&\quad + K\sin(N(\theta - 2\pi/3))i_c - T_L \\
L\dot{i}_a &= -Ri_a + v_a \\
L\dot{i}_b &= -Ri_b + v_b \\
L\dot{i}_c &= -Ri_c + v_c,
\end{aligned}$$

where θ is the shaft angle, ω is the shaft angular rate, and i_a, i_b, i_c are the currents for the three phases. Also J is the moment of inertia, B is the coefficient of viscous friction, K is the motor constant, N is an integer specifying the number of poles, L is the inductance, and R is the resistance per phase. Design a direct adaptive controller for v_a, v_b, v_c so that θ tracks a reference $r(t)$ when the load torque T_L is an unknown constant. Repeat the design when $T_L(\theta)$ is a function of the angular position. What conditions on $T_L(\theta)$ are needed in your design? Hint: $\sin^2(Nx)+\sin^2(N(x+2\pi/3))+\sin^2(N(x-2\pi/3)) = 1.5$.

Exercise 7.14 (Motor Fault) Repeat Exercise 7.13 when one looses the ability to develop torque with phase c so that only phases a and b may be used in the control of the motor.

Exercise 7.15 (Electromagnet Control) A model of a magnetically actuated point mass is defined by

$$m\ddot{x} = f(x) + \frac{k_1 u_1^2}{(G - x_1)^2} - \frac{k_2 u_2^2}{(G + x_1)^2}, \tag{7.77}$$

where x is the position of the mass. The system is actuated by two electromagnets with current inputs u_1 and u_2. Here $G > 0$ is the size of the nominal gap between the point mass and an electromagnet (i.e., when $x = 0$), and k_1, k_2 are electromagnet constants. The actuator u_1 is designed to move the point mass along the $+x$ direction, while u_2 moves the mass along the $-x$ direction. Define a static controller for the case when $f(x)$ is known. Then define a direct adaptive controller for the case when $f(x)$ is approximated by $\mathcal{F}(x, \theta)$.

Chapter 8

Indirect Adaptive Control

8.1 Overview

In the previous chapter we explained how to develop stable direct adaptive controllers of the form $u = \mathcal{F}(z, \hat{\theta})$, where $\mathcal{F}$ is an approximator and $\hat{\theta} \in \mathsf{R}^p$ is a vector of adjustable parameters. The approximator may be defined using knowledge of the system dynamics or using a generic universal approximator. We found that as long as there exists a parameter set for the approximator such that an appropriate static stabilizing controller may be represented, then the parameters of the approximator may be adjusted on-line to achieve stability using either the σ-modification or the ϵ-modification.

In this chapter we will explain how to design indirect adaptive controllers. Unlike the direct adaptive control approach, we will design an indirect adaptive controller by first identifying individual types of uncertainty within the system. A separate adaptive approximator will then be used to compensate for each of the uncertainties. The indirect adaptive control law is then formed by combining the results of each of the approximations.

We will begin our treatment of indirect adaptive control by studying the control of systems which contain uncertainties that are in the span of the input. In this situation, the uncertainties are said to satisfy matching conditions. Both additive and multiplicative uncertainties will be considered so that the error dynamics become

$$\dot{e} = \alpha(t, x) + \beta(x) \left(\Delta(t, x) + \Pi(x) u \right), \tag{8.1}$$

where $\Delta(t, x) \in \mathsf{R}^m$ is a vector of possibly time-varying additive uncertainties, and $\Pi \in \mathsf{R}^{m \times m}$ is a nonsingular matrix of static (time-invariant) multiplicative uncertainties. It will be assumed that the error system is defined to satisfy Assumption 6.1 so that boundedness of e implies boundedness of x. Assuming that a controller may be defined for the case when $\Delta = 0$ and $\Pi = I$, an indirect adaptive scheme will be developed for the case when $\Delta \neq 0$ and $\Pi \neq I$ are unknown.

We will later study the case where the disturbances do not necessarily satisfy matching conditions. A simple example of a system in which uncertainties do not satisfy matching conditions may be defined as

$$\begin{aligned} \dot{x}_1 &= \Delta(x) + x_2 \\ \dot{x}_2 &= u. \end{aligned}$$

Here Δ is an uncertainty that is not in the span of the input. We will later study how to design indirect adaptive controllers for strict-feedback systems that contain possibly time-varying uncertainties which do not satisfy matching conditions.

The purpose of this chapter is not to provide explicit control algorithms suitable for each control application. Rather, it is our intent to provide a set of tools that may be used to design stable controllers for a wide class of nonlinear systems. After reading this chapter, you should be able to design indirect adaptive controllers that are able to compensate for a variety of static and time-varying uncertainties.

8.2 Uncertainties Satisfying Matching Conditions

In this section we will study the adaptive stabilization of uncertain systems in which the uncertainties satisfy a matching condition. For each uncertainty, we will use a separate approximator. Thus unlike the direct adaptive controller which uses a single (possibly large) adjustable approximator, the indirect adaptive controller may use many smaller approximators to compensate for system uncertainties.

8.2.1 Static Uncertainties

Consider the error dynamics

$$\dot{e} = \alpha(t,x) + \beta(x)\left[\Delta(x) + \Pi(x)u\right], \tag{8.2}$$

where $\Delta(x)$ is an additive uncertainty and $\Pi(x)$ is a non-singular multiplicative uncertainty (notice that both uncertainties are time-invariant). The uncertainties satisfy matching conditions since they are in the span of the input u. If $u = \nu_s(t,x)$ is a stabilizing controller for the nominal system ($\Delta \equiv 0$ and $\Pi \equiv I$) and the functions $\Delta(x)$ and $\Pi(x)$ are known, then the control law defined by

$$u = \Pi^{-1}(x)\left(-\Delta(x) + \nu_s(t,x)\right) \tag{8.3}$$

would be a stabilizing controller for (8.2) since it cancels the effects of Δ and Π to render the error dynamics $\dot{e} = \alpha + \beta\nu_s$, which is a stable system by the definition of ν_s. If $-\Delta$ is approximated by $\mathcal{F}_\Delta(z,\hat{\theta})$ and Π by $\mathcal{F}_\Pi(z,\hat{\theta})$, then the control law $u = \nu_a(z,\hat{\theta})$ may be used with

$$\nu_a = \mathcal{F}_\Pi^{-1}(z,\hat{\theta}) \left(\mathcal{F}_\Delta(z,\hat{\theta}) - \eta \left(\frac{\partial V_s}{\partial e} \beta(x) \right)^\top + \nu_s(t,x) \right). \tag{8.4}$$

Here the parameter vector $\hat{\theta}$ is allowed to vary over time. Also, we have used $\mathcal{F}(z,\hat{\theta})$ rather than $\mathcal{F}(x,\hat{\theta})$ since z may only contain a few components of x. Alternatively, z may contain additional signals that are functions of x. Thus we use z as the input to the approximators to help stress that the approximator's inputs may not necessarily be identical to x. The suggested control law $\nu_a(z,\hat{\theta})$ was developed indirectly by first approximating the uncertainties (notice the similarity between (8.3) and (8.4)). Assuming that the approximations are accurate, the controller was developed in an attempt to cancel the effects of the uncertainty so that the performance of the nominally designed closed-loop system is preserved. This is typically referred to as a certainty equivalence approach.

We have included the nonlinear damping term $\eta \left(\frac{\partial V_s}{\partial e}\beta\right)^\top$ to increase closed-loop system robustness. The nonlinear damping term is defined using the definition of the error system and Lyapunov candidate which must satisfy the following assumption:

Assumption 8.1: There exists an error system $e = \chi(t,x)$ satisfying Assumption 6.1 and known static control law $u = \nu_s(z)$ with z measurable, such that for a given radially unbounded, decrescent Lyapunov function $V_s(t,e)$, we find $\dot{V}_s \leq -k_1 V_s + k_2$ along the solutions of (8.2) when $u = \nu_s(z)$, $\Delta \equiv 0$, and $\Pi \equiv I$.

Since the approximator $\mathcal{F}_\Pi$ is a matrix, the Jacobian with respect to its adjustable parameters may not be defined using the familiar notation. Notice, however, for a linearly parameterized approximator

$$\begin{aligned} \mathcal{F}_\Pi(z,\theta)\nu_a &= \begin{bmatrix} \frac{\partial \mathcal{F}_{11}}{\partial \theta}\theta & \cdots & \frac{\partial \mathcal{F}_{1m}}{\partial \theta}\theta \\ \vdots & & \vdots \\ \frac{\partial \mathcal{F}_{m1}}{\partial \theta}\theta & \cdots & \frac{\partial \mathcal{F}_{mm}}{\partial \theta}\theta \end{bmatrix} \nu_a \\ &= \begin{bmatrix} \left(\frac{\partial \mathcal{F}_{11}}{\partial \theta}\nu_{a_1} + \cdots + \frac{\partial \mathcal{F}_{1m}}{\partial \theta}\nu_{a_m} \right)\theta \\ \vdots \\ \left(\frac{\partial \mathcal{F}_{m1}}{\partial \theta}\nu_{a_1} + \cdots + \frac{\partial \mathcal{F}_{mm}}{\partial \theta}\nu_{a_m} \right)\theta \end{bmatrix}. \end{aligned}$$

Thus letting

$$A_{\mathcal{F}}(z) = \begin{bmatrix} \left(\frac{\partial \mathcal{F}_{11}}{\partial \theta}\nu_{a_1} + \cdots + \frac{\partial \mathcal{F}_{1m}}{\partial \theta}\nu_{a_m}\right) \\ \vdots \\ \left(\frac{\partial \mathcal{F}_{m1}}{\partial \theta}\nu_{a_1} + \cdots + \frac{\partial \mathcal{F}_{mm}}{\partial \theta}\nu_{a_m}\right) \end{bmatrix}, \tag{8.5}$$

where $A_{\mathcal{F}} \in \mathrm{R}^{m \times p}$, we find $\mathcal{F}_\Pi(z,\theta)\nu_a = A_{\mathcal{F}}(z)\theta$ and $\mathcal{F}_\Pi(z,\hat{\theta})\nu_a = A_{\mathcal{F}}(z)\hat{\theta}$.

The update law for the indirect controller (8.4) is defined by

$$\dot{\hat{\theta}} = -\Gamma\left[\left(\frac{\partial V_s}{\partial e}\beta(x)\left(\frac{\partial \mathcal{F}_\Delta}{\partial \theta} - A_{\mathcal{F}}(z)\right)\right)^\top + \sigma(\hat{\theta} - \theta^0)\right], \tag{8.6}$$

where Γ is a positive definite symmetric matrix, $\sigma > 0$, and θ^0 are design parameters. This choice for the update law will become apparent in the proof of the following theorem:

Theorem 8.1: *Let Assumption 8.1 hold with $\gamma_{e1}(|e|) \le V_s(e) \le \gamma_{e2}(|e|)$ where γ_{e1} and γ_{e2} are class-$\mathcal{K}_\infty$. If for given linear in the parameter approximators $\mathcal{F}_\Delta(z,\hat{\theta})$ and $\mathcal{F}_\Pi(z,\hat{\theta})$ there exists some θ such that $|\mathcal{F}_\Delta(z,\theta) + \Delta(x)| \le W_\Delta$ and $|\mathcal{F}_\Pi(z,\theta) - \Pi(x)| = 0$ for all $z \in S_z$ where $e \in B_r$ implies $z \in S_z$, then the parameter update law (8.6) with adaptive controller (8.4) guarantee that the solutions of (8.2) are bounded given $B_e \subseteq B_r$ with B_e defined by (8.14).*

Proof: Consider the Lyapunov candidate

$$V_a = V_s + \frac{1}{2}\tilde{\theta}^\top \Gamma^{-1}\tilde{\theta}, \tag{8.7}$$

which has the derivative

$$\dot{V}_a = \frac{\partial V_s}{\partial t} + \frac{\partial V_s}{\partial e}\left[\alpha(t,x) + \beta(x)\left(\Delta(x) + \Pi(x)\nu_a\right)\right] + \tilde{\theta}^\top \Gamma^{-1}\dot{\hat{\theta}}. \tag{8.8}$$

Since $\nu_a = \mathcal{F}_\Pi^{-1}(\mathcal{F}_\Delta - \eta\left(\frac{\partial V_s}{\partial e}\beta(x)\right)^\top + \nu_s(t,x))$, notice that the term in the above equation

$$\begin{aligned} \Delta + \Pi\nu_a &= \Delta + \left(\Pi - \mathcal{F}_\Pi(z,\hat{\theta}) + \mathcal{F}_\Pi(z,\hat{\theta})\right)\nu_a \\ &= \nu_s - \eta\left(\frac{\partial V_s}{\partial e}\beta\right)^\top + \left(\mathcal{F}_\Delta(z,\hat{\theta}) + \Delta\right) - \left(\mathcal{F}_\Pi(z,\hat{\theta}) - \Pi\right)\nu_a. \end{aligned}$$

Using the definition of θ, we find that the third term in this equation is

$$\begin{aligned} \mathcal{F}_\Delta(z,\hat{\theta}) + \Delta &= \mathcal{F}_\Delta(z,\hat{\theta}) - \mathcal{F}_\Delta(z,\theta) + \mathcal{F}_\Delta(z,\theta) + \Delta \\ &= \frac{\partial \mathcal{F}_\Delta(z,\hat{\theta})}{\partial \theta}\tilde{\theta} + w_\Delta, \end{aligned} \tag{8.9}$$

where $w_\Delta = \mathcal{F}_\Delta(z, \theta) + \Delta$ is a representation error with $|w_\Delta| \leq W_\Delta$ for all $z \in S_z$, and the fourth term is

$$\left(\mathcal{F}_\Pi(z, \hat{\theta}) - \Pi\right) \nu_a = \left(\mathcal{F}_\Pi(z, \hat{\theta}) - \mathcal{F}_\Pi(z, \theta) + \mathcal{F}_\Pi(z, \theta) - \Pi\right) \nu_a$$

$$= A_{\mathcal{F}}(z)\tilde{\theta}. \tag{8.10}$$

Using (8.9), (8.10), and Assumption 8.1, we find

$$\dot{V}_a \leq -k_1 V_s + k_2 + \tilde{\theta}^\top \Gamma^{-1} \dot{\hat{\theta}} \tag{8.11}$$
$$+ \frac{\partial V_s}{\partial e} \beta \left[-\eta \left(\frac{\partial V_s}{\partial e} \beta \right)^\top + \frac{\partial \mathcal{F}_\Delta(z, \hat{\theta})}{\partial \theta} \tilde{\theta} + w_\Delta - A_{\mathcal{F}} \tilde{\theta} \right].$$

Using the definition of the update law

$$\dot{V}_a \leq -k_1 V_s + k_2 - \eta \left(\frac{\partial V_s}{\partial e} \beta \right)^2 + \left| \frac{\partial V_s}{\partial e} \beta(x) W_\Delta \right| - \sigma \tilde{\theta}^\top \left(\hat{\theta} - \theta^0 \right). \tag{8.12}$$

Since $-2\tilde{\theta}\left(\hat{\theta} - \theta^0\right) \leq -|\tilde{\theta}|^2 + |\theta - \theta^0|^2$, we obtain

$$\dot{V}_a \leq -k_1 \gamma_{e1}(|e|) + k_2 + \frac{W_\Delta^2}{4\eta} + \frac{\sigma}{2} \left(-|\tilde{\theta}|^2 + |\theta - \theta^0|^2 \right).$$

Let $d = k_2 + \frac{W_\Delta^2}{4\eta} + \frac{\sigma|\theta - \theta^0|^2}{2}$. Then

$$\dot{V}_a \leq -k_1 \gamma_{e1}(|e|) - \frac{\sigma|\tilde{\theta}|^2}{2} + d, \tag{8.13}$$

so it is possible to pick some b_e and $b_{\tilde{\theta}}$ such that $\dot{V}_a \leq 0$ when $|e| \geq b_e$ or $|\tilde{\theta}| \geq b_{\tilde{\theta}}$. In particular, choose

$$b_e = \gamma_{e1}^{-1}\left(\frac{d}{k_1}\right) \qquad b_{\tilde{\theta}} = \sqrt{\frac{2d}{\sigma}}.$$

Using Corollary 7.1, we see that $e \in B_e$, where

$$B_e = \left\{ e \in \mathrm{R}^q : |e| \leq \gamma_{e1}^{-1}\left(\max(V_a(0), V_r)\right) \right\}, \tag{8.14}$$

with

$$V_r = \gamma_{e2} \circ \gamma_{e1}^{-1}\left(\frac{d}{k_1}\right) + \frac{d\lambda_{\max}(\Gamma^{-1})}{\sigma}.$$

Since it is possible to pick k_1, k_2, η, σ, and Γ, we may make V_r arbitrarily small. Thus with a proper choice of the initial conditions, we may always pick $B_e \subseteq B_r$. ■

The assumption that $\nu_a(z,\hat{\theta})$ is well defined for all t is required since it is possible that $\hat{\theta}$ be defined such that $\mathcal{F}_\Pi(z,\hat{\theta})$ is singular. To prevent this, it may be possible to use a projection algorithm which ensures that $\hat{\theta}$ is restricted to a region such that $\mathcal{F}_\Pi$ never becomes singular. The choice of the approximator structure will determine how the projection algorithm is defined. If, for example, fuzzy systems with adjustable output membership function centers are used for a single-input system, then the projection algorithm just needs to ensure that each membership function center is larger than some $c > 0$.

When using Theorem 8.1 to define an indirect adaptive controller, one typically does the following:

1. Place the plant in a canonical representation so that an error system may be defined.

2. Define an error system and Lyapunov candidate V_s for the static problem.

3. Define a static control law $u = \nu_s$ which ensures that $\dot{V}_s \leq -k_1 V_s + k_2$ when $\Delta = 0$ and $\Pi = I$.

4. Choose approximators $\mathcal{F}_\Pi(z,\hat{\theta})$ and $\mathcal{F}_\Delta(z,\hat{\theta})$ such that there exists some θ where $|\mathcal{F}_\Delta(z,\theta) + \Delta(x)| \leq W_\Delta$ and $|\mathcal{F}_\Pi(z,\theta) - \Pi(x)| = 0$ for all $z \in S_z$. Estimate upper bounds for W_Δ and $|\theta - \theta^0|$ where θ^0 may be viewed as a "best guess" of θ.

5. Find some B_r such that $e \in B_r$ implies $z \in S_z$.

6. Choose the initial conditions, control parameters, and update law parameters such that $B_e \subseteq B_r$ with B_e defined by (8.14).

Notice that the design of the indirect adaptive controller is very similar to the design of a direct adaptive controller. Unlike the direct adaptive controller, the design of the indirect adaptive controller does require that some stabilizing control law ν_s be known for the case when $\Delta = 0$ and $\Pi = I$. The approximators and update law are then used only to complement the nominal control law by accounting for the additional system uncertainty.

So far we have assumed that $|\mathcal{F}_\Pi(z,\theta) - \Pi(x)| = 0$. In some cases, this may be a very restrictive assumption since rarely can a fuzzy system or neural network perfectly represent a given function. It should be noted, however, that it is possible to consider the modified control law $\nu_r = \nu_a + \nu_m$ with

$$\nu_m = -\frac{\eta_\pi}{\pi_0}\left(\frac{\partial V_s}{\partial e}\beta(\xi)\right)^\top \nu_a^2, \tag{8.15}$$

where $\eta_\pi > 0$ and $\pi_0 \leq \lambda_{\min}(\pi)$. The above modification is then able to dominate an uncertainty of the form $w_\pi \nu_a \neq 0$ that arrises when

$$\begin{aligned}\left(\mathcal{F}_\Pi(z,\hat{\theta}) - \Pi\right)\nu_a &= \left(\mathcal{F}_\Pi(z,\hat{\theta}) - \mathcal{F}_\Pi(z,\theta) + \mathcal{F}_\Pi(z,\theta) - \Pi\right)\nu_a \\ &= A_{\mathcal{F}}(z)\tilde{\theta} + w_\pi \nu_a,\end{aligned}$$

where $w_\pi = \mathcal{F}_\Pi(z,\theta) - \Pi$ when $z \in S_z$.

Theorem 8.1 only guarantees boundedness of the error trajectory. It is possible to find an ultimate bound for the error since

$$\begin{aligned}k_1 V_s + \frac{\sigma}{2}|\tilde{\theta}|^2 &\geq k_1 V_s + \frac{\sigma}{2}\frac{1}{\lambda_{\max}(\Gamma^{-1})}\tilde{\theta}^\top \Gamma^{-1}\tilde{\theta} \\ &\geq k_m\left(V_s + \frac{1}{2}\tilde{\theta}^\top \Gamma^{-1}\tilde{\theta}\right) \\ &= k_m V_a\end{aligned} \tag{8.16}$$

with $k_m = \min\left(k_1, \frac{\sigma}{\lambda_{\max}(\Gamma^{-1})}\right)$. Then

$$\dot{V}_a \leq -k_m V_a + d, \tag{8.17}$$

where $d = k_2 + \frac{W_\Delta^2}{4\eta} + \frac{\sigma|\theta - \theta^0|^2}{2}$ so that $V_a(t) \leq \frac{d}{k_m} + \left(V_a(0) - \frac{d}{k_m}\right)e^{-k_m t}$. Since $\gamma_{e_1}(|e|) \leq V_a(t)$, we conclude that $|e|$ converges to

$$D_e = \left\{|e| : |e| \leq \gamma_{e_1}^{-1}\left(\frac{d}{k_m}\right)\right\}. \tag{8.18}$$

Since it is possible to make d/k_m arbitrarily small, the ultimate bound may be made arbitrarily small. Notice that this bound is independent of the initial conditions.

We can also find bounds on the RMS error. Using (8.13) notice that

$$\dot{V}_a \leq -k_1 \gamma_{e1}(|e|) + d. \tag{8.19}$$

Rearranging terms and integrating, we see that

$$\int^t \gamma_{e1}(|e|)d\tau \leq \int^t \left(-\frac{\dot{V}_a}{k_1} + \frac{d}{k_1}\right)d\tau. \tag{8.20}$$

Since V_a is bounded, we find

$$\lim_{t\to\infty}\frac{1}{t}\int^t \gamma_{e1}(|e|)d\tau \leq \frac{d}{k_1}. \tag{8.21}$$

Assume that $\gamma_{e1}(s) = \frac{1}{2}s^2$. Then

$$\lim_{t\to\infty}\frac{1}{t}\int^t \frac{1}{2}|e|^2 d\tau \leq \frac{d}{k_1},$$

so the RMS error is bounded by

$$\sqrt{\lim_{t\to\infty}\frac{1}{t}\int^t |e|^2 d\tau} \le \sqrt{\frac{2d}{k_1}}, \tag{8.22}$$

which is again independent of the system initial conditions and may be made arbitrarily small.

Example 8.1 In this example we will use the indirect adaptive approach to design a spacecraft attitude control system. The dynamics of the spacecraft are given by

$$\begin{aligned}
\dot{\phi} &= \omega_x + (\omega_y \sin\phi + \omega_z \cos\phi)\tan\theta \\
\dot{\theta} &= \omega_y \cos\phi - \omega_z \sin\phi \\
\dot{\psi} &= \frac{\omega_y \sin\phi + \omega_z \cos\phi}{\cos\theta} \\
J\begin{bmatrix} \dot{\omega}_x \\ \dot{\omega}_y \\ \dot{\omega}_z \end{bmatrix} &= \begin{bmatrix} 0 & \omega_z & -\omega_y \\ -\omega_z & 0 & \omega_x \\ \omega_y & -\omega_x & 0 \end{bmatrix} J \begin{bmatrix} \omega_x \\ \omega_y \\ \omega_z \end{bmatrix} + \begin{bmatrix} \Delta_x + u_x \\ \Delta_y + u_y \\ \Delta_z + u_z \end{bmatrix},
\end{aligned}$$

where ϕ is roll, θ is pitch, and ψ is yaw in radians. Here, J is the known inertia matrix defined by

$$J = \begin{bmatrix} 100 & -10 & 40 \\ -10 & 150 & 30 \\ 40 & 30 & 150 \end{bmatrix},$$

while u_x, u_y, and u_z are the torques applied by the jet nozzle actuators. The signals Δ_x, Δ_y, and Δ_z represent disturbance torques applied to the spacecraft possibly resulting from a nozzle failure. The goal here is to define an adaptive controller which will force $\phi \to r_\phi$, $\theta \to r_\theta$, and $\psi \to r_\psi$.

We will start by defining the error system. In particular, we will let

$$\begin{aligned}
e_{1,1} &= \phi - r_\phi \\
e_{2,1} &= \theta - r_\theta \\
e_{3,1} &= \psi - r_\psi.
\end{aligned} \tag{8.23}$$

Using the backstepping approach, we will ideally define a controller such that $\dot{e}_{i,1} = -\kappa e_{i,1}$ for $i = 1, 2, 3$ (or $\dot{e}_{i,1} + \kappa e_{i,1} \to 0$). Therefore define $e_{i,2} = \dot{e}_{i,1} + \kappa e_{i,1}$, so that

$$\begin{aligned}
e_{1,2} &= \omega_x + (\omega_y \sin\phi + \omega_z \cos\phi)\tan\theta - \dot{r}_\phi + \kappa e_{1,1} \\
e_{2,2} &= \omega_y \cos\phi - \omega_z \sin\phi - \dot{r}_\theta + \kappa e_{2,1} \\
e_{3,2} &= \frac{\omega_y \sin\phi + \omega_z \cos\phi}{\cos\theta} - \dot{r}_\psi + \kappa e_{3,1}.
\end{aligned}$$

Notice that with this definition, we find $\dot{e}_{i,1} = -\kappa e_{i,1} + e_{i,2}$ for $i = 1, 2, 3$. Also

$$\begin{bmatrix} \dot{e}_{1,2} \\ \dot{e}_{2,2} \\ \dot{e}_{3,2} \end{bmatrix} = \begin{bmatrix} (\omega_y c_\phi - \omega_z s_\phi)\tan\theta\dot{\phi} + \frac{\omega_y s_\phi + \omega_z c_\phi}{c_\theta{}^2}\dot{\theta} - \ddot{r}_\phi + \kappa\dot{e}_{1,1} \\ -(\omega_y s_\phi + \omega_z c_\phi)\dot{\phi} - \ddot{r}_\phi + \kappa\dot{e}_{2,1} \\ \frac{\omega_y c_\phi - \omega_z s_\phi}{c_\theta}\dot{\phi} + \frac{\omega_y s_\phi + \omega_z c_\phi}{c_\theta{}^2} s_\theta\dot{\theta} - \ddot{r}_\psi + \kappa\dot{e}_{3,1} \end{bmatrix}$$
$$+ \begin{bmatrix} 1 & s_\phi \tan\theta & c_\phi \tan\theta \\ 0 & c_\phi & -s_\phi \\ 0 & \frac{s_\phi}{c_\theta} & \frac{c_\phi}{c_\theta} \end{bmatrix} \begin{bmatrix} \dot{\omega}_x \\ \dot{\omega}_y \\ \dot{\omega}_z \end{bmatrix}, \tag{8.24}$$

where we have used the notation $c_\phi = \cos\phi$ and $s_\phi = \sin\phi$. Using the definition for the angular rates, we obtain

$$\begin{bmatrix} \dot{e}_{1,2} \\ \dot{e}_{2,2} \\ \dot{e}_{3,2} \end{bmatrix} = \alpha_2 + \begin{bmatrix} 1 & s_\phi \tan\theta & c_\phi \tan\theta \\ 0 & c_\phi & -s_\phi \\ 0 & \frac{s_\phi}{c_\theta} & \frac{c_\phi}{c_\theta} \end{bmatrix} J^{-1} \begin{bmatrix} \Delta_x + u_x \\ \Delta_y + u_y \\ \Delta_z + u_z \end{bmatrix},$$

where we have grouped the additive terms in $\alpha_2(t, \phi, \theta, \psi, \omega_x, \omega_y, \omega_z)$. Ignoring the uncertainties $\Delta_x, \Delta_y, \Delta_z$, it is possible to define the control law

$$\nu_s = J \begin{bmatrix} 1 & s_\phi \tan\theta & c_\phi \tan\theta \\ 0 & c_\phi & -s_\phi \\ 0 & \frac{s_\phi}{c_\theta} & \frac{c_\phi}{c_\theta} \end{bmatrix}^{-1} \left(-\alpha_2 - \begin{bmatrix} e_{1,1} + \kappa e_{1,2} \\ e_{2,1} + \kappa e_{2,2} \\ e_{3,1} + \kappa e_{3,2} \end{bmatrix} \right),$$

(provided that $c_\theta \neq 0$) so that $\dot{e}_{i,2} = -e_{i,1} - \kappa e_{i,2}$. This control law guarantees that $\dot{V}_s = -2\kappa V_s$, where $V_s = \frac{1}{2}\sum_{i=1}^{3} e_{i,1}^2 + \frac{1}{2}\sum_{i=1}^{3} e_{i,2}^2$. Thus

$$e = [e_{1,1}, e_{2,1}, e_{3,1}, e_{1,2}, e_{2,2}, e_{3,2}]^\top = 0$$

is exponentially stable, and Assumption 8.1 is satisfied with $k_1 = 2\kappa$ and $k_2 = 0$.

We will now define an indirect adaptive controller which uses radial basis neural networks to compensate for the uncertainties. Assume that it is known that one particular (unknown) failure mode causes a torque of

$$\Delta_x = 10 + \frac{400\phi}{1 + 2\dot{\phi}} \tag{8.25}$$

to be applied about the x-axis. To compensate for this type of failure, we will use a normalized radial-basis neural network defined by

$$\mathcal{F}_\Delta(\dot{\phi}, \hat{\theta}) - \frac{\sum_{i=1}^{p} \hat{\theta}_i \mu_i(\dot{\phi})}{\sum_{i=1}^{p} \mu_i(\dot{\phi})},$$

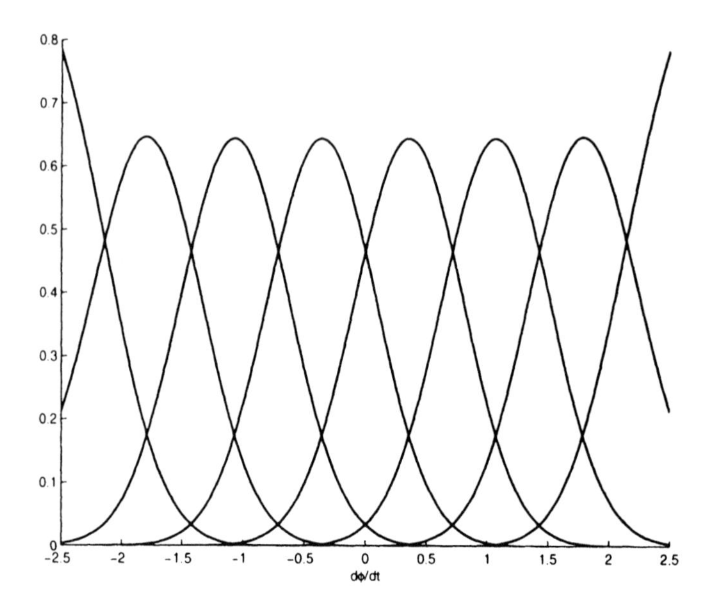

Figure 8.1. Basis functions used to define the neural network in Example 8.1.

where

$$\mu_i = e^{-\left(\frac{\dot\phi - c_i}{\sigma}\right)^2}.$$

Each c_i is used to define the center of the neural network basis function, while σ is chosen to describe the "width" of the basis function. Here we choose to use $p = 8$ basis functions that are evenly distributed between $[-2.5, 2.5]$ as shown in Figure 8.1.

To apply the indirect adaptive control approach, we must make sure that the error trajectory is confined to B_e where $B_e \subseteq B_r$ with $e \in B_r$ implies $z \in S_z$. To do this, one must estimate bounds for the representation error, W, and error in knowledge of the uncertainty, $|\theta - \theta^0|$. Using a least squares approach, it is possible to find a θ such that the representation error, $w = \Delta_x - \mathcal{F}_\Delta$, is bounded by $W_\Delta = 8.48$ and $|\theta| = 460.4$. Since we do not know what type of uncertainty will be applied to the spacecraft (here (8.25) is only one such possible disturbance), we will conservatively choose $b_W = 10^2$ and $b_\theta = 500^2$ to be the parameters used in the design of the control law with $W_\Delta^2 \leq b_W$ and $|\theta|^2 \leq b_\theta$. Thus we will design the controller for disturbances which are characterized by $W_\Delta \leq 10$ and $|\theta| \leq 500$.

Assume that we wish to keep the spacecraft attitude fixed even in the presence of a fault so that $r_\phi = r_\theta = r_\psi = 0$. We may now define B_r such that $e \in B_r$ implies $z \in S_z$. Since the input to the neural

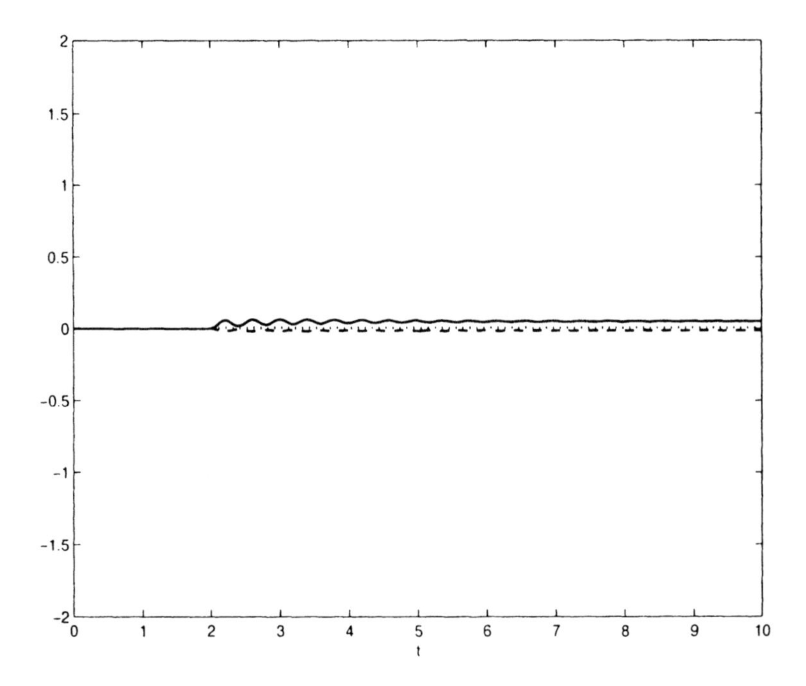

Figure 8.2. Attitude angles ϕ (–), θ ($\cdots$), and ψ (- -) using the indirect adaptive controller (shown in degrees).

network is given by $z = \dot{\phi}$, we must place bounds on $|e|$ to ensure that $\dot{\phi} \in [-2.5, 2.5]$ (i.e., $S_z = [-2.5, 2.5]$). Since $e_{1,2} = \dot{e}_{1,1} + \kappa e_{1,1}$ and $\dot{e}_{1,1} = \dot{\phi}$, we find

$$|\dot{\phi}| \leq |e_{1,2}| + \kappa |e_{1,1}| \leq (1 + \kappa)|e|.$$

Thus $e \in B_r$ with $B_r = \left\{ e \in \mathrm{R}^6 : |e| \leq 2.5/(1 + \kappa) \right\}$ implies that $|\dot{\phi}| \leq 2.5$ so $z \in S_z$.

We are now ready to choose the remaining controller parameters to ensure that $B_e \subseteq B_r$. From (8.14) (with $\gamma_{e_1} = \gamma_{e_2} = \frac{1}{2}|e|^2$) we know that

$$|e| \leq \sqrt{2 \max(V_a(0), V_r)}, \tag{8.26}$$

where

$$V_r = d \left[\frac{1}{k_1} + \frac{\lambda_{\max}(\Gamma^{-1})}{\sigma} \right], \tag{8.27}$$

and

$$d = \frac{W_\Delta^2}{4\eta} + \frac{\sigma |\theta|^2}{2},$$

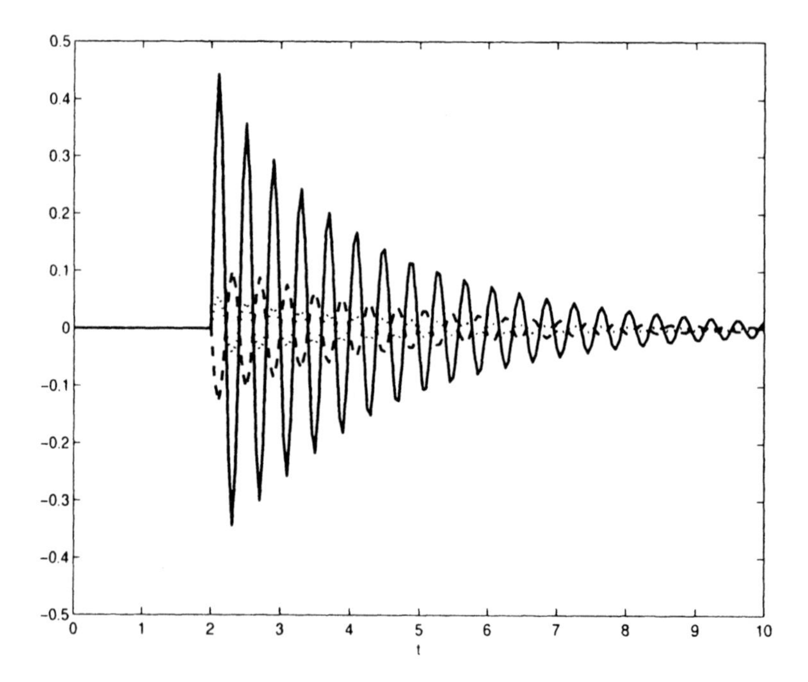

Figure 8.3. Spacecraft rates $\dot{\phi}$ (–), $\dot{\theta}$ (···), and $\dot{\psi}$ (- -) using the indirect adaptive controller (shown in *deg/sec*).

where we have chosen $\theta^0 = 0$ making no assumption about the form of the uncertainty. Choosing $\kappa = 2$, we obtain $k_1 = 4$. This also fixes the size of B_r to be $B_r = \{e \in \mathrm{R}^6 : |e| \leq 0.8334\}$. To ensure that the k_1 term does not dominate the calculation of V_r, we choose $\lambda_{\max}(\Gamma^{-1}) = \sigma/4$ so that $V_r = d/2$ (we will further choose $\Gamma = 4I/\sigma$ to be a diagonal matrix). Choosing $\eta = 2b_W > 2W^2$ and $\sigma = 1/(4b_\theta)$, we find $d \leq 1/8 + 1/8 = 1/4$ so that $V_r \leq 1/8$.

The bound on e (8.26) is also dependent upon $V_a(0) = V_s(0) + \frac{1}{2}\tilde{\theta}^\top(0)\Gamma^{-1}\tilde{\theta}(0)$. Assuming that $\hat{\theta}(0) = 0$ and $V_s(0) = 0$ (no significant pointing errors prior to the fault), we obtain

$$\begin{aligned} V_a(0) &\leq \frac{1}{2}|\theta|^2 \lambda_{\max}(\Gamma^{-1}) \\ &\leq \frac{1}{2} b_\theta \lambda_{\max}(\Gamma^{-1}) = 0.03125. \end{aligned} \tag{8.28}$$

Thus the error is bounded by $|e| \leq \sqrt{2\max(0.03125, 1/8)} = 1/2$. Thus $B_e \subseteq B_r$ since $B_e = \{e \in \mathrm{R}^6 : |e| \leq 1/2\}$.

The spacecraft angles and rates are shown in Figures 8.2 and 8.3, respectively. This example demonstrates the conservativeness sometimes obtained when designing adaptive controllers. Notice that using (8.26) we were able to predict that $|e| \leq 1/2$ so that $|\dot{\phi}| \leq (1+\kappa)|e| \leq 1.5 rad/sec$ ($86 deg/sec$). In Figure 8.3, we see that $|\dot{\phi}|$ never becomes larger than $0.5 deg/sec$ so that our bound is rather conservative. We will find that it is often possible to reduce, e.g., the rate of adaptation Γ and still maintain stable feedback since the stability analysis only provides sufficient and not necessary conditions on stability.

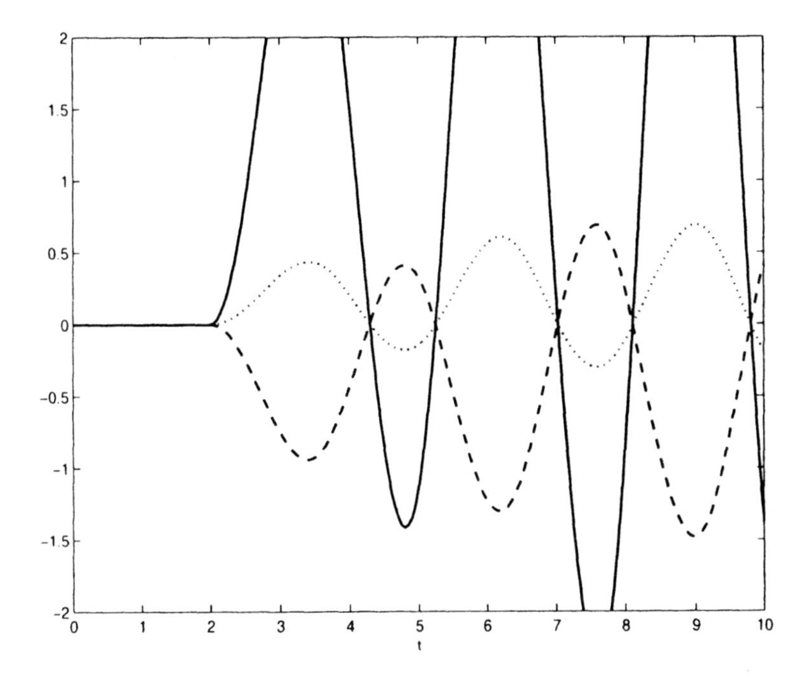

Figure 8.4. Attitude angles ϕ (-), θ ($\cdots$), and ψ (- -) using static feedback (shown in degrees).

To see the effects of the adaptive portion of the controller, another simulation was run where $\Gamma = 0$ so that the adaptation is turned off. In this case, the controller is no longer able to maintain a proper heading as shown in Figure 8.4. $\triangle$

8.2.2 Dynamic Uncertainties

In the previous section we used a feedback linearization approach to define the control law via certainty equivalence and studied the resulting stability. In this section, we will see that systems with additional classes of uncertainties may be stabilized using the same adaptive controller presented in the previous section if we do not restrict our analysis to that based on feedback linearization.

Consider the error dynamics

$$\dot{e} = \alpha(t,x) + \beta(t,x)\left[\Delta(t,x) + \Pi(x)u\right], \tag{8.29}$$

where $\Delta(t,x)$ is now allowed to be a time-varying uncertainty. Since $\Delta(t,x)$ is now dependent upon time which is an unbounded signal, it is not possible to define an approximator $\mathcal{F}_\Delta(t,z,\hat{\theta})$ that may be used to simply cancel the effects of Δ as done in the last section. This is because we cannot find some B_r so that $e \in B_r$ implies $z \in S_z$ with S_z bounded since $t \to \infty$. Instead, we will attempt to approximate a function, $\nu_d(x,\zeta)$, which is able to compensate for its effects. The signal ζ will be defined based upon the choice of the error system and Lyapunov candidate. When Δ is a dynamic uncertainty, we will require the following:

Assumption 8.2: There exists some $\nu_\Delta(x,\zeta)$ such that

$$\zeta^\top\left[\Delta(t,x) + \nu_\Delta(x,\zeta)\right] \le c$$

for all $\zeta \in \mathsf{R}^m$, where $c \ge 0$.

The term ζ will be chosen according to our choice of the Lyapunov candidate as will be shown shortly. Assumption 8.2 ensures us that there exists a controller which is able to compensate for the uncertainty Δ. The following examples show how certain classes of uncertainties may be handled.

Example 8.2 Assume $|\Delta(t,x)| \le \rho\psi(x)$ where ψ is a known non-negative function. Let $\nu_\Delta = -\kappa\psi^2\zeta$, with $\kappa > 0$. Then we find

$$\begin{aligned}\zeta^\top\left[\Delta + \nu_\Delta\right] &= \zeta^\top\left[\Delta - \kappa\psi^2\zeta\right] \\ &\le \frac{\rho^2}{4\kappa},\end{aligned} \tag{8.30}$$

which satisfies Assumption 8.2 with $c = \rho^2/(4\kappa)$. Here we have used the inequality $-x^\top x \pm 2x^\top y \le y^\top y$. $\triangle$

Example 8.3 Assume $|\Delta(t,x)| \le \rho\gamma(|q|)\psi(x)$, where γ is class-$\mathcal{K}$ and ψ is non-negative. Assume that η is a dynamic normalizing signal defined such that $\eta \ge |q|$ for all t. If we let

$$\nu_\Delta = -\kappa\gamma^2(\eta)\psi^2\zeta, \tag{8.31}$$

then $\zeta^\top\left[\Delta + \nu_\Delta\right] \le \rho^2/(4\kappa)$. $\triangle$

We will once again study the behavior of the closed-loop system when using the control law (8.4) and update law (8.6).

Theorem 8.2: *Let Assumption 8.1 and Assumption 8.2 hold with* $\gamma_{e1}(|e|) \leq V_s(e) \leq \gamma_{e2}(|e|)$, *where* γ_{e1} *and* γ_{e2} *are class-*$\mathcal{K}_\infty$ *and*

$$\zeta = \left(\frac{\partial V_s}{\partial e}\beta(x)\right)^\top .$$

If for given linear in the parameter approximators $\mathcal{F}_\Delta(z,\hat{\theta})$ *and* $\mathcal{F}_\Pi(z,\hat{\theta})$ *there exists some* θ *such that*

$$|\mathcal{F}_\Delta(z,\theta) - \nu_\Delta(x,\zeta)| \leq W_\Delta,$$

and $|\mathcal{F}_\Pi(z,\theta) - \Pi(x)| = 0$ *for all* $z \in S_z$ *where* $e \in B_r$ *implies* $z \in S_z$, *then the parameter update law (8.6) with adaptive controller (8.4) guarantee that the solutions of (8.2) are bounded given* $B_e \subseteq B_r$ *with* B_e *defined by (8.35).*

Proof: Notice that

$$\begin{aligned}\mathcal{F}_\Delta(z,\hat{\theta}) + \Delta &= \left(\mathcal{F}_\Delta(z,\hat{\theta}) - \mathcal{F}_\Delta(z,\theta)\right) + (\mathcal{F}_\Delta(z,\theta) - \nu_\Delta) + (\nu_\Delta + \Delta) \\ &= \frac{\partial \mathcal{F}_\Delta(z,\hat{\theta})}{\partial \theta}\tilde{\theta} + w_\Delta + (\nu_\Delta + \Delta), \end{aligned} \tag{8.32}$$

where $w_\Delta = \mathcal{F}_\Delta(z,\theta) - \nu_\Delta$ with $|w_\Delta| \leq W_\Delta$ for all $z \in S_z$. Following the steps up to (8.11) in the proof of Theorem 8.1, we find

$$\begin{aligned}\dot{V}_a \leq\ & -k_1 V_s + k_2 + \tilde{\theta}\Gamma^{-1}\dot{\hat{\theta}} \\ & + \frac{\partial V_s}{\partial e}\beta\left[-\eta\left(\frac{\partial V_s}{\partial e}\beta\right)^\top + \frac{\partial \mathcal{F}_\Delta(z,\hat{\theta})}{\partial\theta}\tilde{\theta} + w_\Delta + \nu_\Delta + \Delta - A_{\mathcal{F}}\tilde{\theta}\right].\end{aligned}$$

Using the definition of the update law and Assumption 8.2

$$\dot{V}_a \leq -k_1 V + k_2 - \eta\left(\frac{\partial V_s}{\partial e}\beta\right)^2 + \left|\frac{\partial V_s}{\partial e}\beta(x) W_\Delta\right| + c - \sigma\tilde{\theta}^\top\left(\hat{\theta} - \theta^0\right). \tag{8.33}$$

Since $-2\tilde{\theta}^\top\left(\hat{\theta} - \theta^0\right) \leq -|\tilde{\theta}|^2 + |\theta - \theta^0|^2$, we obtain

$$\dot{V}_a \leq -k_1\gamma_{e1}(|e|) + k_2 + \frac{W_\Delta^2}{4\eta} + \frac{\sigma}{2}\left(-|\tilde{\theta}|^2 + |\theta - \theta^0|^2\right).$$

Now let $d = k_2 + \frac{W_\Delta^2}{4\eta} + c + \frac{\sigma|\theta - \theta^0|^2}{2}$. Then

$$\dot{V}_a \leq -k_1\gamma_{e1}(|e|) - \frac{\sigma|\tilde{\theta}|^2}{2} + d, \tag{8.34}$$

so it is possible to pick some b_e and $b_{\tilde{\theta}}$ such that $\dot{V}_a \leq 0$ when $|e| \geq b_e$ or $|\tilde{\theta}| \geq b_{\tilde{\theta}}$. In particular, choose

$$b_e = \gamma_{e1}^{-1}\left(\frac{d}{k_1}\right) \qquad b_{\tilde{\theta}} = \sqrt{\frac{2d}{\sigma}}\ .$$

Using Corollary 7.1, we see that $e \in B_e$ with

$$B_e = \left\{e \in \mathrm{R}^q : |e| \leq \gamma_{e1}^{-1}\left(\max(V_a(0), V_r)\right)\right\}, \tag{8.35}$$

where

$$V_r = \gamma_{e2} \circ \gamma_{e1}^{-1}\left(\frac{d}{k_1}\right) + \frac{d\lambda_{\max}(\Gamma^{-1})}{\sigma}. \tag{8.36}$$

By proper choice of the controller parameters and initial conditions, it is possible to choose B_e such that $B_e \subseteq B_r$. ■

By not restricting our analysis to that based on feedback linearization, the indirect controller is not only able to cancel static uncertainties, but is also able to compensate for classes of dynamic uncertainties. Notice that when $\Delta(x)$ is not time-varying, it is possible to let $\nu_\Delta = -\Delta(x)$ in Assumption 8.2 to recover the results of Theorem 8.1.

Notice that the controller form and update law used to stablize the static and dynamic uncertainties are equivalent. Thus if an indirect adaptive controller is designed for a static uncertainty, it is also robust with respect to classes of dynamic uncertainties without modification as long as the approximator is also able to represent an appropriate stabilization term for the dynamic uncertainty.

This approach is also appealing since there are many choices for ν_Δ which may be used to dominate the effects of the dynamic uncertainty. Consider, for example, the case where the additive uncertainty may be bounded by $\Delta(t, x) \leq \rho\psi(x)$. Here it may be possible to dominate the uncertainty by a fuzzy system $\mathcal{F}_{fs}(z, \theta)$ where θ is chosen such that $\mathcal{F}_{fs}(z, \theta) \geq \psi^2(x)$. Then we may let

$$\nu_\Delta = -\eta\mathcal{F}_{fs}(z, \theta)\zeta,$$

which satisfies Assumption 8.2. If the adjustable approximator is then chosen as

$$\mathcal{F}_\Delta(z, \zeta, \hat{\theta}) = -\kappa\mathcal{F}_{fs}(z, \hat{\theta})\zeta, \tag{8.37}$$

then the representation error $w = \mathcal{F}_\Delta(z, \theta) - \nu_\Delta(x)$ may be set to zero so $W_\Delta = 0$. Thus when an approximator is used to represent a dominating term for stabilization, it is typically possible to choose some θ such that the representation error becomes zero. On the other hand, the parameter c in Assumption 8.2 is typically nonzero which increases the size of B_e (defined by (8.35)) in a similar fashion as W_Δ.

The following example demonstrates how to use the indirect adaptive controller to stabilize a system with a dynamic uncertainty.

Example 8.4 In this example, we are to design a controller which is able to accurately point an antenna driven by a permanent magnet DC motor. Here we are primarily interested in overcoming the effects of low-velocity friction. Consider the system defined by

$$\begin{aligned} \dot{q} &= \sigma\omega\left[1 - \frac{q}{T_f}\operatorname{sgn}(\omega)\right] \\ \dot{\theta} &= \omega \\ J\dot{\omega} &= T_c \sin(N\theta + \phi) - q + u, \end{aligned} \tag{8.38}$$

where q is torque caused by friction, θ is the angular position of the antenna, ω is the antenna angular velocity, and u is the commanded torque. Here friction is based on a dynamic friction model proposed by Dahl [35], where σ is used to describe the "stiffness" of the friction, and T_f is the magnitude of the friction torque. The motor cogging torque is described as sinusoidally varying with position, where $T_c = 10$ is the magnitude of the cogging torque, $N = 60$ is the number of motor pole faces, and ϕ is the phase. $J = 20$ is the antenna moment of inertia. We will assume that q is not available for feedback and that J, T_c, and ϕ are unknown. Since $\sin(x+y) = \sin(x)\cos(y) + \cos(x)\sin(y)$, it is possible to express the position and velocity states as

$$\begin{aligned} \dot{x}_1 &= x_2 \\ \dot{x}_2 &= \left(\frac{T_c\cos\phi}{J}\right)\sin(Nx_1) + \left(\frac{T_c\sin\phi}{J}\right)\cos(Nx_1) - \frac{q}{J} + \frac{u}{J}, \end{aligned}$$

where $x_1 = \theta$ and $x_2 = \omega$. We will now design a controller so that $x_1 \to r(t)$.

Define the first error variable as $e_1 = x_1 - r$. Ideally, we will be able to define the controller so that $\dot{e}_1 = -\kappa e_1$ with $\kappa > 0$ so that $e_1 \to 0$. With this in mind, define the second error signal as $e_2 = \dot{e}_1 + \kappa e_1$. This way if $e_2 \to 0$, then $\dot{e}_1 \to -\kappa e_1$. Notice that $\dot{e}_1 = -\kappa e_1 + e_2$. Also

$$\begin{aligned} \dot{e}_2 &= \dot{x}_2 - \ddot{r} + \kappa(e_2 - \kappa e_1) \\ &= -\ddot{r} + \kappa(e_2 - \kappa e_1) + \Delta(t,x) + \Pi u, \end{aligned} \tag{8.39}$$

where $\Delta(t,x)$ is a dynamic uncertainty defined by

$$\Delta(t,x) = \left(\frac{T_c\cos\phi}{J}\right)\sin(Nx_1) + \left(\frac{T_c\sin\phi}{J}\right)\cos(Nx_1) - \frac{q(t)}{J},$$

and $\Pi = 1/J$ is a multiplicative uncertainty. The error dynamics may thus be expressed as $\dot{e} = \alpha(t,x) + \beta(x)[\Delta + \Pi u]$ where

$$\alpha(t,x) = \begin{bmatrix} -\kappa e_1 + e_2 \\ -\ddot{r} + \kappa(e_2 - \kappa e_1) \end{bmatrix},$$

and $\beta = [0, 1]^\top$.

Let $V_s = \frac{1}{2}e_1^2 + \frac{1}{2}e_2^2$. Then choosing

$$\nu_s = \ddot{r} - \kappa(e_2 - \kappa e_1) - \kappa e_2 - e_1$$

renders $\dot{V}_s = -2\kappa V_s$ when $u = \nu_s$, $\Delta = 0$, and $\Pi = 1$, so that Assumption 8.1 is satisfied with $k_1 = 2\kappa$ and $k_2 = 0$.

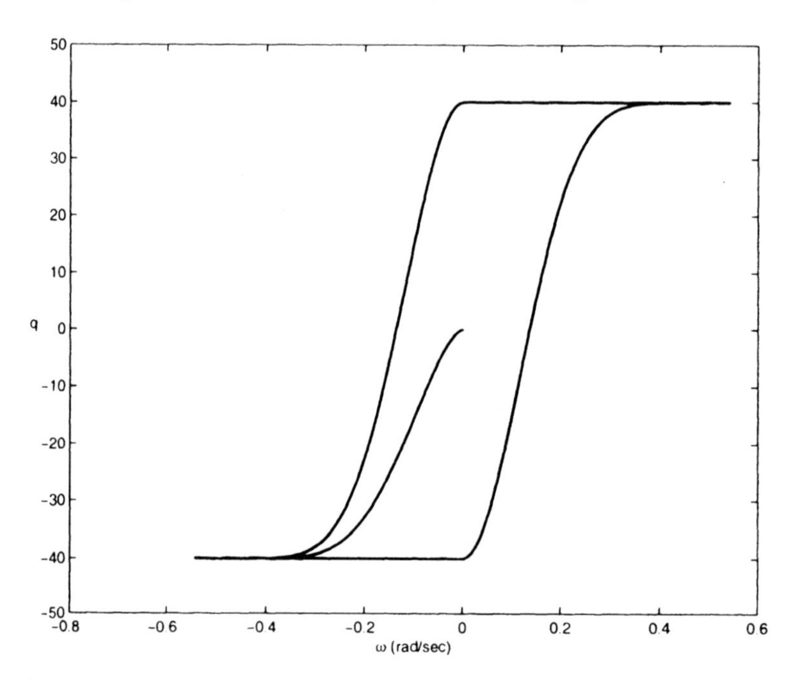

Figure 8.5. A phase plane portrait of q and w when w is a sinusoidal signal.

We will now study the behavior of q so that a dominating control signal may be formed. Letting $V_q = \frac{1}{2}q^2$ we find

$$\dot{V}_q = -\frac{\sigma q^2}{T_f}|\omega| + \sigma\omega q.$$

Thus $\dot{V}_q \leq 0$ when $|q| \geq T_f$, so $M = \{q \in \mathsf{R} : |q| \leq T_f\}$ is an invariant set. If $|q(t_1)| \leq T_f$, then $|q(t)| \leq T_f$ for all $t \geq t_1$. Letting $T_f = 40$, $\sigma = 5000$, and ω a sinusoidal rate with frequency $0.5Hz$, one obtains the hysteretic friction response shown in Figure 8.5. Since the Dahl friction model is discontinuous, we can not be guaranteed that a unique state trajectory exists. Though a unique solution may not be exist, we wee from Figure 8.5 that the friction torque does indeed remain bounded such that $|q| \leq T_f$.

We will now consider the ideal stabilizing signal

$$\nu_\Delta = \theta_1 \sin(Nx_1) + \theta_2 \cos(Nx_1) - \eta_\psi \mathcal{F}_{fs}(V_s, \bar{\theta})\zeta,$$

and

$$\mathcal{F}_\Pi = \theta_3$$

so that $\theta = [\theta_1, \theta_2, \theta_3, \bar{\theta}^\top]^\top$ is the composite parameter vector. From Theorem 8.2, we select $\zeta = \frac{\partial V_s}{\partial e}\beta(x)$. Here $\mathcal{F}_{fs}(V_s, \theta)$ is a fuzzy system with input V_s and 10 fixed output membership functions with centers defined by $\bar{\theta}$. We have shown that $|q| \leq \rho\psi(x)$ with $\rho = T_c$ and $\psi = 1$. Let

$$\theta_1 = -\frac{T_c \cos\phi}{J} \qquad \theta_2 = -\frac{T_c \sin\phi}{J} \qquad \theta_3 = \frac{1}{J}$$

and $\bar{\theta}_i = \frac{T_f^2}{J^2}$ for all i. Even though we are defining θ in terms of the physical parameters of the system, θ is unknown since the physical parameters are not known. Then $\mathcal{F}_{fs}(V_s, \bar{\theta}) = \frac{T_f^2}{J^2}$ so

$$\begin{aligned} \zeta^\top [\Delta + \nu_\Delta] &= \zeta^\top \left[-\frac{q}{J} - \eta_\psi \frac{T_f^2}{J^2}\zeta \right] \\ &\leq -\eta_\psi \frac{T_f^2}{J^2}|\zeta|^2 + \frac{T_f}{J}|\zeta| \leq \frac{1}{4\eta_\psi}. \end{aligned}$$

Assumption 8.2 is thus satisfied with $c = 1/(4\eta_\psi)$. Here ζ is a dummy variable just used to show that Assumption 8.2 is satisfied. Now select the adaptive compensator to be

$$\mathcal{F}_\Delta = \hat{\theta}_1 \sin(Nx_1) + \hat{\theta}_2 \cos(Nx_1) - \eta_\psi \mathcal{F}_{fs}(V_s, \hat{\bar{\theta}})\zeta. \tag{8.40}$$

Since for some θ, $\mathcal{F}_\Delta(z, \theta) = \nu_\Delta(z)$, we find $W_\Delta = 0$.

We now need to find controller parameters that ensure the input to the fuzzy system remains in a valid region. Since V_s is the fuzzy system input, we must find bounds on the state trajectory which ensure that V_s remains in some region. Assume that the fuzzy system is defined with the input range $[0, 7]$. We must then select the control parameters and initial conditions such that $V_s \leq 10$ for all t. We may use $V_s \leq \max(V_a(0), V_r)$ to do this, where

$$V_r = \frac{d}{k_1} + \frac{d\lambda_{\max}(\Gamma^{-1})}{\sigma}$$

is defined by (8.36) in the proof of Theorem 8.2 with $d = c + \sigma|\theta - \theta^0|^2/2$.

Consider the choice $\kappa = 10$ and

$$\Gamma = \frac{1}{\sigma} \begin{bmatrix} 0.1 & & & & & \\ & 0.1 & & & & \\ & & 0.1 & & & \\ & & & 100 & & \\ & & & & \ddots & \\ & & & & & 100 \end{bmatrix}.$$

Then

$$V_r = d\left(\frac{1}{20} + 10\right) \leq 11d$$

since $k_1 = 2\kappa = 20$ and $\lambda_{\max}(\Gamma^{-1}) = 10\sigma$. If we choose $\eta_\psi = 2$, then $c = 1/8$. Also choose $\theta_0 = [0, 0, 1/20, 0, \ldots, 0]^\top$ so that we are assuming $J \approx 20$ based on knowledge of the nominal system. Suppose it is known that $T_c/J \leq 1$, $J \geq 10$ and $T_f/J \leq 2$. Then we may place bounds on $|\theta - \theta_0|$ since

$$|\theta - \theta_0| \leq \begin{vmatrix} 1 \\ 1 \\ 1/10 \\ 4 \\ \vdots \\ 4 \end{vmatrix} + |\theta_0| = 12.8. \tag{8.41}$$

If we choose $\sigma = 1/10000$, then $d \leq 1/8 + 12.8^2/5000 = 0.1332$. Thus

$$V_r \leq 11d \leq 1.46.$$

Also from the proof of Theorem 8.2, we have $V_a(0) = V_s(0) + (\theta - \theta_0)^\top\Gamma^{-1}(\theta - \theta_0)$ assuming that $\hat{\theta}(0) = \theta_0$. If the contribution from $V_s(0)$ may be ignored, then $V_a(0) = 0.0022$. Thus the bound on the trajectory of $V_s(t)$ will be bounded by V_r.

Since we designed the input to the fuzzy system to cover the range $[0, 7]$, the input to the fuzzy system always remains in a valid region so that we may use the indirect update law. According to Theorem 8.2, the update law (8.6) with

$$\frac{\partial \mathcal{F}_\Delta}{\partial \theta} = \begin{bmatrix} \sin(Nx_1) \\ \cos(Nx_1) \\ 0 \\ -\eta_\psi \frac{\partial \mathcal{F}_{fs}}{\partial \theta} \end{bmatrix} \qquad A_{\mathcal{F}} = \begin{bmatrix} 0 \\ 0 \\ 1 \\ 0 \\ \vdots \\ 0 \end{bmatrix}$$

may be used.

To help ensure that $\hat{\theta}_3$ is bounded away from zero, a projection algorithm may be used. If it is known that $J \leq 100$, then is is possible to modify the update routine for $\hat{\theta}_3$ as follows: If $\hat{\theta}_3 \geq 1/100$, then dont modify $\dot{\hat{\theta}}_3$. If $\hat{\theta}_3 \leq 1/100$ and $\dot{\hat{\theta}}_3 < 0$, then set $\dot{\hat{\theta}}_3 = 0$. This will ensure that $\hat{\theta}_3 \geq 1/100$ for all t without affecting the results of Theorem 8.2. This update law does introduce a discontinuity, which may raise concerns about the uniqueness and/or existence of a trajectory for the closed-loop system. In practice, however, a purely discontinuous signal is never generated due to finite slew rates of electronics, etc, so a unique trajectory will exist. For a more rigorous justification, it has been shown in [174] that existence/uniqueness issues associated with discontinuous update laws may be treated using the concept of a Filippov solution to the differential equation. This approach is also commonly used in sliding mode control.

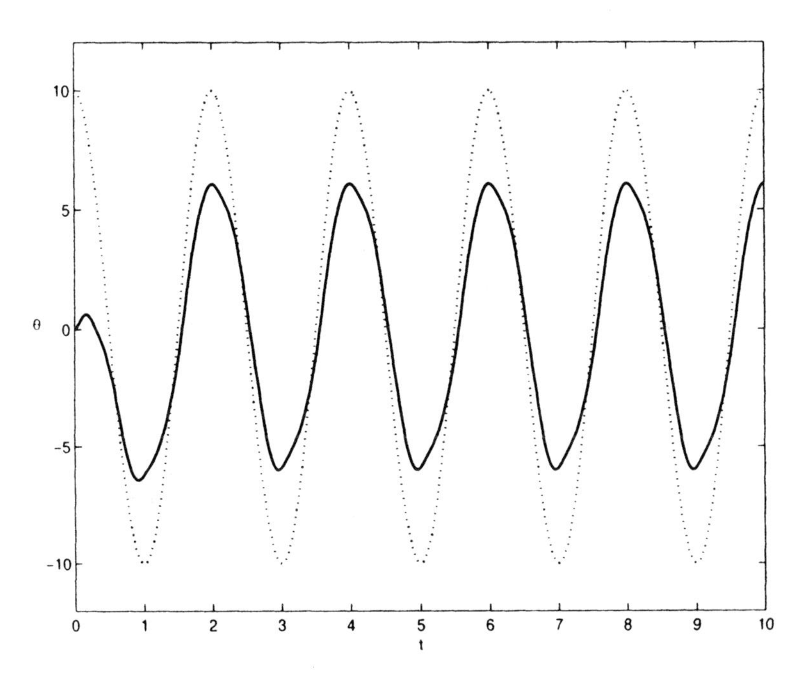

Figure 8.6. Antenna angular position θ (–) and reference signal r ($\cdots$) using a static controller.

To compare the performance obtained with and without the adaptive terms, two simulations were run. Figure 8.6 shows the ability of the static control law $u = J\nu_s$ to track a sinusoidal reference signal. When the adaptive controller is used, the closed-loop performance is

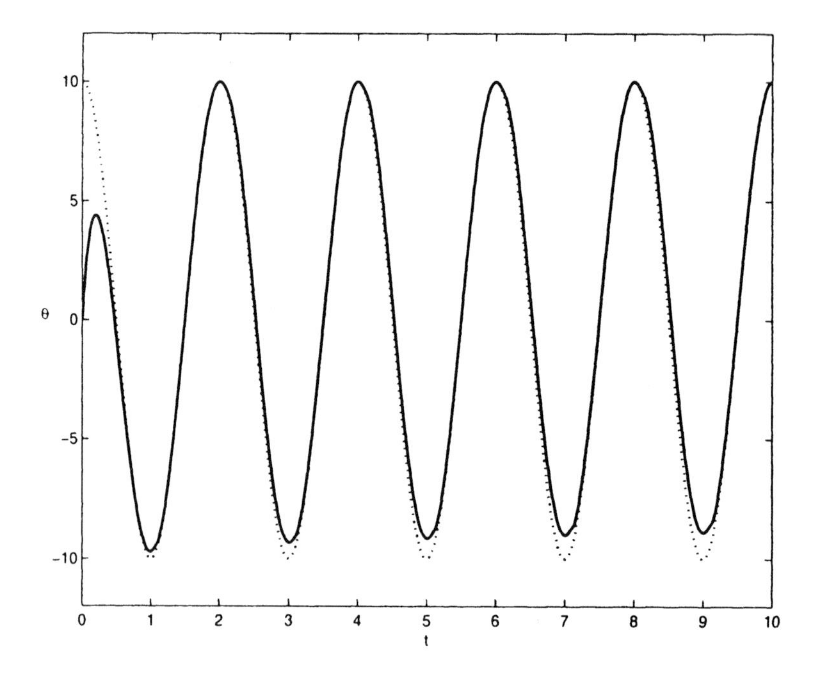

Figure 8.7. Tracking of a reference signal when the adaptive portion of the controller is turned on.

improved as shown in Figure 8.7. △

8.3 Beyond the Matching Condition

The indirect adaptive controllers presented thus far are suitable for systems in which matching conditions are satisfied. Here, via the use of backstepping, we will see how to provide adaptive compensation for uncertainties that do not necessarily satisfy matching conditions.

8.3.1 A Second-Order System

Consider the pure-feedback system defined by

$$\begin{aligned} \dot{x}_1 &= \Delta(y) + x_2 \\ \dot{x}_2 &= u, \end{aligned} \tag{8.42}$$

where Δ is an unknown function with x_1 and x_2 measurable and $y = x_1$ is the plant output which we wish to drive to $r(t)$. Since we want to drive $x_1 \to r$, we will define the first component of the error system as $e_1 = x_1 - r$.

If the plant were simply defined as

$$\dot{x}_1 = \Delta(y) + v$$

with v an input and Δ known, we could define a stabilizing control law. In particular, consider $v = \dot{r} - (\kappa + \nu_1(x_1))e_1 - \Delta(y)$, where $\kappa > 0$ and $\nu_1 \geq 0$ for all $x_1 \in \mathsf{R}$. If $V_1 = \frac{1}{2}e_1^2$, then

$$\begin{aligned} \dot{V}_1 &= e_1(\Delta + v - \dot{r}) \\ &= -2(\kappa + \nu_1(x_1))V_1, \end{aligned} \tag{8.43}$$

so $V_1 \to 0$. The component ν_1 is a nonlinear damping term included to allow for increased closed-loop stability.

Unfortunately, we do not know Δ so the above definition for v cannot be used. Instead we now consider using the control law

$$v(t, x, \hat{\theta}) = \dot{r} - (\kappa + \nu_1(x_1))e_1 + \mathcal{F}(y, \hat{\theta}) \tag{8.44}$$

for the x_1 subsystem, where now $\mathcal{F}(y, \hat{\theta})$ is a linear in the parameter approximation of $-\Delta$. We will assume that there exists some θ such that the representation error $w = \mathcal{F}(y, \theta) + \Delta(x)$ may be bounded by $|w(y)| \leq W$ for all $y \in S_y$. Using this definition for v, we find that

$$\begin{aligned} \dot{V}_1 &= e_1\left(\Delta - (\kappa + \nu_1(x_1))e_1 + \mathcal{F}(y, \hat{\theta}) - \mathcal{F}(y, \theta) + \mathcal{F}(y, \theta)\right) \\ &= -(\kappa + \nu_1(x_1))e_1^2 + e_1\left(\frac{\partial \mathcal{F}}{\partial \theta}\tilde{\theta} + w(x)\right), \end{aligned}$$

where $\tilde{\theta} = \hat{\theta} - \theta$ is the parameter error.

Ideally, we will be able to define an indirect adaptive control law $u = \nu_a(z, \hat{\theta})$ that will force $x_2 \to v$ so that $x_1 \to r$. With this in mind, we define the error system for (8.42) as

$$\begin{aligned} e_1 &= x_1 - r \\ e_2 &= x_2 - v. \end{aligned}$$

After some manipulation it is possible to show that the error dynamics are defined by

$$\begin{aligned} \dot{e}_1 &= -(\kappa + \nu_1(x_1))e_1 + \frac{\partial \mathcal{F}}{\partial \theta}\tilde{\theta} + w + e_2 \\ \dot{e}_2 &= -\frac{\partial v}{\partial t} - \frac{\partial v}{\partial x_1}(\Delta + x_2) - \frac{\partial v}{\partial \hat{\theta}}\dot{\hat{\theta}} + u \end{aligned}$$

when $\mathcal{F}$ is linear with respect to the adjustable parameters. Here $\partial v/\partial t$ is used to account for terms defined explicitly in t. From (8.44) we define $\partial v/\partial t = \ddot{r}$.

We will now consider the Lyapunov candidate

$$V = \frac{1}{2}e^\top e + \frac{1}{2}\tilde{\theta}^\top \Gamma^{-1}\tilde{\theta},$$

where Γ is a symmetric positive definite matrix. Taking the derivative of V, we find

$$\begin{aligned} \dot{V} &= e_1\left(-(\kappa + \nu_1(x_1))e_1 + \frac{\partial \mathcal{F}}{\partial \theta}\tilde{\theta} + w + e_2\right) \\ &+ e_2\left(-\frac{\partial v}{\partial t} - \frac{\partial v}{\partial x_1}(\Delta + x_2) - \frac{\partial v}{\partial \hat{\theta}}\dot{\hat{\theta}} + u\right) + \tilde{\theta}^\top \Gamma^{-1}\dot{\hat{\theta}} \end{aligned} \quad (8.45)$$

Consider the indirect adaptive control law $u = \nu_a(z, \hat{\theta})$ defined by

$$\nu_a = \frac{\partial v}{\partial t} + \frac{\partial v}{\partial x_1}\left(-\mathcal{F}(y, \hat{\theta}) + x_2\right) + \frac{\partial v}{\partial \hat{\theta}}\dot{\hat{\theta}} - e_1 - (\kappa + \nu_2)e_2, \quad (8.46)$$

where $\nu_2 \geq 0$ is another nonlinear damping term that will be defined shortly. With this control law, we find

$$\begin{aligned} \dot{V} &\leq -\kappa|e|^2 + |e_1|W - \nu_1|e_1|^2 + \left|\frac{\partial v}{\partial x_1}e_2\right|W - \nu_2|e_2|^2 \\ &+ \left(e_1 - \frac{\partial v}{\partial x_1}e_2\right)\frac{\partial \mathcal{F}(y, \hat{\theta})}{\partial \theta}\tilde{\theta} + \tilde{\theta}^\top \Gamma^{-1}\dot{\hat{\theta}}. \end{aligned}$$

Notice that choosing the nonlinear damping terms as $\nu_1 = \eta$ and $\nu_2 = \eta\left|\frac{\partial v}{\partial x_1}\right|$ with $\eta > 0$ renders (after completing the square twice with terms 2-5 in the above equation)

$$\dot{V} \leq -\kappa|e|^2 + \frac{W^2}{2\eta} + \left(e_1 - \frac{\partial v}{\partial x_1}e_2\right)\frac{\partial \mathcal{F}(y, \hat{\theta})}{\partial \theta}\tilde{\theta} + \tilde{\theta}^\top \Gamma^{-1}\dot{\hat{\theta}}. \quad (8.47)$$

To cancel the term containing $\tilde{\theta}$, we now consider the update law

$$\dot{\hat{\theta}} = -\Gamma\left[\left(e_1 - \frac{\partial v}{\partial x_1}e_2\right)\left(\frac{\partial \mathcal{F}(y, \hat{\theta})}{\partial \theta}\right)^\top + \sigma\left(\hat{\theta} - \theta^0\right)\right], \quad (8.48)$$

where $\sigma > 0$ is added to improve closed-loop system robustness. Using this update law, we find

$$\dot{V} \leq -\kappa|e|^2 + \frac{W^2}{2\eta} - \sigma\tilde{\theta}\left(\hat{\theta} - \theta^0\right). \quad (8.49)$$

Since $-2\tilde{\theta}\left(\hat{\theta} - \theta^0\right) \leq -|\tilde{\theta}|^2 + |\theta - \theta^0|^2$, we find

$$\dot{V} \leq -\kappa|e|^2 + \frac{W^2}{2\eta} - \frac{\sigma}{2}|\tilde{\theta}|^2 + \frac{\sigma}{2}|\theta - \theta^0|^2, \quad (8.50)$$

so we may define b_e and $b_{\tilde{\theta}}$ such that $\dot{V} \leq 0$ when $|e| > b_e$ or $|\tilde{\theta}| > b_{\tilde{\theta}}$. This ensures that the closed-loop system is stable assuming that the controller parameters are chosen such that y never leaves S_y where the representation error is guaranteed to be bounded.

As seen here, the steps required to develop an indirect adaptive controller for a simple second-order system in which an uncertainty does not satisfy matching conditions are much more involved than the case where matching conditions are satisfied. We will now extend the above results to a more general class of nth-order nonlinear systems which contain uncertainties that do not satisfy matching conditions.

8.3.2 Strict-Feedback Systems with Static Uncertainties

The adaptive backstepping procedure may be applied to the single-input strict-feedback system defined by

$$\begin{aligned} \dot{x}_1 &= f_1(x_1) + g_1(x_1)\left[\Delta_1(y) + \Pi_1(y)x_2\right] \\ &\vdots \\ \dot{x}_{n-1} &= f_{n-1}(\bar{x}_{n-1}) + g_{n-1}(\bar{x}_{n-1})\left[\Delta_{n-1}(y) + \Pi_{n-1}(y)x_n\right] \\ \dot{x}_n &= f_n(x) + g_n(x)\left[\Delta_n(y) + \Pi_n(y)u\right], \end{aligned} \tag{8.51}$$

where $y = x_1$ is the output. The notation $\bar{x}_i$ is defined as $\bar{x}_i = [x_1, \ldots, x_i]^\top$. For the above system, each f_i and g_i are known, while Δ_i and Π_i represent system uncertainties. It is assumed that the product $g_i\Pi_i$ is bounded away from zero.

To simplify the analysis to follow, we will consider the case where $\Pi_i(y) = 1$ for $i = 1, \ldots, n$. Here we will be interested in the tracking problem where $x_1 \to r(t)$. Using the backstepping methodology, we define the error system as

$$e_i = x_i - v_{i-1}(t, \bar{x}_{i-1}, \hat{\theta}) \tag{8.52}$$

for $i = 1, \ldots, n$ with $v_0 = r(t)$. Here $\hat{\theta}$ will be a vector of adjustable approximator parameters to be defined shortly. Notice that $e_1 = x_1 - r$ is independent of $\hat{\theta}$. We will then define

$$\begin{aligned} g_i v_i &= \frac{\partial v_{i-1}}{\partial r^{(i-1)}} r^{(i)} - (\kappa_i + \nu_i(\bar{x}_i))e_i - g_{i-1}e_{i-1} - f_i + g_i\mathcal{F}_i(y, \hat{\theta}) \\ &\quad + \sum_{j=1}^{i-1}\left[\frac{\partial v_{i-1}}{\partial x_j}\left(f_j + g_j[x_{j+1} - \mathcal{F}_j(y, \hat{\theta})]\right) + \frac{\partial v_{i-1}}{\partial r^{(j-1)}} r^{(j)}\right] \\ &\quad + \tau_i(\bar{x}_i, \hat{\theta}) \end{aligned}$$

for $i = 1, \ldots, n$ (where for convenience we let $g_0 = e_0 = 0$). The functions ν_i will be used to improve the closed-loop system robustness, while each τ_i will be used to help place the error dynamics into a special form. Notice that the time dependence in the definition of $v_i(t, \bar{x}_i)$ comes about indirectly from the time dependence in the definition of the reference signal, $r(t)$. The control is then chosen as

$$u = v_n. \tag{8.53}$$

Here each $\mathcal{F}_i(y, \hat{\theta})$ is used to approximate $-\Delta_i(y)$ for $i = 1, \ldots, n$. It is assumed that there exists some ideal parameter set $\theta \in \mathsf{R}^p$ such that the representation error

$$w_i = \mathcal{F}_i(y, \theta) + \Delta_i \tag{8.54}$$

is bounded by $|w_i| \leq W_i$ for $i = 1, \ldots, n$ when $y \in S_y$. The size of S_y will once again be dependent upon the controller parameters as will be defined shortly. For convenience, we will omit various arguments of the functions when we do not need to emphasize the dependence.

Notice that

$$g_1 v_1 = \dot{r} - (\kappa_1 + \nu_1) e_1 - f_1 + g_1 \mathcal{F}_1(\hat{\theta}) + \tau_1.$$

Thus

$$\begin{aligned} \dot{e}_1 &= f_1 + g_1[\Delta_1 + x_2] - \dot{r} \\ &= f_1 + g_1[\Delta_1 + e_2 + v_1] - \dot{r} \\ &= -(\kappa_1 + \nu_1)e_1 + g_1 e_2 + g_1 \left(\Delta_1 + \mathcal{F}_1(\hat{\theta})\right) + \tau_1. \end{aligned} \tag{8.55}$$

The derivative of the error terms $e_2, \ldots, e_n$ may be calculated in a similar fashion. Notice that

$$\begin{aligned} \dot{e}_i &= f_i + g_i[\Delta_i + x_{i+1}] - \dot{v}_{i-1} \\ &= f_i + g_i[\Delta_i + e_{i+1} + v_i] - \dot{v}_{i-1}. \end{aligned} \tag{8.56}$$

Since

$$\dot{v}_i = \sum_{j=0}^{i} \frac{\partial v_i}{\partial r^{(j)}} r^{(j+1)} + \sum_{j=1}^{i} \frac{\partial v_i}{\partial x_j} (f_j + g_j(\Delta_j + x_{j+1})) + \frac{\partial v_i}{\partial \hat{\theta}} \dot{\hat{\theta}},$$

we may use the approach for $\dot{e}_1$ above. Thus it is possible to use the definition of v_i to obtain

$$\begin{aligned} \dot{e}_i &= -(\kappa_i + \nu_i)e_i - g_{i-1}e_{i-1} + g_i e_{i+1} + g_i \left(\Delta_i + \mathcal{F}_i(\hat{\theta})\right) \\ &\quad - \frac{\partial v_{i-1}}{\partial \hat{\theta}} \dot{\hat{\theta}} - \sum_{j=1}^{i-1} \left[\frac{\partial v_{i-1}}{\partial x_j} g_j \left(\Delta_j + \mathcal{F}_j(\hat{\theta})\right)\right] + \tau_i, \end{aligned} \tag{8.57}$$

for $i = 2, \ldots, n-1$, and

$$\begin{aligned} \dot{e}_n &= -(\kappa_n + \nu_n(x))e_n - g_{n-1}e_{n-1} + g_n \left(\Delta_n + \mathcal{F}_n(\hat{\theta})\right) \\ &\quad - \frac{\partial v_{n-1}}{\partial \hat{\theta}} \dot{\hat{\theta}} - \sum_{j=1}^{n-1} \left[\frac{\partial v_{n-1}}{\partial x_j} g_j \left(\Delta_j + \mathcal{F}_j(\hat{\theta})\right)\right] + \tau_n. \end{aligned} \tag{8.58}$$

Using the definition of the representation error, (8.54), we find

$$\begin{aligned} \dot{e}_i &= -(\kappa_i + \nu_i)e_i - g_{i-1}e_{i-1} + g_i e_{i+1} + g_i\left(w_i + \mathcal{F}_i(\hat{\theta}) - \mathcal{F}_i(\theta)\right) \\ &\quad - \frac{\partial v_{i-1}}{\partial \hat{\theta}}\dot{\hat{\theta}} - \sum_{j=1}^{i-1}\left[\frac{\partial v_{i-1}}{\partial x_j} g_j \left(w_j + \mathcal{F}_j(\hat{\theta}) - \mathcal{F}_j(\theta)\right)\right] + \tau_i. \end{aligned} \quad (8.59)$$

Using linear in the parameter approximators, (8.59) becomes

$$\begin{aligned} \dot{e}_i &= -(\kappa_i + \nu_i)e_i - g_{i-1}e_{i-1} + g_i e_{i+1} + g_i\left(w_i + \frac{\partial \mathcal{F}_i}{\partial \hat{\theta}}\tilde{\theta}\right) \\ &\quad - \frac{\partial v_{i-1}}{\partial \hat{\theta}}\dot{\hat{\theta}} - \sum_{j=1}^{i-1}\left[\frac{\partial v_{i-1}}{\partial x_j} g_j \left(w_j + \frac{\partial \mathcal{F}_j}{\partial \hat{\theta}}\tilde{\theta}\right)\right] + \tau_i, \end{aligned} \quad (8.60)$$

where $\tilde{\theta} = \hat{\theta} - \theta$ is the parameter error.

Let

$$z_{i,j} = -\frac{\partial v_{i-1}}{\partial x_j} g_j. \quad (8.61)$$

Then define

$$\begin{aligned} q_i &= g_i \frac{\partial \mathcal{F}_i}{\partial \hat{\theta}} + \sum_{j=1}^{i-1} z_{i,j}\frac{\partial \mathcal{F}_j}{\partial \hat{\theta}} \\ h_i &= g_i w_i + \sum_{j=1}^{i-1} z_{i,j} w_j, \end{aligned} \quad (8.62)$$

and

$$D_i = -\frac{\partial v_{i-1}}{\partial \hat{\theta}}. \quad (8.63)$$

It is now possible to group terms so that

$$\dot{e} = A_e e + \begin{bmatrix} q_1 \\ \vdots \\ q_n \end{bmatrix}\tilde{\theta} + \begin{bmatrix} D_1 \\ \vdots \\ D_n \end{bmatrix}\dot{\hat{\theta}} + \begin{bmatrix} h_1 + \tau_1 \\ \vdots \\ h_n + \tau_n \end{bmatrix}, \quad (8.64)$$

where

$$A_e = \begin{bmatrix} -\kappa_1 - \nu_1 & g_1 & 0 & \cdots & 0 \\ -g_1 & -\kappa_2 - \nu_2 & g_2 & & \vdots \\ 0 & -g_2 & -\kappa_3 - \nu_3 & & \\ \vdots & & & \ddots & g_{n-1} \\ 0 & \cdots & & -g_{n-1} & -\kappa_n - \nu_n \end{bmatrix}. \quad (8.65)$$

For now, consider the error system defined by

$$\dot{e} = A_e e + \begin{bmatrix} q_1 \\ \vdots \\ q_n \end{bmatrix} \tilde{\theta} + \begin{bmatrix} h_1 \\ \vdots \\ h_n \end{bmatrix}, \tag{8.66}$$

so that we are ignoring the contribution from each D_i, and τ_i in (8.64). Define a Lyapunov candidate as

$$V = \frac{1}{2} e^\top e + \frac{1}{2} \tilde{\theta}^\top \Gamma^{-1} \tilde{\theta}, \tag{8.67}$$

where Γ is a positive definite diagonal matrix. Now consider the parameter update law defined by

$$\dot{\hat{\theta}} = -\Gamma \left(\sum_{i=1}^{n} q_i{}^\top e_i + \sigma(\hat{\theta} - \theta^0) \right), \tag{8.68}$$

where $\sigma > 0$ and θ^0 is a best guess of θ. The derivative of the Lyapunov candidate now becomes

$$\dot{V} = -\sum_{i=1}^{n} (\kappa_i + \nu_i) e_i^2 + \sum_{i=1}^{n} h_i e_i - \sigma \tilde{\theta}^\top (\hat{\theta} - \theta^0). \tag{8.69}$$

Since $-2\tilde{\theta}^\top (\hat{\theta} - \theta^0) \leq -|\tilde{\theta}|^2 + |\theta - \theta^0|^2$ we find

$$\dot{V} \leq -\sum_{i=1}^{n} (\kappa_i + \nu_i) e_i^2 + \sum_{i=1}^{n} h_i e_i - \frac{\sigma}{2} |\tilde{\theta}|^2 + \frac{\sigma}{2} \left|\theta - \theta^0\right|^2. \tag{8.70}$$

The nonlinear damping terms may then be defined to account for the h_i terms in (8.70). Let

$$\nu_i(\bar{x}_i) = \eta_i \left(g_i^2 + \sum_{j=1}^{i-1} \left|z_{i,j}\right|^2 \right), \tag{8.71}$$

where $\eta_i > 0$ for $i = 1, \ldots, n$. Notice that

$$-\sum_{i=1}^{n} \eta_i \left(g_i^2 + \sum_{j=1}^{i-1} \left|z_{i,j}\right|^2 \right) e_i^2 + \sum_{i=1}^{n} h_i \leq \sum_{i=1}^{n} \left(\frac{\sum_{j=1}^{i} W_j^2}{4\eta_i} \right)$$

when $y \in S_y$. Using this inequality,

$$\dot{V} \leq -\sum_{i=1}^{n} \kappa_i e_i^2 - \frac{\sigma}{2} |\tilde{\theta}|^2 + \frac{\sigma}{2} \left|\theta - \theta^0\right|^2 + \sum_{i=1}^{n} \left(\frac{\sum_{j=1}^{i} W_j^2}{4\eta_i} \right)$$

so that there exists b_e and $b_{\tilde{\theta}}$ such that $|e| \geq b_e$ and $|\tilde{\theta}| > b_{\tilde{\theta}}$ imply that $\dot{V} \leq 0$. This then ensures that $|e|$ and $|\tilde{\theta}|$ are bounded with a proper choice of controller parameters and initial conditions so that $y \in S_y$ for all t.

But (8.66) is not the true error system. We will see, however, it is now possible to choose the functions τ_i to maintain stability. Using the definition of the proposed update law (8.68), notice that

$$
\begin{aligned}
\begin{bmatrix} D_1 \\ \vdots \\ D_n \end{bmatrix} \dot{\hat{\theta}} &= -\begin{bmatrix} D_1 \\ \vdots \\ D_n \end{bmatrix} \Gamma \sum_{i=1}^{n} {q_i}^\top e_i - \sigma \begin{bmatrix} D_1 \\ \vdots \\ D_n \end{bmatrix} \Gamma \left(\hat{\theta} - \theta_0\right) \\
&= \begin{bmatrix} 0 & \cdots & 0 \\ m_{2,1} & \cdots & m_{2,n} \\ \vdots & & \vdots \\ m_{n,1} & \cdots & m_{n,n} \end{bmatrix} e - \sigma \begin{bmatrix} D_1 \\ \vdots \\ D_n \end{bmatrix} \Gamma \left(\hat{\theta} - \theta_0\right),
\end{aligned}
$$

where

$$
m_{i,j} = -\frac{\partial v_{i-1}}{\partial \hat{\theta}} \Gamma q_j. \tag{8.72}
$$

Substituting this into (8.64) we obtain

$$
\begin{aligned}
\dot{e} &= A_e e + \begin{bmatrix} q_1 \\ \vdots \\ q_n \end{bmatrix} \tilde{\theta} + \begin{bmatrix} 0 & \cdots & 0 \\ m_{2,1} & \cdots & m_{2,n} \\ \vdots & & \vdots \\ m_{n,1} & \cdots & m_{n,n} \end{bmatrix} e \\
&\quad - \sigma \begin{bmatrix} D_1 \\ \vdots \\ D_n \end{bmatrix} \Gamma(\hat{\theta} - \theta^0) + \begin{bmatrix} h_1 + \tau_1(x_1) \\ \vdots \\ h_n + \tau_n(x_n) \end{bmatrix}.
\end{aligned}
$$

We are now ready to define each $\tau_i(t, \bar{x}_i, \hat{\theta})$. Notice that τ_i is not dependent upon $x_{i+1}, \ldots, x_n$. Consider the case were $\dot{e} = f + Ge$. If $V = \frac{1}{2} e^\top e$, then

$$
\dot{V} = e^\top f + \frac{1}{2} e^\top \left(G + G^\top\right) e.
$$

If G is skew-symmetric so that $G + G^\top = 0$, then the effects of Ge may be ignored in the stability analysis. Therefore we will define each τ_i to cancel the effects of the terms not included in (8.66). That is, we will choose τ_i to cancel the effects of the $D_i(\hat{\theta} - \theta^0)$ terms of the update law and also the $m_{i,j}$ terms. Notice that

$$
\begin{bmatrix} 0 & \cdots & 0 \\ m_{2,1} & \cdots & m_{2,n} \\ \vdots & & \vdots \\ m_{n,1} & \cdots & m_{n,n} \end{bmatrix} - \begin{bmatrix} 0 & 0 & \cdots & 0 \\ m_{2,1} & m_{2,2} & \ddots & \vdots \\ \vdots & \vdots & \ddots & 0 \\ m_{n,1} & m_{n,2} & \cdots & m_{n,n} \end{bmatrix}
$$

$$-\begin{bmatrix} 0 & & \cdots & & 0 \\ 0 & 0 & & & \\ 0 & m_{2,3} & 0 & & \vdots \\ \vdots & \vdots & \ddots & \ddots & \\ 0 & m_{2,n} & \cdots & m_{n-1,n} & 0 \end{bmatrix}$$

is a skew-symmetric matrix defined by

$$G = \begin{bmatrix} 0 & 0 & 0 & \cdots & 0 \\ 0 & 0 & m_{2,3} & \cdots & m_{2,n} \\ 0 & -m_{2,3} & 0 & \ddots & \vdots \\ \vdots & \vdots & \ddots & \ddots & m_{n-1,n} \\ 0 & -m_{2,n} & \cdots & -m_{n-1,n} & 0 \end{bmatrix}. \tag{8.73}$$

Letting

$$\begin{aligned} \tau &= -\begin{bmatrix} 0 & 0 & \cdots & 0 \\ m_{2,1} & m_{2,2} & \ddots & \vdots \\ \vdots & \vdots & \ddots & 0 \\ m_{n,1} & m_{n,2} & \cdots & m_{n,n} \end{bmatrix} e + \sigma \begin{bmatrix} D_1 \\ \vdots \\ D_n \end{bmatrix} \Gamma(\hat{\theta} - \theta^0) \\ &\quad - \begin{bmatrix} 0 & & \cdots & & 0 \\ 0 & 0 & & & \\ 0 & m_{2,3} & 0 & & \vdots \\ \vdots & \vdots & \ddots & \ddots & \\ 0 & m_{2,n} & \cdots & m_{n-1,n} & 0 \end{bmatrix} e \end{aligned} \tag{8.74}$$

we obtain the desired result

$$\dot{e} = A_e e + \begin{bmatrix} q_1 \\ \vdots \\ q_n \end{bmatrix} \tilde{\theta} + \begin{bmatrix} h_1 \\ \vdots \\ h_n \end{bmatrix} + Ge. \tag{8.75}$$

Notice that each τ_i is defined in terms of $m_{k,l}$ with $k \leq i$ and $l \leq i$ so that it is not dependent upon $x_{i+1}, \ldots, x_n$. The following theorem summarizes the resulting closed-loop behavior:

Theorem 8.3: *Assume for given linear in the parameter approximators $\mathcal{F}_i(y, \hat{\theta})$ there exists some θ such that $|\mathcal{F}_i(y, \theta) + \Delta_i(y)| \leq W_i$ for all $y \in S_y$ where $e \in B_r$ implies $y \in S_y$. Then the parameter update law (8.68) with adaptive controller (8.53) guarantee that the solutions of (8.51) are bounded given $B_e \subseteq B_r$ with B_e defined by (8.78).*

Proof: Defining the Lyapunov candidate as

$$V = \frac{1}{2}e^\top e + \frac{1}{2}\tilde{\theta}^\top \Gamma^{-1}\tilde{\theta},$$

we obtain

$$\dot{V} \leq -2\bar{\kappa}|e|^2 - \frac{\sigma}{2}|\tilde{\theta}|^2 + d, \tag{8.76}$$

where $\bar{\kappa} = \min \kappa_i$ and

$$d = \frac{\sigma}{2}\left|\theta - \theta^0\right|^2 + \sum_{i=1}^{n}\left(\frac{\sum_{j=1}^{i} W_j^2}{4\eta_i}\right). \tag{8.77}$$

If $|e| \geq b_e$ or $|\tilde{\theta}| \geq b_{\tilde{\theta}}$ where

$$b_e = \sqrt{\frac{d}{2\bar{\kappa}}} \qquad b_{\tilde{\theta}} = \sqrt{\frac{2d}{\sigma}},$$

then $\dot{V} \leq 0$.

Corollary 7.1 may now be used to establish that $e \in B_e$ for all t with

$$B_e = \left\{e \in \mathsf{R}^p : |e| \leq \sqrt{2\left(\max(V(0), V_r)\right)}\right\}, \tag{8.78}$$

where $V_r = \frac{1}{2}b_e^2 + \frac{\lambda_{\max}(\Gamma^{-1})b_{\tilde{\theta}}^2}{2}$. By proper choice of the controller parameters and initial conditions, it is always possible to make B_e arbitrarily small (so that $B_e \subseteq B_r$). ■

As with the other adaptive techniques presented thus far, it is necessary that there exists some θ such that the representation errors are bounded for all $y \in S_y$. When designing an adaptive controller for a strict-feedback system, one typically knows the desired range over which the state variables are allowed to vary. Using this knowledge, an appropriate approximator structure may be defined so that it is able to compensate for system uncertainties when properly tuned. The remaining controller parameters (such as the rate of adaptation) are then chosen so that we are ensured that the state trajectories will remain bounded in such a manner that the inputs to the approximator(s) will remain in a valid input space.

Here we considered uncertainties which are only dependent upon the output y. This was done since given some bound on $e^\top e$ it is possible to place a bound on $|y|$ given r is also bounded. Thus we can place restrictions on the maximum allowable error to ensure that $y \in S_y$. Since $e_2, \ldots, e_n$ are dependent upon $\hat{\theta}$, it is not as easy to place bounds on $x_2, \ldots, x_n$ given bounds on $|e|$.

In Theorem 8.3 we showed that it is possible to guarantee that the closed-loop system is stable when using the adaptive controller with strict feedback systems. It was not shown, however, what one may expect for an ultimate bound on the output error, $e_1 = y - r$. Since

$$\dot{V} \leq -2\bar{\kappa}|e|^2 - \frac{\sigma}{2}|\tilde{\theta}|^2 + d \tag{8.79}$$

$$\leq -4\bar{\kappa}\left(\frac{1}{2}|e|^2\right) - \frac{\sigma}{\lambda_{\max}(\Gamma^{-1})}\left(\frac{1}{2}\tilde{\theta}^\top\Gamma^{-1}\tilde{\theta}\right) + d \tag{8.80}$$

we find

$$\dot{V} \leq -kV + d,$$

where $k = \min(4\bar{\kappa}, \sigma/\lambda_{\max}(\Gamma^{-1}))$. Thus an ultimate bound on $e_1 = y - r$ is given by $\sqrt{2d/k}$ since $e_1^2 \leq 2V$ and V converges to a ball of radius d/k. Since k may be made arbitrarily large by proper choice of κ, σ and Γ, it is possible to make the ultimate bound arbitrarily small. Though this is an appealing result, one must keep in mind that it is not possible to make the feedback gains arbitrarily large in practice due to unmodeled quantities such as structural dynamics and time delays.

The following example demonstrates how to use the above technique to control a simple strict-feedback system fitting the form defined by (8.51).

Example 8.5 Consider the strict-feedback system defined by

$$\begin{aligned} \dot{x}_1 &= \Delta_1(x_1) + x_2 \\ \dot{x}_2 &= u, \end{aligned} \tag{8.81}$$

where $\Delta_1(x_1) \approx \theta x_1^2$ when $|x_1| \leq 10$. Here it will be assumed that $\theta = 2$ is an unknown constant and that $|\Delta_1(x_1) - \theta x_1^2| \leq 0.1$ when $|x_1| \leq 10$. If $\mathcal{F}_1(x_1, \hat{\theta}) = -\theta x_1^2$ is used to approximately cancel the uncertainty Δ_1, then $W_1 = 0.1$. We will be interested in designing an adaptive controller such that $x_1 \to 0$. Based on this objective, the first error state is defined by $e_1 = x_1$.

Defining the second error state According to (8.52), the second error state is defined as $e_2 = x_2 - v_1$, where

$$v_1 = -(\kappa + \nu_1)e_1 + \mathcal{F}(x_1, \hat{\theta}) + \tau_1, \tag{8.82}$$

and $\kappa > 0$. Using (8.61) for $z_{1,j}$, (8.71) for ν_1, (8.62) for q_1, (8.63) for D_1, and (8.72) for $m_{1,j}$ we find

$$\begin{aligned} z_{1,1} = z_{1,2} &= 0 \\ \nu_1 &= \eta \\ q_1 &= -x_1^2 \\ D_1 &= 0 \\ m_{1,1} = m_{1,2} &= 0, \end{aligned}$$

where $\eta > 0$ is a design variable. Using the above definitions, we use (8.74) to obtain $\tau_1 = \sigma D_1 \Gamma(\hat{\theta} - \theta^0) = 0$. Thus $v_1 = -(\kappa + \eta)e_1 + \mathcal{F}(x_1, \hat{\theta})$.

Defining the control input With $u = v_2$ we now obtain

$$u = -(\kappa + \nu_2)e_2 - e_1 - \left(\kappa + \eta + 2\hat{\theta}x_1\right)\left(x_2 - \mathcal{F}(x_1, \hat{\theta})\right) + \tau_2. \quad (8.83)$$

Using the definitions above,

$$\begin{aligned} z_{2,1} &= -(\kappa + \eta) - 2\hat{\theta}x_1 \\ z_{2,2} &= 0 \\ \nu_2 &= \eta(1 + z_{2,1}^2) \\ q_2 &= -z_{2,1}x_1^2 \\ D_2 &= x_1^2 \\ m_{2,1} &= x_1^2 \Gamma q_1 \\ m_{2,2} &= x_1^2 \Gamma q_2 \end{aligned}$$

so that $\tau_2 = -(m_{2,1}e_1 + m_{2,2}e_2) + \sigma D_2 \Gamma \left(\hat{\theta} - \theta^0\right)$. We are now ready to define the update and control parameters.

Choosing the controller parameters According to (8.68), the parameter update law is defined by

$$\dot{\hat{\theta}} = -\Gamma\left(q_1 e_1 + q_2 e_2 + \sigma\left(\hat{\theta} - \theta^0\right)\right), \quad (8.84)$$

where $\Gamma, \sigma > 0$ and θ^0 is our best guess of the value of θ.

We will now use (8.78) to place bounds on the tracking error. If we choose $\theta^0 = 0$, $\sigma = 0.01$, and $\eta = 1$, then we may use (8.77) to find $d = 0.0225$ (since $W_1 = 0.1$ and $\theta = 2$). Since we want $|e| \leq 10$ (so that $|e_1| \leq 10$), we should ensure that $V_r \leq 10^2/2$ according to (8.78). This is accomplished choosing $\Gamma = 2$ and $\kappa = 1$ so that $V_r = 1.13$.

The performance of the resulting adaptive controller is shown by the solid line in Figure 8.8. As a comparison to the case where no adaptation is used, consider the controller defined by

$$u = -(\kappa + \nu_2)e_2 - e_1 - (\kappa + \eta)x_2, \quad (8.85)$$

which is similar to (8.83) with the terms related to the adaptive approximation removed. The resulting closed-loop performance is shown by the dotted line in Figure 8.8. As seen in the figure, when the nonlinear uncertainty is not compensated the system is unstable. △

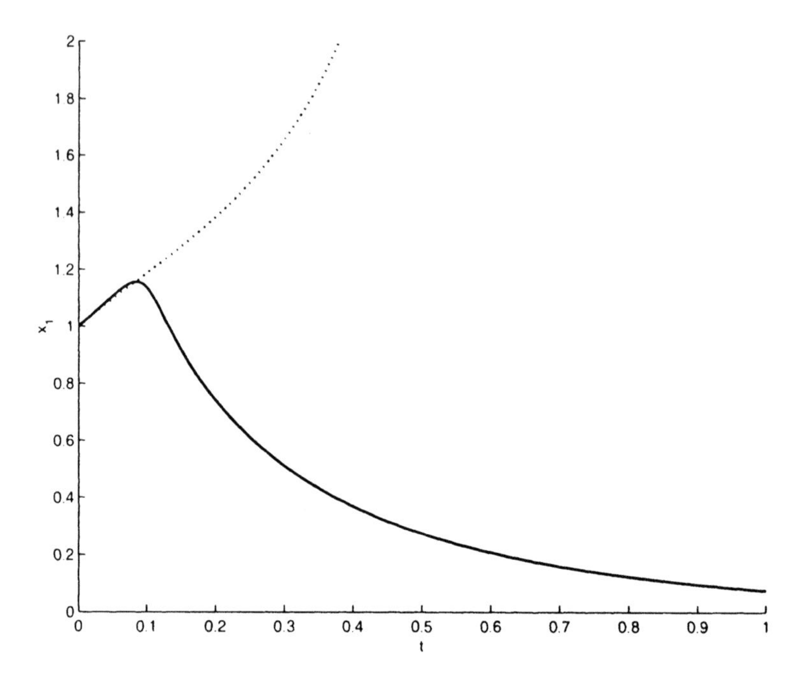

Figure 8.8. The plant output when using the adaptive controller (-), and when adaptation is turned off $(\cdots)$.

Even for the simple example above, we see that defining the control law for a system which does not satisfy matching conditions may become rather involved.

8.3.3 Strict-Feedback Systems with Dynamic Uncertainties

We will now consider the control of strict-feedback systems with possibly time-varying uncertainties. In particular, we will consider the system defined by

$$\begin{aligned} \dot{x}_1 &= f_1(x_1) + g_1(x)\left[\Delta_1(t,y) + x_2\right] \\ &\vdots \\ \dot{x}_{n-1} &= f_{n-1}(\bar{x}_{n-1}) + g_{n-1}(\bar{x}_{n-1})\left[\Delta_{n-1}(t,y) + x_n\right] \\ \dot{x}_n &= f_n(x) + g_n(x)\left[\Delta_n(t,y) + u\right], \end{aligned} \tag{8.86}$$

where $y = x_1$ is the output. For the above system, each f_i and g_i are known, while each Δ_i is a dynamic uncertainty. Again it is assumed that each g_i is bounded away from zero.

Notice that each Δ_i may be a time-varying uncertainty. We will therefore make the following assumption:

Assumption 8.3: There exists some $\nu_{\Delta_i}(y,\zeta)$ such that

$$\zeta^\top \left[\Delta_i(t,y) + \nu_{\Delta_i}(y,\zeta)\right] \le c_i$$

for each $i = 1, \ldots, m$ and all $\zeta \in \mathsf{R}^m$, where $c_i \ge 0$.

Notice that this is similar to the requirement made for uncertainties which do satisfy matching conditions. As in the case where matching conditions were satisfied, we will approximate the compensating terms ν_{Δ_i} for $i = 1, \ldots, n$. The representation error will be defined as $w_i = \mathcal{F}_i(y,\zeta,\theta) - \nu_{\Delta_i}$ for some ideal θ. It is additionally assumed that there exists some ideal parameter set $\theta \in \mathsf{R}^p$ such that $|\mathcal{F}_i(y,\zeta,\theta) - \nu_{\Delta_i}| \le W_i$ for all $\zeta \in \mathsf{R}^m$ and $y \in S_y$. The definition of ζ will be based upon our choice of the error system as will be demonstrated shortly.

To develop a controller for (8.86) based on the backstepping method we define $e_1 = x_1 - r$ and

$$e_i = x_i - v_{i-1}(t, \bar{x}_{i-1}), \tag{8.87}$$

where

$$\begin{aligned} g_i v_i &= \frac{\partial v_{i-1}}{\partial r^{(i-1)}} r^{(i)} - (\kappa_i + \nu_i(\bar{x}_i))e_i - g_{i-1}e_{i-1} - f_i \\ &\quad + g_i \mathcal{F}_i(y, g_i e_i, \hat{\theta}) + \tau_i(\bar{x}_i) \\ &\quad + \sum_{j=1}^{i-1} \left[\frac{\partial v_{i-1}}{\partial x_j} \left(f_j + g_j[x_{j+1} - \mathcal{F}_j(y, z_{i,j} e_i, \hat{\theta})] \right) + \frac{\partial v_{i-1}}{\partial r^{(j-1)}} r^{(j)} \right] \end{aligned} \tag{8.88}$$

for $i = 2, \ldots, n$, and we define

$$z_{i,j} = -\frac{\partial v_{i-1}}{\partial x_j} g_j.$$

The control is then chosen as

$$u = v_n \tag{8.89}$$

as in the case with static uncertainties.

The error derivatives may be expressed as

$$\begin{aligned} \dot{e}_i &= -(\kappa_i + \nu_i)e_i - g_{i-1}e_{i-1} + g_i e_{i+1} + g_i(\Delta_i + \mathcal{F}_i(y, g_i e_i, \hat{\theta})) \\ &\quad - \frac{\partial v_{i-1}}{\partial \hat{\theta}} \dot{\hat{\theta}} - \sum_{j=1}^{i-1} \left[\frac{\partial v_{i-1}}{\partial x_j} g_j \left(\Delta_j + \mathcal{F}_j(y, z_{i,j} e_i, \hat{\theta}) \right) \right] + \tau_i, \end{aligned} \tag{8.90}$$

for $i = 2, \ldots, n-1$. Adding and subtracting ν_{Δ_i} we obtain

$$\dot{e}_i = -(\kappa_i + \nu_i)e_i - g_{i-1}e_{i-1} + g_i e_{i+1} - \frac{\partial v_{i-1}}{\partial \hat{\theta}} \dot{\hat{\theta}}$$

$$+ g_i(\Delta_i + \mathcal{F}_i(y, g_i e_i, \hat{\theta}) + \nu_{\Delta_i}(\bar{x}_i, g_i e_i) - \nu_{\Delta_i}(\bar{x}_i, g_i e_i)) + \tau_i$$
$$+ \sum_{j=1}^{i-1} \left[z_{i,j} \left(\Delta_j + \mathcal{F}_j(y, z_{i,j} e_i, \hat{\theta}) + \nu_{\Delta_j}(\bar{x}_j, z_{i,j} e_i) - \nu_{\Delta_j}(\bar{x}_j, z_{i,j} e_i) \right) \right].$$

Letting

$$q_i = g_i \frac{\partial \mathcal{F}_i(y, g_i e_i, \hat{\theta})}{\partial \hat{\theta}} + \sum_{j=1}^{i-1} z_{i,j} \frac{\partial \mathcal{F}_j(y, z_{i,j} e_i, \hat{\theta})}{\partial \hat{\theta}} \tag{8.91}$$

$$h_i = g_i w_i + \sum_{j=1}^{i-1} z_{i,j} w_j \tag{8.92}$$

$$b_i = g_i(\Delta_i + \nu_{\Delta_i}) + \sum_{j=1}^{i-1} z_{i,j}(\Delta_j + \nu_{\Delta_j}), \tag{8.93}$$

and

$$D_i = -\frac{\partial v_{i-1}}{\partial \hat{\theta}}, \tag{8.94}$$

we may group terms and express the error system as

$$\dot{e} = A_e e - \begin{bmatrix} q_1 \\ \vdots \\ q_n \end{bmatrix} \tilde{\theta} + \begin{bmatrix} D_1 \\ \vdots \\ D_n \end{bmatrix} \dot{\hat{\theta}} + \begin{bmatrix} h_1 + b_1 + \tau_1 \\ \vdots \\ h_n + b_n + \tau_n \end{bmatrix}, \tag{8.95}$$

where A_e is defined by (8.65).

Using the same parameter update law used with the static uncertainty case

$$\dot{\hat{\theta}} = -\Gamma \left(\sum_{i=1}^{n} q_i{}^\top e_i + \sigma(\hat{\theta} - \theta^0) \right), \tag{8.96}$$

we once again find

$$\begin{bmatrix} D_1 \\ \vdots \\ D_n \end{bmatrix} \dot{\hat{\theta}} = \begin{bmatrix} 0 & \cdots & 0 \\ m_{2,1} & \cdots & m_{2,n} \\ \vdots & & \vdots \\ m_{n,1} & \cdots & m_{n,n} \end{bmatrix} e - \sigma \begin{bmatrix} D_1 \\ \vdots \\ D_n \end{bmatrix} \Gamma \left(\hat{\theta} - \theta_0 \right), \tag{8.97}$$

where $m_{i,j}$ is defined in (8.72), Γ is diagonal with positive elements, $\sigma > 0$ and θ^0 is a best guess of θ.

Rather than forming a skew-symmetric matrix as before, here will use the properties of a diagonally dominant system to overcome the effects of Me in (8.97) with $M = [m_{i,j}]$ defined element-wise. It is possible to show that given $A, K \in \mathbb{R}^{n\times n}$, then $x^\top(A+K)x \geq 0$ holds for all x with $K = \text{diag}(k_1, \ldots, k_n)$ if

$$k_i \geq n\left(1 + a_{1,i}^2 + \cdots + a_{n,i}^2\right)$$

defined along the columns of A, or

$$k_i \geq n\left(1 + a_{i,1}^2 + \cdots + a_{i,n}^2\right)$$

defined along the rows of A with $A = [a_{i,j}]$ (this result is proved in Theorem 14.1).

Now choose

$$\tau = -Ke + \sigma \begin{bmatrix} D_1 \\ \vdots \\ D_n \end{bmatrix} \Gamma\left(\hat{\theta} - \theta_0\right), \tag{8.98}$$

where

$$K = \text{diag}\left(\begin{bmatrix} n\left(1 + m_{1,1}^2\right) \\ n\left(1 + (m_{2,1} + m_{1,2})^2 + m_{2,2}^2\right) \\ \vdots \\ n\left(1 + (m_{n,1} + m_{1,n})^2 + (m_{n,2} + m_{2,n})^2 + \cdots + m_{n,n}^2\right) \end{bmatrix}\right).$$

We will now show that the above choice for K allows one to dominate the terms in M.

Define the Lyapunov candidate as

$$V = \frac{1}{2}e^\top e + \frac{1}{2}\tilde{\theta}^\top \Gamma^{-1}\tilde{\theta}, \tag{8.99}$$

where Γ is a positive definite diagonal matrix. Using the definition of the update law and τ, we find

$$\dot{V} = \frac{1}{2}e^\top\left[M + M^\top - 2K\right]e - \sum_{i=1}^{n}(\kappa_i + \nu_i)e_i^2 + \sum_{i=1}^{n}(h_i + b_i)e_i - \sigma\tilde{\theta}^\top(\hat{\theta} - \theta^0).$$

Notice that $M + M^\top = M_1 + M_2$, where

$$M_1 = \begin{bmatrix} m_{1,1} & m_{2,1} + m_{1,2} & \cdots & m_{n,1} + m_{1,n} \\ 0 & \ddots & \ddots & \vdots \\ \vdots & \ddots & & m_{n,n-1} + m_{n-1,n} \\ 0 & \cdots & 0 & m_{n,n} \end{bmatrix}$$

is upper triangular and

$$M_2 = \begin{bmatrix} m_{1,1} & 0 & \cdots & 0 \\ m_{2,1} + m_{1,2} & & \ddots & \vdots \\ \vdots & \ddots & \ddots & 0 \\ m_{n,1} + n_{1,n} & \cdots & m_{n,n-1} + m_{n-1,n} & m_{n,n} \end{bmatrix}$$

is lower triangular. Also notice that

$$\begin{aligned} e^\top \left[M + M^\top - 2K\right] e &= e^\top \left[M_1 + M_2 - 2K\right] e \\ &= e^\top \left[M_1 - K\right] e + e^\top \left[M_2 - K\right] e \leq 0 \end{aligned}$$

since K is diagonally dominating. Thus

$$\dot{V} \leq -\sum_{i=1}^{n} (\kappa_i + \nu_i) e_i^2 + \sum_{i=1}^{n} (h_i + b_i) e_i - \sigma \tilde{\theta}^\top (\hat{\theta} - \theta^0). \tag{8.100}$$

Since $-2\tilde{\theta}^\top (\hat{\theta} - \theta^0) \leq -|\tilde{\theta}|^2 + |\theta - \theta^0|^2$ we find

$$\dot{V} \leq -\sum_{i=1}^{n} (\kappa_i + \nu_i) e_i^2 + \sum_{i=1}^{n} (h_i + b_i) e_i - \frac{\sigma}{2} |\tilde{\theta}|^2 + \frac{\sigma}{2} \left|\theta - \theta^0\right|^2 .$$

Using Assumption 8.3 we find each

$$\begin{aligned} b_i e_i &= g_i (\Delta_i + \nu_{\Delta_i}) e_i + \sum_{j=1}^{i-1} z_{i,j} e_i (\Delta_j + \nu_{\Delta_j}) \\ &\leq c_i + \sum_{j=1}^{i-1} c_j , \end{aligned}$$

so

$$\dot{V} \leq -\sum_{i=1}^{n} (\kappa_i + \nu_i) e_i^2 + \sum_{i=1}^{n} h_i e_i - \frac{\sigma}{2} |\tilde{\theta}|^2 + \frac{\sigma}{2} \left|\theta - \theta^0\right|^2 + \sum_{i=1}^{n} \sum_{j=1}^{i} c_j . \tag{8.101}$$

The nonlinear damping terms may then be defined to account for the h_i terms in (8.101). Let

$$\nu_i(\bar{x}_i) = \eta_i \left(g_i^2 + \sum_{j=1}^{i-1} \left|z_{i,j}\right|^2 \right), \tag{8.102}$$

where $\eta_i > 0$ for $i = 1, \ldots, n$, so that

$$\dot{V} \leq -\sum_{i=1}^{n} \kappa_i e_i^2 - \frac{\sigma}{2} |\tilde{\theta}|^2 + d,$$

where

$$d = \frac{\sigma}{2}\left|\theta - \theta^0\right|^2 + \sum_{i=1}^{n}\left(\frac{\sum_{j=1}^{i} W_j^2}{4\eta_i}\right) + \sum_{i=1}^{n}\sum_{j=1}^{i} c_j \tag{8.103}$$

so that there exists b_e and $b_{\tilde{\theta}}$ such that $|e| \geq b_e$ and $|\tilde{\theta}| > b_{\tilde{\theta}}$ imply that $\dot{V} \leq 0$. It is now possible to state the following theorem:

Theorem 8.4: *Let Assumption 8.3 hold, and assume that for given linear in the parameter approximators $\mathcal{F}_i(y, \hat{\theta})$ there exists some θ such that $|\mathcal{F}_i(y, \theta) + \Delta_i(y)| \leq W_i$ for all $y \in S_y$, where $e \in B_r$ implies $y \in S_y$. Then the parameter update law (8.96) with adaptive controller (8.89) guarantee that the solutions of (8.86) are bounded given $B_e \subseteq B_r$ with B_e defined by (8.104).*

The proof of Theorem 8.4 follows the one for Theorem 8.3. Following these steps, it is possible to show that if one ignores the effect of y leaving the space S_y, then $e \in B_e$ for all t with

$$B_e = \left\{e \in \mathbb{R}^p : |e| \leq \sqrt{2\left(\max(V(0), V_r)\right)}\right\}, \tag{8.104}$$

where $V_r = \frac{1}{2}b_e^2 + \frac{\lambda_{\max}(\Gamma^{-1})b_{\tilde{\theta}}^2}{2}$. By properly choosing $B_e \subseteq B_r$ so that $y \in S_y$, then one may conclude that the closed-loop system is stable. As is the case with static uncertainties, one may further show that

$$\dot{V} \leq -kV + d,$$

where $k = \min(2\bar{\kappa}, \sigma/\lambda_{\max}(\Gamma^{-1}))$, $\bar{\kappa} = \min \kappa_i$, and d is defined by (8.103). Since $|e_1| \leq 2V$, the ultimate bound on e_1 is given by $\sqrt{2d/k}$.

When using the above procedure to define compensating approximators, it is suggested that each approximator take the form

$$\mathcal{F}_i(y, \zeta, \hat{\theta}) = -\eta_i \mathcal{F}_{d_i}(y, \hat{\theta})\zeta, \tag{8.105}$$

where $\mathcal{F}_{d_i}$ is a dominating term to be approximated. If $|\Delta_i| \leq \rho_i\psi_i$, then we may define θ such that $\mathcal{F}_{d_i}(y, \theta) \geq \psi_i^2(y)$. This way

$$\begin{aligned}\zeta\left[\Delta_i + \mathcal{F}_i(y, \zeta, \theta)\right] &\leq \rho_i|\psi_i(\bar{y})\zeta| - \eta_i\psi_i^2(\bar{y})\zeta^2 \\ &\leq \frac{\rho_i^2}{4\eta_i},\end{aligned} \tag{8.106}$$

which satisfies Assumption 8.3. Also $\mathcal{F}_{d_i}$ is not dependent upon ζ, which makes it easier to ensure that the inputs to $\mathcal{F}_{d_i}$ remain in a valid input space.

8.4 Summary

In this chapter, we learned how to develop indirect adaptive controllers to compensate for various system uncertainties. It was shown that it is possible to define adjustable approximators which are able to compensate for either static or dynamics uncertainties. In general the indirect adaptive controller for the case where matching conditions are satisfied is defined as follows:

1. Place the plant in a canonical representation so that an error system may be defined.
2. Define an error system and Lyapunov candidate V_s for the static problem.
3. Define a static control law $u = \nu_s$ which ensures that $\dot{V}_s \leq -k_1 V_s + k_2$ when $\Delta = 0$ and $\Pi = I$.
4. Choose approximators $\mathcal{F}_\Pi(z, \hat{\theta})$ and $\mathcal{F}_\Delta(z, \hat{\theta})$ such that there exists some θ where $|\mathcal{F}_\Delta(x, \theta) + \Delta(x)| \leq W_\Delta$ and $|\mathcal{F}_\Pi(x, \theta) - \Pi(x)| = 0$ for all $z \in S_z$. Estimate upper bounds for W_Δ and $|\theta - \theta^0|$ where θ^0 may be viewed as a "best guess" of θ.
5. Find some B_r such that $e \in B_r$ implies $z \in S_z$.
6. Choose the initial conditions, control parameters, and update law parameters such that $B_e \subseteq B_r$ with B_e defined as the ball to which e is confined ignoring the effects of the case when z leaves S_z.

Using an adaptive backstepping methodology, it was shown how to design indirect adaptive controllers for systems in which uncertainties do not satisfy matching conditions.

8.5 Exercises and Design Problems

Exercise 8.1 (Imperfect Approximation) Prove that the assumption that $|\mathcal{F}_\Pi(x, \theta) - \Pi(x)| = 0$ in Theorems 8.1 and 8.2 can be relaxed if a lower bound $\pi_0 \leq \lambda_{\min}(\Pi)$ is known, and the control $\nu_r = \nu_a + \nu_m$ is used, with

$$\nu_m = -\frac{\eta_\pi}{\pi_0}\left(\frac{\partial V_s}{\partial e}\beta(x)\right)^\top \nu_a^2$$

and $\eta_\pi > 0$.

Exercise 8.2 (Controller-Identifier Separation) Consider the error dynamics $\dot{e} = \alpha(t, x) + \beta(x)\left[\Delta(x) + u\right]$, where $\Delta(x)$ may be approximated by the known linear in the parameter approximator $\mathcal{F}(x, \hat{\theta})$.

Assume that there exists some θ such that $|\mathcal{F}(x,\theta) + \Delta(x)| \leq W$ for all $x \in \mathbb{R}^n$ and a control law $\nu_s(t,x)$ such that the Lyapunov function $V_s(e)$ satisfies $\dot{V}_s \leq -k_1 V_s + k_2$ when $u = \nu_s$ and $\Delta = 0$.

Now assume that a parameter update routine

$$\dot{\hat{\theta}} = \phi(t, x, \hat{\theta})$$

has been defined such that $\tilde{\theta} \in \mathcal{L}_\infty$. Use the Lyapunov function $V_a = V_s$ and nonlinear damping to define a control law $u = \nu_a(t, x, \hat{\theta})$ such that the closed-loop system is stable. Find the ultimate bound on $|e|$.

Exercise 8.3 (ϵ-Modification) Modify Theorem 8.1 and the associated assumptions to allow the case where an ϵ-modification is used in the update law.

Exercise 8.4 (Strict-Feedback Extension I) Modify the controller associated with Theorem 8.3 to cover the case when $\Pi_i \neq 1$ is unknown.

Exercise 8.5 (Strict-Feedback Extension I) Develop an indirect adaptive controller for the system

$$\begin{aligned}
\dot{x}_1 &= x_2 \\
&\vdots \\
\dot{x}_{i-1} &= x_i \\
\dot{x}_i &= \Delta(x_i) + x_{i+1} \\
&\vdots \\
\dot{x}_n &= \Delta(x_i) + u,
\end{aligned}$$

where $\Delta(x_i)$ is an uncertainty that may be approximated by the linear in the parameter approximator $\mathcal{F}(x_i, \theta)$ such that $|\mathcal{F}(x_i,\theta) + \Delta| \leq W$ for all x_i in B_r where $B_r = \{x_i \in \mathbb{R} : |x_i| \leq r\}$.

Exercise 8.6 (Control of a Double Integrator) Consider the system defined by

$$\begin{aligned}
\dot{x}_1 &= x_2 \\
\dot{x}_2 &= \Delta(x) + u,
\end{aligned} \tag{8.107}$$

where the output $y = x_1$ is to be regulated to the value r_0. Assuming that $\Delta \equiv 0$, define a static stabilizing controller using the following approaches :

1. A stable manifold defined by $e = x_2 + \kappa x_1$ with a Lyapunov candidate $V_s = \frac{1}{2}e^2$.

2. An error system defined by $e = x$ and Lyapunov candidate $V_s = e^\top P e$ with P a symmetric positive definite matrix.
3. Backstepping with $e_1 = x_1$ and $e_2 = x_2 + \kappa e_1$ and Lyapunov candidate $V = \frac{1}{2}e_1^2 + \frac{1}{2}e_2^2$.

For each of these approaches, define an indirect adaptive controller to compensate for the uncertainty $\Delta(x)$.

Exercise 8.7 (Stepper Motor) The model of a two-phase permanent-magnet stepper motor [119] is given by

$$\begin{aligned} \dot{\theta} &= \omega \\ J\dot{\omega} &= -K_m i_a \sin(N\theta) + K_m i_b \cos(N\theta) - B\omega - T_L \\ L\dot{i}_a &= -Ri_a + K_b\omega \sin(N\theta) + v_a \\ L\dot{i}_b &= -Ri_b - K_b\omega \cos(N\theta) + v_b, \end{aligned} \tag{8.108}$$

where θ is the angular position, ω is the angular rate, and i_a, i_b are the currents for phase A and B. Here J is the moment of inertia, K_m is the motor torque constant, N is the number of teeth on the rotor, B is the viscous friction coefficient, L is the inductance, R is the resistance, K_b is the back EMF constant, and T_L is the load torque. Define an indirect adaptive controller for the voltages v_a and v_b such that $\theta \to r(t)$ when J, K_m, B, R, L, and T_L are unknown.

Exercise 8.8 (Fuzzy Control of a Surge Tank) Consider the surge tank defind in Exercise 7.11. Design an indirect adaptive controller using a fuzzy system with adjustable output membership centers to approximate the cross-sectional area when $A(x)$ is defined as

- $A(x) = 1 + |x|$
- $A(x) = 2 + (x - 1)(x + 1)$

so that $x \to r$, where $r > 0$ is a constant. Is it possible to develop an indirect adaptive controller that is stable for both cases?

Exercise 8.9 (Neural Control of a Surge Tank) Repeat Exercise 8.8 using a radial-basis neural network.

Chapter 9

Implementations and Comparative Studies

9.1 Overview

In Chapters 6, 7 and 8 we discussed certain important classes of continuous time nonlinear systems, and presented general methods based on state feedback for control of such systems, including in particular direct and indirect adaptive control methods. Here, we will illustrate these adaptive approaches by applying them to the problem of controlling a rotational inverted pendulum apparatus. Moreover, we will compare the performance of these methods with that of "conventional" adaptive control techniques.

The direct and indirect adaptive control approaches of Chapters 7 and 8 are general enough that they can be applied to a wide class of nonlinear systems. The main requirement is that a Lyapunov function that implies closed-loop stability be known, from which the analysis can be carried out. On the other hand, for the purposes of this application chapter we do not require such generality: As will be shown, the inverted pendulum fits a particular class of systems, that of input-output feedback linearizable plants, and this knowledge may be used to more fully exploit the power of the adaptive techniques presented earlier.

We will start by showing how the stability results of Chapters 7 and 8 can be strengthened by restricting the class of systems under consideration and assuming the systems to have a particular structure (i.e., an input-output state feedback linearizable structure). Then we will give a detailed explanation about how the modified techniques can be implemented, paying special attention to details which may help the control designer to avoid common problems and construct effective adaptive designs.

9.2 Control of Input-Output Feedback Linearizable Systems

We will begin our presentation of the implementation issues associated with the direct and indirect adaptive controllers by first giving some additional background material related to input-output feedback linearizable systems. In particular, we will see that input-output feedback linearizable systems have a special form that allows stronger stability results to be obtained than seen in the previous chapters since we can ensure asymptotic stability of the error dynamics rather than just ultimate boundedness.

9.2.1 Direct Adaptive Control

Recall from Chapter 6 the class of input-output feedback linearizable systems

$$\begin{aligned} \dot{\xi} &= f_\xi(\xi) + g_\xi(\xi)u \\ y &= h_\xi(\xi), \end{aligned} \tag{9.1}$$

where $\xi \in \mathsf{R}^{n+d}$ with $d \geq 0$. By virtue of the change of coordinates $[q^\top, x^\top]^\top = T(\xi)$, this system may be transformed into

$$\begin{aligned} \dot{q} &= \phi(q, x) \\ \dot{x}_1 &= x_2 \\ &\vdots \\ \dot{x}_{n-1} &= x_n \\ \dot{x}_n &= f(q, x) + g(q, x)u \end{aligned} \tag{9.2}$$

and $y = x_1$. This system has strong relative degree n, which may also be determined by differentiating the output y until the input u appears for the first time and it is multiplied by a function $g(q, x)$ which does not vanish for any $q \in \mathsf{R}^d$, $x \in \mathsf{R}^n$. In this way, we may write the input-output behavior of (9.1) as

$$y^{(n)} = f(q, x) + g(q, x)u. \tag{9.3}$$

We will require the function g to become neither arbitrarily small nor arbitrarily large, that is, $0 < g_0 \leq g(q, x) \leq g_1 < \infty$ for some known g_0 and g_1 within a compact set. Moreover, we will assume the existence and knowledge of some piecewise continuous and bounded function $g_d(q, x)$ such that $|\dot{g}(q, x)| \leq g_d(q, x)$ within a compact set. As in the previous chapters, we will use function approximators in the adaptive controllers. This means that somehow the state must be confined to a compact set, which in turn implies that boundedness of $|\dot{g}|$ is automatically satisfied within this compact set as long as all functions in the system are piecewise continuous and bounded for bounded arguments.

We also assume the zero dynamics of the system, $\dot{q} = \phi(q, 0)$, to be input-to-state stable with x as the input. We will see that the pendulum does not satisfy this condition; however, it will be possible to stabilize the entire state of this system by properly choosing the adaptive controller. We are interested in having system (9.1) track a reference trajectory $r(t)$. We will operate under the assumption that r and its derivatives up to the nth one are bounded and can be measured.

Now we define an error system for tracking using a stable manifold $e = \chi(t, x)$ where

$$\chi(t,x) = k_1(r - x_1) + \ldots + k_{n-1}(r^{(n-2)} - x_{n-1}) + (r^{(n-1)} - x_n) \tag{9.4}$$

It was shown in Example 6.6 that if $L(s) = s^{n-1} + k_{n-1}s^{n-2} + \ldots + k_2 s + k_1$ has its roots in the open left half plane (it is Hurwitz), then the error system using the stable manifold (9.4) satisfies Assumption 6.1. Now, let $\bar{\chi}(t,x) = k_1(\dot{r} - x_2) + \ldots + k_{n-1}(r^{(n-1)} - x_n)$, and note that

$$\begin{aligned} \dot{e} &= \bar{\chi} - f(q,x) - g(q,x)u + r^{(n)} \\ &= \alpha(t,q,x) + \beta(q,x)u, \end{aligned} \tag{9.5}$$

where $\alpha(t,q,x) = \bar{\chi} - f(q,x) + r^{(n)}$ and $\beta(q,x) = -g(q,x)$. When no dynamic uncertainty exists, we may consider the radially unbounded Lyapunov candidate $V_s = \frac{1}{2g}e^2$ (assuming $g(q,x) > 0$; when g is negative the Lyapunov candidate must be modified accordingly), so that

$$\dot{V}_s = \frac{1}{g}(\alpha + \beta u) - \frac{\dot{g}}{2g^2}e^2. \tag{9.6}$$

We will study the following static control law that contains a feedback linearizing term and a stabilizing term,

$$u = \nu_s(z) = \frac{1}{\beta}(-\alpha - c_1 e) + \frac{g_d}{2g_0^2}|e|\text{sgn}(e), \tag{9.7}$$

where $z = [q^\top, x^\top, r, \ldots, r^{(n)}]^\top$ and $c_1 > 0$.

Note that we are using a discontinuous term that includes a sgn(e) function in our control law. Such a discontinuous term may potentially create problems of existence and uniqueness of solutions of (9.5), which is the reason why the sgn function is usually replaced by a continuous approximation (see Exercise 9.1). In practice, however, the discontinuous term has distinct advantages, such as robustness in the presence of noise and unmodeled dynamics. At the same time, it may also reduce the life of the actuators due to chattering, so the control designer must evaluate the possible benefits of a discontinuous controller versus its disadvantages. In our pendulum application we chose to use the discontinuous term because we were interested in very precise tracking and robustness, both of which the discontinuous term provides.

With the static law (9.7) we obtain

$$\begin{aligned}\dot{V}_s &= -\frac{c_1}{g}e^2 - \frac{\dot{g}}{2g^2}e^2 - \frac{g_d}{2g_0^2}e|e|\text{sgn}(e) \\ &\leq -\frac{c_1}{g}e^2 + \frac{|\dot{g}|}{2g^2}|e|^2 - \frac{g_d}{2g_0^2}|e|^2 \\ &\leq -\frac{c_1}{g}e^2 \end{aligned} \tag{9.8}$$

so that $e = 0$ is a stable equilibrium point. Given that we are restricting the class of systems studied in Chapter 7, we are slightly modifying the overall approach in the stability proof while maintaining the fundamental concepts of the direct adaptive control methodology. For this reason, instead of the control law in Theorem 7.2, consider

$$\nu_a = \nu_s + W\text{sgn}(e), \tag{9.9}$$

where W will be defined shortly. Since the term $\frac{1}{\beta}(-\alpha - c_1 e)$ is actually unknown due to uncertainty or poor plant knowledge, this part of ν_a will need to be approximated, and the direct adaptive control law $u = \nu_a$ becomes

$$\nu_a = \mathcal{F}(z, \hat{\theta}) = \frac{\partial \mathcal{F}}{\partial \theta}\hat{\theta} + \left(\frac{g_d}{2g_0^2}|e| + W\right)\text{sgn}(e), \tag{9.10}$$

where we assume the existence of some θ so that the mismatch $w = \mathcal{F}(z, \theta) - \nu_a$ is of known, finite size, that is, $|w| \leq W$ for all $z \in S_z$, where $e \in B_r$ implies $z \in S_z$. Using (9.10) in (9.5) we obtain

$$\begin{aligned}\dot{e} &= \alpha + \beta\left[\frac{\partial \mathcal{F}}{\partial \theta}\hat{\theta} + \left(\frac{g_d}{2g_0^2}|e| + W\right)\text{sgn}(e) + \mathcal{F}(z,\theta) - \mathcal{F}(z,\theta)\right] \\ &= \alpha + \beta\left[\frac{\partial \mathcal{F}}{\partial \theta}\tilde{\theta} + \nu_s + W\text{sgn}(e) + w\right] \\ &= -c_1 e + \beta\left(\frac{g_d}{2g_0^2}|e|\text{sgn}(e) + W\text{sgn}(e) + \frac{\partial \mathcal{F}}{\partial \theta}\tilde{\theta} + w\right)\end{aligned} \tag{9.11}$$

with $\tilde{\theta} = \hat{\theta} - \theta$. Now, define the Lyapunov candidate

$$V_a = V_s + \frac{1}{2}\tilde{\theta}^\top \Gamma^{-1}\tilde{\theta}, \tag{9.12}$$

where Γ is positive definite and symmetric. Then, computing the derivative we obtain

$$\dot{V}_a \leq -\frac{c_1}{g}e^2 - \frac{\partial \mathcal{F}}{\partial \theta}\tilde{\theta}e + \tilde{\theta}^\top \Gamma^{-1}\dot{\hat{\theta}}. \tag{9.13}$$

Consider the adaptation law

$$\dot{\hat{\theta}} = \Gamma \frac{\partial \mathcal{F}}{\partial \theta} e, \tag{9.14}$$

which results in

$$\dot{V}_a \le -\frac{c_1}{g} e^2 \le -\frac{c_1}{g_1} e^2. \tag{9.15}$$

By applying Barbalat's Lemma one can show asymptotic stability of the error, so that $\lim_{t \to \infty} e = 0$. This implies that the tracking objective is satisfied. Note, however, that the result we obtain here is unlike Theorem 7.2, where application of Corollary 7.1 yields explicit bounds on e which can in turn be used to ensure $z \in S_z$. In this case, we have chosen not to use the σ-modification in the adaptation law (9.14) and resort to Barbalat's Lemma instead. A high gain bounding control term may be used to impose explicit bounds on e to keep it within some B_r that implies $z \in S_z$. The use of such a bounding term will be illustrated in the inverted pendulum application.

9.2.2 Indirect Adaptive Control

We again consider the class of input-output feedback linearizable systems (9.1) and the error dynamics (9.5). Notice that when the functions f and g are unknown, (9.5) is a special case of the dynamic uncertainty error dynamics (8.29). In particular, letting $\alpha = \alpha_\Delta + \alpha_k$ and $\beta = \beta_\Pi + \beta_k$ we may rewrite (9.5) as

$$\dot{e} = \alpha_\Delta(t, q, x) + \alpha_k(t, q, x) + (\beta_\Pi(q, x) + \beta_k(q, x))u \tag{9.16}$$

where α_k and β_k represent any known part of the plant dynamics, and α_Δ, β_Π are uncertainties. Whereas the error system (8.29) is set in terms of multiplicative uncertainties, here we concentrate on additive uncertainties. Both cases can be made equivalent by using the appropriate functional substitutions. If no knowledge is available to the designer about the plant dynamics, α_k and β_k may be set to zero. The only constraints we impose on the known functions is that they be piecewise continuous and bounded for bounded arguments, and, as before, that $-g(q, x) = \beta_\Pi(q, x) + \beta_k(q, x)$ be bounded away from zero.

We start by assuming no uncertainty, though. In this case, let the vector of inputs z be defined as before and consider the feedback linearizing control law

$$u = \nu_s(z) = \frac{1}{\beta}(-\alpha - c_1 e) \tag{9.17}$$

and the Lyapunov candidate

$$V_s = \frac{1}{2} e^2. \tag{9.18}$$

For the indirect adaptive case, the only assumption we make on the plant (other than input-to-state stability of the zero dynamics) is that $g(q,x) \geq g_0 > 0$ (or bounded by a negative constant in the corresponding negative case). Then,

$$\dot{V}_s = e(\alpha + \beta u) = -c_1 e^2. \tag{9.19}$$

As in the direct adaptive case, we are applying the ideas put forth in Chapter 8 to a more particular class of systems. Thus, by slightly modifying the analysis in Theorem 8.2 we will obtain stronger stability results, albeit restricted to the class of input-output feedback linearizable systems. Since many systems of practical interest belong to this class, the loss of generality is acceptable.

When dynamic uncertainty is taken into account we use function approximators such as fuzzy systems or neural networks to represent the unknown parts of α and β (specific examples will be given in Section 9.7). Letting $w_\alpha = \mathcal{F}_\alpha(z,\theta) - \alpha$ and $w_\beta = \mathcal{F}_\beta(z,\theta) - \beta$ be the mismatch between approximators and unknown functions, consider the control law

$$u = \nu_a = \frac{1}{\mathcal{F}_\beta(z,\hat{\theta})}(-\mathcal{F}_\alpha(z,\hat{\theta}) - c_1 e) + \frac{1}{g_0}(W_\alpha + W_\beta|\nu_s|)\text{sgn}(e), \tag{9.20}$$

where ν_s is redefined to include $\mathcal{F}_\alpha(z,\hat{\theta})$ and $\mathcal{F}_\beta(z,\hat{\theta})$ instead of α and β, and there exists some θ (whose estimate $\hat{\theta}$ we will update) such that $|w_\alpha| \leq W_\alpha$ and $|w_\beta| \leq W_\beta$ for $z \in S_z$. The nonlinear damping term in (8.4) has been replaced by a discontinuous term. The same potential problems and advantages of using such a term outlined in Section 9.2.1 apply here.

We now study the Lyapunov candidate

$$V_a = V_s + \frac{1}{2}\tilde{\theta}^\top \Gamma^{-1}\tilde{\theta}, \tag{9.21}$$

where $\tilde{\theta} = \hat{\theta} - \theta$. Notice that

$$\begin{aligned}
\alpha + \beta\nu_a &= \alpha + \frac{\beta}{\mathcal{F}_\beta(z,\hat{\theta})}(-\mathcal{F}_\alpha(z,\hat{\theta}) - c_1 e) + \frac{\beta}{g_0}(W_\alpha + W_\beta|\nu_s|)\text{sgn}(e) \\
&\quad + c_1 e + \mathcal{F}_\alpha(z,\hat{\theta}) - c_1 e - \mathcal{F}_\alpha(z,\hat{\theta}) \\
&= \left(\frac{\beta}{\mathcal{F}_\beta(z,\hat{\theta})} - 1\right)(-\mathcal{F}_\alpha(z,\hat{\theta}) - c_1 e) + (\alpha - \mathcal{F}_\alpha(z,\hat{\theta})) \\
&\quad - c_1 e + \frac{\beta}{g_0}(W_\alpha + W_\beta|\nu_s|)\text{sgn}(e) \\
&= -c_1 e - (\mathcal{F}_\alpha(z,\hat{\theta}) - \alpha) - (\mathcal{F}_\beta(z,\hat{\theta}) - \beta)\nu_s \\
&\quad + \frac{\beta}{g_0}(W_\alpha + W_\beta|\nu_s|)\text{sgn}(e).
\end{aligned} \tag{9.22}$$

Then,

$$\dot{V}_a \leq -c_1 e^2 + \left(\frac{\partial \mathcal{F}_\alpha}{\partial \theta} \tilde{\theta} + \frac{\partial \mathcal{F}_\beta}{\partial \theta} \tilde{\theta} \nu_s \right) e + \tilde{\theta}^\top \Gamma^{-1} \dot{\hat{\theta}}. \tag{9.23}$$

Finally, with the adaptation law

$$\dot{\hat{\theta}} = -\Gamma \left(\frac{\partial \mathcal{F}_\alpha}{\partial \theta} + \frac{\partial \mathcal{F}_\beta}{\partial \theta} \nu_s \right) e \tag{9.24}$$

we obtain

$$\dot{V}_a \leq -c_1 e^2. \tag{9.25}$$

Similar to the direct adaptive controller of Section 9.2.1, asymptotic convergence of e to zero can be shown using Barbalat's Lemma, with the provision that a bounding term be used to guarantee $e \in B_r$, which implies $z \in S_z$.

9.3 The Rotational Inverted Pendulum

We now turn our attention to the application of the direct and indirect adaptive techniques. We chose a rotational inverted pendulum as a platform to illustrate the methods. The rotational inverted pendulum is an unstable and under-actuated system (i.e., it has fewer inputs than degrees of freedom). It presents considerable control design challenges and is therefore appropriate for testing the performance of different control techniques. Moreover, this experiment allows us to highlight some of the main points and pitfalls involved in the design of adaptive controllers. The experimental setup used in this book was developed in [240, 249], where a nonlinear mathematical model of the system was obtained via physics and system identification techniques, and four different control methods were applied: proportional-derivative control, linear quadratic regulation, direct fuzzy control, and auto-tuned fuzzy control.

The hardware setup of the system is shown in Figure 9.1. It consists of three principal parts: the pendulum itself (controlled object), interface circuits, and the controller, implemented by means of a C program in a digital computer. The controller can actuate the pendulum by means of a DC motor. The motor has an optical encoder on its base that allows the measurement of its angle (with respect to the starting position), which we will refer to as θ_0. The shaft of the motor has a fixed arm attached to it at a right angle. The pendulum can rotate freely about the arm, and its angle, θ_1, is also measured with an optical encoder[1]. Measurement of θ_1 is performed with respect to the pendulum's stable equilibrium point, where it is assumed to have a value of π radians. The system was built in such a way that the base of the pendulum is not allowed to turn more than ± 10

[1] Recall that we use θ to denote the vector of adjustable parameters, which should not be confused with the pendulum angles θ_0 and θ_1.

radians from the starting position, in order to protect the wires from being ripped off. The input voltage to the DC motor amplifier is constrained to a range of ±5 Volts.

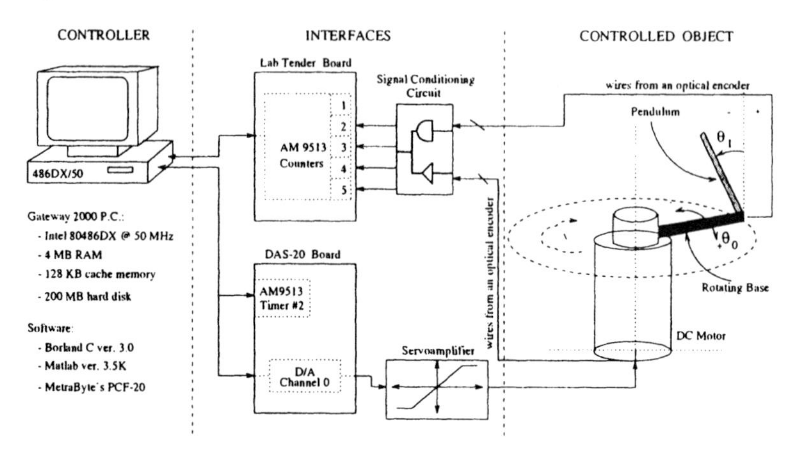

Figure 9.1. Hardware setup of the inverted pendulum system (taken from [35]).

The rotational pendulum system presents two somewhat separate problems: first, a controller needs to be designed that is able to balance the pendulum. Second, an adequate algorithm has to be used to swing up the pendulum so that when it reaches an upright position (i.e., where $\theta_1 \approx 0$) its angular velocity ($\dot{\theta}_1$) is close to zero. This facilitates the job of the controller which "catches" the pendulum and tries to balance it. In this work we will not be concerned with swing-up details, and will concentrate only on the balancing control of the pendulum. The so called "simple energy pumping" swing-up algorithm developed in [249] will be used without changes in all the experiments and simulations, only with minor tunings depending on the nature of the test. This algorithm is just a proportional controller which takes as input the error between a maximum swing angle (a tuning parameter) and the base angle θ_0.

In implementation, a sampling time of 0.01 seconds was used. All the simulation and experimental plots include the swing-up phase, and show the first 6 seconds only, since this time was considered enough to show the representative aspects of the results.

9.4 Modeling and Simulation

The rotational inverted pendulum can be represented with a four-state nonlinear model. The states are θ_0, $\dot{\theta}_0$, θ_1 and $\dot{\theta}_1$. Of them, only θ_0 and θ_1 are directly available for measurement; the other two states have to be estimated. To do this we use a first-order backward difference approximation of the derivative. This estimation method turns out to be reliable and accurate enough. Thus, for the rest of the chapter, it will be assumed that all states are directly available for the controllers without need of further estimation.

With the techniques in Chapter 10 it will be possible to remove the need to measure all the states by using state observers. The simple approximation to the derivatives of θ_0 and θ_1 used here performs, essentially, the same function as a state observer, but without any theoretical performance guarantee. Thus, the use of such simple approximators is motivated here by experience and good performance observed in practice.

The differential equations that describe the dynamics of the pendulum system (note that $\theta_1 = 0$ is the unstable equilibrium point) are given by

$$\ddot{\theta}_0 = -a_p\dot{\theta}_0 + K_p u \tag{9.26}$$

$$\ddot{\theta}_1 = -\frac{C_1}{J_1}\dot{\theta}_1 + \frac{m_1 g l_1}{J_1}\sin\theta_1 + \frac{K_1}{J_1}\ddot{\theta}_0, \tag{9.27}$$

where $m_1 = 8.6184 \times 10^{-2}$ Kg is the mass of the pendulum, $l_1 = 0.113$m is the distance from the center of mass of the pendulum, $g = 9.81 \frac{m}{s^2}$ is the acceleration due to gravity, $J_1 = 1.301 \times 10^{-3}$ N-m-s^2 is the inertia of the pendulum, $C_1 = 2.979 \times 10^{-3} \frac{N-m-s}{rad}$ is the frictional constant between the pendulum and the rotating base, $K_1 = \pm 1.9 \times 10^{-3}$ is a proportionality constant, and u is the control input (voltage applied to the motor). The numerical values of the constants were determined experimentally in [249].

A linear model of the DC motor is given by

$$\frac{\Theta_0(s)}{U(s)} = \frac{K_p}{s(s+a_p)} \tag{9.28}$$

with $a_p = 33.04$ and $K_p = 74.89$. Note that the sign of K_1 depends on whether the pendulum is in the inverted or the non-inverted position, that is, for $\frac{\pi}{2} < \theta_1 < \frac{3\pi}{2}$ we have $K_1 = 1.9 \times 10^{-3}$, and $K_1 = -1.9 \times 10^{-3}$ otherwise. When simulating the system, a conditional statement is used to determine the sign of K_1 according to the relation above.

Let $\xi_1 = \theta_0$, $\xi_2 = \dot{\theta}_0$, $\xi_3 = \theta_1$ and $\xi_4 = \dot{\theta}_1$. Then a state variable representation of the plant is given by:

$$\begin{aligned}
\dot{\xi}_1 &= \xi_2 \\
\dot{\xi}_2 &= a_1\xi_2 + b_1 u \\
\dot{\xi}_3 &= \xi_4 \\
\dot{\xi}_4 &= a_2\xi_2 + a_3\sin\xi_3 + a_4\xi_4 + b_2 u,
\end{aligned} \tag{9.29}$$

where $a_1 = -a_p$, $a_2 = -\frac{K_1 a_p}{J_1}$, $a_3 = \frac{m_1 g l_1}{J_1}$, $a_4 = -\frac{C_1}{J_1}$, $b_1 = K_p$, and $b_2 = \frac{K_1 K_p}{J_1}$. Since we are only interested in balancing the pendulum, we take the output of the system as $y = \xi_3$.

For simulation of the system, a fourth-order Runge-Kutta numerical method was used in all cases, with an integration step size of 0.001 seconds. The controllers are assumed to operate in continuous time; therefore, the sampling time of the controller was set equal to the integration step size. Also, the initial conditions were kept identical in all simulations: $\xi_1(0) = 0\,rad$, $\xi_2(0) = 0\,\frac{rad}{s}$, $\xi_3(0) = \pi\,rad$, and $\xi_4(0) = 0\,\frac{rad}{s}$. Under these conditions, the pendulum is in the downward position. When the simulations start, the pendulum is first swung up with the same swing-up algorithm used for implementation, and then "caught" by the balancing controller currently being tested, in order to resemble experimental conditions as accurately as possible. The balancing controller begins to act when $|\xi_3| \leq 0.3\,rad$; at the same time, the swing-up controller is shut down.

The heuristic "energy pumping" swing-up scheme seeks to "pump" energy from the base of the system to the pendulum in such a way that the magnitude of each swing increases until the pendulum reaches its inverted position. The first design parameter in this algorithm is M, which determines the maximum amplitude of each swing, in radians. For this study, M was varied between 1.1 and 1.4 radians, depending on whether a disturbance was applied to the pendulum or not (we encourage the reader to experiment with different values of M in simulation to see how it affects the swing-up algorithm). A small M is preferable to a large one, because it is more likely that the pendulum reaches its inverted position with almost zero velocity if M is small. The second parameter involved is a gain, k_{sw}, which for all cases will be fixed at 0.75. Then, taking u_{sw} as the swing-up control input, the algorithm works as follows:

If $\theta_0 - \pi < 0$
  then $\theta_{0ref} = -M$
  else $\theta_{0ref} = M$

$$e_{sw} = \theta_{0ref} - \theta_0$$
$$u_{sw} = k_{sw} e_{sw} \ ,$$

where the subscript sw is used to denote "swing up."

9.5 Two Non-Adaptive Controllers

In this section two non-adaptive controllers for the inverted pendulum will be introduced and these will serve as a base-line for comparison to the results to follow. We will start with a linear quadratic regulator, which provided the best experimental results for nominal conditions. Second, a feedback linearizing control law will be used; as shown later, there is no guarantee of boundedness using this technique, and the results obtained here corroborate this theoretical prediction.

9.5.1 Linear Quadratic Regulator

To design a linear quadratic regulator (LQR) for the pendulum we linearize the system model by using the approximation $\sin \xi_3 \approx \xi_3$, which is valid for small angles, in (9.29). The resulting system can be shown to be controllable; thus, an LQR can be constructed. For the design, the greatest penalty was assigned to the error in states ξ_3 and ξ_4, since the primary objective of the controller is to balance the pendulum. The state-feedback gain obtained was tested experimentally, and after some fine tuning, the gain vector used was:

$$K = [-0.7, 1, 10.8, 0.7]^\top, \tag{9.30}$$

where the state error $e = r - \xi$ ($r = 0$ is a reference state trajectory) is used, and $\xi = [\xi_1, \xi_2, \xi_3, \xi_4]^\top$.

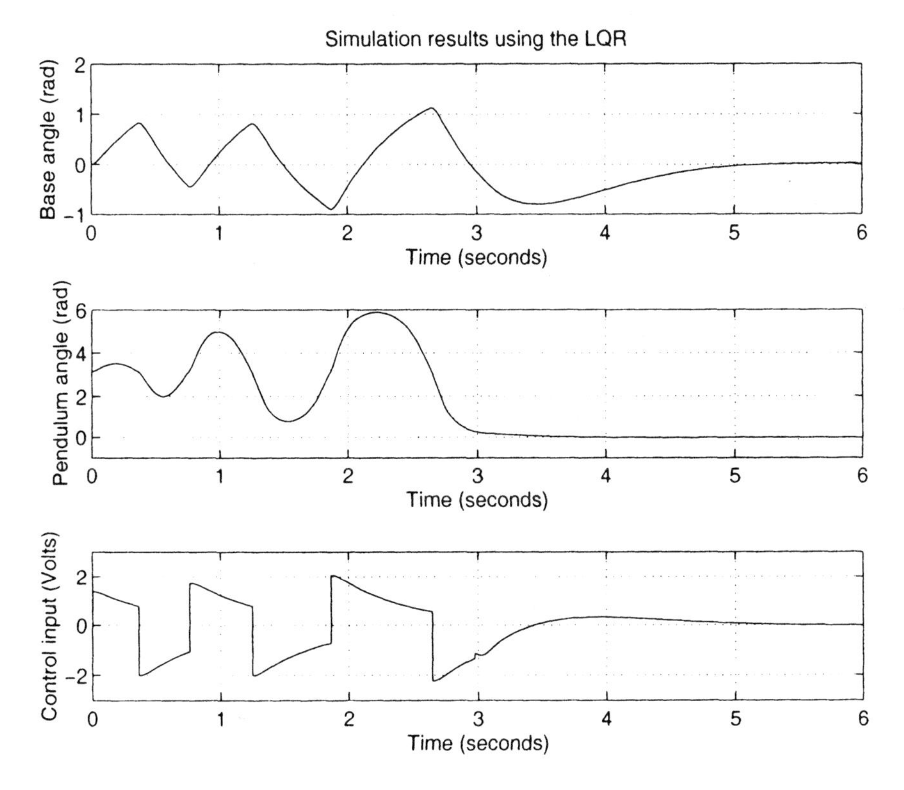

Figure 9.2. LQR simulation.

As shown in Figures 9.2 and 9.3, the LQR performs very well in both simulation and implementation; it successfully balances the pendulum and drives all the system states to zero. The control input ideally goes to zero

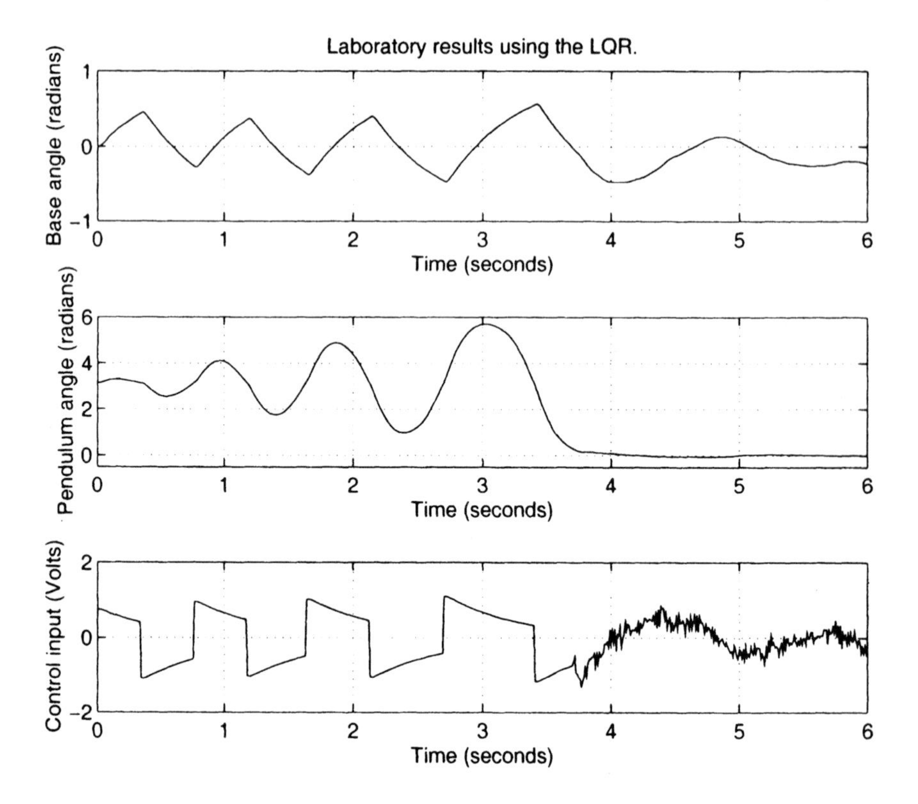

Figure 9.3. LQR implementation.

when equilibrium is reached (Figure 9.2), although in practice it does not (Figure 9.3) since the unmodeled aspects of the system (e.g., sensor noise, sampling time, nonlinear characteristics, etc.) prevent the controller from behaving perfectly.

9.5.2 Feedback Linearizing Controller

We find that the inverted pendulum has a strong relative degree $n = 2$ because, after differentiating the output twice, we obtain

$$\ddot{y} = a_2\xi_2 + a_3 \sin \xi_3 + a_4\xi_4 + b_2 u. \tag{9.31}$$

We can define a feedback linearizing control law as

$$\nu_f = \frac{1}{g_\xi(\xi)}(-f_\xi(\xi) + \nu), \tag{9.32}$$

where $f_\xi(\xi) = a_2\xi_2 + a_3 \sin \xi_3 + a_4\xi_4$, $g_\xi(\xi) = b_2$, and ν is a signal that we will define later. Since the plant has a strong relative degree, it is possible to find a mapping, T, such that $T(\xi)$ is a diffeomorphism that transforms the system to the form (9.2). Then, $[q^\top, x^\top]^\top = T(\xi)$ is a new, linear state variable representation of the system if we use the control law (9.32). Such a mapping $T(\xi)$ can be found to be

$$\begin{aligned} q_1 &= T_1(\xi) = \xi_1 + \xi_3 \\ q_2 &= T_2(\xi) = \xi_2 - \frac{b_1}{b_2}\xi_4 \\ x_1 &= T_3(\xi) = \xi_3 \\ x_2 &= T_4(\xi) = \xi_4. \end{aligned} \tag{9.33}$$

Given this new set of states, the output of the plant is given by $y = x_1$, and $\dot{y} = x_2$.

If we restrict the output y to be identically zero for all time, and given that the origin is an (unstable) equilibrium point of the undriven system, then the zero-dynamics of the pendulum are given by

$$\begin{aligned} \dot{q}_1 &= q_2 \\ \dot{q}_2 &= \left(a_1 - \frac{a_2 b_1}{b_2}\right) q_2. \end{aligned} \tag{9.34}$$

In order for the system to be minimum-phase, its zero-dynamics have to be input-to-state stable, i.e. $a_1 - \frac{a_2 b_1}{b_2} < 0$ for this choice of $T(\xi)$. However, a simple computation shows that this is not the case; indeed, the zero-dynamics of the pendulum are *unstable*, because $a_1 - \frac{a_2 b_1}{b_2} = 0$. This causes two states of the system (ξ_1 and ξ_2, since ξ_3 and ξ_4 are bounded when the output is bounded because we assume the state ξ to be uniformly continuous) to be potentially unbounded under feedback linearization. As will be seen later, this prediction is corroborated both in simulation and experimentation.

The instability of the zero-dynamics implies that, although the control law (9.32) will yield a stable input-output behavior under the right choice of ν, if the initial conditions of the system are distinct from zero, then a subset of the states will be unbounded; in particular, by solving (9.34) for nonzero initial conditions, we find that the state ξ_1 (which represents the base angle θ_0 of the rotational pendulum) will be given by $\xi_1(t) = ct$ in steady-state, where t represents time, and c is an integration constant.

It is worth noting that the unboundedness of ξ_1 can be tolerated in this experiment, since all it means is that the pendulum base keeps rotating, while the pendulum is being balanced. Of course, in practice a limit is imposed on the rate and amount of this rotation to protect the machinery, but in principle the instability of the zero-dynamics can effectively be dealt with.

In order to simulate and implement the feedback linearizing controller, it is necessary to specify the signal ν. Let $r(t)$ be a pre-specified reference trajectory (set equal to zero for all cases in this work), at least twice differentiable, and take $e_o = x_1 - r$ (following the notation of [192]). Then ν may be defined as

$$\nu = \ddot{r} - 2\dot{e}_o - 8e_o, \tag{9.35}$$

which, together with (9.32), yields a stable closed-loop system in the input-output sense with poles at $s = -1 \pm 2.65j$. This choice of the closed-loop poles was made because of practical considerations: experience with feedback linearization on the pendulum experiment indicates that attempting to bring the output error to zero too fast can sometimes result in failure or a degraded performance; therefore, these somewhat slow poles were chosen.

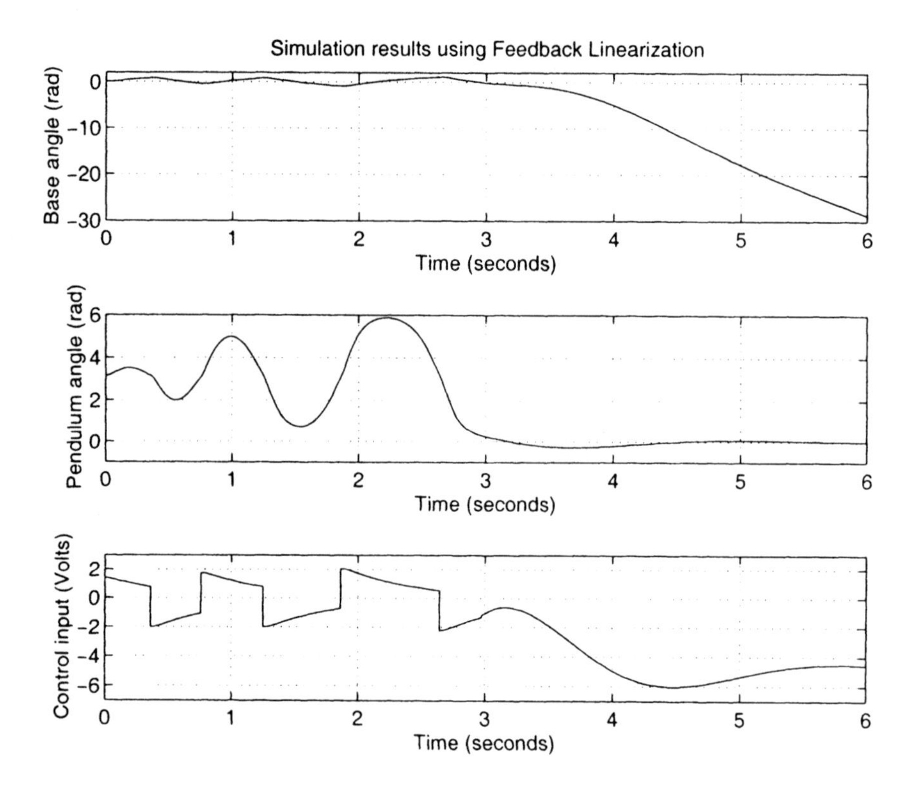

Figure 9.4. Linearization simulation.

Figure 9.4 shows that the simulation results are as expected: the pendulum is balanced, and the base keeps rotating at an almost constant velocity. Note that this controller settles the control input at a nonzero value, which in turn introduces energy into the system, and causes its base to rotate; this is easily explained by the theoretical analysis of the zero-dynamics. A comparison with the experimental results in Figure 9.5 shows significant similarities: after the swing-up phase, the states ξ_1 and ξ_3, as well as the

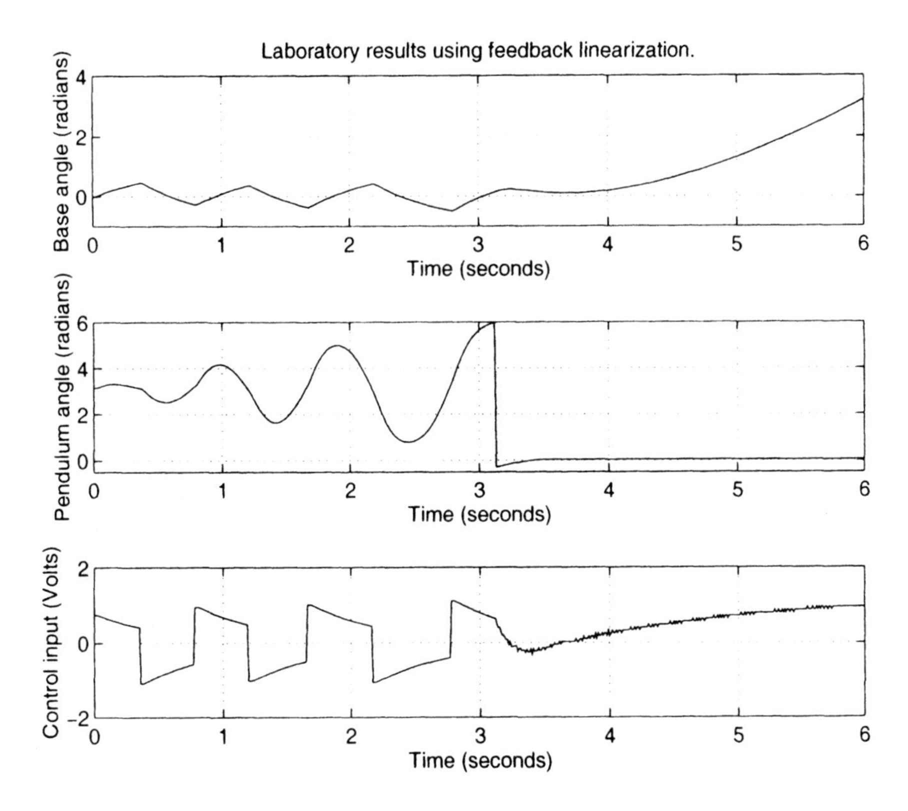

Figure 9.5. Feedback linearization implementation.

control input, behave as in the simulation. Observe that in simulation the feedback linearizing controller reaches a peak value of $-6V$ to balance the pendulum; although in implementation the control input is limited to $\pm 5V$ as explained above, no such bound was used for simulation, since it was desired to preserve ideal circumstances regarding control action.

9.6 Adaptive Feedback Linearization

Both the feedback linearizing controller and the LQR, as well as almost any other non-adaptive technique used on the pendulum fail (or in the case of non-adaptive fuzzy control have degraded performance [249]) when the nominal system (i.e., the pendulum without any mass changes, or added disturbances) is altered in an *unknown manner*. It is in such a situation that adaptive control plays a central role, since it is, at least in principle, able to deal with significant plant changes. In this investigation, two types of plant alterations were used:

- A container half filled with metal bolts fixed at the tip of the pendulum, and

- A container half filled with water fixed at the tip of the pendulum.

The added weight (not accounted for in the design of the controllers) not only shifts the pendulum's center of mass away from the pivot point (which in turn decreases the natural frequency of the pendulum) and makes the effects of friction less dominant, but also introduces random disturbances, which vary in nature with the bolts and the water. In the case of the water container, a "sloshing liquid" effect is created, which strongly affects the dynamics of the system. As will be seen from the results below, the bolts have a different type of effect that is caused by their "rattling" during balancing.

We note that the adaptive controllers in this section and the remainder of this chapter are based on the assumption that the plant is minimum-phase. Since this assumption does not hold for the pendulum, some consequences of the techniques are no longer guaranteed; specifically, it is not to be expected that the state ξ_1 and possibly the control input are bounded. In particular, the behavior of ξ_1 will depend on the initial conditions of the system. However, the study of the adaptive techniques on such a system is of theoretical and practical relevance, because, in the first place, it provides a good example of practical results very well predicted by theoretical analysis of nonlinear systems. Secondly, it can give insight into how to overcome the limitations of the adaptive controllers inherited by their underlying assumptions. As will be shown, it is indeed possible to obtain not only stability and boundedness, but also good robustness to unmodeled plant changes.

The adaptive fuzzy techniques we illustrate here are based on feedback linearization; therefore, adaptive feedback linearization (AFL) seemed the most natural choice for reference and comparison to conventional adaptive control. For the design of this controller the technique described in [192, 191] was used as we discuss next.

We first rewrite the system equation as a linear combination of known, fixed nonlinear functions

$$\begin{aligned} \dot{\xi} &= f(\xi) + g(\xi)u \\ y &= h(\xi), \end{aligned} \tag{9.36}$$

where

$$f(\xi) = \theta_1^{(1)*} \begin{bmatrix} \xi_2 \\ 0 \\ \xi_4 \\ 0 \end{bmatrix} + \theta_2^{(1)*} \begin{bmatrix} 0 \\ \xi_2 \\ 0 \\ 0 \end{bmatrix} + \theta_3^{(1)*} \begin{bmatrix} 0 \\ 0 \\ 0 \\ \xi_2 \end{bmatrix} + \theta_4^{(1)*} \begin{bmatrix} 0 \\ 0 \\ 0 \\ \sin \xi_3 \end{bmatrix}$$

$$+ \theta_5^{(1)^*} \begin{bmatrix} 0 \\ 0 \\ 0 \\ \xi_4 \end{bmatrix} \tag{9.37}$$

$$g(\xi) = \theta_1^{(2)^*} \begin{bmatrix} 0 \\ 1 \\ 0 \\ 0 \end{bmatrix} + \theta_2^{(2)^*} \begin{bmatrix} 0 \\ 0 \\ 0 \\ 1 \end{bmatrix} \tag{9.38}$$

$$h(\xi) = \xi_3. \tag{9.39}$$

We shall *estimate* $f(\xi)$ and $g(\xi)$ by searching for the optimum vectors $\theta^{(1)^*} = [\theta_1^{(1)^*}, \ldots, \theta_5^{(1)^*}]^\top$ and $\theta^{(2)^*} = [\theta_1^{(2)^*}, \theta_2^{(2)^*}]$. We use $\theta^{(1)}(t)$ and $\theta^{(2)}(t)$ to denote the estimates of the optimum parameter vectors at time t. Then, the adaptive control law is given by

$$\nu_{af} = \frac{1}{(L_g L_f h)_e} \left[-(L_{f^2} h)_e + \nu \right], \tag{9.40}$$

where $(L_f h)_e$ stands for the estimated Lie derivative of h with respect to f, and the variable ξ has been dropped for convenience. To allow for tracking, we take

$$\nu = \ddot{r} + \alpha_2(\dot{r}_m - x_2) + \alpha_1(r - x_1). \tag{9.41}$$

A simple computation gives

$$\begin{aligned} (L_{f^2} h)_e &= \theta_1^{(1)}(t)\theta_3^{(1)}(t)\xi_2 + \theta_1^{(1)}(t)\theta_4^{(1)}(t)\sin\xi_3 + \theta_1^{(1)}(t)\theta_5^{(1)}(t)\xi_4 \\ (L_g L_f h)_e &= \theta_1^{(1)}(t)\theta_2^{(2)}(t). \end{aligned} \tag{9.42}$$

Following [192], define $\theta \in \mathbf{R}^{32}$ as a vector containing all the combinations $\theta_i^{(1)}(t)$, $\theta_i^{(2)}(t)$, $\theta_i^{(1)}(t)\theta_j^{(1)}(t)$ and $\theta_i^{(1)}(t)\theta_j^{(2)}(t)$. For adaptation, define an error signal of the form $e_1 = -\beta_2 \dot{e}_o - \beta_1 e_o$, with $e_o = r - x_1$, where the transfer function

$$\frac{\beta_2 s + \bar{g}_\xi}{s^2 + \alpha_2 s + \alpha_1} \tag{9.43}$$

is strictly positive real. Then, the adaptation law, given by a normalized gradient approach, is

$$\dot{\theta} = -\frac{e_1 w}{1 + w^\top w}, \tag{9.44}$$

where w is the *regressor* vector obtained by computing the output error equation $\ddot{e}_o + \alpha_2 \dot{e}_o + \alpha_1 e_o = 0$ (see [191]).

In order to start the search for the optimum vectors $\theta^{(1)^*}$ and $\theta^{(2)^*}$ at the best known point in the search space, their estimates $\theta^{(1)}(t)$ and $\theta^{(2)}(t)$ were initialized using the parameters obtained from the system model,

$$\theta^{(1)}(0) = [1, a_1, a_2, a_3, a_4]^\top \qquad \theta^{(2)}(0) = [b_1, b_2]^\top. \tag{9.45}$$

For simulation and implementation, the following set of parameters was used: $\alpha_1 = 8$, $\alpha_2 = 2$, $\bar{g}_\xi = 0.08$, and $\beta_2 = 0.1$. In this way, the same equation for ν is used as for the non-adaptive feedback linearizing controller (see equation (9.35)). The values for $\bar{g}_\xi$ and β_2 were determined via manual tuning; after several simulation and experimentation trials, they were the choices that, as far as we could determine, made the controller work at its best and still maintain the strictly positive real condition.

Figure 9.6 has characteristics similar to those of the non-adaptive feedback linearization in Figure 9.4, although the required control input stays within implementation bounds. For the nominal plant, the controller exhibits a very good behavior, as seen in Figure 9.7: the pendulum is perfectly balanced, and the control input settles at a value close to zero, so that the base rotates slowly. In Figure 9.8 we observe that the controller manages to balance the pendulum with bolts, but only for a short time; the base is turning rapidly, and at about the fifth second the control input reaches the its limit of $5V$. The controller had the greatest problems with water, and was not able to maintain equilibrium, as seen in Figure 9.9. In the next section we will show how the adaptive fuzzy control techniques can significantly outperform this conventional adaptive controller.

9.7 Indirect Adaptive Fuzzy Control

Here, an indirect adaptive fuzzy controller (IAFC) will be developed for the inverted pendulum; two possible configurations will be presented and experimentally tested. First, a controller that does not make explicit use of any plant dynamics knowledge will be used. Second, it will be illustrated how to incorporate the knowledge of the model (9.29) in the design. It will be shown experimentally that such an enhanced controller has, in the case of the pendulum, a noticeable advantage over the previous techniques and provides an increased robustness in the presence the induced disturbances.

9.7.1 Design Without Use of Plant Dynamics Knowledge

As previously shown, the pendulum model has a relative degree of two. Substituting the numerical values of the parameters, we obtain from (9.31)

$$f_\xi(\xi) \approx 48.2521\xi_2 + 73.4085 \sin \xi_3 - 2.2898\xi_4 \tag{9.46}$$

$$g_\xi(\xi) \approx -109.3705. \tag{9.47}$$

In these equations we use "approximately equal" signs because the numerical parameters of the equations are not expected to represent the pendulum's input-output dynamics *exactly*; rather, the right-hand side of (9.46) and (9.47) are simply our ***best known approximations*** to $f_\xi(\xi)$ and $g_\xi(\xi)$, respectively. Note that $\xi = T^{-1}(q, x)$, which can be easily computed, and

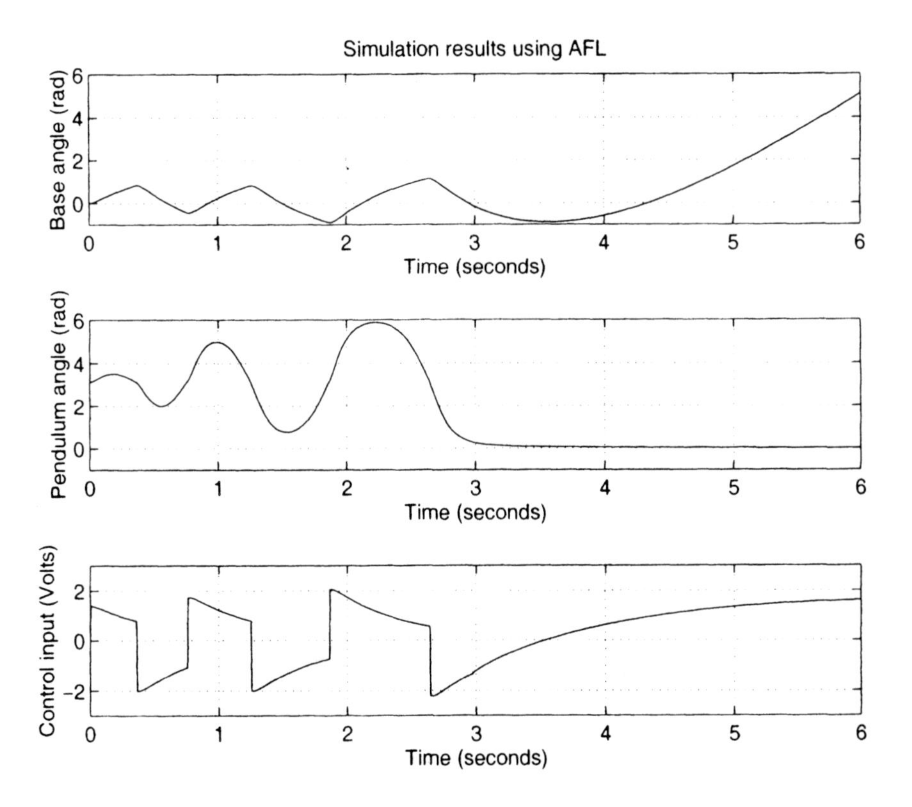

Figure 9.6. AFL simulation.

$f(q,x) = f_\xi(T^{-1}(q,x))$. Moreover, $g_\xi(\xi) = g(q,x) = g < 0$, so there exists a $g_0 < 0$ (take, for instance, $g_0 = -100$, which gives us a safe margin of error) such that $g \leq g_0$ for all $t \geq 0$; thus, g is bounded away from zero,

Recalling that $e_o = r - x_1$, consider the stable manifold $\chi(t,x) = k_1 e_o + \dot{e}_o$ and the error system $e = \chi(t,x)$ (shown before to satisfy Assumption 6.1). Then, letting $\bar{\chi}(t,x) = k_1 \dot{e}_0$,

$$\begin{aligned} \dot{e} &= \bar{\chi} + \ddot{r} - f(q,x) - g(q,x)u \\ &= \alpha(t,q,x) + \beta(q,x)u \end{aligned} \tag{9.48}$$

with $\alpha(t,q,x) = \bar{\chi} + \ddot{r} - f(q,x)$ and $\beta(q,x) = -g(q,x) = -b_2$. Since the reference we are attempting to track is $r(t) = 0$ for all t, the problem reduces to one of regulation to the origin, and $\dot{e} = \alpha(q,x) + \beta(q,x)u$ with $\alpha(q,x) = \bar{\chi} - f(q,x)$ and $\bar{\chi}(x) = -k_1 x_1$.

It is possible to represent (9.46) and (9.47) using a special form of Takagi-Sugeno fuzzy systems. To briefly present the notation, take a fuzzy

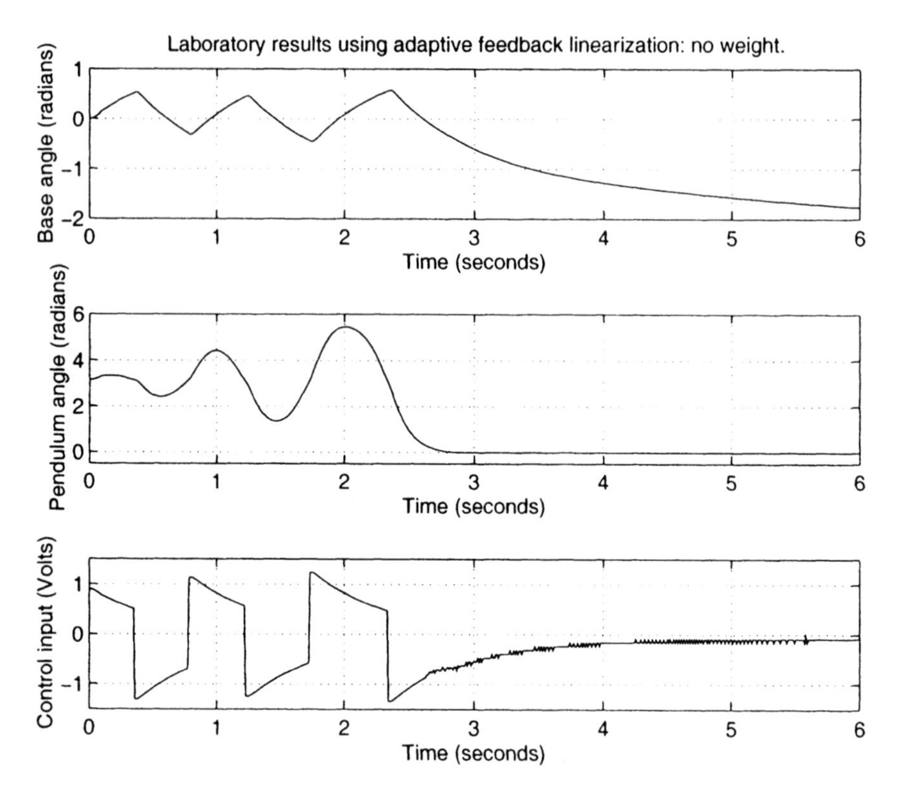

Figure 9.7. Experimental results of AFL with nominal plant.

system denoted by $\mathcal{F}_{fs}(x)$. Then,

$$\mathcal{F}_{fs}(\xi) = \frac{\sum_{i=1}^{p} c_i \mu_i}{\sum_{i=1}^{p} \mu_i}. \tag{9.49}$$

Here, singleton fuzzification of the input $x = [x_1, \ldots, x_n]^\top$ is assumed; the fuzzy system has p rules, and μ_i is the value of the membership function for the antecedent of the ith rule given the input ξ. It is assumed that the fuzzy system is constructed in such a way that $\sum_{i=1}^{p} \mu_i \neq 0$ for all $\xi \in \mathrm{R}^n$. The parameter c_i is the consequent of the ith rule, which in this work will be taken as a linear combination of Lipschitz continuous functions $d_k(\xi) \in \mathrm{R}$, $k = 1, \ldots, m-1$, so that

$$c_i = \theta_{i,0} + \theta_{i,1} d_1(\xi) + \ldots + \theta_{i,m-1} d_{m-1}(\xi)\ , \quad i = 1, \ldots, p. \tag{9.50}$$

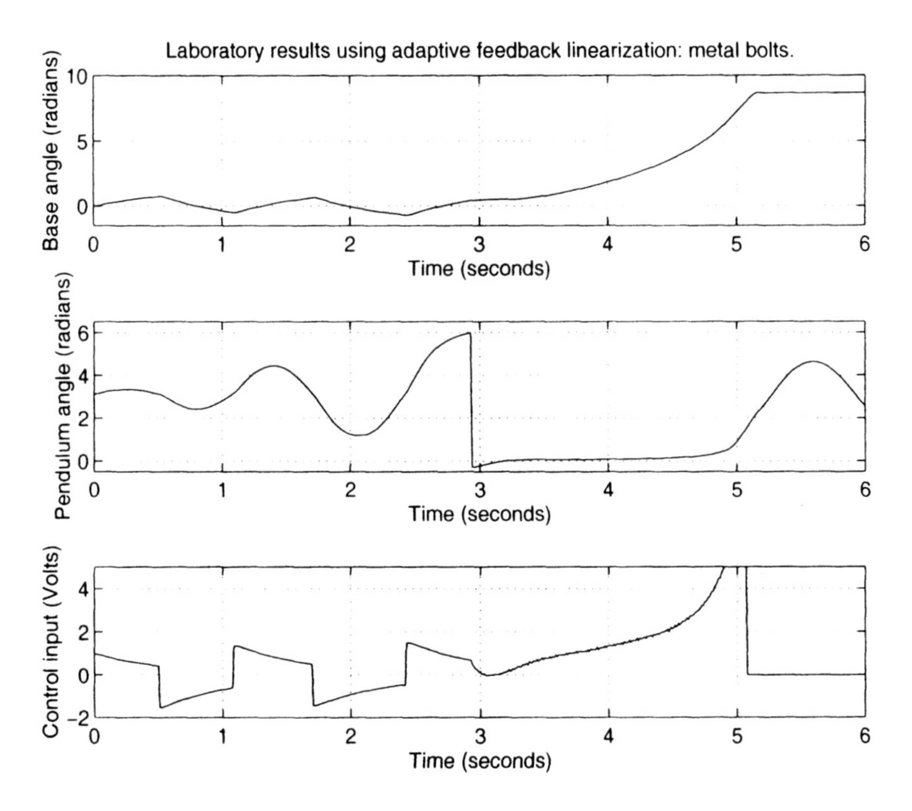

Figure 9.8. Experimental results of AFL with disturbance: metal bolts.

Define $d = [1, d_1(\xi), \dots, d_{m-1}(\xi)]^\top \in \mathrm{R}^m$, $\zeta^\top = \frac{[\mu_1, \dots, \mu_p]}{\sum_{i=1}^{p} \mu_i}$, and

$$\theta = \begin{bmatrix} \theta_{1,0} & \cdots & \theta_{p,0} \\ \vdots & \ddots & \vdots \\ \theta_{1,m-1} & \cdots & \theta_{p,m-1} \end{bmatrix}. \tag{9.51}$$

Then, the nonlinear equation that describes the fuzzy system can be written as

$$\mathcal{F}_{fs}(\xi) = d^\top \theta \zeta \tag{9.52}$$

and $\frac{\partial \mathcal{F}_{fs}}{\partial \theta} = d\zeta^\top$. Even though so far the approximators have been set up using a parameter vector, the notation also allows for our a parameter matrix as in our case here.

Given this notation, we can write

$$\begin{aligned} \mathcal{F}_\alpha(z, \theta) &= \bar{\chi}(x) - d_\alpha^\top \theta_\alpha \zeta_\alpha & (9.53) \\ \mathcal{F}_\beta(z, \theta) &= -d_\beta^\top \theta_\beta \zeta_\beta, & (9.54) \end{aligned}$$

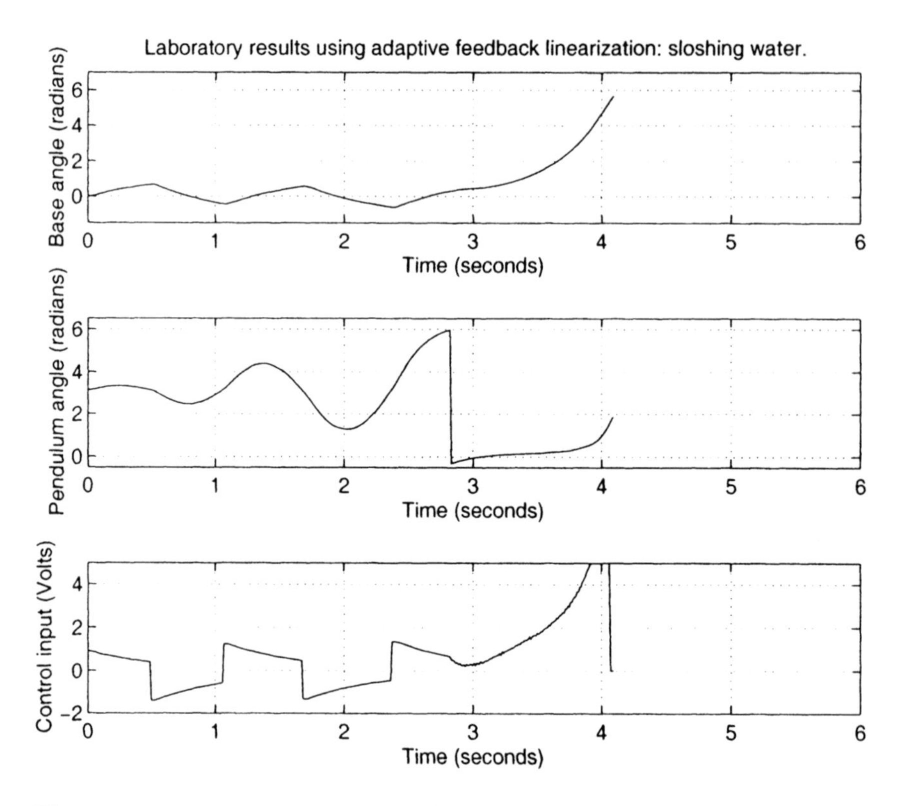

Figure 9.9. Experimental results of AFL with disturbance: sloshing water.

where we have partitioned θ in θ_α and θ_β for convenience, and we let $z = [q^\top, x^\top]$. Moreover, let $w_\alpha = \mathcal{F}_\alpha(z, \theta) - \alpha$ and $w_\beta = \mathcal{F}_\beta(z, \theta) - \beta$, and we assume the existence of some θ_α, θ_β such that $|w_\alpha| \leq W_\alpha$ and $w_\beta \leq W_\beta$ whenever $z \in S_z$ for some known $W_\alpha > 0$ and $W_\beta > 0$. The set S_z will be defined below with the aid of a bounding control term. Since θ_α and θ_β are unknown, we use $\hat{\theta}_\alpha$ and $\hat{\theta}_\beta$, which will be modified on line with an adaptation law.

In simulation we took $W_\alpha = 0.5$ and $W_\beta = 1.1$, since the representation mismatches were expected to be small. In the laboratory, however, it was necessary to increase these bounds to $W_\alpha = 5$ and $W_\beta = 8$, because apparently the complexities of the real plant were more difficult to represent than the model. Note that these values were chosen as the result of a tuning process, where rough, intuitive estimates of the values were used to start with and then tuned in order to improve the performance of the controller. The functional effect of increasing the error bounds W_α and W_β is to increase the magnitude of the discontinuous term.

Based upon the general form of the system model, (9.29), we take the following set of equations for both $\mathcal{F}_\alpha(z,\hat{\theta})$ and $\mathcal{F}_\beta(z,\hat{\theta})$ (also used for the direct adaptive fuzzy controller in the next section):

$$\begin{aligned} d^\top &= [1, \xi_1, \xi_2, \sin\xi_3, \xi_4] \\ &= [1, q_1 - x_1, q_2 + \frac{b_1}{b_2}x_2, \sin x_1, x_2]. \end{aligned} \tag{9.55}$$

Since the original states ξ are readily available, we will generally use these, instead of $[q^\top, x^\top]$, keeping in mind that the transformation $[q^\top, x^\top] = T(\xi)$ makes both choices equivalent.

The fuzzy systems use five rules each, of the form

$$\mathbf{If} y \mathrm{is} F_i \mathbf{Then} c_i = f_i(d), \quad i = 1, \ldots, 5 \tag{9.56}$$

where each $f_i(d)$ is, respectively, a row of the matrices $d^\top\hat{\theta}_\alpha$ and $d^\top\hat{\theta}_\beta$, and we initialize the system with

$$\hat{\theta}_\alpha(0) = \begin{bmatrix} 0 & 0 & 0 & 0 & 0 \\ 0 & 0 & 0 & 0 & 0 \\ a_2 & a_2 & a_2 & a_2 & a_2 \\ a_3 & a_3 & a_3 & a_3 & a_3 \\ a_4 & a_4 & a_4 & a_4 & a_4 \end{bmatrix}, \quad \hat{\theta}_\beta(0) = \begin{bmatrix} b_2 & b_2 & b_2 & b_2 & b_2 \\ 0 & 0 & 0 & 0 & 0 \\ 0 & 0 & 0 & 0 & 0 \\ 0 & 0 & 0 & 0 & 0 \\ 0 & 0 & 0 & 0 & 0 \end{bmatrix}. \tag{9.57}$$

Note that this fuzzy system design is over-specified, mainly in the case of $\mathcal{F}_\beta(z,\hat{\theta})$, because, from the system model, this function is not expected to depend on the state. However, this choice was made to allow for a greater adaptation flexibility. The initialization (9.57) gives the system the best known starting point in the search space. The input fuzzy sets F_i are as described in Figure 9.10.

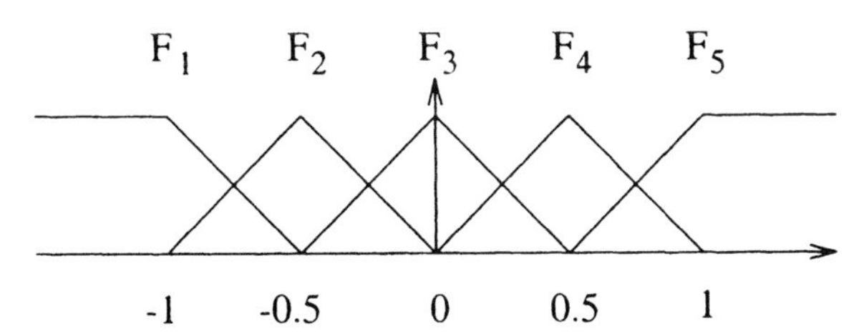

Figure 9.10. Input membership functions.

We may now define the indirect adaptive control law from (9.20) as

$$u = \nu_a = \nu_s + \frac{1}{g_0}(W_\alpha + W_\beta|\nu_s|)\mathrm{sgn}(e) + \nu_{bi}, \tag{9.58}$$

where $\nu_s = \frac{1}{\mathcal{F}_\beta}(-\mathcal{F}_\alpha - c_1 e)$. We let $c_1 = 1$ and $k_1 = 8$ (with these choices, the poles of the error transfer function are at $s = -1$ and $s = -9$, which produce a small error settling time). The term ν_{bi} is a bounding control term (see Exercise 9.2). To define it, we first need to determine bounding functions $\bar{f}_\xi(\xi) \geq |f_\xi(\xi)|$ and $\bar{g}_\xi(\xi) \geq |g_\xi(\xi)|$ whenever $|e| \geq M_e$. Based on the numerical values of (9.46) and (9.47), the bounding functions were empirically determined to be

$$\bar{f}_\xi(\xi) = 70\xi_2 + 75\xi_3 + 10\xi_4 \tag{9.59}$$

$$\bar{g}_\xi(\xi) = 140. \tag{9.60}$$

Then, let $\nu_{bi} = 0$ whenever $|e| \leq M_e$, and otherwise

$$\nu_{bi} = \left[\frac{1}{g_0} \left(|d_\alpha^\top \theta_\alpha \zeta_\alpha| + \bar{f}_\xi(\xi) + (|d_\beta^\top \theta_\beta \zeta_\beta| + \bar{g}_\xi(\xi))|\nu_s| \right) - |\nu_{si}| \right] \operatorname{sgn}(e), \tag{9.61}$$

where $\nu_{si} = \frac{1}{g_0}(W_\alpha + W_\beta |\nu_s|)\operatorname{sgn}(e)$. The parameter M_e defines a bounded, closed subset of the error state space, within which the error is guaranteed to stay. Since we are dealing with a regulation problem, by definition of the error system boundedness of e implies boundedness of the state ξ and of z, which is therefore confined to a compact set S_z whose size is a function of M_e. For simulation, we took $M_e = 0.4$; again, a larger margin had to be used in implementation, and the smallest acceptable value was $M_e = 3$. Note that although it is possible in principle to take an arbitrarily small M_e, in practice it is often the case that the bounding control acts "too much" with a small M_e, and the unavoidable limits in the control input signal cause the system to become unstable. Thus, the values shown here were first tuned, and were chosen because they gave us the best overall results.

To define the adaptation equations, let I_5 be a 5×5 identity matrix, and let $\Gamma = 0.05\, I_5$ for simulation, and $\Gamma = 0.1\, I_5$ for implementation, and take

$$\begin{aligned} \dot{\hat{\theta}}_\alpha &= -\Gamma d \zeta_\alpha^\top e \\ \dot{\hat{\theta}}_\beta &= -\Gamma d \zeta_\beta^\top e \nu_s. \end{aligned} \tag{9.62}$$

Note that these adaptation equations are equivalent to (9.24). A projection algorithm is used to ensure that $\hat{\theta}_\alpha$ and $\hat{\theta}_\beta$ remain within reasonable limits; specifically, it is sufficient to ensure that $\mathcal{F}_\beta(z, \hat{\theta}_\beta)$ is bounded away from zero, and that its value is such that the assumption $g(q, x) \leq g_0$ holds.

Note that the indirect adaptive control adaptation algorithm outlined in Section 9.2.2 guarantees that the parameter error matrices $\tilde{\theta}_\alpha = \hat{\theta}_\alpha - \theta_\alpha$ and $\tilde{\theta}_\beta = \hat{\theta}_\beta - \theta_\beta$ will at least stay bounded. Note also that the equations (9.46) and (9.47) are themselves only approximations, based on our best knowledge of the plant. Thus, it is possible that the approximators used do not represent f and g accurately, in spite of which closed-loop stability is achieved.

One of the assumptions in Section 9.2.2 is not satisfied, namely, that the zero-dynamics of the plant be input-to-state stable; thus, as happened with the adaptive and non-adaptive feedback linearizing controllers, boundedness of the state ξ_1, and possibly of the control input, are not expected. However, in principle, the controller should be able to achieve output regulation (i.e., keep the pendulum balanced).

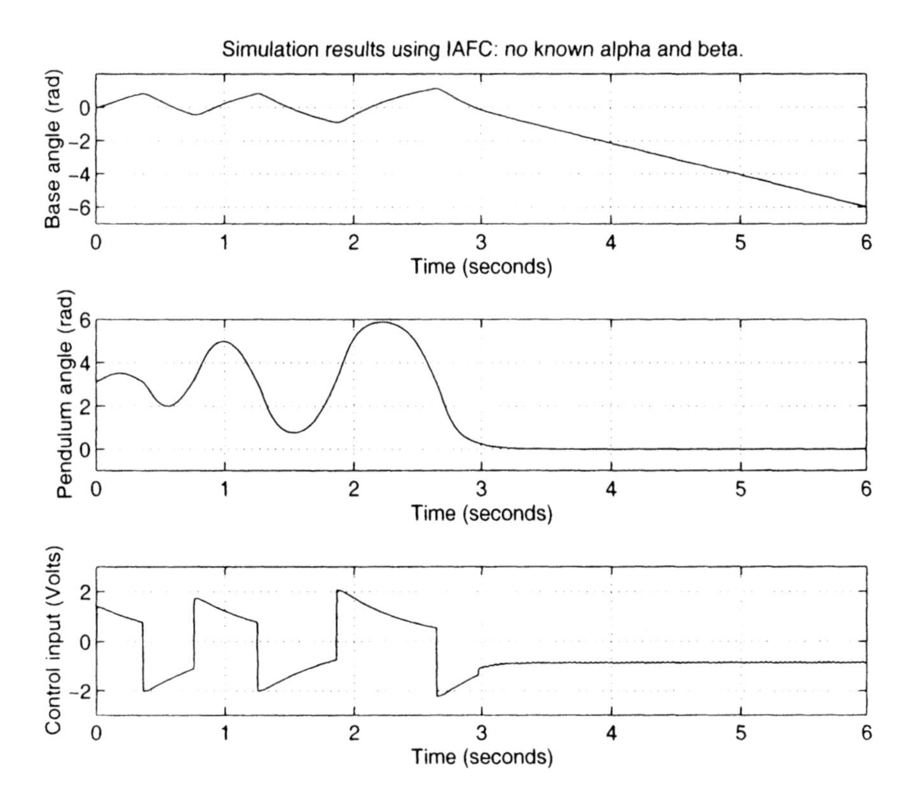

Figure 9.11. No plant dynamics knowledge used: IAFC simulation.

We see in Figure 9.11 that this is, indeed, the case. The pendulum is successfully balanced, and the settling time of the controller is smaller than in the case of any of the previous controllers, since here the output error's closed-loop poles are set at $s = -1$ and $s = -9$; the reason why these poles were not used for the (adaptive and non-adaptive) feedback linearizing controllers is that with these choices the performance of both algorithms degraded in simulation and experimentation, in terms of error convergence and robustness.

For implementation, we see in Figure 9.12 that the pendulum is balanced, using the nominal plant, although the output error is not exactly zero. When the bolts disturbance is used (Figure 9.13), the controller has trouble, similar to AFL (see Figure 9.13), because the control input reaches its lower limit of $-5V$. We see in Figure 9.14 that, with sloshing water, the controller performs better, although it is apparent that the control input limit is about to be reached. Thus, the performance of this IAFC design is roughly similar to that of adaptive feedback linearization.

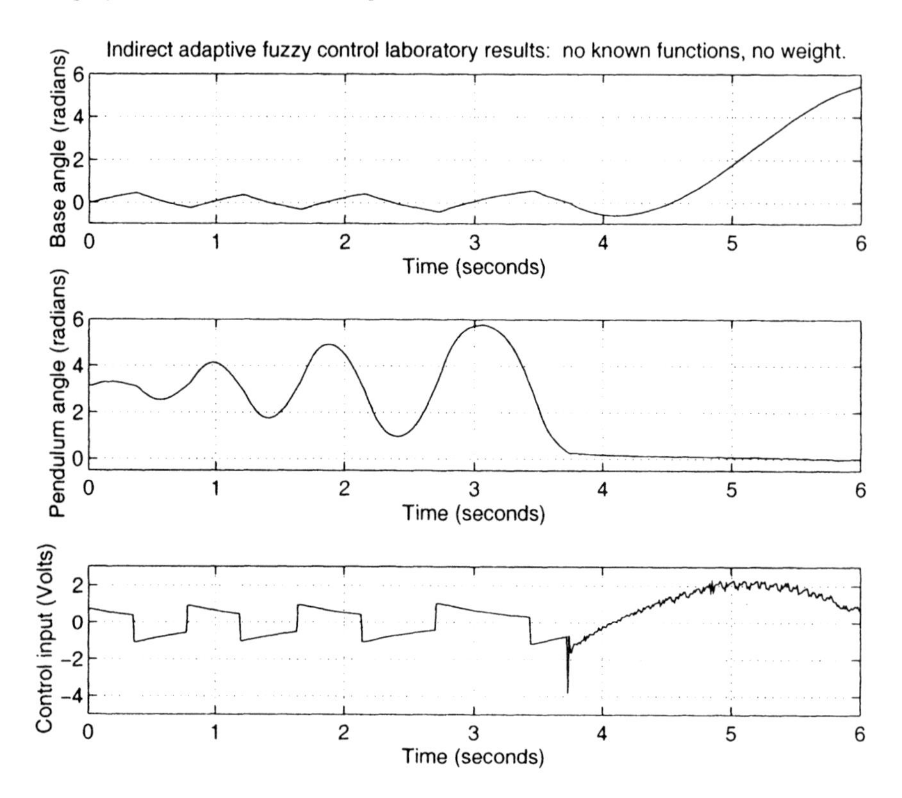

Figure 9.12. No plant dynamics knowledge used: Experimental results of IAFC with nominal plant.

9.7.2 Incorporation of Plant Dynamics Knowledge

In order to improve the robustness characteristics of the IAFC, we will now take a slightly different design approach, and will make explicit use of the knowledge we have of the plant, that is., the nonlinear model (9.29). By comparing the simulation and experimental results so far, we see that, although very useful for theoretical analysis and design, the model is nevertheless a relatively poor approximation of the rotational inverted pendulum. In spite of this fact, it can effectively be incorporated into the indirect adaptive scheme, and thus provide it with an improved disturbance rejection ability.

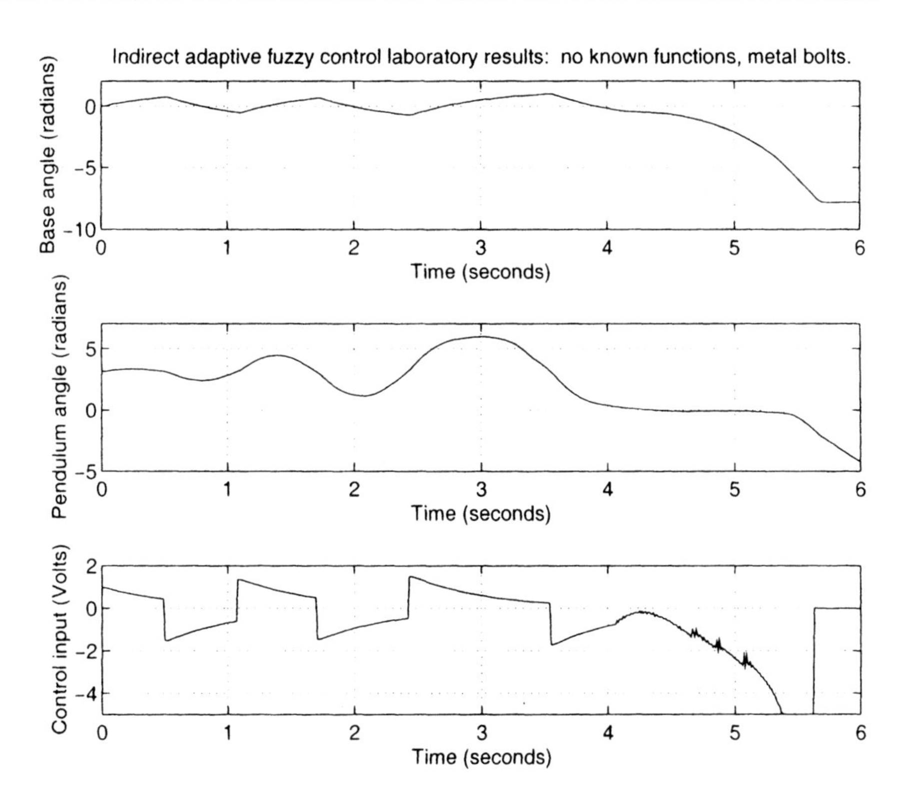

Figure 9.13. No plant dynamics knowledge used: Experimental results of IAFC with disturbance: metal bolts.

Consider the input-output equation (9.31), which we now rewrite as

$$\ddot{y} = \dot{x}_2 = (f_k(t) + \tilde{f}(q,x)) + (g_k(t) + \tilde{g}(q,x))u. \tag{9.63}$$

We are now assuming that the pendulum input-output equation can be represented by some known, non-zero, fixed functions $f_k(t)$ and $g_k(t)$, and unknown functions $\tilde{f}(q,x)$ and $\tilde{g}(q,x)$, which are to be identified on-line by the IAFC adaptation mechanism. That is, $f = f_k + \tilde{f}$ and $g = g_k + \tilde{g}$, and we assume $g_k(q,x) + \tilde{g}(q,x) \leq g_0 < 0$. Since the model (9.29) represents all our knowledge about the pendulum, we use it to specify the known functions as

$$\begin{aligned} f_k(t) &= a_2\xi_2 + a_3 \sin\xi_3 + a_4\xi_4 \\ g_k &= b_2. \end{aligned} \tag{9.64}$$

The functions f_k and g_k are computable, since we can measure the entire state of the plant. We will use these functions to rewrite the error system (9.48), where we let $\alpha(t,q,x) = \alpha_\Delta(t,q,x) + \alpha_k(t)$ and $\beta(q,x) = \beta_\Pi(q,x) + \beta_k(t)$, and we choose $\alpha_k = f_k$ and $\beta_k = g_k$ (strictly speaking, α_k and β_k

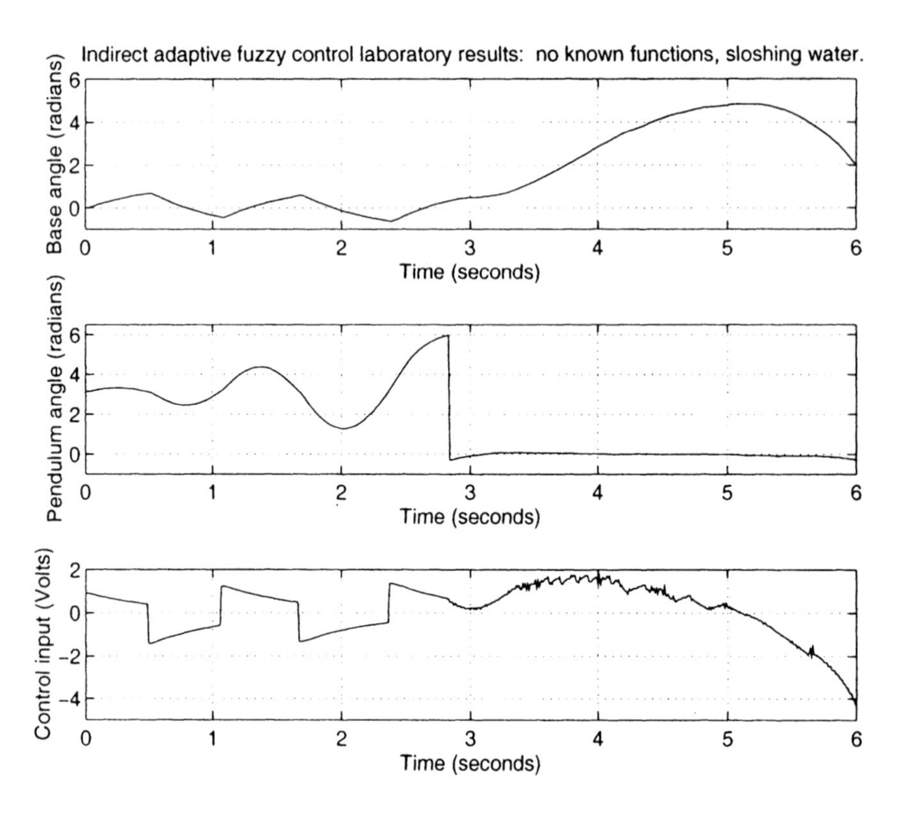

Figure 9.14. No plant dynamics knowledge used: Experimental results of IAFC with disturbance: sloshing water.

are not explicitly functions of time, but of the plant states; however, we use this notation for simplicity).

The unknown functions $\tilde{f}(q, x)$ and $\tilde{g}(q, x)$ can be represented using fuzzy systems as in equations (9.53), where we use the same vector of functions (9.55). If we know nothing about $\tilde{f}(q, x)$ and $\tilde{g}(q, x)$, a possible way to initialize their fuzzy system approximation is by letting $\theta_\alpha(0) = 0$ and $\theta_\beta(0) = 0$. In this manner, the adaptation mechanism will attempt to identify the plant by introducing variations to the functions defined in (9.64). Notice the fundamental difference that this design has with respect to the previous one: above, the input-output dynamics of the system were represented entirely by $f(q, x)$ and $g(q, x)$, which were estimated by the adaptation mechanism. Here, we let the IAFC estimate *perturbations* off $f_k(t)$ and $g_k(t)$ (recall that $f_k(t)$ and $g_k(t)$ contain all our knowledge about the plant). As we will see, the characteristics of the adaptive process change based upon how we define and initialize the system.

The approximation bounds W_α and W_β need not be reset with this configuration, because the representation errors w_α and w_β using (9.64) are expected to be of the same order of magnitude as in the previous case, and possibly less or at most equal; therefore, they are taken as defined before, with their respective values for simulation and implementation.

Following a similar reasoning, the bounding functions $\bar{f}_\xi(\xi)$ and $\bar{g}_\xi(\xi)$ are expected to be less than or equal in magnitude to (9.59) and (9.60), respectively; thus, although they could be redefined and made smaller, it is certainly still valid that the required conditions $\bar{f}_\xi(\xi) \geq |\tilde{f}(T(\xi))|$ and $\bar{g}_\xi(\xi) \geq |\tilde{g}(T(\xi))|$ are satisfied using (9.59) and (9.60).

To incorporate the plant knowledge into the design we use the same control law (9.20), but now

$$\begin{aligned}\mathcal{F}_\alpha(z,\hat{\theta}) &= \bar{\chi} - \alpha_k - d_\alpha^\top \hat{\theta}_\alpha \zeta_\alpha \\ \mathcal{F}_\beta(z,\hat{\theta}) &= -\beta_k - d_\beta^\top \hat{\theta}_\beta \zeta_\beta .\end{aligned}$$

The discontinuous and bounding control terms are taken without change, as well as the adaptation law (9.62).

Observation of the simulation results in Figure 9.15 using this modified IAFC shows little difference with Figure 9.11; apparently, the basic characteristics of the controller remain the same, that is, the state ξ_1 is still unbounded, and the pendulum is balanced with a very similar control input. Implementation of the controller on the nominal plant, as seen in Figure 9.16 shows a slightly faster error convergence to zero. The most notable differences arise when the plant is disturbed; in Figure 9.17 we see that the controller is able to handle the bolts disturbance effectively, and without saturation of the control input. The same good behavior is observed in Figure 9.18, where the plant is under the effects of sloshing water dynamics; initially, the pendulum is not perfectly balanced, but eventually the error converges to zero.

9.8 Direct Adaptive Fuzzy Control

Now we turn our attention to direct adaptive fuzzy control (DAFC) for the inverted pendulum. Using IAFC, a controller is constructed that seeks to identify the plant dynamics and use its best estimate to produce an approximation to a feedback linearizing law. Here, the approach is to search for an unknown control law that provides (at least) asymptotically stable tracking and is able to compensate for disturbances and maintain stability. As was the case with IAFC, the DAFC methodology allows the designer to use previous knowledge or experience with the plant in various ways. Here we will illustrate two representative possibilities, and will study how the results change depending on how the design is chosen; in fact, it will be seen that it is possible to obtain significantly different control results, depending on the approach taken.

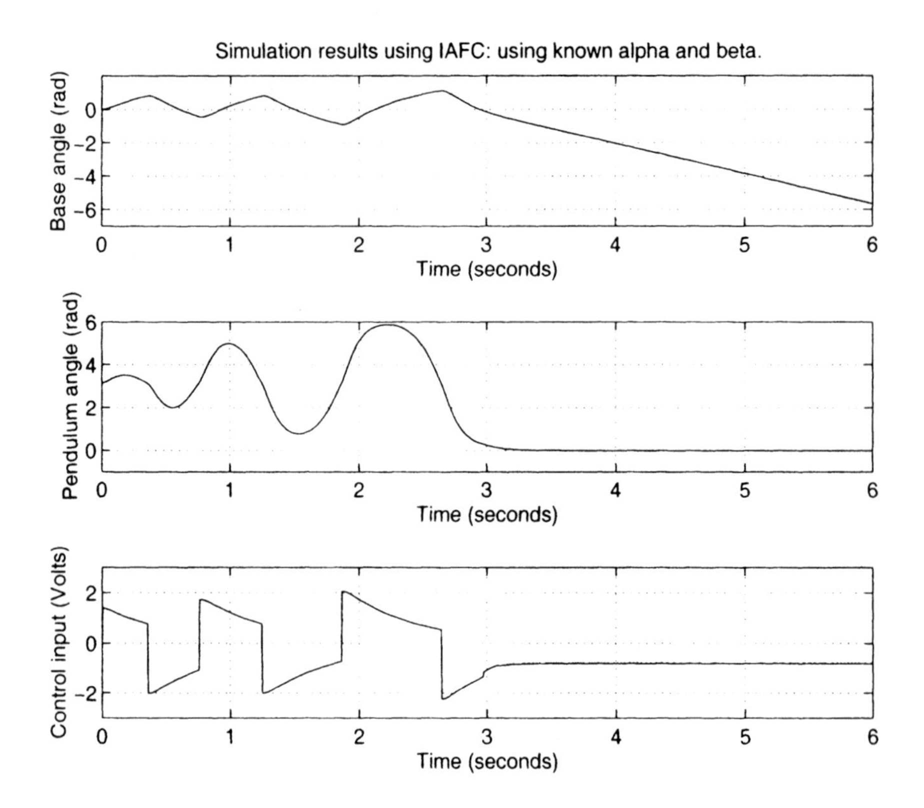

Figure 9.15. Plant dynamics knowledge used: IAFC simulation.

In IAFC, it was possible to use a known part of the plant dynamics, represented by α_k and β_k, in the control design. We saw that for the pendulum application it was beneficial to include the known dynamics, because it increased the robustness of the design. DAFC provides the designer with a method to incorporate a best guess of what the controller should be (below we will call this the "known controller", denoted by ν_k). The algorithm then adaptively tunes a fuzzy controller to compensate for inaccuracies in our choice of this known controller.

9.8.1 Using Feedback Linearization as a Known Controller

As described in Section 9.2.1, DAFC is a somewhat more restrictive technique than its indirect counterpart, since in addition to the assumption that the plant is minimum-phase, it is assumed that $g(q,x)$ is bounded by two finite constants, g_0 and g_1. For the pendulum this assumption holds, since $-\infty < g_1 \leq g(q,x) \leq g_0 < 0$, where for instance we take, as before, $g_0 = -100$ and $g_1 = -140$. The last plant assumption needed is that, for some $g_d(q,x) \geq 0$, $|\dot{g}(q,x)| \leq g_d(q,x)$. Since g is expected to be a constant,

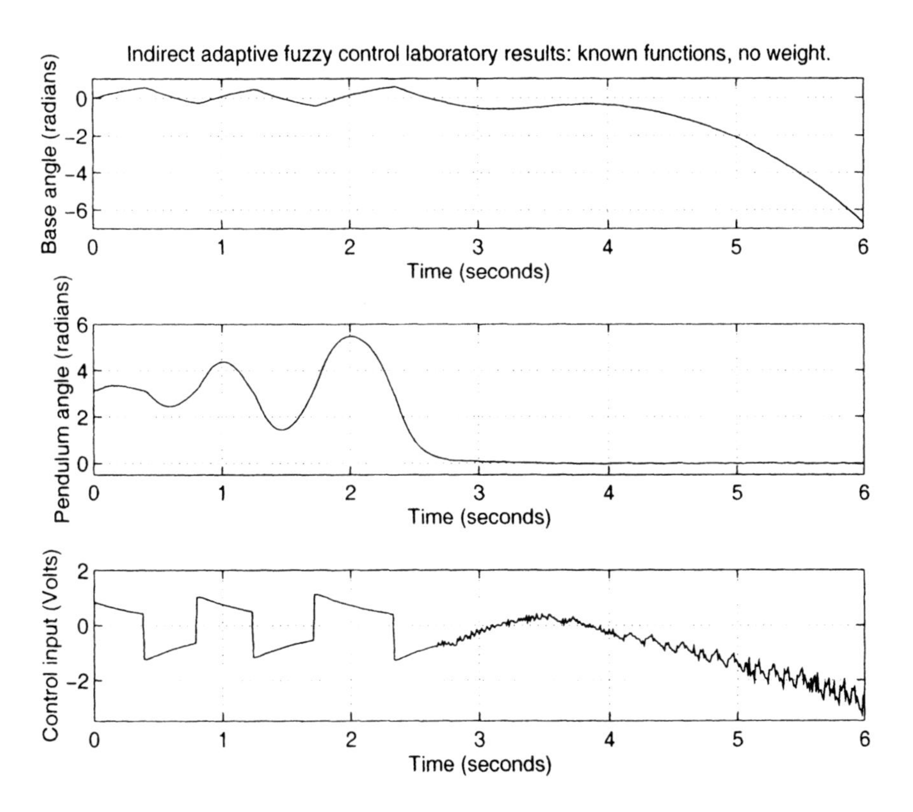

Figure 9.16. Plant dynamics knowledge used: experimental results of IAFC with nominal plant.

we can safely set $g_d(q, x) = 0$, and the assumption holds.

Note that the control equations derived in Section 9.2.1 are based on the premise that $g(q, x)$ is positive. However, the negative case can be easily accommodated by making appropriate changes. Specifically, the adaptation differential equation and the discontinuous term will have, each, a small but crucial sign change.

Recall from Section 9.2.1 the static control law (9.7). The feedback linearizing portion of (9.7) cannot be implemented due to uncertainty; however, in some cases a designer may know some controller ν_k with good closed-loop performance that may be derived from experience or from a successful design model. Then it may be desirable to include such a known term in the adaptive scheme. Then, instead of (9.10) we will select the direct adaptive control law

$$u = \mathcal{F}(z, \hat{\theta}) = d_u^\top \hat{\theta} \zeta_u + \nu_k - W\text{sgn}(e) + \nu_{bd} \tag{9.65}$$

with $\zeta_u \in \mathbb{R}^5$ is as defined for the IAFC with the fuzzy sets of Figure 9.10.

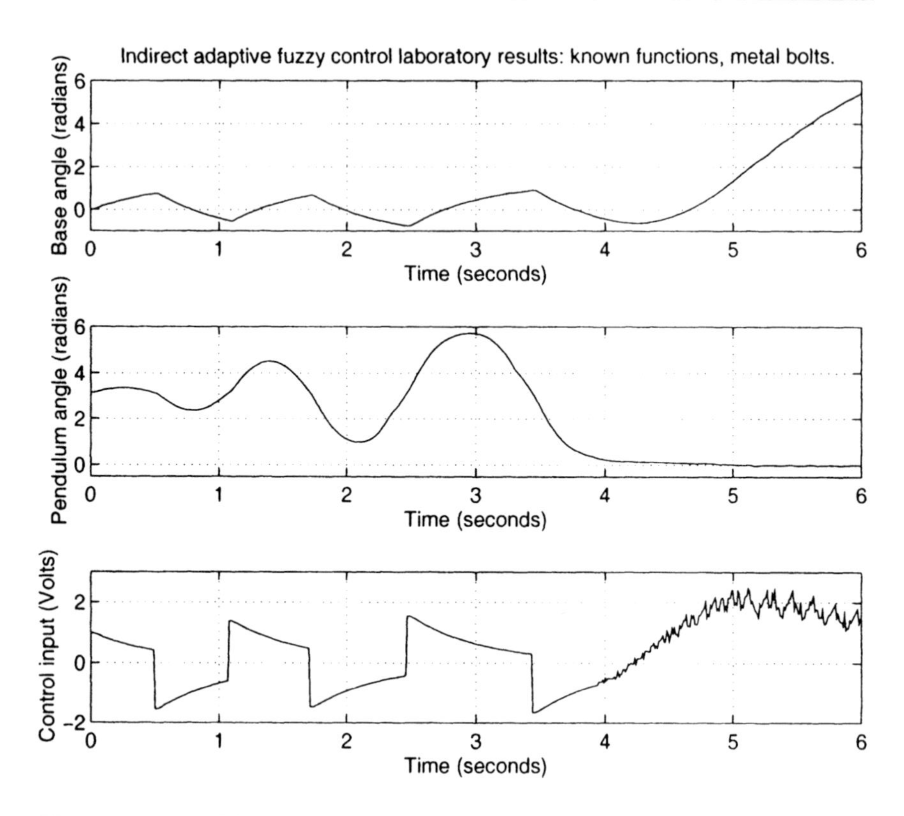

Figure 9.17. Plant dynamics knowledge used: experimental results of IAFC with disturbance: metal bolts.

Notice the sign change of the discrete term, which is due to the fact that g is negative for the pendulum, and recall that $g_d = 0$. In this section we will use ν_k, and we will set it to zero in the next.

Letting $w = \mathcal{F}(z, \theta) - \nu_a$ (with ν_a redefined as $\nu_a = \nu_s - W\text{sgn}(e)$ and the discontinuous term in ν_s also has negative sign), we assume some θ exists such that $|w| \leq W$ for $z \in S_z$. The compact set S_z will be determined by the bounding control term ν_{bd}. In practice, it is often hard to have a concrete idea about the magnitude of W because the relation between ν_a and its fuzzy representation might be difficult to characterize; however, it is much easier to begin with a rough, intuitive idea about this bound, and then iterate the design process and adjust it, until the performance of the controller indicates that one is close to the right value. For simulation, we found that $W = 0.01$ gave us good results, and in the laboratory we increased it to $W = 0.1$. These bounds are both relatively small, which indicates that the fuzzy system we used, although a simple one, could represent the unknown controller ν_a with sufficient accuracy.

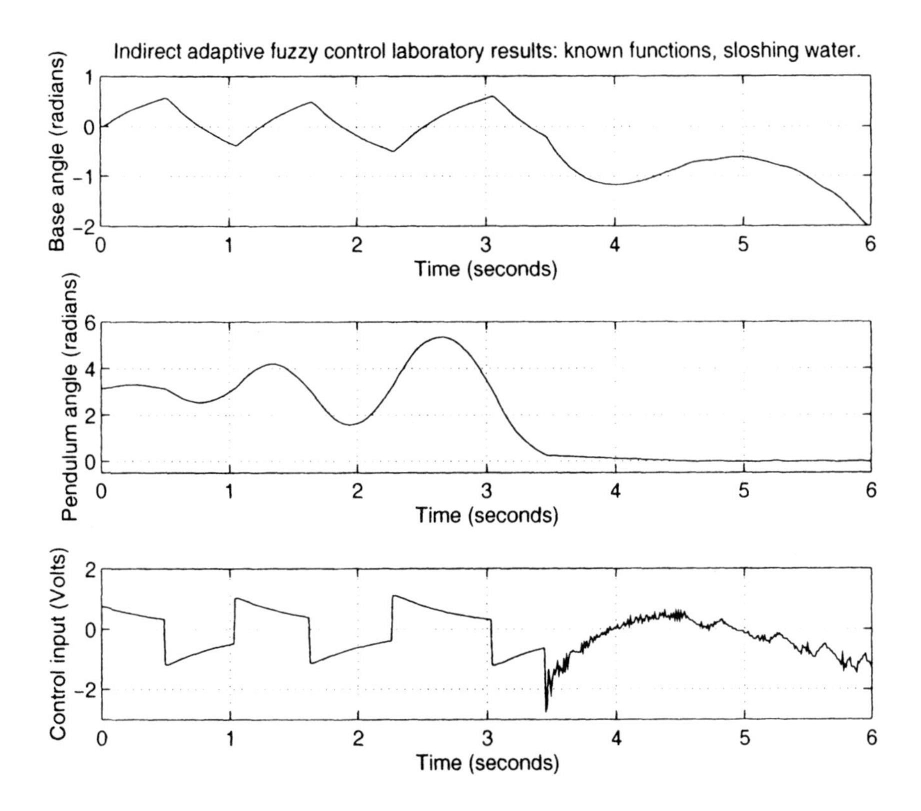

Figure 9.18. Plant dynamics knowledge used: experimental results of IAFC with disturbance: sloshing water.

The matrix $\hat{\theta} \in \mathsf{R}^{5\times 5}$ is adaptively updated on-line, and the function vector d_u is taken as defined in (9.55). The fuzzy system again uses only five rules, as given by (9.56), and now each $f_i(\xi)$ is a row of the matrix $d_u^\top \hat{\theta}$. *In order to approximate a feedback linearizing controller* we will define ν_k as in (9.32), and we will take ν as in (9.35). Further, following the same line of reasoning as in Section 9.7.2, we initialize the fuzzy system with $\hat{\theta}(0) = 0$.

The bounding control term ν_{bd} (see Exercise 9.2) needs the assumption that $f_\xi(\xi)$ is bounded, with $|f_\xi(\xi)| \leq \bar{f}_\xi(\xi)$. We take $\bar{f}_\xi(\xi)$ as defined in (9.59); then, if $e > M_e$,

$$\nu_{bd} = -\left(|d_u^\top \hat{\theta}\zeta_u| + \frac{\bar{f}_\xi + |\bar{\chi} + r^{(n)}|}{|g_0|}\right) \operatorname{sgn}(e) \tag{9.66}$$

and $\nu_{bd} = 0$ otherwise. As in the IAFC case, M_e defines a ball to which e converges and within which it stays afterwards – an invariant set.

For simulation, we used $M_e = 0.6$, and increased it to $M_e = 2.5$ in implementation. Please refer to the discussion on IAFC for an explanation on how we determined these values.

The last part of the DAFC mechanism is the adaptation law, which is chosen in such a way that the output error converges asymptotically to zero, and the parameter error remains at least bounded. To account for the negative sign of g, instead of (9.14) we use

$$\dot{\theta} = -\Gamma d_u \zeta_u^\top e. \tag{9.67}$$

For simulation, we used $\Gamma = 0.9I_5$, and in experimentation we decreased the gain slightly to $\Gamma = 0.5I_5$. With these choices the algorithm was able to adapt fast enough to perform well and compensate for disturbances, but without inducing oscillations typical of a too high adaptation rate.

Figure 9.19 shows the simulation results with this controller. It has a behavior typical of feedback linearizing controllers on this plant: the control input settles and oscillates around a non-zero value, thus keeping the pendulum base rotating. Observe in Figure 9.20 the performance of the DAFC design on the nominal plant: the error is effectively decreased to zero, and the behavior of the base is similar to the previous cases. Again, the advantages given by the adaptive capability of this algorithm appear most distinctively in the presence of strong disturbances: the controller is quite successful with both the metal bolts (Figure 9.21) and the sloshing water (Figure 9.22). The pendulum is kept balanced, and the control input remains within small bounds around zero. Thus, this design proved to be robust and reliable, although it still has the weakness that all the other controllers presented in this work until now share: it is not able to deal with the instability condition of the system's zero-dynamics. Therefore, as a last and, in our opinion, best adaptive fuzzy control design example, we will now describe a DAFC that can not only compensate for the induced disturbances (and, in fact, it does it with greater ease than all the previous controllers), but is also able to keep state boundedness, even though the theoretical analysis of Section 9.2.1 does not predict it (recall that such analysis does not preclude it).

9.8.2 Using the LQR to Obtain Boundedness

Although the theoretical analysis in Section 9.2.1 uses the assumption that the unknown control law ν_a which the DAFC tries to identify contains a feedback linearizing law, it was found experimentally that this does not need to be the case. If the right known controller is used, and/or the adaptation mechanism is initialized appropriately, then the adaptation algorithm will converge to a controller that might behave in a very different manner, because this mechanism seems to try to find a (local) optimum controller closest to its starting point in the search space, and this optimum does not necessarily have to be a feedback linearizing controller.

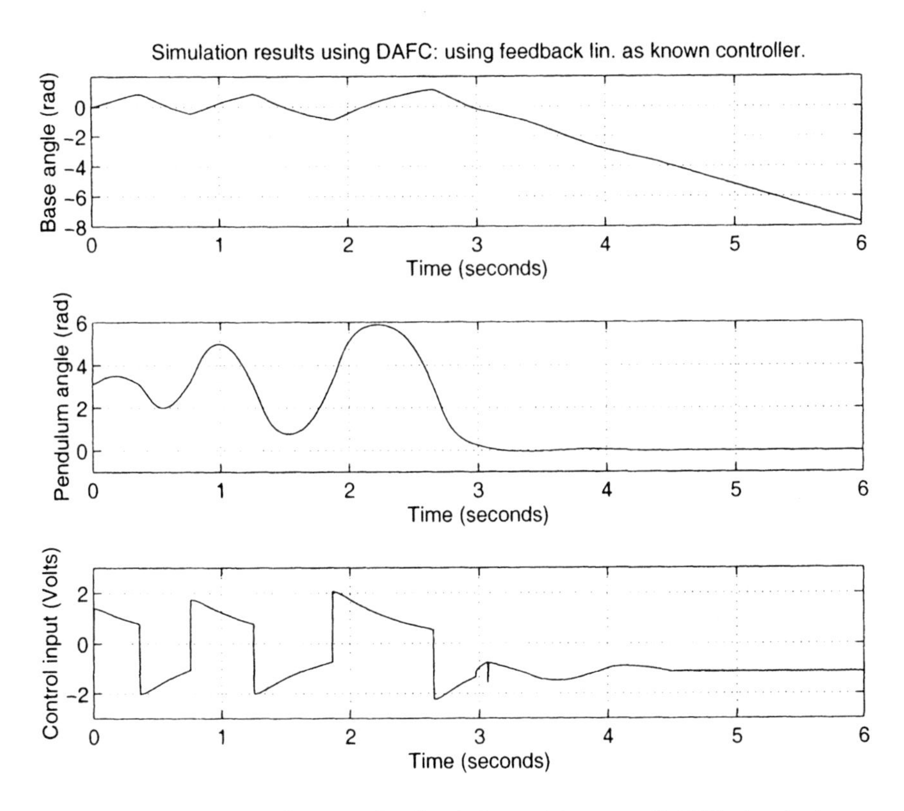

Figure 9.19. DAFC using feedback linearizing u_k: DAFC simulation.

This finding is of special importance when the control design task involves dealing with a non-minimum phase plant like the pendulum, for which feedback linearization based adaptive techniques have the limitation of being unable to maintain complete state boundedness. As stated before, the unboundedness of the state ξ_1 is admissible for the pendulum, but it might not be for other systems.

Consider, for instance, that a non-adaptive controller is available that can control the non-minimum phase plant with state boundedness. Then, it is possible that the desirable boundedness characteristics of this controller can be incorporated into the DAFC design, and enhanced by the robustness that the adaptive method provides. This is precisely the point of view taken in Chapter 7, where we assume the existence of a Lyapunov function and a corresponding controller that is able to guarantee stability. This analysis is independent of the plant's dynamic structure (e.g., whether it is feedback linearizable non-minimum phase), and rather relies on the existence of a stabilizing controller. Clearly, this stabilizing controller will need to take the plant's characteristics into account, but if it is conceived appropriately,

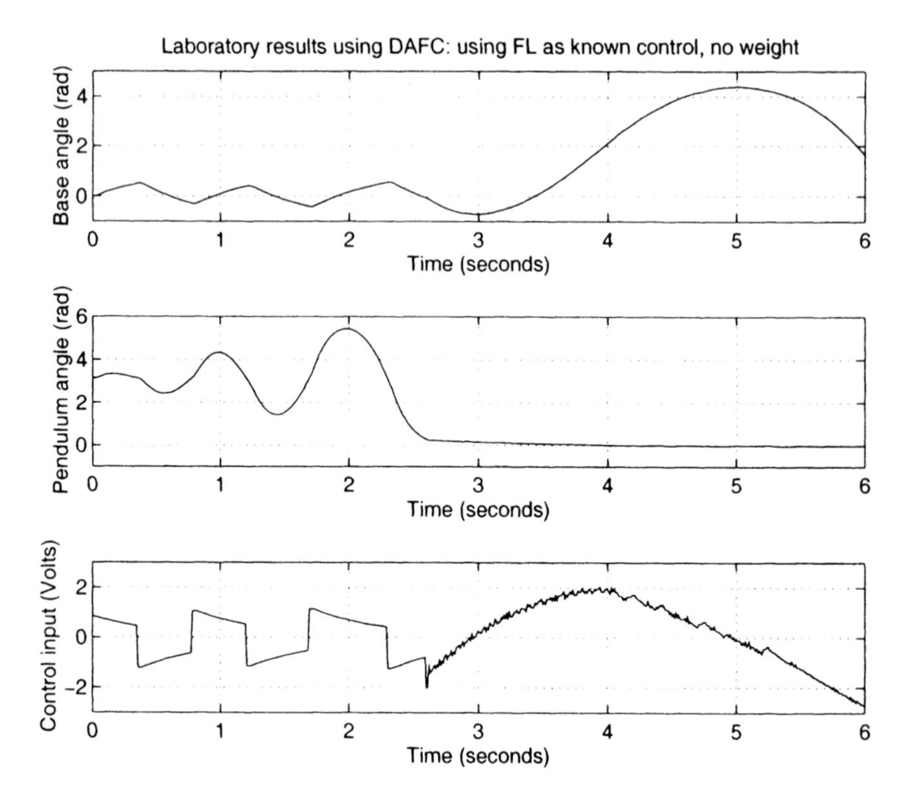

Figure 9.20. DAFC using feedback linearizing u_k: experimental results of DAFC with nominal plant.

it may be able to get around problems such as the unstable zero dynamics. Although the stability analysis is less general, we will attempt to apply the concepts in Chapter 7 to design a better controller.

For our present study, a most natural and intuitive choice for this purpose is the LQR. This controller implements a linear function of the plant states, and is able to drive the state error to zero for the nominal plant while maintaining state boundedness. Observe in Figures 9.2 and 9.3 that all the plant states are indeed kept bounded. The LQR was shown to have very good performance using the nominal, undisturbed system. Nevertheless, it fails immediately when significant disturbances are introduced.

A DAFC will be designed based on the LQR, so that its good behavior in terms of state boundedness can be kept, and its weakness regarding plant disturbances eliminated. Two different, and functionally equivalent ways were found to accomplish this. The first makes use of the term ν_k, as illustrated above. The second uses an appropriate initialization of the matrix $\hat{\theta}$. Since the use of ν_k has already been shown, only the second

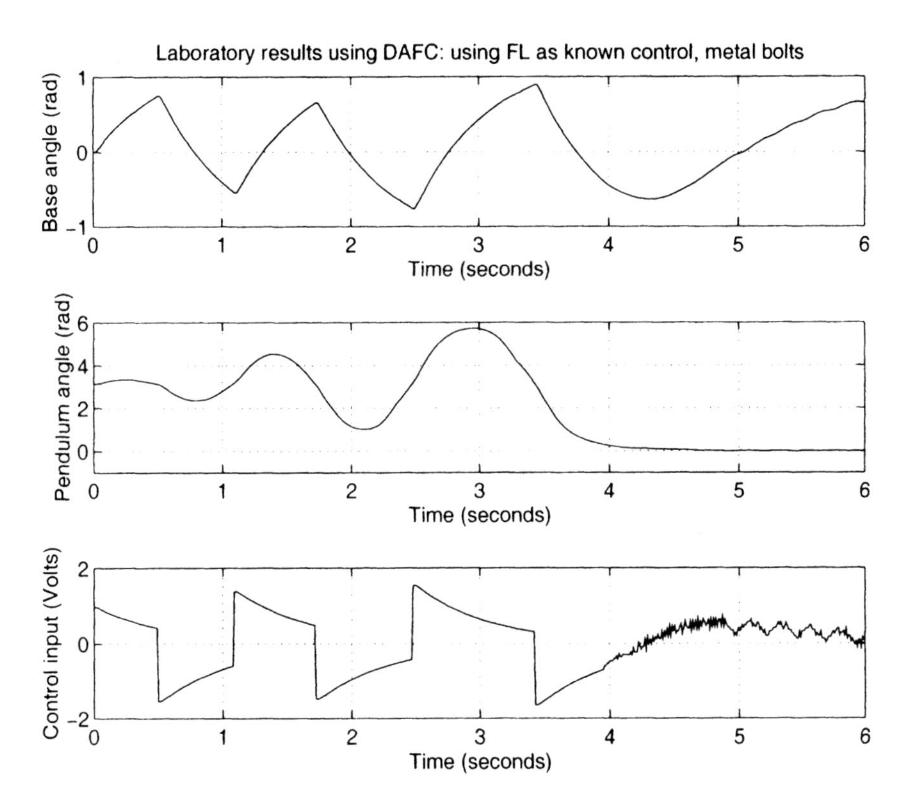

Figure 9.21. DAFC using feedback linearizing u_k: experimental results of DAFC with disturbance: metal bolts.

approach will be described here.

Take, again, the adaptive control law (9.65), now with a smaller gain $\Gamma = 0.005I_5$ (i.e., we slow adaptation down) for simulation and implementation purposes. This adaptation gain was chosen via tuning of the controller. We found that higher gains tended to produce a more oscillatory behavior.

The fundamental difference between this and the previous design lies in the unknown stabilizing control that we aim to identify. Before, the adaptive search was configured in such a way that the mechanism converged to a feedback linearizing law; now, we want it to identify a control input that behaves basically like an LQR, that is, we want to implement a design that behaves like an *adaptive LQR*. To do this, we start the adaptation algorithm at a point in the search space in the proximity of the LQR controller. That

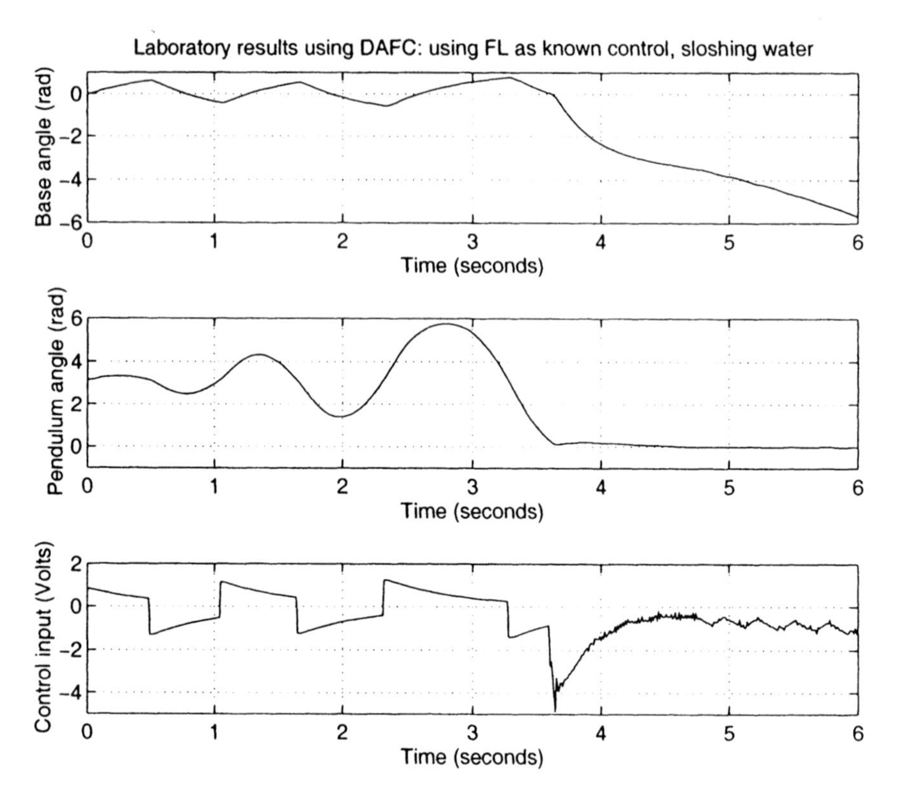

Figure 9.22. DAFC using feedback linearizing u_k: experimental results of DAFC with disturbance: sloshing water.

is, we will use it to initialize the parameter matrix $\hat{\theta}$ as

$$\hat{\theta}(0) = \begin{bmatrix} 0 & 0 & 0 & 0 & 0 \\ 0.7 & 0.7 & 0.7 & 0.7 & 0.7 \\ 1 & 1 & 1 & 1 & 1 \\ 10.8 & 10.8 & 10.8 & 10.8 & 10.8 \\ 0.7 & 0.7 & 0.7 & 0.7 & 0.7 \end{bmatrix}. \tag{9.68}$$

Notice that the sign of the gains has been reversed, since in this case we do not use the state error $r - \xi$, but rather the vector (9.55). It is worth mentioning that an alternate, similar way of implementing this design consists of using the control term $\nu_k = K^\top \xi$ (i.e., we set ν_k equal to the LQR state feedback law) and letting $\hat{\theta}(0) = 0$.

The design is now complete, and the results obtained corroborate our expectations about it. We see in Figure 9.23 the behavior of the controller in simulation. Observe that it closely resembles the performance of the LQR in Figure 9.2, both in terms of the states and the control input it produces.

Figure 9.24 shows the experimental results of the modified DAFC on the nominal plant. The pendulum is balanced with a control input that approaches zero on average (which means that the state ξ_1 is not going to grow without bound), and the performance is similar to that of the LQR in Figure 9.3, although the output error is not exactly zero. The most interesting results are found in Figure 9.25 and Figure 9.26. We note that using both the metal bolts and the sloshing water disturbances the controller is able to maintain convergence, and in addition it has a behavior much like that of an LQR re-tuned for the disturbed system: the pendulum base does not keep rotating, but lightly oscillates around a constant position, and the pendulum is balanced with a control input that has an average value close to zero. We observe in both cases, and most distinctly in the case of the water, how the controller adapts to the system with random disturbances; the control input oscillations are relatively large at first, and after a couple of seconds decrease in amplitude, as the DAFC approximates the ideal controller more and more. At the same time, the error converges to zero, and the base movement decreases.

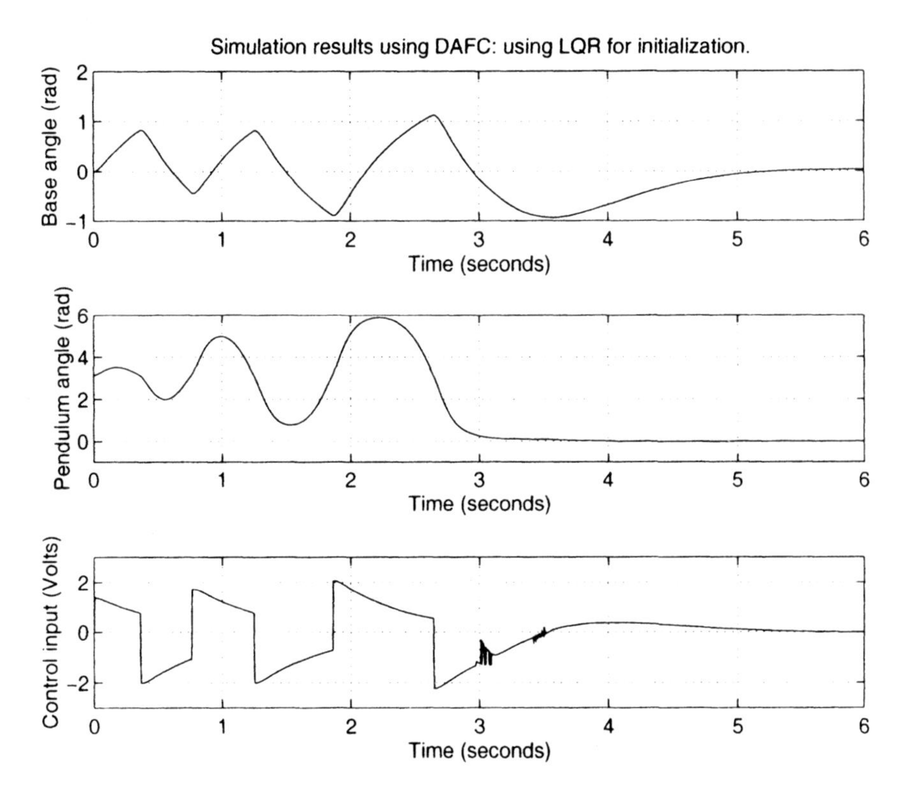

Figure 9.23. DAFC initialized as LQR: DAFC simulation.

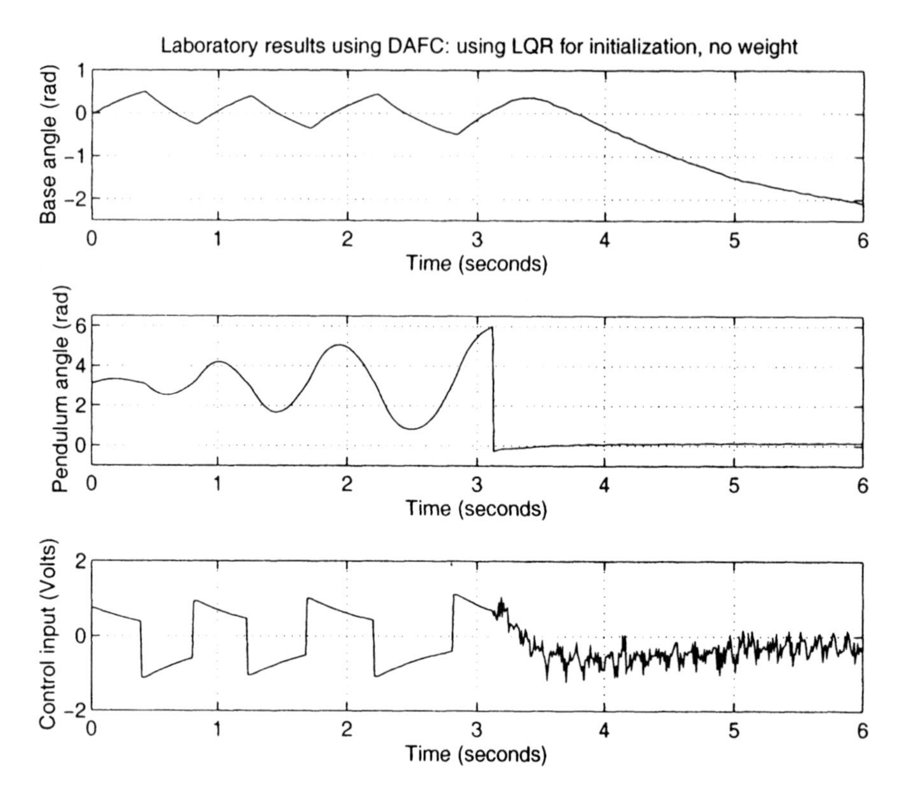

Figure 9.24. DAFC initialized as LQR: experimental results of DAFC with nominal plant.

9.8.3 Other Approaches

We tried several more DAFC configurations, which gave us more insight into the technique and showed it to be very flexible and reliable. Here we summarize three of these alternative approaches and the results we obtained. All these tests were done only in simulation; since the results in this chapter show a good match between simulation and implementation, we are confident that the following designs would also have similar simulation and experimental performance.

First, to test the ability of the algorithm to adapt and maintain stability, we started the controller without providing it with any information at all about the system, that is, we let $\nu_k = 0$ and $\hat{\theta}(0) = 0$. The performance of this controller eventually resembled that of the first design in this section (see Figure 9.19), where we tried to approach a feedback linearizing law: stability was attained with unboundedness of the state ξ_1. However, the control input it required was outside the implementation bounds.

Second we tested a *destabilizing* known controller ν_k. For the inverted

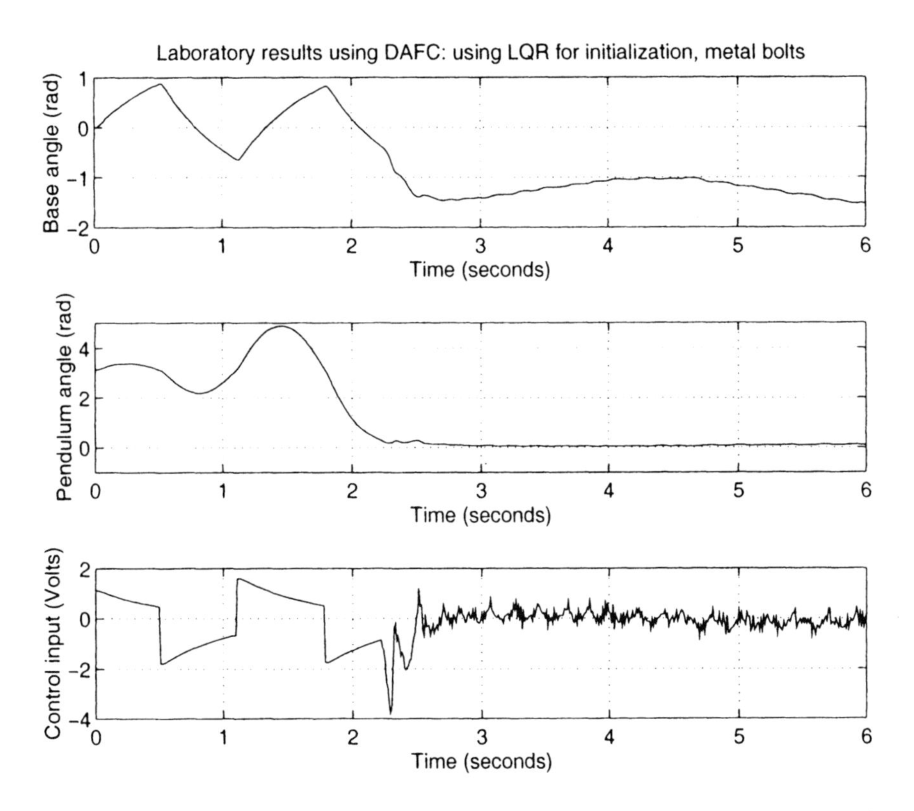

Figure 9.25. DAFC initialized as LQR: experimental results of DAFC with disturbance: metal bolts.

pendulum, the simplest controller that makes the system unstable is a nonzero constant input; hence we set $\nu_k = c$, with c a constant (which we kept within $\pm 5V$ to be consistent with implementation constraints), and let $\hat{\theta}(0) = 0$. Then, as predicted by the theoretical analysis, the controller was able to compensate for the destabilizing ν_k and kept input-output stability, again with ξ_1 unbounded, although the control input it required was beyond implementation bounds.

The third design attempted to find the simplest way to attain state boundedness. Partial state feedback seemed at first to be a possible solution. We let $\nu_k = k_1\xi_1 + k_2\xi_2$, with k_1 and k_2 positive constants, and $\hat{\theta}(0) = 0$. This controller was able to drive the output error to zero and to keep the states bounded, but only when the initial conditions were such that the pendulum was close to its unstable equilibrium. In our experiments the only way to obtain reliable and consistent results that can also be implemented is to use complete state feedback that resembles the LQR.

As mentioned before, four control methods have been previously used

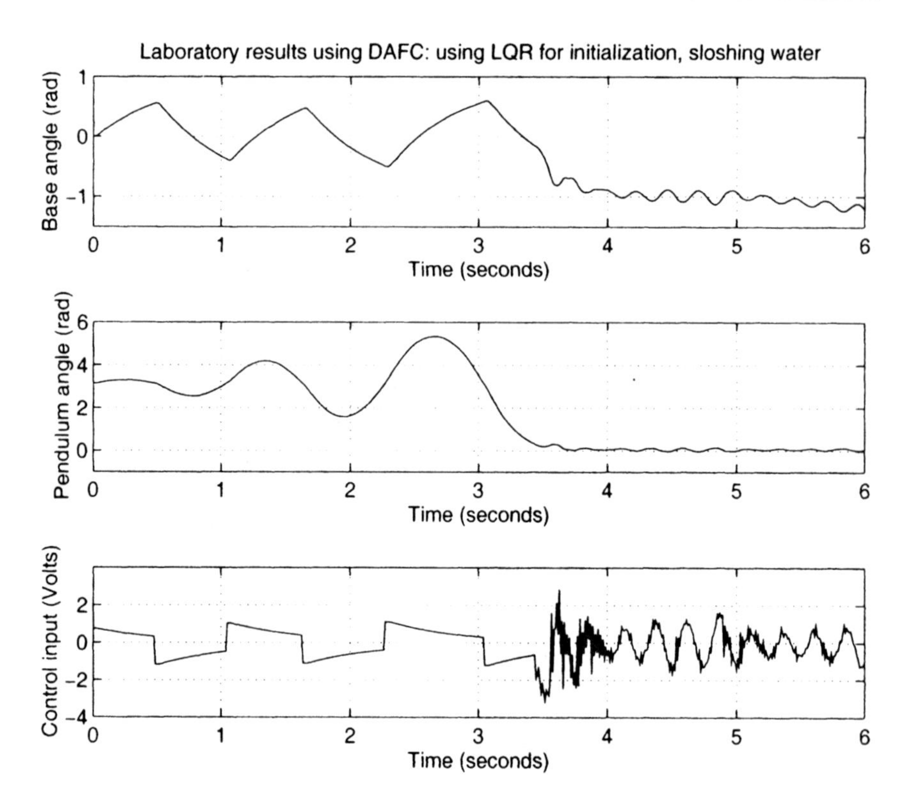

Figure 9.26. DAFC initialized as LQR: experimental results of DAFC with disturbance: sloshing water.

on the inverted pendulum in [240, 249]. Of them, only direct fuzzy control and the so-called "auto tuned fuzzy control" were able to control the system in the presence of the induced random disturbances (this is not surprising, since they are both nonlinear controllers, and therefore have an advantage in this case over linear methods). However, the performance of the fuzzy controller was severely degraded, and thus the auto-tuned method was the one that best compensated for disturbances. The auto-tuned fuzzy control method in [240, 249] dynamically expands or shrinks the universes of discourse for the inputs (it uses the four states) of a direct fuzzy controller, by tuning the input gains at every fixed time interval (50 samples), based on the magnitude of the error. The DAFC shown in this section performs in a similar way to the auto-tuned controller, although the DAFC is actually able to balance the pendulum with somewhat less base and pendulum oscillations, and tends to use a control input of a lesser magnitude. The first DAFC design and the modified IAFC do almost equally well, and somewhat better than the auto-tuned and the DAFC based on

LQR; these controllers are able to balance the pendulum with surprising ease (the control input they use has fewer high-frequency oscillations, and with a smaller magnitude), but present the disadvantage of keeping the base rotating. The conventional adaptive technique described here, adaptive feedback linearization, also has this disadvantage; it proved to have an acceptable performance on the nominal system, but was unsuccessful when the plant presented disturbances.

9.9 Summary

In this chapter, we studied various approaches to the implementation of both the direct and indirect adaptive controllers. Although all the experiments were performed using fuzzy systems for ease of comparison, similar results may be obtained using neural networks.

In general it was found that the results obtained in implementation corresponded very well with those obtained using simulation. In implementation, however, we typically had to allow for more conservative bounds in the controller design to achieve the desired performance. This may be attributed to the additional uncertainty associated with our experimental setup. For example, implementation had delays associated with the sampled-data nature of the computer system in which the controller was implemented, and also had unmodeled dynamics such as friction and structural dynamics. The issue of delay associated with a computer-based implementation will be addressed in more detail in Chapter 13. In general, if one plans to use the continuous-time approaches discussed thus far, then the sampling rate should be chosen to be at least twice as high as the highest frequency used in the model of the system and a factor of 10 greater than the desired closed-loop system bandwidth. This will help ensure a valid continuous-time model (though not necessarily guarantee stability). When the sampling rate may not be set high enough, then directly working in the discrete-time framework is appropriate.

We also found that there are a number of ways to design either a direct or indirect adaptive controller for the same system. In general it is typically advantageous to use as much knowledge about the plant as possible to design the control algorithm. If, for example, the nonlinearities of a system are known except for a single term, it is often better to design the static portion of the controller to compensate for the known nonlinearities, and just use the adaptive portion of the controller to compensate for the unknown term. This way the adaptive controller will have an easier time trying to approximate the uncertainties of the system. On the other hand, designing the adaptive portion to account for more uncertainty may often result in a more robust closed-loop system.

9.10 Exercises and Design Problems

Exercise 9.1 (Continuous Approximation to the Sign Function) Re-derive the stability proof of the direct and indirect adaptive controllers, but replace the sgn function with a continuous approximation. Some possibilities include the saturation function ($\text{sat}(y) = \text{sgn}(y)$ if $|y| \geq 1$ and $\text{sat}(y) = y$ if $-1 < y < 1$) and the hyperbolic tangent function. Show that, when appropriately defined, such approximations yield convergence of the error system to an ϵ-neighborhood of the origin, whose size can be modified by the choice of a design constant.

Exercise 9.2 (Bounding Control) Derive the bounding control terms ν_{bi} and ν_{bd} used for the indirect and direct adaptive methods, respectively. Consider, for the indirect case, the Lyapunov candidate $V_{bi} = \frac{1}{2}e^2$, and $V_{bd} = \frac{1}{2g}e^2$ for the direct case. Show that for both cases the bounding terms make $e \leq M_e$ into an invariant set to which the error converges exponentially fast.

Exercise 9.3 (Choosing an Approach) Consider the system described by

$$\dot{x} = -k_1 x + k_2 x^2 + \sin(x + k_3) + u,$$

where each k_i is unknown and we wish to drive the error $e = x - r$ to zero if possible with r a constant. Discuss the advantages and disadvantages of each of the following approaches to developing an appropriate controller:

1. Use the nonlinear damping terms

$$\nu_d = -\eta\left(1 + x^2 + x^4\right)e$$

in the controller design with the Lyapunov candidate $V = \frac{1}{2}e^2$. Thus in this case we are just dominating the uncertainty using a static controller.

2. Use the relationship

$$\sin(x + k_3) = \cos(k_3)\sin(x) + \sin(k_3)\cos(x)$$

to develop an approximator with unknown parameters

$$\theta = [k_1, k_2, \cos(k_3), \sin(k_3)]^\top$$

that may be used with the adaptive control approaches.

3. Use a fuzzy system or neural network in an adaptive controller to approximate $-k_1 x + k_2 x^2 + \sin(x + k_3)$ directly.

Exercise 9.4 (High Gain Systems) Discuss and show with a simulation why controllers using high feedback gain may cause a problem when the following occur:

- There are unmodeled delays associated with a controller.
- The sensors are noisy.
- There are unmodeled dynamics.

In light of these issues, discuss why an adaptive controller may be more robust or achieve better performance than a controller designed using nonlinear damping.

Exercise 9.5 (Indirect Adaptive Control for the Pendulum) Design and implement the indirect adaptive control method discussed in this chapter in simulation (using MATLAB or a custom-written computer program) for the inverted pendulum.

1. Implement the method without use of any available knowledge on the plant dynamics in the design.
2. Use the known functions (9.64) in your design. Investigate if the design remains viable in the presence of noise in the plant dynamics or at the actuator.
3. Is it possible to design the indirect adaptive controller in such a manner that it does not act like a feedback linearizing controller?
4. Discuss possible shortcomings and advantages of the indirect adaptive methodology. When would such an approach be most beneficial? When could one prefer to use a direct adaptive approach instead?

Exercise 9.6 (Direct Adaptive Control for the Pendulum) Design and implement the direct adaptive control method discussed in this chapter in simulation (using MATLAB or a custom-written computer program) for the inverted pendulum.

1. Implement the method without use of any previous design knowledge, and use a fuzzy system or a neural network to approximate the unknown controller.
2. Re-design an LQR controller for the pendulum, and use it as your known controller. Can you improve on the results in this chapter?
3. Are there other suitable approaches to designing a fixed controller for your design? Is there a clear advantage in using such a previously available design in a direct adaptive controller?

Exercise 9.7 (Adaptive Control of a Ball and Beam System [163]) Consider the ball and beam system in Figure 9.27. The ball is allowed to roll (without sliding) along the beam, and its position relative to the left edge of the beam is denoted as r. The beam tilts about its center point, thus causing the ball to roll from one position to another. The control problem consists in designing a controller that tilts the beam in such a way that the ball is brought from its initial position to another desired position. The beam is driven by a DC motor whose shaft is attached to the center of the beam.

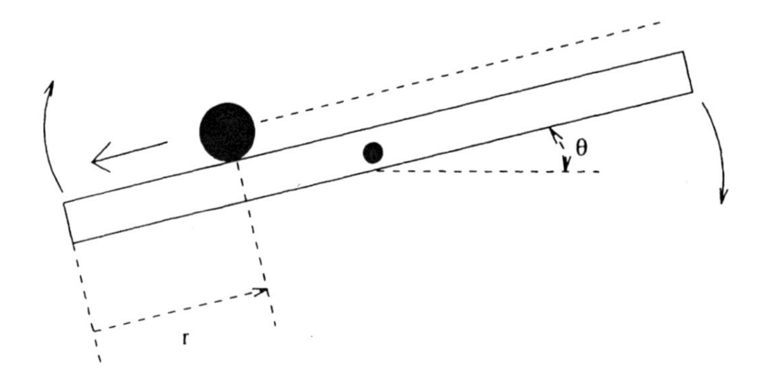

Figure 9.27. Ball and beam system.

Consider Figure 9.28 for a block-diagram description of the system. Let i_a be the input armature current to the motor, θ the angle of the beam and r the position of the ball on the beam. A simple Proportional-Integral-Derivative (PID) controller is used to drive the motor and to position the beam at any desired angle. This controller takes as an input the error Θ_e between an angle reference Θ_r and the beam angle θ. The signal Θ_r is produced by the ball position controller (which seeks to achieve our primary objective). By means of appropriate tuning of the PID controller it is possible to achieve very good angle tracking, and since the inner loop has much faster dynamics than the outer loop, it can be considered virtually invisible to the ball position controller.

Let $x_1 = \theta$ and $u = i_a$. Then, a linear state-space model of the motor is given by: $\dot{x}_1 = x_2, \dot{x}_2 = x_3 + b_1 u, \dot{x}_3 = a_1 x_2 + a_2 x_3 + b_2 u$, where $a_1 = -87885.84$, $a_2 = -1416.4$, $b_1 = 280.12$, and $b_2 = -18577.14$. If we now let $x_4 = r$ we can obtain two more equations which represent the ball and beam dynamics when the beam angle is taken as the input, using Newton's second law. Here we are using the approximation $\sin x_1 \approx x_1$ (valid because the beam angle varies within a small range around zero), in order to have the input enter

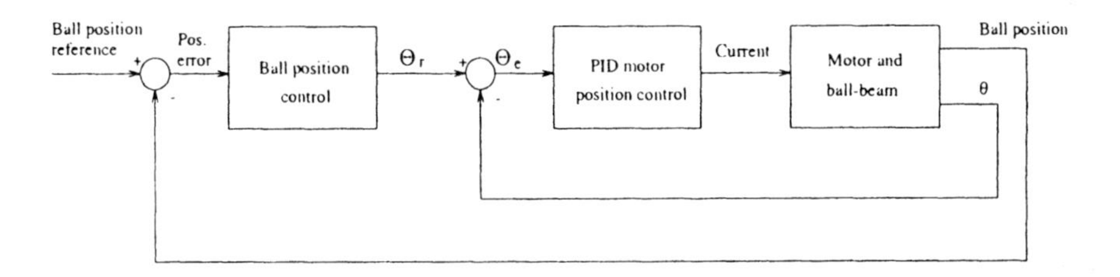

Figure 9.28. Motor-ball-beam control scheme. Θ_r is the angle reference input, Θ_e is the angle error, and θ is the beam angle.

linearly. A reasonably good model of the ball and beam system which has well-defined relative degree is given by

$$\begin{aligned}\dot{x}_4 &= x_5 \\ \dot{x}_5 &= a_3 x_1 + a_4 \tan^{-1}(100x_5)(e^{-10^4 x_5^2} - 1),\end{aligned} \tag{9.69}$$

where $a_3 = -514.96$ and $a_4 = 9.84$, and the system output is $y = x_4$. The numerical values take into account the acceleration due to gravity and the friction constant between the ball and the beam. The function $\tan^{-1}(100x_5)(e^{-10^4 x_5^2} - 1)$ is an approximation to the acceleration due to friction that the ball experiences on the beam.

1. Design the inner-loop PID controller for the motor, keeping in mind that it must have small settling time and little overshoot.
2. Design a direct adaptive controller for the ball-and-beam system. Can you improve its performance by including some fixed controller as your "known controller" term?
3. Simulate your design and discuss your results. In particular, compare the performance of your adaptive design with that of your fixed controller alone. Does the adaptive element provide any improvement?
4. Perform an indirect adaptive design for this system. Do you need to consider the motor subsystem in your design?

Part III

Output-Feedback Control

Chapter 10

Output-Feedback Control

10.1 Overview

In Chapter 6 we introduced non-adaptive control design tools for certain classes of nonlinear systems. All of them were based on the assumption that the state of the plant is available for feedback. The scope of this chapter is to remove this restriction by dealing with the case when the state of the system is not available for feedback but, rather, only the output can be measured. We will, in other words, introduce a set of techniques to perform output-feedback control. Recall, from Chapter 6, the structure of the system dynamics

$$\begin{aligned} \dot{x} &= f(x, u) \\ y &= h(x) \end{aligned} \tag{10.1}$$

with state $x \in \mathbb{R}^n$, input $u \in \mathbb{R}^m$, and output $y \in \mathbb{R}^p$. As in Chapter 6, we will assume that f is piecewise continuous in t and locally Lipschitz in x to ensure that there exists a unique solution to (10.1) defined on a compact time interval $[0, t_1]$, for some $t_1 > 0$.

Throughout this chapter we will try to find controllers that drive the state trajectories $x(t)$ to the origin $x = 0$ by only using the information provided by y (this is referred to as the output feedback stabilization problem). Additionally, we will investigate the problem of finding a control law forcing $y \to r(t)$, where $r(t)$ is a reference signal, by using only y (this is referred to as an output-feedback tracking problem). In both cases, the controllers we will find, rather than being static with $u = \nu(t, y)$, will be dynamic

$$\begin{aligned} \dot{x}_c &= f_c(x_c, y, r) \\ u &= h_c(x_c, y), \end{aligned} \tag{10.2}$$

with a state x_c and sufficiently smooth functions f_c and h_c. Note that, when solving a stabilization problem, the dependence of (10.2) on r will be dropped.

When seeking an output-feedback tracking controller, in analogy to what we have done in Chapter 6 we will define an error system $e = \chi(t, x)$

with $e \in \mathbb{R}^n$, which provides a measure of the closed-loop system tracking performance and, subsequently, we will study the stability of the error dynamics, $\dot{e} = \alpha(t, x, u)$. Here, however, we will relax the assumption made in previous chapters that the function χ measuring the tracking performance is analytically known, and will introduce tools to estimate this function on-line. On the other hand, when solving a stabilization problem, we will not need to define an error system (we will directly work with the plant dynamics (10.1)), and hence there will be no need to estimate the function χ.

We will first describe a technique to solve the stabilization and tracking problems for a particular class of nonlinear systems (namely, systems in output-feedback form). Following that, we will use the notion of *uniform complete observability* (UCO) to find output-feedback controllers for more general classes of nonlinear systems. Specifically, in the spirit of a separation principle, given any state-feedback controller (e.g., designed using the tools described in Chapter 6), we will estimate the state of the plant by means of a nonlinear observer. The state estimate will then be employed to recover the performance of the state-feedback controller. The theory will be initially developed for the output-feedback stabilization of SISO systems and will be successively extended to the robust output-feedback stabilization of MIMO systems. After that, we will introduce the concept of *practical internal model* to derive a solution of the tracking problem with the on-line estimation of the function $\chi(t, x)$.

10.2 Partial Information Framework

Recall from Section 6.2 that, given a smooth reference signal $r(t)$, a sufficient condition for the existence of an error system $e = \chi(t, x)$ that satisfies Assumption 6.1 is the existence of two sufficiently smooth and bounded functions $x^r(t)$ and $c^r(t)$ satisfying

$$\begin{aligned} \dot{x}^r &= f(x^r, c^r) \\ r &= h(x^r, c^r). \end{aligned} \tag{10.3}$$

Once the functions $x^r(t)$ and $c^r(t)$ have been calculated, the error system is simply given by $e = \chi(t, x) = x - x^r(t)$. The pair $(x^r(t), c^r(t))$ is referred to as the *stable inverse of the plant.* For particular classes of nonlinear systems, such as input-output feedback linearizable systems with full relative degree or strict-feedback systems, the solution of (10.3) is rather straightforward (see Chapter 6) and amounts to knowing $n - 1$ derivatives of the reference signal $r(t)$. For more general classes of systems, however, the solution of (10.3) may be difficult to calculate.

Example 10.1 Consider the second-order system

$$\dot{x}_1 = x_2 + u$$
$$\dot{x}_2 = -x_1 + kx_2^2$$
$$y = x_1,$$

and suppose that, given the reference signal $r(t) = \exp\{-(t\cos t)^2\}$, we want to find an error system satisfying Assumption 6.1. In order to do that, we seek to find two functions $x^r(t)$ and $c^r(t)$ satisfying

$$\dot{x^r}_1 = x^r{}_2 + c^r$$
$$\dot{x^r}_2 = -x^r{}_1 + kx^r{}_2^2$$
$$\exp(-(t\cos t)^2) = x^r{}_1.$$

Assume, for now, that $k = 0$ and note that $x^r{}_1(t) = \exp\{-(t\cos t)^2\}$, $c^r(t) = -2t\left(\cos^2 t + t\sin t\cos t\right)\exp\{-(t\cos t)^2\} - x^r{}_2(t)$, and $\dot{x^r}_2 = -\exp\{-(t\cos t)^2\}$. In this case, the function $x^r{}_2$ is given by the integral

$$x^r{}_2(t) = \int_{-\infty}^{t} -\exp\{-(t\cos t)^2\}dt\ ,$$

which is well defined for all $t \in \mathsf{R}$ but cannot be calculated explicitly. One could resort to a numerical off-line approximation of the integral above and achieve approximate tracking with arbitrary accuracy. Note, however, that the controller so obtained would yield tracking of the reference signal $r(t) = \exp\{-(t\cos t)^2\}$ but could not be employed to make the system follow different reference inputs. A more practical solution of the problem would be to estimate the function $x^r{}_2(t)$ on-line.

Now set $k = 1$ and consider the problem of tracking the reference signal $r(t) = \sin^2 t - \cos t$. This time we have that $x^r{}_1(t) = \sin^2 t - \cos t$, $c^r(t) = \sin t + 2\sin t\cos t - x^r{}_2(t)$, and $x^r{}_2(t)$ is a bounded solution (if it exists), of the differential equation

$$\dot{x}^r{}_2 = x^r{}_2^2 + \cos t - \sin^2 t. \tag{10.4}$$

Note that (10.4) is an unstable nonlinear system with a bounded time-varying input $\cos t - \sin^2 t$. Hence, unless an appropriate initial condition $x^r{}_2(0)$ that guarantees that $x^r{}_2(t)$ exists and is bounded for all $t \geq 0$ is found, one cannot numerically calculate the solution to (10.4). Given a general nonlinear system, such a "stable solution" may not exist and may be difficult to calculate. In this case, the choice of initial condition $x^r{}_2(0) = 0$ yields a "stable solution" to (10.4) given by $x^r{}_2(t) = \sin t$, yielding $c^r(t) = 2\sin t\cos t$. For all other choices of $x^r{}_2(0)$, the solution to (10.4) grows unbounded in finite time. $\triangle$

Throughout Chapters 6, 7, and 8 we assumed the tracking performance measure $e = \chi(t,x)$ to be available for feedback. The previous example illustrates that this assumption may be, in practice, too restrictive. In this part of the book we will develop control design tools in a *partial information framework*, that is, a framework where the state x of the plant and the tracking performance measure $e = \chi(t,x)$ are not directly available for feedback. Within a partial information framework one seeks to design a controller achieving tracking of bounded reference signals using exclusively the information provided by the output of the system and the reference signals (and not their time derivatives). In contrast to that, a *full information framework* is one where the state of the system and the tracking performance measure $e = \chi(t,x)$ are directly available for feedback. The previous part of this book was devoted to studying robust adaptive control design tools in a full information framework. Here we will establish the foundations for developing similar tools in a partial information framework. The control design problem becomes more involved but the techniques introduced here have greater practical relevance.

We will start, in the next section, by solving the output-feedback stabilization and tracking problems for the special class of systems in output-feedback form. In this instance, given the strong assumption we will make on the structure of the plant, we will be able to derive a systematic procedure to define an appropriate error system $e = \chi(t,x)$. Furthermore, we will rely on the knowledge of the time derivatives of the reference signals to have e directly available for feedback, and thus the approach will not entirely follow a partial information philosophy. In later sections we will depart from this idea and follow a more general approach.

10.3 Output-Feedback Systems

A single-input single-output system is said to be in output-feedback form if its dynamics can be written as

$$\begin{aligned}
\dot{x}_1 &= x_2 + g_1(y) \\
\dot{x}_2 &= x_3 + g_2(y) \\
&\vdots \\
\dot{x}_{r-1} &= x_r + g_{r-1}(y) \\
\dot{x}^r &= x_{r+1} + g_r(y) + d_m\sigma(y)u \\
&\vdots \\
\dot{x}_{n-1} &= x_n + g_{n-1}(y) + d_1\sigma(y)u \\
\dot{x}_n &= g_n(y) + d_0\sigma(y)u \\
y &= x_1,
\end{aligned} \tag{10.5}$$

where each g_i and $\sigma(y)$ are locally Lipschitz functions, $g_i(0) = 0$, $\sigma(y) \neq 0$ for all $y \in \mathsf{R}$, and $r = n - m$ is the relative degree of the system. The scalars d_i, $i = 1, \ldots m$ are assumed to be such that the polynomial $p(s) = d_m s^m + \ldots + d_1 s + d_0$ is Hurwitz. Notice that the system nonlinearities are only allowed to depend on the output y and that the zero dynamics of the system are linear and exponentially stable (this comes from the fact that $p(s)$ is Hurwitz).

Defining

$$A = \begin{bmatrix} 0 & & \\ \vdots & I & \\ 0 & \ldots & 0 \end{bmatrix}, \; d = \begin{bmatrix} 0 \\ \vdots \\ 0 \\ d_m \\ \vdots \\ d_0 \end{bmatrix}, \; g(y) = \begin{bmatrix} g_1(y) \\ \vdots \\ g_n(y) \end{bmatrix}, \; C^{\top} = \begin{bmatrix} 1 \\ 0 \\ \vdots \\ 0 \end{bmatrix} \tag{10.6}$$

(10.5) can be rewritten in vector form as

$$\begin{aligned} \dot{x} &= Ax + g(y) + d\sigma(y)u \\ y &= Cx, \end{aligned} \tag{10.7}$$

where (A, C) are an observable pair. Assume that the reference signal and its first r time derivatives are bounded for all $t \geq 0$ and available for feedback, and note that the special form (10.7) lends itself easily to the design of an estimator for the state x (also referred to as an observer) using y

$$\begin{aligned} \dot{\hat{x}} &= A\hat{x} + g(y) + d\sigma(y)u + L(y - \hat{y}) \\ \hat{y} &= C\hat{x}. \end{aligned} \tag{10.8}$$

By letting $\tilde{x} = x - \hat{x}$ and subtracting (10.8) from (10.7) we get the error dynamics

$$\dot{\tilde{x}} = (A - LC)\tilde{x}, \tag{10.9}$$

which are exponentially stable provided the vector L is chosen so that the matrix $(A - LC)$ is Hurwitz (this can be always achieved because (A, C) is an observable pair). Let P be the positive definite solution to the Lyapunov equation $P(A - LC) + (A - LC)^{\top} P = -I$. Next, it is fairly easy to design controllers solving the stabilization and the tracking problems by using the tools introduced in Chapter 6. In particular, by treating the estimation error $\tilde{x}(t)$ as an exponentially decaying disturbance, one can employ nonlinear damping and backstepping to solve the tracking problem using the feedback given by y and the observer states $\hat{x}$.

Example 10.2 Consider system (10.5) in output-feedback form. We will see that backstepping and nonlinear damping can be employed to solve the tracking problem. The stabilization problem is solved by setting $r(t) = \dot{r}(t) = \ldots r^{(r)}(t) = 0$.

x_1 subsystem In analogy to the procedure outlined in Chapter 6, we start by considering the subsystem defined by

$$\dot{x}_1 = x_2 + g_1(y) = \hat{x}_2 + g_1(y) + \tilde{x}_2 \tag{10.10}$$

and letting $\hat{x}_2$ be the virtual input v, so that

$$\dot{x}_1 = v + g_1(y) + \tilde{x}_2. \tag{10.11}$$

Note that we choose the virtual input to be $\hat{x}_2$ because the variable x_2 is not measurable, hence (10.11) contains a term $\tilde{x}_2$ which will be treated as a disturbance to reject. Let the first component of the error system be $e_1(t) = y - r$. Using (10.11) we find the error dynamics to be

$$\dot{e}_1 = v + g_1(y) + \tilde{x}_2 - \dot{r}. \tag{10.12}$$

Choose a Lyapunov function candidate for this subsystem as $V_1 = \frac{1}{2}e_1^2 + \frac{1}{c_1}\tilde{x}^\top P\tilde{x}$. Its derivative along the trajectories of (10.12) is

$$\begin{aligned}\dot{V}_1 &= e_1\left(v + g_1(y) + \tilde{x}_2 - \dot{r}\right) - \tfrac{1}{c_1}\tilde{x}^\top\tilde{x} \\ &\leq e_1\left(v + g_1(y) + \tilde{x}_2 - \dot{r}\right) - \tfrac{1}{c_1}(\tilde{x}_1^2 + \tilde{x}_2^2),\end{aligned}$$

where the inequality comes from the fact that some of the negative definite terms have been dropped. Choose $v = \nu_1(z_1) = -g_1(y) + \dot{r} - \kappa e_1 - c_1 e_1$, where $z_1 = [y, r, \dot{r}]^\top$ is a vector containing measurable variables. This yields

$$\dot{V}_1 \leq -\kappa e_1^2 - c_1 e_1^2 + e_1\tilde{x}_2 - \frac{1}{c_1}\tilde{x}_1^2 - \frac{1}{c_1}\tilde{x}_2^2$$

Using Young's inequality we have that $e_1\tilde{x}_2 \leq c_1 e_1^2 + \frac{1}{4c_1}\tilde{x}_2^2$, so that

$$\dot{V}_1 \leq -\kappa e_1^2 - \frac{1}{c_1}\tilde{x}_1^2 - \frac{3}{4c_1}\tilde{x}_2^2,$$

from which we get asymptotic stability of $e_1 = 0$.

x_2 subsystem Now consider the new subsystem

$$\begin{aligned}\dot{x}_1 &= \hat{x}_2 + g_1(y) + \tilde{x}_2 \\ \dot{\hat{x}}_2 &= \hat{x}_3 + g_2(y) + L_2(y - \hat{x}_1) \\ &= v + g_2(y) + L_2(y - \hat{x}_1),\end{aligned} \tag{10.13}$$

where $v = \hat{x}_3$ is the new virtual control input. The second component of the error system is $e_2 = \hat{x}_2 - \nu_1(z_1)$. The time derivative of $e = [e_1, e_2]^\top$ is now given by

$$\begin{aligned}
\dot{e}_1 &= \hat{x}_2 - \nu_1(z_1) + \nu_1(z_1) + g_1(y) + \tilde{x}_2 - \dot{r} \\
&= -\kappa e_1 - c_1 e_1 + e_2 + \tilde{x}_2 \\
\dot{e}_2 &= v + g_2(y) + L_2(y - \hat{x}_1) - \frac{\partial \nu_1}{\partial y}\left[\hat{x}_2 + \tilde{x}_2 + g_1(y)\right] \\
&\quad - \frac{\partial \nu_1}{\partial r}\dot{r} - \frac{\partial \nu_1}{\partial \dot{r}}\ddot{r}.
\end{aligned} \tag{10.14}$$

A new Lyapunov function candidate is now defined as

$$V_2 = V_1 + \frac{1}{2}e_2^2 + \frac{1}{c_2}\tilde{x}^\top P \tilde{x}.$$

Its time derivative along the trajectories of (10.14) is bounded by

$$\begin{aligned}
\dot{V}_2 &\leq -\kappa e_1^2 - c_1 e_1^2 + e_1 e_2 + e_1 \tilde{x}_2 + e_2 \Big\{ v + g_2(y) + L_2(y - \hat{x}_1) \\
&- \frac{\partial \nu_1}{\partial y}\left[\hat{x}_2 + \tilde{x}_2 + g_1(y)\right] - \frac{\partial \nu_1}{\partial r}\dot{r} - \frac{\partial \nu_1}{\partial \dot{r}}\ddot{r} \Big\} - \left(\tfrac{1}{c_1} + \tfrac{1}{c_2}\right)(\tilde{x}_1^2 + \tilde{x}_2^2).
\end{aligned}$$

Let

$$\begin{aligned}
v = \nu_2(z_2) = \quad & -\kappa e_2 - c_2\left(\frac{\partial \nu_1}{\partial y}\right)^2 e_2 - e_1 - g_2(y) - L_2(y - \hat{x}_1) \\
& + \frac{\partial \nu_1}{\partial y}\left[\hat{x}_2 + g_1(y)\right] + \frac{\partial \nu_1}{\partial r}\dot{r} + \frac{\partial \nu_1}{\partial \dot{r}}\ddot{r},
\end{aligned}$$

where $z_2 = [y, \hat{x}_1, \hat{x}_2, r, \dot{r}]^\top$ is a vector formed by measurable variables. Next,

$$\begin{aligned}
\dot{V}_2 &\leq -\kappa e_1^2 - c_1 e_1^2 + e_1 \tilde{x}_2 - \kappa e_2^2 - c_2\left(\frac{\partial \nu_1}{\partial y}\right)^2 e_2^2 - \frac{\partial \nu_1}{\partial y}\tilde{x}_2 e_2 \\
&- \left(\tfrac{1}{c_1} + \tfrac{1}{c_2}\right)(\tilde{x}_1^2 + \tilde{x}_2^2).
\end{aligned}$$

By applying Young's inequality to the sign-indefinite terms we have

$$e_1\tilde{x}_2 \leq c_1 e_1^2 + \frac{1}{4c_1}\tilde{x}_2^2, \qquad \frac{\partial \nu_1}{\partial y}\tilde{x}_2 e_2 \leq c_2\left(\frac{\partial \nu_1}{\partial y}\right)^2 e_2^2 + \frac{1}{4c_2}\tilde{x}_2^2,$$

and thus $\dot{V}_2 \leq -\kappa(e_1^2 + e_2^2) - \frac{3}{4}\left(\frac{1}{c_1} + \frac{1}{c_2}\right)(\tilde{x}_1^2 + \tilde{x}_2^2)$. The procedure at the next step will be repeated by considering the x_3 subsystem and setting $e_3 = \hat{x}_3 - \nu_2(z_2)$. The associated error dynamics become

$$\begin{aligned}
\dot{e}_1 &= -\kappa e_1 - c_1 e_1 + e_2 + \tilde{x}_2 \\
\dot{e}_2 &= \hat{x}_3 - \nu_2(z_2) + \nu_2(z_2) + g_2(y) + L_2(y - \hat{x}_1) \\
&\quad - \frac{\partial \nu_1}{\partial y}\left[\hat{x}_2 + \tilde{x}_2 + g_1(y)\right]\frac{\partial \nu_1}{\partial r}\dot{r} - \frac{\partial \nu_1}{\partial \dot{r}}\ddot{r} \\
&= -\kappa e_2 - e_1 + e_3 - c_2\left(\frac{\partial \nu_1(z_1)}{\partial y}\right)^2 e_2 - \frac{\partial \nu_1(z_1)}{\partial y}x_2.
\end{aligned} \tag{10.15}$$

x_i subsystem ($1 \le i < r$) Repeating the procedure above $i-1$ times, one gets the x_i subsystem

$$\begin{aligned} \dot{x}_1 &= \hat{x}_2 + g_1(y) + \tilde{x}_2 \\ &\vdots \\ \dot{\hat{x}}_i &= \hat{x}_{i+1} + g_i(y) = v + g_i(y), \end{aligned} \tag{10.16}$$

where $v = \hat{x}_{i+1}$ is the new virtual control input. As we did before, we define $e_{i+1} = \hat{x}_{i+1} - \nu_i(z_i)$. After noticing that $e_i = \hat{x}_i - \nu_{i-1}(z_{i-1})$, where ν_{i-1} is the virtual control found at the last step and $z_{i-1} = [y, \hat{x}_1, \dots, \hat{x}_{i-1}, r, \dots, r^{(i-1)}]^\top$ is a vector formed by measurable variables, we calculate the error dynamics

$$\begin{aligned} \dot{e}_1 &= -\kappa e_1 - c_1 e_1 + e_2 + \tilde{x}_2 \\ &\vdots \\ \dot{e}_i &= v + g_i(y) - \frac{\partial \nu_{i-1}}{\partial y}[\hat{x}_2 + \tilde{x}_2 + g_1(y)] - \textstyle\sum_{j=1}^{i-1} \frac{\partial \nu_{i-1}}{\partial \hat{x}_j}[\hat{x}_{j+1} \\ &\quad + g_j(y) + L_j(y - \hat{x}_1)] - \textstyle\sum_{j=0}^{i-1} \frac{\partial \nu_{i-1}}{\partial r^{(j)}} r^{(j+1)}. \end{aligned} \tag{10.17}$$

The choice of the Lyapunov function candidate $V_i = V_{i-1} + \frac{1}{c_i}\tilde{x}^\top P \tilde{x}$ and of the virtual control

$$\begin{aligned} v = \nu_i(z_i) &= -\kappa e_i - e_{i-1} - c_i \left(\frac{\partial \nu_{i-1}}{\partial y}\right)^2 e_i - L_i(y - \hat{x}_1) - g_i(y) \\ &+ \frac{\partial \nu_{i-1}}{\partial y}[\hat{x}_2 + g_1(y)] + \sum_{j=1}^{i-1} \frac{\partial \nu_{i-1}}{\partial \hat{x}_j}[\hat{x}_{j+1} + g_j(y) + L_j(y - \hat{x}_1)] \\ &+ \sum_{j=0}^{i-1} \frac{\partial \nu_{i-1}}{\partial r^{(j)}} r^{(j+1)}, \end{aligned} \tag{10.18}$$

yields $\dot{V}_i \le -\sum_{j=1}^{i}\left[\kappa e_j^2 + \frac{3}{4c_j}\tilde{x}^\top \tilde{x}\right]$. Note that the virtual control has arguments $z_i = [y, \hat{x}_1, \dots, \hat{x}_i, r, \dots, r^{(i)}]^\top$. At the next step we let $e_{i+1} = \hat{x}_{i+1} - \nu_i(z_i)$ and obtain the error dynamics

$$\begin{aligned} \dot{e}_1 &= -\kappa e_1 - c_1 e_1 + e_2 + \tilde{x}_2 \\ &\vdots \\ \dot{e}_i &= -\kappa e_i - e_{i-1} + e_{i+1} - c_i \left(\frac{\partial \nu_{i-1}}{\partial y}\right)^2 e_i - \frac{\partial \nu_{i-1}}{\partial y}\tilde{x}_2. \end{aligned} \tag{10.19}$$

x_r subsystem The procedure can be iterated until step r, where a controller for system (10.5) is found to be

$$u = \frac{1}{d_m \sigma(y)}\left[\nu_r - \hat{x}_{r+1} - r^{(r)}\right]$$

yielding the error dynamics

$$\begin{aligned}
\dot{e}_1 &= -\kappa e_1 - c_1 e_1 + e_2 + \tilde{x}_2 \\
&\vdots \\
\dot{e}_i &= -\kappa e_i - e_{i-1} + e_{i+1} - c_i \left(\frac{\partial \nu_{i-1}}{\partial y}\right)^2 e_i - \frac{\partial \nu_{i-1}}{\partial y}\tilde{x}_2. \\
&\vdots \\
\dot{e}_r &= -\kappa e_r - e_{r-1} - c_r \left(\frac{\partial \nu_{r-1}}{\partial y}\right)^2 e_r - \frac{\partial \nu_{r-1}}{\partial y}\tilde{x}_2.
\end{aligned} \tag{10.20}$$

At step r, the Lyapunov function candidate becomes

$$V_r = V_{r-1} + \frac{1}{c_r}\tilde{x}^\top P\tilde{x} = \sum_{j=1}^{r} \frac{1}{2}e_j^2 + \frac{1}{c_j}\tilde{x}^\top P\tilde{x}$$

and its time derivative is negative definite

$$\dot{V}_r \leq -\sum_{j=1}^{r}\left[\kappa e_j^2 + \frac{3}{4c_j}\tilde{x}^\top \tilde{x}\right],$$

thus implying that $e_1, \ldots, e_r$ and $\tilde{x}$ are bounded and tend to zero asymptotically (and thus the tracking error tends to zero). The boundedness of $e_1, \ldots, e_r$ and $\tilde{x}$ implies the boundedness of the states $x_1, \ldots, x_r$ and estimates $\hat{x}_1, \ldots, \hat{x}_r$. Using the fact that the zero dynamics are globally exponentially stable one can easily show that $x_{r+1}, \ldots, x_n$ are bounded as well.

△

The above example demonstrates how one may use the backstepping procedure to develop output-feedback controllers for systems when the system dynamics are known. The next example demonstrates how one may similarly design an observer when there is uncertainty in the system

Example 10.3 Consider the system defined by

$$\begin{aligned}
\dot{x}_1 &= x_2 + g(y) \\
\dot{x}_2 &= u
\end{aligned} \tag{10.21}$$

with output $y = x_1$. In this example, we will assume that $g(y)$ may be approximated by the fuzzy system or neural network $\mathcal{F}(y, \theta)$ over all $y \in \mathrm{R}$. We will also assume that θ is a known vector of parameters such that $|g(y) - \mathcal{F}(y, \theta)| < W$ for all $y \in \mathrm{R}$, and that we want to develop an observer to create an estimate of x_2.

Consider the observer

$$\begin{aligned} \dot{\hat{x}} &= A\hat{x} + \begin{bmatrix} \mathcal{F}(y,\theta) \\ u \end{bmatrix} + L(y - \hat{y}) \\ \hat{y} &= C\hat{x}, \end{aligned} \tag{10.22}$$

where

$$A = \begin{bmatrix} 0 & 1 \\ 0 & 0 \end{bmatrix} \qquad C = [\ 1 \quad 0\ ].$$

Notice that the dynamics of this observer are based upon the system dynamics with the feedback term, $L(y - \hat{y})$, included to stabilize the error dynamics. Define the error in the state estimate as $\tilde{x} = x - \hat{x}$ and choose L so that $P(A - LC) + (A - LC)^\top P = -2I$, where P is a positive definite symmetric matrix.

Now define the positive definite function $V_o = \tilde{x}^\top P \tilde{x}$. The derivative of this decays according to

$$\dot{V}_o = -2\tilde{x}^\top \tilde{x} + 2\tilde{x}^\top P \begin{bmatrix} g - \mathcal{F} \\ 0 \end{bmatrix}. \tag{10.23}$$

Since $|g - \mathcal{F}| < W$, we may use $-|\tilde{x}|^2 + 2|\tilde{x}||P|W \leq W^2|P|^2$ to show that

$$\begin{aligned} \dot{V}_o &\leq -\tilde{x}^\top \tilde{x} + W^2|P|^2 \\ &\leq -\frac{V_o}{\lambda_{\max}(P)} + W^2|P|^2 \end{aligned} \tag{10.24}$$

so that V_o and $\tilde{x}$ are UUB. The size of the ultimate bound may be made arbitrarily small by making $|P|$ small which will in turn set the observer gain via L. △

As seen from the previous examples, it is possible to design observers and output-feedback controllers for systems in output-feedback form using the backstepping approach. Very often, however, the system nonlinearities are not simply functions of the plant output. In these cases, it is desirable to use other approaches to design output-feedback controllers. In this book we will show how to use the separation principle to design output-feedback controllers for systems when the plant dynamics are known. We will then extend these results to cases when there is uncertainty in the system dynamics so that approximations of the plant dynamics (possibly due to using a fuzzy system or neural network) may be used.

10.4 Separation Principle for Stabilization

Return to the general problem of stabilizing the origin $x = 0$ of the nonlinear system

$$\begin{aligned} \dot{x} &= f(x, u) \\ y &= h(x, u) \end{aligned} \tag{10.25}$$

by means of a dynamic output-feedback controller

$$\begin{aligned} \dot{x}_c &= f_c(x_c, y, r) \\ u &= h_c(x_c, y). \end{aligned} \tag{10.26}$$

Assume for now that $m = \rho = 1$, i.e., the plant is single-input single-output (later we will consider the general multi-input multi-output case). Throughout this section we will develop a methodology to solve this problem by assigning a particular structure to the dynamics of the controller. Specifically, assume there exists a smooth state-feedback controller $u = \bar{u}(x)$ which stabilizes the origin of (10.25), i.e, such that the origin of

$$\begin{aligned} \dot{x} &= f(x, \bar{u}(x)) \\ y &= h(x, u) \end{aligned} \tag{10.27}$$

is asymptotically stable (or globally asymptotically stable). Then, we will try to find a controller which estimates the state x on-line and employs this estimate to recover the performance of the state-feedback controller $\bar{u}(x)$. This approach, which is based on the so-called separation principle, has the advantage of decoupling the state-feedback control design, for which well-established tools like the ones introduced in Chapter 6 exist, from the state estimation problem, thus making the overall output-feedback control design easier. Before going into the details of this approach, we need to introduce a definition of observability for nonlinear systems which will be useful to develop a general class of nonlinear observers to estimate $x(t)$.

10.4.1 Observability and Nonlinear Observers

Consider the following mapping

$$y_e \triangleq \begin{bmatrix} y \\ \vdots \\ y^{(n-1)} \end{bmatrix} = \mathcal{H}\left(x, u, \ldots, u^{(n_u-1)}\right) \triangleq \begin{bmatrix} h(x,u) \\ \varphi_1(x, u, u^{(1)}) \\ \vdots \\ \varphi_{n-1}\left(x, u, \ldots, u^{(n_u-1)}\right) \end{bmatrix}, \tag{10.28}$$

where

$$\begin{aligned}
\varphi_1(x,u,u^{(1)}) &= \frac{\partial h}{\partial x} f(x,u) + \frac{\partial h}{\partial u} u^{(1)} \\
\varphi_2(x,u,u^{(1)},u^{(2)}) &= \frac{\partial \varphi_1}{\partial x} f(x,u) + \frac{\partial \varphi_1}{\partial u} u^{(1)} + \frac{\partial \varphi_1}{\partial u^{(1)}} u^{(2)} \\
&\vdots \\
\varphi_{n-1}\left(x,u,\ldots,u^{(n_u-1)}\right) &= \frac{\partial \varphi_{n-2}}{\partial x} f(x,u) + \sum_{j=0}^{n_u-2} \frac{\partial \varphi_{n-2}}{\partial u^j} u^{(j+1)}.
\end{aligned} \tag{10.29}$$

$\mathcal{H}$ is thus the mapping relating the first $n-1$ derivatives of the output y to the state of the system and a number n_u of control input derivatives (note that $n_u \leq n$). When $\mathcal{H}$ does not depend on u we will set $n_u = 0$. Now assume that (10.25) is *uniformly completely observable* (UCO), i.e., the mapping $\mathcal{H}$ is invertible with respect to x and its inverse $x = \mathcal{H}^{-1}(y_e, u, \dot{u}, \ldots, u^{(n_u-1)})$ is smooth (in other words, $\mathcal{H}$ is assumed to be a diffeomorphism), for all $x \in \mathsf{R}^n$, $[u, \dot{u}, \ldots, u^{(n_u-1)}] \in \mathsf{R}^{n_u}$. Later on, we will relax this assumption by not requiring it to hold globally on $\mathsf{R}^n \times \mathsf{R}^{n_u}$.

Example 10.4 In the particular case when (10.25) is a linear system

$$\begin{aligned}
\dot{x} &= Ax + Bu, \\
y &= Cx + Du
\end{aligned} \tag{10.30}$$

$\mathcal{H}(x, u, \ldots, u^{(n-1)})$ is given by

$$\mathcal{H} = \underbrace{\begin{bmatrix} C \\ CA \\ \vdots \\ CA^{n-1} \end{bmatrix}}_{\mathcal{H}_1} x + \underbrace{\begin{bmatrix} Du \\ CBu + D\dot{u} \\ \vdots \\ \left(\sum_{j=0}^{n-2} CA^j B u^{(n-j-3)}\right) + Du^{(n-1)} \end{bmatrix}}_{\mathcal{H}_2}. \tag{10.31}$$

In this case, $\mathcal{H}$ is invertible with respect to x if and only if the $n \times n$ constant matrix $\mathcal{H}_1$ is invertible. This corresponds to the well-known observability condition for linear time invariant SISO systems. △

Definition 10.1: The system

$$\begin{aligned}
\dot{\hat{x}} &= \hat{f}(\hat{x}, u, y) \\
\hat{y} &= \hat{h}(\hat{x}, u)
\end{aligned} \tag{10.32}$$

is an observer for (10.25) if the following two conditions are satisfied

(i) $\hat{x}(0) = x(0)$ implies that $\hat{x}(t) = x(t)$ for all $t \geq 0$.

(ii) $\hat{x}(t) \to x(t)$ as $t \to \infty$ whenever $\hat{x}(0)$ and $x(0)$ belong to some suitable subset of R^n.

Thus, in our definition, an observer is a dynamical system which estimates the state of the plant by only using the information given by the control input u and the system output y.

Next, we will illustrate how to design observers for the general class of systems in (10.25). From the observability assumption we have that

$$x = \mathcal{H}^{-1}(y_e, u, \ldots, u^{(n_u - 1)}) \tag{10.33}$$

(where $\mathcal{H}^{-1}$ denotes the smooth inverse of $\mathcal{H}$) and thus, if the first $n_u -$ 1 derivatives of u were known, one could estimate x by estimating the first first $n - 1$ derivatives of y (vector y_e) and inverting the mapping $\mathcal{H}$. However, in practice the derivatives of u are not available and the inverse of $\mathcal{H}$ may be difficult (if not impossible) to calculate analytically.

Example 10.5 The following nonlinear system

$$\begin{aligned} \dot{x}_1 &= x_2 \\ \dot{x}_2 &= (1 + x_1)\exp(x_1^2) + u - 1 \\ y &= x_2 \end{aligned} \tag{10.34}$$

is uniformly completely observable since $\mathcal{H}$ is given by

$$y_e = \mathcal{H}(x, u) = \begin{bmatrix} y \\ \dot{y} \end{bmatrix} = \begin{bmatrix} x_2 \\ (1 + x_1)\exp(x_1^2) + u - 1 \end{bmatrix}, \tag{10.35}$$

which is invertible for all x and u. To see this, observe that from the first equation we get $x_2 = y$, and the second equation is invertible with respect to x_1 because the function $(1 + x_1)\exp(x_1^2)$ is strictly increasing. However, one cannot *analytically* find the inverse $\mathcal{H}^{-1}$ other than by numerical approximation. △

To remove the first of the two obstructions above (the fact that the derivatives of u are not available for feedback), we add n_u integrators at the input side of the system (see Figure 10.1),

$$\dot{x} = f(x, s_1), \; \dot{s}_1 = s_2, \; \ldots, \; \dot{s}_{n_u} = v \tag{10.36}$$

and, using integrator backstepping (see Theorem 6.2), we employ $\bar{u}(x)$ to design a stabilizing controller $\bar{v}(x, s_1, \ldots, s_{n_u})$ for the augmented system (10.36). In what follows, to simplify our notation we will let $s = [s_1, \ldots, s_{n_u}]^\top$ and $x_a = [x^\top, s^\top]^\top$ so that (10.36) can be rewritten as

$$\begin{aligned} \dot{x}_a &= f_a(x_a, v) \\ y &= h_a(x_a, s_1), \end{aligned} \tag{10.37}$$

where $f_a(x_a, v) = [f(x, s_1)^\top, s_2, \ldots, v]^\top$ and $h_a(x_a, s_1) = h(x, u)$. The subscript a is used to denote the fact that (10.37) is an augmented system. Now note that, by definition, $u = s_1, \dot{u} = s_2, \ldots, u^{(n_u-1)} = s_{n_u}$, and thus we can rewrite (10.33) as

$$x = \mathcal{H}^{-1}(y_e, s). \tag{10.38}$$

Since the state of the chain of integrators is part of the controller, s is now available for feedback and can be employed to estimate x.

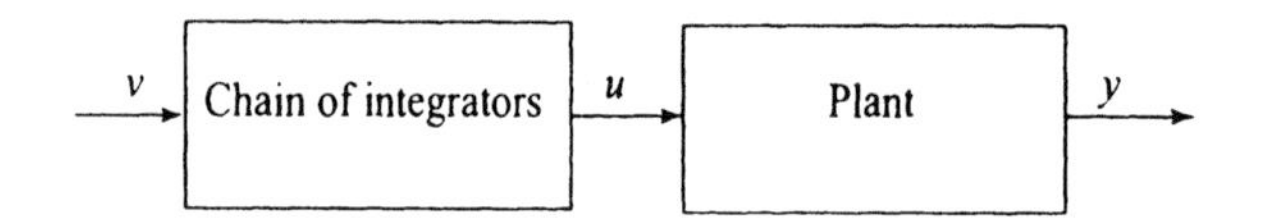

Figure 10.1. Adding integrators at the input side of the system.

In order to avoid the calculation of $\mathcal{H}^{-1}$ (the second obstruction mentioned earlier), rather than estimating the derivatives of y (vector y_e) and calculating x from (10.38), we will estimate x directly by means of the following nonlinear observer

$$\begin{aligned} \dot{\hat{x}} &= f(\hat{x}, s_1) + \left[\frac{\partial \mathcal{H}(\hat{x}, s)}{\partial \hat{x}}\right]^{-1} \mathcal{E}^{-1} L\, [y(t) - \hat{y}(t)] \\ \hat{y}(t) &= h(\hat{x}). \end{aligned} \tag{10.39}$$

where $\mathcal{E} = \text{diag}\left[\eta, \eta^2, \ldots, \eta^n\right]$, η is a design parameter such that $0 < \eta \leq 1$, and $L = [l_1, l_2, \ldots, l_n]^\top$ is such that $s^n + l_1 s^{n-1} + \ldots + l_n$ is a Hurwitz polynomial. ¿From the fact that the plant is uniformly completely observable and, hence, that $\mathcal{H}$ is a diffeomorphism, we have that its Jacobian with respect to x, $\partial \mathcal{H}(x, s)/\partial x$, is nonsingular for all x and s, and thus its inverse in (10.39) is well-defined. Notice that, while calculating the inverse of the mapping $\mathcal{H}$ may be a difficult or impossible task, the calculation of the inverse of the state-dependent matrix $\frac{\partial \mathcal{H}}{\partial \hat{x}}$ is straightforward.

In order to study the stability properties of the observer we have just introduced, we need to assume that $x_a(t)$, the state of the augmented system (10.37), is bounded for all $t \geq 0$. Specifically, we will assume that there exists a compact set Ω such that $x_a(t) \in \Omega$ for all $t \geq 0$. The controller we will define in the next section will then guarantee that these conditions are automatically satisfied. We want to show that the observer estimation error dynamics $\tilde{x}(t) = x - \hat{x}$ are asymptotically stable. To this end, consider the filtered transformation

$$y_e = \mathcal{H}(x, s) = \mathcal{H}(x_a) = \begin{bmatrix} h(x, s_1) \\ \varphi_1(x, s_1, s_2) \\ \vdots \\ \varphi_{n-1}(x, s_1, \ldots, s_{n_u}) \end{bmatrix}, \tag{10.40}$$

which by assumption is invertible with smooth inverse $x = \mathcal{H}^{-1}(y_e, s)$, and express system (10.25) in new coordinates. By definition, the extended output is $y_e = [y,\, \dot{y},\, \ldots,\, y^{(n-1)}]^{\mathsf{T}}$ and, with φ_{n-1} defined in (10.29),

$$\begin{aligned} y^{(n)} &= \left[\frac{\partial \varphi_{n-1}}{\partial x} f\left(\mathcal{H}^{-1}(y_e, s), s\right) + \sum_{k=1}^{n_u - 1} \frac{\partial \varphi_{n-1}}{\partial s_k}\left(\mathcal{H}^{-1}(y_e, s), s\right) s_{k+1}\right] \\ &\quad + \left[\frac{\partial \varphi_{n-1}}{\partial s_{n_u}}\left(\mathcal{H}^{-1}(y_e, s), s\right)\right] v \\ &\stackrel{\triangle}{=} \alpha(y_e, s) + \beta(y_e, s) v. \end{aligned}$$

Hence, in the new coordinates (10.25) becomes

$$\dot{y}_e = A y_e + B\left[\alpha(y_e, s) + \beta(y_e, s) v\right], \tag{10.41}$$

where

$$A = \begin{bmatrix} 0 & & \\ \vdots & I & \\ 0 & \ldots & 0 \end{bmatrix}, \; B = \begin{bmatrix} 0 \\ \vdots \\ 0 \\ 1 \end{bmatrix}. \tag{10.42}$$

Next, transform the observer (10.39) to new coordinates $\hat{y}_e = \mathcal{H}(\hat{x}, s) = [\hat{y}_{e_1}, \hat{y}_{e_2}, \ldots, \hat{y}_{e_n}]^{\mathsf{T}}$ where $\hat{y}_{e_1} = \hat{y}$, so that

$$\begin{aligned} \dot{\hat{y}}_{e_1} &= \frac{\partial h}{\partial \hat{x}} f(\hat{x}, s_1) + \frac{\partial h}{\partial \hat{x}}\left[\frac{\partial \mathcal{H}}{\partial \hat{x}}\right]^{-1} \mathcal{E}^{-1} L\left[y - h(\hat{x}, s_1)\right] + \frac{\partial h}{\partial s_1} \dot{s}_1 \\ &= \hat{y}_{e_2} + \frac{\partial h}{\partial \hat{x}}\left[\frac{\partial \mathcal{H}}{\partial \hat{x}}\right]^{-1} \mathcal{E}^{-1} L\left[y - h(\hat{x}, s_1)\right]. \end{aligned} \tag{10.43}$$

Similarly, for $i = 2, \ldots, n-1$

$$\dot{\hat{y}}_{e_i} = \frac{\partial \varphi_{i-1}}{\partial \hat{x}}(\hat{x}, s_1, \ldots, s_i) \left\{ f(\hat{x}, s_1) + \left[\frac{\partial \mathcal{H}}{\partial \hat{x}}\right]^{-1} \mathcal{E}^{-1} L\left(y - h(\hat{x}, s_1)\right) + \sum_{k=1}^{i} \frac{\partial \varphi_{i-1}}{\partial s_k} s_{k+1} \right\}.$$

By definition,

$$\hat{y}_{e_{i+1}} = \varphi_i(\hat{x}, s_1, \ldots, s_{i+1}) = \frac{\partial \varphi_{i-1}}{\partial \hat{x}} f(\hat{x}, s_1) + \sum_{k=1}^{i} \frac{\partial \varphi_{i-1}}{\partial s_k} s_{k+1}.$$

Hence, we conclude that

$$\dot{\hat{y}}_{e_i} = \hat{y}_{e_{i+1}} + \frac{\partial \varphi_{i-1}}{\partial \hat{x}} \left[\frac{\partial \mathcal{H}}{\partial \hat{x}}\right]^{-1} \mathcal{E}^{-1} L\left(y - h(\hat{x}, s_1)\right), \tag{10.44}$$

for $i = 2, \ldots, n-1$. Finally,

$$\dot{\hat{y}}_{e_n} = \alpha(\hat{y}_e, s) + \beta(\hat{y}_e, s)v + \frac{\partial \varphi_{n-1}}{\partial \hat{x}} \left[\frac{\partial \mathcal{H}}{\partial \hat{x}}\right]^{-1} \mathcal{E}^{-1} L\left[y - h(\hat{x}, s_1)\right]. \tag{10.45}$$

By using (10.43), (10.44), and (10.45) we can write, in compact form,

$$\begin{aligned} \dot{\hat{y}}_e &= A\hat{y}_e + B[\alpha(\hat{y}_e, s) + \beta(\hat{y}_e, s)v] + \left[\frac{\partial \mathcal{H}}{\partial \hat{x}}\right] \left[\frac{\partial \mathcal{H}}{\partial \hat{x}}\right]^{-1} \mathcal{E}^{-1} L\left[y - h(\hat{x}, s_1)\right] \\ &= A\hat{y}_e + B[\alpha(\hat{y}_e, s) + \beta(\hat{y}_e, s)v] + \mathcal{E}^{-1} L\left[y_{e1} - \hat{y}_{e1}\right]. \end{aligned} \tag{10.46}$$

Define the observer error in the new coordinates, $\tilde{y_e} = \hat{y}_e - y_e$. Let $C \in \mathbb{R}^n$ be defined as $C = [1, \ldots, 0]$, then the observer error dynamics are given by

$$\dot{\tilde{y}}_e = (A - \mathcal{E}^{-1} L\, C)\tilde{y}_e + B\left[\alpha(\hat{y}_e, s) + \beta(\hat{y}_e, s)v - \alpha(y_e, s) - \beta(y_e, s)v\right]. \tag{10.47}$$

Now notice that the pair (C, A) is observable and hence one can choose L such that the eigenvalues of $A - LC$ have negative real parts or, in other words, the roots of the polynomial $s^n + l_1 s^{n-1} + \ldots + l_n$ have negative real parts. Defining the coordinate transformation

$$\tilde{\nu} = \mathcal{E}' \tilde{y}_e, \; \mathcal{E}' \triangleq \operatorname{diag}\left[\frac{1}{\eta^{n-1}}, \frac{1}{\eta^{n-2}}, \ldots, 1\right], \tag{10.48}$$

in the new domain the observer error equation becomes

$$\dot{\tilde{\nu}} = \frac{1}{\eta}(A - LC)\tilde{\nu} + B\left[\alpha(\hat{y}_e, s) + \beta(\hat{y}_e, s)v - \alpha(y_e, s) - \beta(y_e, s)v\right], \tag{10.49}$$

where, by our choice of L, $A - LC$ is Hurwitz. Note that the form of A was used to go from (10.47) to (10.49). Let P be the solution to the Lyapunov equation

$$P(A - LC) + (A - LC)^{\top} P = -I \tag{10.50}$$

and consider the Lyapunov function candidate $V_o(\tilde{\nu}) = \tilde{\nu}^{\top} P \tilde{\nu}$. Calculate the time derivative of V_o along the $\tilde{\nu}$ trajectories

$$\dot{V}_o = -\frac{\tilde{\nu}^{\top}\tilde{\nu}}{\eta} + 2\tilde{\nu}^{\top} PB\left[\alpha(\hat{y}_e, s) + \beta(\hat{y}_e, s)v - \alpha(y_e, s) - \beta(y_e, s)v\right]. \tag{10.51}$$

Assume that $[\hat{x}(0)^{\top}, s(0)^{\top}]^{\top} \in \Omega$ (i.e., the estimate of x_a is initialized in the same compact set) and define the compact set

$$\mathcal{K}_{\tilde{y}_e} \triangleq \{\tilde{y}_e \in \mathrm{R}^n \mid \hat{y}_e, y_e \in \mathcal{H}(\Omega)\}.$$

By definition, $\mathcal{K}_{\tilde{y}_e}$ contains the initial condition $\tilde{y}_e(0)$. Assume now that the following time signal is bounded as follows

$$\begin{aligned} |\alpha(\hat{y}_e(t), s(t)) + \beta(\hat{y}_e(t), s(t))v(t) &- \alpha(y_e(t), s(t)) \\ &- \beta(y_e(t), s(t))v(t)| \leq k^* |\tilde{y}_e(t)|, \end{aligned} \tag{10.52}$$

for some $\gamma > 0$, for all $t \geq 0$, and for all $\hat{y}_e(0) \in \mathcal{H}(\Omega)$, $y_e(t) \in \mathcal{H}(\Omega)$. This condition requires that the function $\alpha(\hat{y}_e, s) + \beta(\hat{y}_e, s)v - \alpha(y_e, s) - \beta(y_e, s)v$ satisfies a Lipschitz-type inequality at any time instant with a fixed Lipschitz constant k^*. Notice that the boundedness of the control input $v(t)$ and the smoothness of the functions α and β are, in general, not sufficient to fulfill requirement (10.52) since, while $x_a(t)$, and hence $y_e(t) = \mathcal{H}(x_a(t))$, is assumed to be bounded for all $t \geq 0$, nothing can be said about the behavior of $\hat{y}_e(t)$ (this point is made clearer in the proof to follow). If α and β are globally Lipschitz functions and $v(t)$ is a bounded function of time, then requirement (10.52) is automatically satisfied. We will see in the following that, without requiring α and β to be globally Lipschitz or any other additional assumption, (10.52) is always fulfilled by applying to the observer a suitable dynamic projection onto a fixed compact set or, in some cases, by using a saturation. By virtue of (10.52), if $\hat{y}_e(0) \in \mathcal{H}(\Omega)$ and $y_e(t) \in \mathcal{H}(\Omega)$, there exists a fixed scalar $k^* > 0$, independent of η, such that the bracketed term in (10.51) can be bounded as follows

$$\left[\alpha(\hat{y}_e, s) + \beta(\hat{y}_e, s)v - \alpha(y_e, s) - \beta(y_e, s)v\right] \leq k^* |\hat{y}_e - y_e|, \tag{10.53}$$

and thus the time derivative of V_o can be bounded as follows

$$\dot{V}_o \leq -\frac{|\tilde{\nu}|^2}{\eta} + 2k^*|P||\tilde{y}_e||\tilde{\nu}| \leq -\frac{|\tilde{\nu}|^2}{\eta} + 2k^*|P||\tilde{\nu}|^2, \tag{10.54}$$

where we have used the fact that $|\tilde{y_e}| \leq |\tilde{\nu}|$.

As we mentioned earlier, the smoothness of α and β and the boundedness of $v(t)$ for all $t \geq 0$ are not sufficient to guarantee that a bound of the type (10.53) hold for the bracketed term in (10.51). This is seen by noticing that the level sets of the Lyapunov function V_o, expressed in $\tilde{\xi}$ coordinates,

$$\Lambda_c = \{\tilde{\xi} \in \mathsf{R}^n \mid V_o(\mathcal{E}'\tilde{\xi}) \leq c\},$$

are parameterized by η and become larger as η is decreased. Thus, letting $\tilde{\xi}$ range over Λ_c, a straightforward application of Lipschitz inequality would result in a bound like (10.53) where k^*, rather than being constant, is a function of η. Defining $\bar{\eta} = \min\{1/(2|P|k^*), 1\}$, we conclude that, for all $\eta < \bar{\eta}$, the $\tilde{y_e}$ trajectories starting in $\mathcal{K}_{\tilde{y_e}}$ will converge asymptotically to the origin.

We have thus proved that, provided the state x_a of the augmented system (10.37) is contained in some compact set Ω and (10.52) is satisfied, one can choose a sufficiently small value of η in the observer (10.39) guaranteeing that $\tilde{y}_e(t) = \hat{y}_e(t) - y_e(t) \to 0$ as $t \to \infty$. Recalling that $\mathcal{H}$ is a diffeomorphism, we conclude that $\hat{x}(t) \to x(t)$. Finally notice that $\hat{x}(t) = x(t)$ is the unique solution of (10.39) when $\hat{x}(0) = x(0)$. Hence, (10.39) is an observer for (10.25) in the sense of Definition 10.1.

Besides proving that the estimation error vanishes asymptotically, we are also interested in assessing how *fast* does it vanish. Specifically, given any positive scalars ϵ and T, we now show that one can pick η guaranteeing that $\hat{x} - x < \epsilon$ for all $t \geq T$, and thus the convergence of the observer can be made arbitrarily fast. To this end, note that

$$\lambda_{min}(\mathcal{E}'P\mathcal{E}') \geq \lambda_{min}(\mathcal{E}')^2\lambda_{min}(P) = \lambda_{min}(P),$$

since $\lambda_{min}(\mathcal{E}') = 1$. Next,

$$\lambda_{max}(\mathcal{E}'P\mathcal{E}') \leq \lambda_{max}(\mathcal{E}')^2\lambda_{max}(P) = 1/(\eta^{2(n-1)})\lambda_{max}(P),$$

since $\lambda_{max}(\mathcal{E}') = 1/\eta^{(n-1)}$. Therefore,

$$\lambda_{min}(P)|\tilde{y_e}|^2 \leq V_o = \tilde{y_e}^{\top}\mathcal{E}'P\mathcal{E}'\tilde{y_e} \leq \frac{1}{\eta^{2(n-1)}}\lambda_{max}(P)|\tilde{y_e}|^2. \qquad (10.55)$$

Define $\bar{\epsilon}$ so that $|\tilde{y_e}| \leq \bar{\epsilon}$ implies that $|\hat{x} - x| \leq \epsilon$ (the smoothness of $\mathcal{H}^{-1}$ guarantees that $\bar{\epsilon}$ is well defined). By inequality (10.55) we have that $V_o \leq \bar{\epsilon}^2\lambda_{min}(P)$ implies that $|\tilde{y_e}| \leq \bar{\epsilon}$, and $V_o(0) \triangleq V_o(\tilde{\nu}(0)) \leq (1/\eta^{2(n-1)})\lambda_{max}(P)|\tilde{y_e}(0)|^2$. Moreover, from (10.54)

$$\dot{V}_o(t) \leq -\left(\frac{1}{\eta} - 2|P|k^*\right)|\tilde{\nu}|^2 \leq -\frac{1}{\lambda_{max}(P)}\left(\frac{1}{\eta} - 2|P|k^*\right)V_o(t).$$

Therefore, by Lemma 2.1, $V_o(t)$ satisfies the following inequality

$$\begin{aligned} V_o(t) &\le V_o(0)\exp\left\{-\frac{1}{\lambda_{max}(P)}\left(\frac{1}{\eta}-2|P|k^*\right)t\right\} \qquad (10.56) \\ &\le \frac{1}{\eta^{2(n-1)}}\lambda_{max}(P)|\tilde{y}_e(0)|^2\exp\left\{-\frac{1}{\lambda_{max}(P)}\left(\frac{1}{\eta}-2|P|k^*\right)t\right\}, \end{aligned}$$

which, for sufficiently small η, can be written as

$$V_o(t) \le \frac{a_1}{\eta^{2n}}\exp\left\{-\frac{a_2}{\eta}t\right\},\ a_1, a_2 > 0.$$

An upper estimate of the time T such that $|\hat{x}-x| \le \epsilon$ for all $t \ge T$, is calculated as follows :

$$\frac{a_1}{\eta^{2n}}\exp\left\{-\frac{a_2}{\eta}t\right\} \le \bar{\epsilon}^2\lambda_{min}(P) \text{ for all } t \ge T = \frac{2n\eta}{a_2}\log\left(\frac{a_1}{\bar{\epsilon}\eta}\right).$$

Noticing that since $T \to 0$ as $\eta \to 0$, we conclude that T can be made arbitrarily small by choosing a sufficiently small η^*. The results above are summarized in the following theorem.

Theorem 10.1: *Assume that the plant (10.25) is uniformly completely observable, the state x_a of the augmented system (10.37) belongs to a compact invariant set Ω, and that the Lipschitz-type condition (10.52) is satisfied. Choose L such that the roots of the polynomial $s^n + l_1 s^{n-1} + \ldots + l_n$ have negative real part.*

Under these conditions, using observer (10.39), for all $[\hat{x}(0)^\top, s(0)^\top]^\top \in \Omega$ the following two properties hold

(i) There exists $\bar{\eta}$, $0 < \bar{\eta} \le 1$, such that for all $\eta \in (0, \bar{\eta})$, $\hat{x} \to x$ as $t \to +\infty$.

(ii) For each positive T, ϵ, there exists η^*, $0 < \eta^* \le 1$, such that for all $\eta \in (0, \eta^*]$, $|\hat{x} - x| \le \epsilon\ \forall t \ge T$.

10.4.2 Peaking Phenomenon

Using inequality (10.56), we can calculate an upper bound for the estimation error in $y_e(t)$-coordinates

$$|\tilde{y}_e| \le \sqrt{\frac{\lambda_{max}(P)}{\lambda_{min}(P)}}\frac{1}{\eta^{n-1}}|\tilde{y}_e(0)|\exp\left\{-\frac{1}{2\lambda_{min}(P)}\left(\frac{1}{\eta}-2|P|k^*\right)t\right\}. \qquad (10.57)$$

Notice that the magnitude of the upper bound above during the initial transient (around $t = 0$) grows larger as η is made smaller. Thus, while decreasing η has the beneficial effect of improving the rate of convergence of the observer, it may also yield an undesirable peak in the observer estimation error. When employing the state estimate $\hat{x}$ in place of x (i.e., $v = \bar{v}(\hat{x}, s)$) to stabilize the augmented system (10.37), the peak in the observer estimation error may destroy the stability properties of the state-feedback controller, no matter how fast the rate of convergence of the observer is made. This effect, known as the *peaking phenomenon*, is essentially due to the fact that the peak of the observer estimation error during transient may generate a large control action which may drive the closed-loop system trajectories to instability. This is seen in the next example.

Example 10.6 Consider the second-order system

$$\begin{aligned} \dot{x}_1 &= x_2 \\ \dot{x}_2 &= x_2^2 + \exp(-x_2^2)u \\ y &= x_1, \end{aligned} \tag{10.58}$$

for which a globally stabilizing controller is easily found to be

$$u = -\exp(x_2^2)\left(x_2^2 + x_1 + x_2\right). \tag{10.59}$$

In this case the mapping $\mathcal{H}$ is simply given by

$$\mathcal{H}(x, s) = [x_1, x_2]^\top,$$

which is a diffeomorphism on $\mathbb{R}^2$ thus showing that the system is uniformly completely observable. Furthermore, $n_u = 0$ (i.e., $\mathcal{H}$ does not depend on s), and thus we do not need to add integrators at the input side of the system. Noting that $\partial\mathcal{H}/\partial\hat{x} = I$, the observer equation (10.39) takes the form

$$\dot{\hat{x}}_1 = \hat{x}_2 + \frac{l_1}{\eta}(y - \hat{x}_1) \tag{10.60}$$

$$\dot{\hat{x}}_2 = \hat{x}_2^2 + \exp(-\hat{x}_2^2)u + \frac{l_2}{\eta^2}(y - \hat{x}_1). \tag{10.61}$$

Following Theorem 10.1, we choose $l_1 = l_2 = 1$ and we pick η sufficiently small. In practice, a convenient way to choose η is to run a few simulations and decrease its value until a satisfactory rate of convergence for the observer is found. Figure 10.2 shows the state trajectories $x_1(t)$ and $x_2(t)$ and the observer estimates $\hat{x}_1(t)$, $\hat{x}_2(t)$ when the state-feedback control law (10.59) is employed. The result confirms the theoretical predictions of Theorem 10.1: the observer estimation error vanishes and its rate of convergence is improved by decreasing the value of η. However, when comparing the performance of the observer when $\eta = 0.1$ and $\eta = 0.01$, the undesirable peaking phenomenon appears to be evident; as in the latter case $\hat{x}(t)$ exhibits

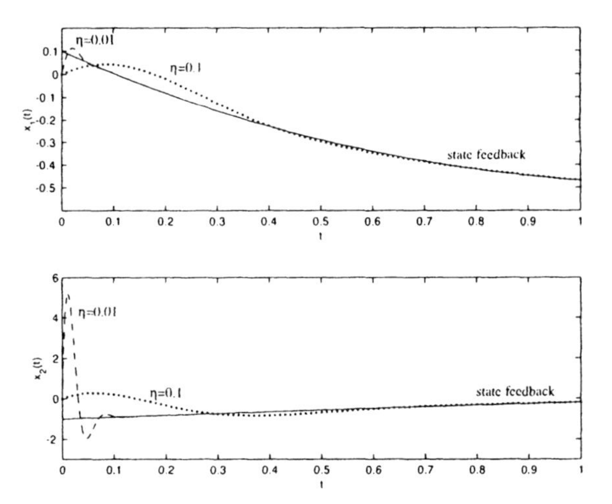

Figure 10.2. Peaking phenomenon in the observer estimate.

a significantly higher peak. Next, the destabilizing effect of the peaking phenomenon is illustrated by the phase plot in Figure 10.3, where the state-feedback controller (10.59) is replaced by

$$u = \exp(\hat{x}_2^2)\left(\hat{x}_2^2 + \hat{x}_1 + \hat{x}_2\right).$$

Now the presence of the peak in the observer estimates yields a large control input generating a deviation in the closed-loop system trajectories which becomes larger as η is decreased. When $\eta = 0.004$ the closed-loop system is unstable. △

10.4.3 Dynamic Projection of the Observer Estimate

The separation principle for linear time-invariant systems states that one can find an output-feedback controller by first independently designing a state-feedback controller and an observer, and then replacing x by $\hat{x}$ in the controller. Unfortunately, as seen in Example 10.6, such an interconnection does not guarantee closed-loop stability in the nonlinear setting. Specifically, we have seen that the main obstruction to achieving a separation principle is the presence of the peaking phenomenon in the observer. One way to eliminate the destabilizing effect of the peak in the observer is to project the observer states onto a suitable compact set so that the resulting output-feedback controller is globally bounded, with a bound independent of η (hence, the peaking phenomenon is *eliminated*). If this compact set is chosen to be larger than the set containing the state trajectories, then one can recover the stability properties of the state-feedback controller and thus

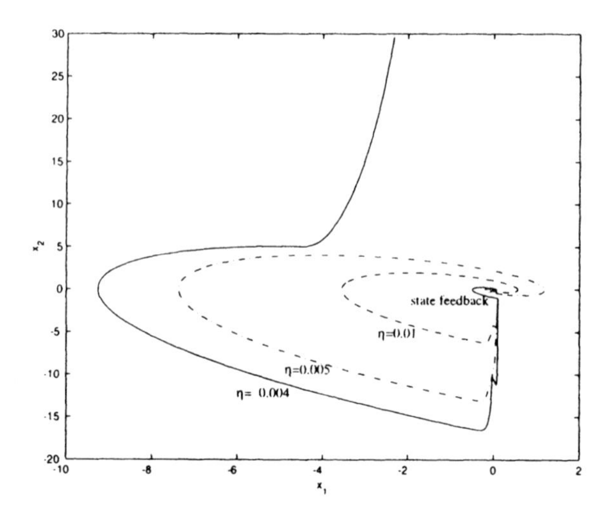

Figure 10.3. The peaking of the observer destabilizes the closed-loop system.

achieve a separation principle for nonlinear systems (this statement will be proved in the next section). Next, we introduce the dynamic projection and analyze its main properties.

Let $\mathcal{F} : \mathsf{R}^{n+n_u} \to \mathsf{R}^{n+n_u}$ be the mapping defined by

$$\mathcal{Y} = \mathcal{F}(x_a) = \begin{bmatrix} y_e \\ s \end{bmatrix} = \begin{bmatrix} \mathcal{H}(x, s) \\ s \end{bmatrix} \tag{10.62}$$

and notice that $\mathcal{F}$ is a diffeomorphism on R^{n+n_u} (this is due to the fact that its inverse is given by $x_a = [\mathcal{H}^{-1}(\mathcal{Y})^\top, s^\top]^\top$, which is smooth). Similarly, let $\hat{\mathcal{Y}} = \mathcal{F}(\hat{x}_a) = [\hat{y}_e^\top, s]^\top$, where $\hat{x}_a = [\hat{x}^\top, s]^\top$. Next, let $\mathcal{C}$ be a set in $\mathcal{Y}$ coordinates and denote by $N(\mathcal{Y})$ the normal vector to the boundary of $\mathcal{C}$ at $\mathcal{Y}$. In practice, the set $\mathcal{C}$ can always be expressed by an inequality

$$\mathcal{C} = \{\mathcal{Y} \,|\, g(\mathcal{Y}) \leq 0\},$$

where g is a smooth function. The boundary of $\mathcal{C}$ is then the set

$$\partial\mathcal{C} = \{\mathcal{Y} \,|\, g(\mathcal{Y}) = 0\}$$

and the normal vector $N(\mathcal{Y})$ is then calculated as

$$N(\mathcal{Y}) = \frac{\partial g(\mathcal{Y})}{\partial \mathcal{Y}}^\top,$$

with the convention that N calculated at any point of the boundary $\partial\mathcal{C}$ points outside of $\mathcal{C}$. Let $N_{y_e}(y_e, s)$ and $N_s(y_e, s)$ denote the y_e and s

components of N, that is,

$$N(\mathcal{Y}) = \begin{bmatrix} N_{y_e}(y_e, s) \\ N_s(y_e, s) \end{bmatrix} = \begin{bmatrix} \dfrac{\partial g(y_e, s)}{\partial y_e}^{\top} \\ \dfrac{\partial g(y_e, s)}{\partial s}^{\top} \end{bmatrix}.$$

Assumption 10.1: Assume that $\mathcal{C}$ has the following properties

(i) $\mathcal{C}$ has a smooth boundary.

(ii) Each slice $\mathcal{C}^{\bar{s}}$ of $\mathcal{C}$ obtained by holding s constant at $\bar{s}$, i.e.,

$$\mathcal{C}^{\bar{s}} = \{y_e \in \mathsf{R}^n \,|\, \mathcal{Y} \in \mathcal{C}\}$$

is convex, for all $\bar{s} \in \mathsf{R}^{n_u}$.

(iii) The vector $N_{y_e}(y_e, s)$ does not vanish anywhere on $\partial\mathcal{C}$.

(iv) $\bigcup_{\bar{s}} \mathcal{C}^{\bar{s}}$ is a compact set.

Consider the following *dynamic projection* applied to the observer dynamics:

$$\dot{\hat{x}}^P = \begin{cases} \dot{\hat{x}} - \left[\dfrac{\partial \mathcal{H}}{\partial \hat{x}}\right]^{-1} \left\{ \Gamma \dfrac{N_{y_e}(\hat{\mathcal{Y}}) N_{y_e}(\hat{\mathcal{Y}})^{\top}}{N_{y_e}(\hat{\mathcal{Y}})^{\top} \Gamma N_{y_e}(\hat{\mathcal{Y}})} \dot{\hat{y}}_e + \Gamma \dfrac{N_{y_e}(\hat{\mathcal{Y}}) N_s(\hat{\mathcal{Y}})^{\top}}{N_{y_e}(\hat{\mathcal{Y}})^{\top} \Gamma N_{y_e}(\hat{\mathcal{Y}})} \dot{s} \right\} \\ \qquad \text{if } N_{y_e}(\hat{\mathcal{Y}})^{\top} \dot{\hat{y}}_e + N_s(\hat{\mathcal{Y}})^{\top} \dot{s} \geq 0 \text{ and } \hat{\mathcal{Y}} \in \partial\mathcal{C} \\ \dot{\hat{x}} \qquad \text{otherwise} \end{cases} \tag{10.63}$$

where $\Gamma = (S\mathcal{E}')^{-1}(S\mathcal{E}')^{-1}$, $S = S^{\top}$ denotes the matrix square root of P (defined in (10.50)), $\partial\mathcal{C}$ denotes the boundary of $\mathcal{C}$, and

$$\dot{\hat{y}}_e = \frac{\partial \mathcal{H}}{\partial \hat{x}} \dot{\hat{x}} + \frac{\partial \mathcal{H}}{\partial s} \dot{s}.$$

The following lemma shows that (10.63) guarantees boundedness and preserves convergence for $\hat{x}$.

Lemma 10.1: *Let $\mathcal{C}$ a set in $\mathcal{Y}$ coordinates satisfying Assumption 10.1. The observer (10.39) with dynamic projection (10.63) has the following properties*

(i) Boundedness: if $\hat{x}^P(0) \in \mathcal{H}^{-1}(\mathcal{C})$, then $\hat{x}^P(t) \in \mathcal{H}^{-1}(\mathcal{C})$ for all t.

If, in addition, $x_a(t) \in \Omega$ for all $t \geq 0$, where Ω is a compact set, then the following holds

(ii) *Preservation of original convergence characteristics: properties (i) and (ii) established by Theorem 10.1 remain valid for* $\hat{x}^P$.

(iii) *Requirement (10.52) is guaranteed to hold provided* $\sup_{t\geq 0} v(t)$ *is bounded.*

Proof: In order to prove part (i) we need another coordinate transformation, $\zeta = S\mathcal{E}'y_e$, (similarly, let $\hat{\zeta} = S\mathcal{E}'\hat{y}_e$, $\tilde{\zeta} = S\mathcal{E}'\tilde{y}_e$). Define the linear map $\mathcal{G} = \mathrm{diag}\,[S\mathcal{E}', I_{n_u \times n_u}]$ and consider the set $\mathcal{C}'$, image of $\mathcal{C}$ under the map $\mathcal{G}$, i.e., $\mathcal{C}' \triangleq \{[\zeta^\top, s^\top]^\top \in \mathrm{R}^{n+n_u} \,|\, \mathcal{G}^{-1}[\zeta^\top, s^\top]^\top \in \mathcal{C}\}$ ($\mathcal{C}'$ is convex because of the linearity of $\mathcal{G}$). Let $N'_\zeta(\zeta, s)$, $N'_s(\zeta, s)$ be the ζ and s components of the normal vector to the boundary of $\mathcal{C}'$. In order to prove part (i) of the Lemma, it is sufficient to show that the dynamic projection (10.63) renders the set $\mathcal{C}'$ positively invariant which in turn guarantees that $\hat{x} = \mathcal{H}^{-1}(\hat{y}_e, s)$ is contained in the compact set $\mathcal{H}^{-1}(\mathcal{C})$. In $\hat{y}_e$ coordinates we have that

$$\dot{\hat{y}}_e^P = \frac{d}{dt}\{\mathcal{H}(\hat{x}^P, s)\} = \left[\frac{\partial \mathcal{H}}{\partial \hat{x}}\dot{\hat{x}}^P + \frac{\partial \mathcal{H}}{\partial s}\dot{s}\right]$$

$$= \begin{cases} \dot{\hat{y}}_e - \Gamma \dfrac{N_{y_e}N_{y_e}^\top \dot{\hat{y}}_e + N_{y_e}N_s^\top \dot{s}}{N_{y_e}^\top \Gamma N_{y_e}} & \\ \qquad \text{if } N_{y_e}(\hat{\mathcal{Y}})^\top \dot{\hat{y}}_e + N_s(\hat{\mathcal{Y}})^\top \dot{s} \geq 0 \text{ and } \hat{\mathcal{Y}} \in \partial\mathcal{C} & \\ \dot{\hat{y}}_e \qquad \text{otherwise.} & \end{cases} \tag{10.64}$$

In order to relate $N'_\zeta(\hat{\zeta}, s)$, $N'_s(\hat{\zeta}, s)$ to $N_{y_e}(\hat{y}_e, s)$, $N'_s(\hat{\zeta}, s)$, recall that that $\partial\mathcal{C} = \{[y_e^\top, s^\top]^\top \,|\, g(y_e, s) = 0\}$ and $N_{y_e}(\hat{y}_e, s) = (\partial g(\hat{y}_e, s)/\partial \hat{y}_e)^\top$, $N_s(\hat{y}_e, s) = (\partial g(\hat{y}_e, s)/\partial s)^\top$.

The boundary of $\mathcal{C}'$ is the set $\partial\mathcal{C}' = \{[\zeta^\top, s^\top]^\top \in \mathrm{R}^{n+n_u} \,|\, g((S\mathcal{E}')^{-1}\zeta, s) = 0\}$ and

$$N'_\zeta(\hat{\zeta}, s) = (S\mathcal{E}')^{-1}(\partial g(\hat{y}_e, s)/\partial \hat{y}_e)^\top = (S\mathcal{E}')^{-1}N_{y_e}(\hat{y}_e, s),$$

$$N'_s(\hat{\zeta}, s) = N_s(\hat{y}_e, s).$$

The expression of the projection (10.63) in ζ coordinates is found by using the definition of Γ and noting that

$$\dot{\hat{\zeta}}^P = S\mathcal{E}'\dot{\hat{y}}_e^P = \begin{cases} S\mathcal{E}'\dot{\hat{y}}_e - (S\mathcal{E}')^{-1}\dfrac{N_{y_e}\left(N_{y_e}^\top \dot{\hat{y}}_e + N_s^\top \dot{s}\right)}{N_{y_e}^\top \Gamma N_{y_e}} & \\ \qquad \text{if } N_{y_e}^\top \dot{\hat{y}}_e + N_s^\top \dot{s} \geq 0 \text{ and } \hat{\mathcal{Y}} \in \partial\mathcal{C} & \\ S\mathcal{E}'\dot{\hat{y}}_e \qquad \text{otherwise} & \end{cases} \tag{10.65}$$

and then substituting $N'_\zeta = (S\mathcal{E}')^{-1}N_{y_e}$, $N'_s = N_s$, and $\dot{\hat{y}}_e = (S\mathcal{E}')^{-1}\dot{\hat{\zeta}}$, to find that

$$\dot{\hat{\zeta}}^P = \begin{cases} \dot{\hat{\zeta}} - \dfrac{N'_\zeta\left(N'^{\top}_\zeta\dot{\hat{\zeta}} + N'^{\top}_s\dot{s}\right)}{N'^{\top}_\zeta N'_\zeta} & \text{if } N'^{\top}_\zeta\dot{\hat{\zeta}} + N'^{\top}_s\dot{s} \geq 0 \\ & \text{and } [\hat{\zeta}^\top, s^\top]^\top \in \partial\mathcal{C}' \\ \dot{\hat{\zeta}} & \text{otherwise .} \end{cases} \tag{10.66}$$

Next, we show that the domain $\mathcal{C}'$ is positively invariant for (10.66). In order to do that, consider the continuously differentiable function

$$V_{\mathcal{C}'} = \frac{1}{2}b^2\left((S\mathcal{E}')^{-1}\zeta, s\right)$$

and calculate its time derivative along the trajectory of (10.66) when $\hat{\zeta}^P \in \partial\mathcal{C}'$,

$$\begin{aligned} \dot{V}_{\mathcal{C}'} &= b\left((S\mathcal{E}')^{-1}\hat{\zeta}^P, s\right)\left[N'_\zeta(\hat{\zeta}^P, s)^\top\dot{\hat{\zeta}}^P + N'_s(\hat{\zeta}^P, s)^\top\dot{s}\right] \\ &= b\left((S\mathcal{E}')^{-1}\hat{\zeta}^P, s\right)\left[N'^{\top}_\zeta\dot{\hat{\zeta}} - \frac{N'^{\top}_\zeta N'_\zeta\left(N'^{\top}_\zeta\dot{\hat{\zeta}} + N'^{\top}_s\dot{s}\right)}{N'^{\top}_\zeta N'_\zeta} + N'^{\top}_s\dot{s}\right] \\ &= 0. \end{aligned} \tag{10.67}$$

Since $\dot{V}_{\mathcal{C}'} = 0$ on the boundary of $\mathcal{C}'$ the solutions $[\hat{\zeta}^{P\top}(t)^\top, s^\top(t))]^\top$ of (10.66) can not cross $\partial\mathcal{C}'$ and hence $\mathcal{C}'$ is positively invariant. This in turn implies that $[\hat{y}_e^{P\top}(t), s^\top(t)]^\top$ cannot cross $\partial\mathcal{C}$ and, thus, $\hat{x}^P(t) \in \mathcal{H}^{-1}(\mathcal{C})$ for all t.

The proof of part (ii) is based on the knowledge of a Lyapunov function for the observer in $\tilde{\nu}$ coordinates (see (10.48)). Notice that $\tilde{\zeta} = S\tilde{\nu}$, and $V_o = \tilde{\nu}^\top P\tilde{\nu} = (\tilde{\nu}^\top S)(S\tilde{\nu}) = \tilde{\zeta}^\top\tilde{\zeta}$. We want to show that, in $\tilde{\zeta}$ coordinates, $\dot{V}_o < 0$ and $\dot{\hat{\zeta}}^P = S\mathcal{E}'\dot{\hat{y}}_e^P$ implies that $\dot{V}_o^P \leq \dot{V}_o$, where $\tilde{\zeta}^P = \hat{\zeta}^P - \zeta$, and $V_o^P = \tilde{\zeta}^{P\top}\tilde{\zeta}^P$.

Recall that, by assumption, $x_a(t) \in \mathcal{F}^{-1}(\mathcal{C})$ for all $t \geq 0$ or, equivalently, $[y_e(t)^\top, s(t)^\top]^\top \in \mathcal{C}$ or, what is the same, $[\zeta(t)^\top, s(t)^\top]^\top \in \mathcal{C}'$ for all $t \geq 0$. From (10.66), when $[\hat{\zeta}^\top, s^\top]^\top$ is in the interior of $\mathcal{C}'$, or $[\hat{\zeta}^\top, s^\top]^\top$ is on the boundary of $\mathcal{C}'$ and $N'^{\top}_\zeta\dot{\hat{\zeta}} + N'^{\top}_s\dot{s} < 0$ (i.e., the update is pointed to the interior of $\mathcal{C}'$), we have that $V_o = V_o^P$. Let us consider all the remaining cases and, since $\hat{\zeta}^P = \hat{\zeta}$ and $\tilde{\zeta}^P = \tilde{\zeta}$ (the projection is dynamic, it only operates on $\dot{\hat{\zeta}}$), we have

$$\dot{V}_o^P = 2\tilde{\zeta}^P\dot{\tilde{\zeta}}^P = 2\tilde{\zeta}^\top\dot{\tilde{\zeta}}^P = 2\tilde{\zeta}^\top\left[\dot{\hat{\zeta}} - \dot{\zeta} - p(\hat{\zeta}, \dot{\hat{\zeta}}, s, \dot{s})N'_\zeta(\hat{\zeta}, s)\right], \tag{10.68}$$

where

$$p(\hat{\zeta}, \dot{\hat{\zeta}}, s, \dot{s}) = \frac{N_{\zeta}'^{\top} \dot{\hat{\zeta}} + N_s'^{\top} \dot{s}}{N_{\zeta}'(\hat{\zeta}, s)^{\top} N_{\zeta}'(\hat{\zeta}, s)}$$

is nonnegative since, by assumption, $N_{\zeta}'^{\top} \dot{\hat{\zeta}} + N_s'^{\top} \dot{s} \geq 0$. Thus,

$$\dot{V}_o^P = \dot{V}_o - 2p\tilde{\zeta}^{\top} N_{\zeta}'. \tag{10.69}$$

Using the fact that $[\zeta^{\top}, s^{\top}]^{\top} \in \mathcal{C}'$ and that $[\hat{\zeta}^{\top}, s^{\top}]^{\top}$ lies on the boundary of $\mathcal{C}'$, we have that the vector $[\hat{\zeta} - \zeta, 0^{\top}]^{\top}$ points outside of $\mathcal{C}'$ which, by the convexity of $\mathcal{C}'$, implies that $\tilde{\zeta}^{\top} N_{\zeta}' + 0^{\top} N_s' = \tilde{\zeta}^{\top} N_{\zeta}' \geq 0$, thus concluding the proof of part (ii).

Next, to prove part (iii), we want to show that if $x_a \in \Omega$ for all $t \geq 0$, then inequality (10.52) holds for all $t \geq 0$, with $\hat{y}_e(t)$ replaced by $\hat{y}_e^P$, provided that $v(t)$ is uniformly bounded. We start by noting that $y_e(t) = \mathcal{H}(x(t), s(t))$ is contained in the compact set $\mathcal{H}(\Omega)$ for all $t \geq 0$ and $s(t)$ is contained in the compact set $\Omega^s = \{s \in \mathsf{R}^{n_u} \,|\, x_a(t) \in \Omega\}$ for all $t \geq 0$. Furthermore, using part (i) of this lemma and property (iv) in Assumption 10.1 we have that

$$[\hat{y}_e^{P\top}, s^{\top}]^{\top} \in \bar{\mathcal{C}} = \left(\bigcup\nolimits_{s \in \Omega^s} \mathcal{C}^s\right) \times \Omega^s, \text{ for all } t \geq 0,$$

where $\bar{\mathcal{C}}$ is a compact set. Now, part (iii) is proved by noticing that inequality (10.52) follows directly from the facts above, the boundedness of $v(t)$, and the local Lipschitz continuity of α and β. ■

Putting the results of Theorem 10.1 and Lemma 10.1 together, we have that if the x_a trajectories of the system are contained in some fixed compact set Ω for all time and the plant is UCO, then the observer (10.39) together with the dynamic projection (10.63) provide an asymptotically convergent and arbitrarily fast estimate of the state x, without any additional assumption. Note that there is no restriction on the size of the compact set Ω.

In the particular case when the plant (10.25) has the form

$$\begin{aligned} \dot{x}_i &= x_{i+1}, \qquad i = 1, \ldots, n-1 \\ \dot{x}_n &= f(x, u) \\ y &= x_1 \end{aligned} \tag{10.70}$$

and hence the observability mapping $\mathcal{H}$ is the identity, i.e., $y_e = \mathcal{H}(x) = x$, the observer (10.39) is simply given by

$$\begin{aligned} \dot{\hat{x}}_i &= \hat{x}_{i+1} + \frac{l_i}{\eta^i}(y - \hat{x}_1), \qquad i = 1, \ldots, n-1 \\ \dot{\hat{x}}_n &= f(\hat{x}, u) + \frac{l_n}{\eta^n}(y - \hat{x}_1). \end{aligned} \tag{10.71}$$

Observer (10.71) is commonly referred to as a high-gain observer. For the particular class of systems (10.70), in order to satisfy requirement (10.52) and eliminate the destabilizing effects of the peaking phenomenon, one can remove the dynamic projection (10.63) by saturating $\hat{x}$ in (10.71) over the same compact set $\mathcal{F}^{-1}(\mathcal{C})$

$$
\begin{aligned}
\dot{\hat{x}}_i &= \hat{x}_{i+1} + \frac{l_i}{\eta^i}(y - \hat{x}_1), \qquad i = 1, \ldots, n-1 \\
\dot{\hat{x}}_n &= f(\hat{x}^s, u) + \frac{l_n}{\eta^n}(y - \hat{x}_1),
\end{aligned}
\tag{10.72}
$$

where

$$
\hat{x}^s = \mathcal{V}\,\mathrm{sat}(\mathcal{V}^{-1}\hat{x}), \tag{10.73}
$$

with

$$
\mathrm{sat}\{\hat{x}\} = [\mathrm{sat}(\hat{x}_1), \ldots, \mathrm{sat}(\hat{x}_n)]^{\top}, \; \mathrm{sat}(\hat{x}_i) = \begin{cases} 1 & \text{if } \hat{x}_i > 1 \\ \hat{x}_i & \text{if } |\hat{x}_i| \le 1 \\ -1 & \text{if } \hat{x}_i < -1 \end{cases}
$$

$$
\mathcal{V} = \mathrm{diag}[\mathcal{V}_1, \ldots, \mathcal{V}_n], \; \mathcal{V}_i = \sup_{x \in \mathcal{H}^{-1}(\mathcal{C})} \{x_i\},
$$

and using $\hat{x}^s$ in place of x in the stabilizing control law.

10.4.4 Output-Feedback Stabilizing Controller

Recall that for the augmented system

$$
\begin{aligned}
\dot{x}_a &= f_a(x_a, v) \\
y &= h_a(x_a, s_1),
\end{aligned}
\tag{10.74}
$$

where $x_a = [x^{\top}, s^{\top}]^{\top}$, we may design a controller $\bar{v}(x_a)$ which makes the origin $x_a = 0$ asymptotically stable with some domain of attraction $\mathcal{D}$. Clearly, when the controller $\bar{v}(x_a)$ globally asymptotically stabilizes the origin $x_a = 0$, we will have $\mathcal{D} = \mathsf{R}^{n+n_u}$. Using the converse Lyapunov theorem found in [121], we know that there exists a continuously differentiable function V defined on $\mathcal{D}$ satisfying, for all $x_a \in \mathcal{D}$,

$$
\alpha_1(|x_a|) \le V(x_a) \le \alpha_2(|x_a|) \tag{10.75}
$$

$$
\lim_{x_a \to \partial\mathcal{D}} \alpha_1(|x_a|) = \infty \tag{10.76}
$$

$$
\frac{\partial V}{\partial x_a} f_a(x_a, \bar{v}(x_a)) \le -\alpha_3(|x_a|), \tag{10.77}
$$

where α_i, $i = 1, 2, 3$ are class $\mathcal{K}$ functions, and $\partial\mathcal{D}$ stands for the boundary of the set $\mathcal{D}$. Given any positive scalar c, define the corresponding level set of V as

$$
\Omega_c \triangleq \{x_a \in \mathsf{R}^{n+n_u} \mid V(x_a) \le c\}.
$$

Clearly, $\Omega_c \subset \mathcal{D}$ for all $c > 0$ and, from (10.76), Ω_c becomes arbitrarily close to $\mathcal{D}$ as $c \to \infty$. In what follows, we will introduce an output-feedback controller of the form (10.26) yielding the following two properties:

1. The origin of the closed-loop system is asymptotically stable.
2. The set which estimates the domain of attraction using the output feedback controller is contained in $\mathcal{D}$, and it can be made arbitrarily close to it by appropriately choosing a design parameter in the controller.

Consider now two arbitrary positive scalars c_1 and c_2, where $c_1 < c_2$, and the associated compact sets Ω_{c_1} and Ω_{c_2}. Note that, from the properties of the Lyapunov function $V(x_a)$, $\Omega_{c_1} \subset \Omega_{c_2} \subset \mathcal{D}$. Recall that $\hat{x}_a = [\hat{x}^\top, s^\top]^\top$, define $x_a^P = [\hat{x}^{P\top}, s^\top]^\top$, and consider the output-feedback controller

$$\hat{v} = \bar{v}(\hat{x}_a^P), \tag{10.78}$$

where $\hat{x}^P$ is the state of the observer (10.39) with the dynamic projection (10.63) operating on the set $\mathcal{C}$ which we choose so that it satisfies Assumption 10.1 and

$$\mathcal{F}(\Omega_{c_2}) \subset \mathcal{C}\,.$$

Notice that, due to the projection, the controller (10.78) does not exhibit peaking. When the plant takes the form (10.70), then saturation can be used in place of the projection and $\hat{x}^P$ in (10.78) can be replaced by $\hat{x}^s$, defined in (10.73). The following theorem states that the output-feedback controller (10.78) guarantees asymptotic stability of the origin $x_a = 0$ with domain of attraction containing the set Ω_{c_1}.

Theorem 10.2: *Consider the closed-loop system formed by a plant augmented by a chain of n_u integrators (10.74), the observer (10.39), and the controller (10.78). Then, for any pair of scalars c_1 and c_2 such that $0 < c_1 < c_2$, there exists a scalar $\eta^*, 0 < \eta^* \leq 1$, such that, for all $\eta \in (0, \eta^*]$, the set Ω_{c_1} is contained in the region of attraction of the origin $x_a = 0$.*

This theorem states that any compact set Ω_{c_1} contained in $\mathcal{D}$ can be made a domain of attraction for the origin $x_a = 0$ of the augmented plant provided η is chosen sufficiently small. From the property that Ω_{c_1} approaches $\mathcal{D}$ as $c_1 \to \infty$, we deduce that the proposed output-feedback controller recovers the domain of attraction achieved with the original state-feedback controller. In particular, when the state-feedback controller globally asymptotically stabilizes the origin $x_a = 0$ (i.e., $\mathcal{D} = \mathbb{R}^{n+n_u}$), the output feedback controller stabilizes the origin *semiglobally* or, in other words, the domain of attraction of the origin with the output-feedback controller is *any* compact set in $\mathbb{R}^{n+n_u}$.

We next proceed to proving the theorem. To this end, we divide the proof in three parts.

1. (Lemma 10.2). *Invariance of Ω_{c_2} and ultimate boundedness:* Using the arbitrarily fast rate of convergence of the observer (see part (ii) in Theorem 10.1), we show that any trajectory originating in Ω_{c_1} cannot exit the set Ω_{c_2} and converges in finite time to an arbitrarily small neighborhood of the origin. Here, the proof is based on the fact that the projection eliminates the peaking phenomenon.

2. (Lemma 10.3). *Asymptotic stability of the origin:* By using the exponential stability of the observer estimate and Lemma 10.2, we prove that the origin of the closed-loop system is asymptotically stable.

3. *Closed-loop stability:* Finally, by putting together the results obtained in Lemmas 10.2 and 10.3, we conclude the closed-loop stability proof.

Before stating the first lemma, notice that the smoothness of f_a and the control law $\bar{v}$ implies that there exists a positive scalar $\bar{\gamma}$ such that

$$|f_a(x_a, \bar{v}(\hat{x}_a^P)) - f_a(x_a, \bar{v}(x_a))| \leq \bar{\gamma}|\hat{x}^P - x| \tag{10.79}$$

for all $x_a, \hat{x}_a \in \mathcal{F}^{-1}(\mathcal{C}) \supset \Omega_{c_2}$. Furthermore, from the fact that V is continuously differentiable we have that there exists a positive scalar A such that $|\partial V/\partial x_a| \leq A$ for all x_a in Ω_{c_2}.

Lemma 10.2: *Suppose that the initial condition $x_a(0)$ is contained in Ω_{c_1} and consider the set Ω_{d_ϵ}, where $d_\epsilon = \alpha_2 \circ \alpha_3^{-1}(\mu A \bar{\gamma} \epsilon)$, and choose $\epsilon > 0$ and $\mu > 1$ such that $d_\epsilon < c_1$. Then, there exists a positive scalar $\eta^*, 0 < \eta^* \leq 1$, such that, for all $\eta \in (0, \eta^*]$, the solution of (10.74) remains confined in Ω_{c_2}, the set $\Omega_{d_\epsilon} \subset \Omega_{c_1}$ is positively invariant, and is reached in finite time.*

Proof: Since $V(x_a(0)) \leq c_1 < c_2$, there exists a time $T_1 > 0$ such that $V(x_a(t)) \leq c_2$, for all $t \in [0, T_1)$. Choose T_0 such that $0 < T_0 < T_1$. Then, since $x_a(t) = [x(t)^\top, s(t)^\top]^\top \in \Omega_{c_2} \subset \mathcal{H}^{-1}(\mathcal{C})$ for all $t \in [0, T_1)$ and, by Lemma 10.1, part (i), $\hat{x}^P(t) \in \mathcal{H}^{-1}(\mathcal{C})$ for all $t \geq 0$, we conclude that, for all $t \in [0, T_1)$,

$$[\hat{x}^{P\top}, s^\top]^\top \in \mathcal{F}^{-1}\left(\bigcup_{s \in \Omega_{c_2}^s} \mathcal{C}^s, \Omega_{c_2}^s\right)$$

which, from part (iv) in Assumption 10.1, is a compact set. Thus, from the smoothness of $\bar{v}$, $v(t)$ is bounded by a constant independent of η, and hence we can apply part (ii) in Theorem 10.1 and parts (ii), (iii) in Lemma 10.1 and conclude that for any positive ϵ there exists a positive $\eta^*, 0 < \eta^* \leq 1$ such that, for all $\eta \leq \eta^*$, $|\tilde{x}^P| \triangleq |\hat{x}^P - x| \leq \epsilon, \forall t \in [T_0, T_1)$. Hence, for all $t \in [T_0, T_1)$, we have that $V(x_a(t)) \leq c_2$ and $|\tilde{x}(t)| \leq \epsilon$. In order for part (ii) in Theorem 10.1 to hold for all $t \geq T_0$, x_a must belong to Ω_{c_2} for all $t \geq 0$. So far we can only guarantee that $x_a \in \Omega_{c_2}$ for all $t \in [0, T_1)$ and hence the result of Theorem 10.1 applies in this time interval, only. Next, we will show that $T_1 = \infty$, i.e., Ω_{c_2} is a positively invariant set, so that the result of Theorem 10.1 will be guaranteed to hold for all $t \geq 0$.

Consider the Lyapunov function candidate $V(x_a)$ defined in (10.75)-(10.77). Taking its derivative with respect to time and using (10.79) we get

$$
\begin{aligned}
\dot{V} &= \frac{\partial V}{\partial x_a} f_a(x_a, \hat{v}) = \frac{\partial V}{\partial x_a} f_a(x_a, \bar{v}) + \frac{\partial V}{\partial x_a}\left[f_a(x_a, \hat{v}) - f_a(x_a, \bar{v})\right] \\
&\leq -\alpha_3(|x_a|) + \left|\frac{\partial V}{\partial x_a}\right| \left|f_a(x_a, \bar{v}(\hat{x}_a^P)) - f_a(x_a, \bar{v}(x_a))\right| \\
&\leq -\alpha_3(|x_a|) + A\bar{\gamma}|\tilde{x}^P| \\
&\leq -\alpha_3(|x_a|) + A\bar{\gamma}\epsilon \\
&\leq -\alpha_3 \circ \alpha_2^{-1}(V) + A\bar{\gamma}\epsilon
\end{aligned}
$$

for all $t \in [T_0, T_1)$. When $V \geq d_\epsilon$ we have that

$$\dot{V} \leq -(\mu - 1)A\bar{\gamma}\epsilon,$$

hence V decays linearly, which in turn implies that $x_a(t) \in \Omega_{c_2}$ for all $t \geq 0$ (i.e., $T_1 = \infty$), and Ω_ϵ is reached in finite time. ■

The use of the dynamic projection plays a crucial role in the proof of Lemma 10.2. As η is made smaller, the observer peak may grow larger, thus generating a large control input, which in turn might drive the system states x_a outside of Ω_{c_1} in shorter time. The projection eliminates the peaking and makes sure that the exit time T_1 is independent of ϵ, since the maximum size of the $\hat{v}$ will not depend on ϵ, thus allowing us to choose ϵ independently of T_1.

Lemma 10.2 proves that all the trajectories starting in Ω_{c_1} will remain confined within Ω_{c_2} and converge to an arbitrarily small neighborhood of the origin in finite time. Now, in order to complete the stability analysis, it remains to show that the origin of the output-feedback closed-loop system is asymptotically stable, so that if Ω_{d_ϵ} is small enough all the closed-loop system trajectories converge to it.

Lemma 10.3: *There exists a positive scalar ϵ^* such that for all $\epsilon \in (0, \epsilon^*]$ all the trajectories starting inside the compact set*

$$\Delta_\epsilon \triangleq \{[x_a^\top, \tilde{x}^{P\top}]^\top \mid V(x_a) \leq d_\epsilon \text{ and } |\tilde{x}^P| \leq \epsilon\}$$

converge asymptotically to the origin $x_a = 0, \tilde{x}^P = 0$.

Proof: Recall that $\tilde{x}^P = \mathcal{H}^{-1}(\hat{y}_e^P, s) - \mathcal{H}^{-1}(y_e, s)$. From the uniform complete observability of the plant we have that the mapping $\mathcal{H}^{-1}$ is locally Lipschitz. Hence, there exists a neighborhood $N_{\tilde{y}_e}$ such that $|\tilde{x}^P| \leq l|\hat{y}_e^P - y_e|$, for all $\hat{y}_e^P - y_e \in N_{\tilde{y}_e}$, and for some positive constant l, which, by

(10.57), implies that the origin of the $\tilde{x}^P$ system is exponentially stable. By the converse Lyapunov theorem we conclude that there exists a Lyapunov function $V_o'(\tilde{x}^P)$ and positive constants $\bar{c}_1, \bar{c}_2, \bar{c}_3$ such that

$$\begin{aligned} \bar{c}_1|\tilde{x}^P|^2 \leq V_o' \leq \bar{c}_2|\tilde{x}^P|^2 \\ \dot{V}_o' \leq -\bar{c}_3|\tilde{x}^P|^2. \end{aligned} \tag{10.80}$$

Define the positive scalar ϵ^* such that $|\tilde{x}^P| \leq \epsilon^*$ implies $\hat{y}_e^P - y_e \in N_{\bar{y}_e}$ (the existence of ϵ^* is a direct consequence of the fact that $\mathcal{H}$ is locally Lipschitz). Next, define the following composite Lyapunov function candidate

$$V_c(x_a, \tilde{x}^P) = V(x_a) + \lambda\sqrt{V_o'(\tilde{x}^P)}, \qquad \lambda > \frac{2\sqrt{\bar{c}_2}\bar{\gamma}A}{\bar{c}_3}$$

then,

$$\begin{aligned} \dot{V}_c \quad & \leq -\alpha_3(|x_a|) + A\bar{\gamma}|\tilde{x}^P| - \frac{\lambda}{2\sqrt{V_o'(\tilde{x}^P)}}\bar{c}_3|\tilde{x}^P|^2 \\ & \leq -\alpha_3(|x_a|) - \left(\frac{\bar{c}_3\lambda}{2\sqrt{\bar{c}_2}} - A\bar{\gamma}\right)|\tilde{x}^P| < 0, \end{aligned}$$

where we have used the fact that $[x_a^\top, \tilde{x}^{P\top}]^\top \in \Delta_\epsilon$ implies that $x_a \in \Omega_{c_2}$ (provided ϵ is small enough), and hence $\left|\frac{\partial V}{\partial x_a}\right| \leq A$. Since $\dot{V}_c$ is negative definite, all the $[x_a^\top, \tilde{x}^{P\top}]^\top$ trajectories starting in Δ_ϵ will converge asymptotically to the origin. ■

We are now ready to prove Theorem 10.2.

Proof of Theorem 10.2: By Lemma 10.3, there exists $\epsilon^* > 0$ such that, for all $\epsilon \in (0, \epsilon^*]$, Δ_ϵ is a region of attraction for the origin. Use Lemma 10.2 and the fact that $x_a(0) \in \Omega_{c_1}$ to find $\eta^*, 0 < \eta^* \leq 1$, so that for all $\eta \in (0, \eta^*]$ the state trajectories enter Δ_ϵ in finite time. This concludes the proof of the theorem. ■

10.5 Extension to MIMO Systems

If the input u of the plant has dimension m and the output y has dimension ρ, with m and ρ integers greater than one, the output-feedback controller introduced in the previous section remains essentially unchanged. The only difference is introduced in the definition of the observability mapping $\mathcal{H}$ and the related observability assumption. Specifically, when $u \in \mathbb{R}^m$ and

$y \in \mathbb{R}^\rho$, we define the mapping

$$y_e \stackrel{\triangle}{=} \begin{bmatrix} y_1 \\ \dot{y}_1 \\ \vdots \\ y_1^{(k_1-1)} \\ y_2 \\ \vdots \\ y_2^{(k_2-1)} \\ \vdots \\ y_\rho \\ \vdots \\ y_\rho^{(k_\rho-1)} \end{bmatrix} = \begin{bmatrix} h_1(x,u) \\ \varphi_1^1(x,u,u^{(1)}) \\ \vdots \\ \varphi_1^{k_1-1}\left(x,u,\ldots,u^{(k_1-1)}\right) \\ h_2(x,u) \\ \vdots \\ \varphi_2^{k_2-1}\left(x,u,\ldots,u^{(k_2-1)}\right) \\ \vdots \\ h_\rho(x,u) \\ \vdots \\ \varphi_\rho^{k_\rho-1}\left(x,u,\ldots,u^{(k_\rho-1)}\right) \end{bmatrix} \stackrel{\triangle}{=} \mathcal{H}(x,s), \quad (10.81)$$

where $s \stackrel{\triangle}{=} [u_1,\ldots,u_1^{(n_1-1)},\ldots,u_m,\ldots,u_m^{(n_m-1)}]^\top \in \mathbb{R}^{n_u}$, $\sum_{i=1}^{\rho} k_i = n$, $n_u \stackrel{\triangle}{=} n_1 + \ldots + n_m$, $0 \le n_i \le \max\{k_1,\ldots,k_\rho\}$, (when $\mathcal{H}$ does not depend on u_i then we set $n_i = 0$). Similarly to the SISO case, the functions φ_i^j are defined as follows:

$$\begin{aligned} \varphi_i^j(x,u,\ldots,u^{(j)}) &= \frac{\partial \varphi_i^{j-1}}{\partial x} f(x,u) + \sum_{k=0}^{j-1} \frac{\partial \varphi_i^{j-1}}{\partial u^{(k)}} u^{(k+1)} \\ &\quad \text{for } i = 1,\ldots,\rho,\ j = 1,\ldots,k_i - 1 \end{aligned} \quad (10.82)$$

$$\varphi_i^0(x,u) = h_i(x,u), \text{ for } i = 1,\ldots,\rho.$$

Instead of augmenting the plant with one chain of integrators, we now augment it with m such chains

$$\begin{aligned} \dot{s}_{i,j} &= s_{i,j+1}, & j &= 1,\ldots,n_i - 1 \\ \dot{s}_{i,n_i} &= u_i', & i &= 1,\ldots,m \\ u_i &= s_{i,1}. \end{aligned} \quad (10.83)$$

In complete analogy to the SISO case, we say that the MIMO system is uniformly completely observable if there exists a set of indices $\{k_1,\ldots,k_\rho\}$ such that the mapping $\mathcal{H}$ in (10.81) is a diffeomorphism for all $x \in \mathbb{R}^n$ and $s \in \mathbb{R}^{(n_u)}$.

The matrix $\mathcal{E}$ and the vector L in (10.39) are modified as follows:

$$L = block\text{-}diag[L^1,\ldots,L^\rho], \mathcal{E} = block\text{-}diag[\mathcal{E}_1,\ldots,\mathcal{E}_\rho],$$

where L_i and $\mathcal{E}_i = \operatorname{diag}[\eta,\eta^2,\ldots,\eta^{k_i}]$ have dimension $n_j \times 1$ and $k_i \times k_i$, respectively. The assumption that L is Hurwitz in Theorem 10.1 (i.e., the polynomial $s^n + l_1 s^{n-1} + \ldots + l_n$ is Hurwitz) is replaced by the requirement that each L_i is Hurwitz.

The output-feedback controller is now given by

$$\hat{v} = \bar{v}(\hat{x}_a^P), \tag{10.84}$$

where, as before, $\hat{x}_a^P = [\hat{x}^{P\top}, s^\top]^\top$ and $\hat{x}^P$ is the state of the observer (10.39) with projection (10.63). In the particular case when the observability mapping is the identity and the plant takes the form (10.70), the projection can be eliminated by employing the observer (10.72) and replacing $\hat{x}^P$ by $\hat{x}^s$, as defined in (10.73).

10.6 How to Avoid Adding Integrators

Thus far we have seen that, given a state-feedback stabilizing controller $\bar{u}(x)$, in order to find an output-feedback controller based on a separation principle, one has to augment the plant with n_u integrators at the input side and design a stabilizing controller $\bar{v}(x_a)$ for the augmented system. Even though one can always perform this step in a systematic fashion by, e.g., employing integrator backstepping, on the other hand the resulting control law may become rather complex. This is seen in the next example.

Example 10.7 Consider the simple system

$$\begin{aligned} \dot{x}_1 &= x_2 + u \\ \dot{x}_2 &= x_1 x_2 - x_2. \end{aligned} \tag{10.85}$$

Using the Lyapunov function candidate $V_1 = (1/2)x_1^2 + (1/2)x_2^2$, we get

$$\dot{V}_1 = x_1(x_2 + u) + x_2(x_1 x_2 - x_2). \tag{10.86}$$

The controller $\bar{u}(x) = -x_1 - x_2 - x_2^2$ yields $\dot{V}_1 = -x_1^2 - x_2^2$ and hence globally asymptotically stabilizes the origin of (10.85). Now suppose we augment the system with two integrators

$$\dot{s}_1 = s_1, \qquad \dot{s}_2 = v, \qquad u = s_1,$$

and consider the problem of designing a controller $\bar{v}(x, s)$ for the augmented system. Following the recursive procedure outlined in Example 6.9, we start by considering the x subsystem, for which we have already designed a virtual control input $\nu_1(z_1) = \bar{u}(x)$, with $z_1 = x$. Then, we consider the subsystem

$$\begin{aligned} \dot{x}_1 &= x_2 + s_1 \\ \dot{x}_2 &= x_1 x_2 - x_2 \\ \dot{s}_1 &= \nu_2(z_2), \end{aligned} \tag{10.87}$$

we define $e_1 = s_1 - \nu_1(z_1)$, and we employ the Lyapunov function candidate $V_2 = V_1 + \frac{1}{2}e_1^2$ to stabilize (10.87)

$$\dot{V}_2 = \dot{V}_1 + e_1\left(\nu_2(z_2) - \dot{\nu}_1(z_1)\right) = -x_1^2 - x_2^2 + x_1 e_1 + e_1\left(\nu_2(z_2) - \dot{\nu}_1(z_1)\right).$$

Choosing $\nu_2(z_2) = \dot{\nu}_1(z_1) - x_1 - e_1$, where $z_2 = [x^\top, s_1]^\top$, we get $\dot{V}_2 = x_1^2 - x_2^2 - e_1^2$. Next, consider the full augmented system, define $e_2 = s_2 - \nu_2(z_2)$, and employ the Lyapunov function candidate $V_3 = V_1 + V_2 + \frac{1}{2}e_2^2$ to find the desired stabilizing controller $\bar{v}(x, s)$

$$\dot{V}_3 = -x_1^2 - x_2^2 - e_1^2 + e_1 e_2 + e_2\left(v - \dot{\nu}_2(z_2)\right).$$

Setting $v = \dot{\nu}_2(z_2) - e_1 - e_2 \stackrel{\triangle}{=} \bar{v}(x, s)$ we obtain $\dot{V}_3 = -x_1^2 - x_2^2 - e_1^2 - e_2^2$ and thus the origin $x_a = [x^\top, s^\top]^\top = 0$ of the augmented system is globally asymptotically stable. Notice that the control input we have just derived employs the time derivative of the function

$$\nu_2 = \dot{\nu}_1 - x_1 - e_1 = -s_1 - x_1 x_2 - 2x_2^2 - 2x_1 x_2^2 - x_2 - s_1 + \nu_1(z_1)$$

along the trajectories of the augmented system. The reader may easily see that such a time derivative contains a large number of nonlinear terms. Thus, even though the original controller $\bar{u}(x)$ is quite simple, the stabilizing controller for the augmented system $\bar{v}(x_1)$ becomes rapidly too complex as n_u grows large. △

This simple example shows the undesirable feature of the output feedback controller in Theorem 10.2, namely, the exponential explosion, as n_u grows, of nonlinear terms in the controller. In order to understand how to avoid this problem it is useful to recall that the reason for adding a chain of integrators at the input side of the plant is to employ the control input derivatives $\dot{u}, \ldots, u^{(n_u - 1)}$ in the observer (10.39). Thus, the knowledge of the time derivatives of u is really only employed for estimation, rather than control. If, instead of augmenting the system with n_u integrators (vector s), one generates *estimates* of s to be used in the observer, there will be no need to augment the plant and thus no need to redesign a controller.

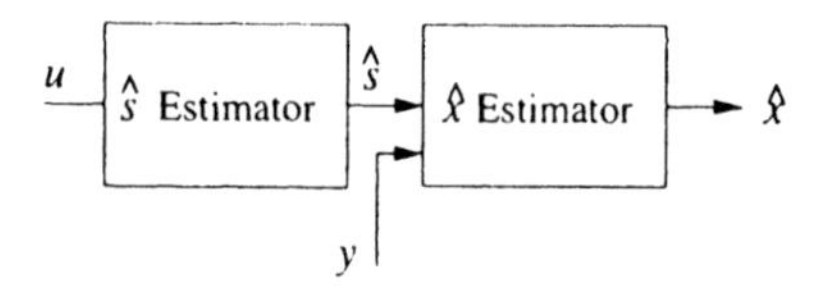

Figure 10.4. Cascade estimation scheme.

The new estimation scheme is illustrated in Figure 10.4, where two estimators in cascade are employed to calculate $\hat{x}$. The first estimator approximates the derivatives of $u(t)$ (vector $\hat{s}$), which are then employed by the second estimator to approximate x. Observer (10.39) is modified as follows

$$\begin{aligned}\dot{\hat{x}} &= f(\hat{x},\hat{u}) + \left[\frac{\partial \mathcal{H}(\hat{x},\hat{s})}{\partial \hat{x}}\right]^{-1}\Big[(\mathcal{E}^x)^{-1}L(y-\hat{y}) \\ &\qquad - \frac{\partial \mathcal{H}(\hat{x},\hat{s})}{\partial \hat{s}}(\mathcal{E}^s)^{-1}KC(u-\hat{u})\Big] \\ \hat{y} &= h(\hat{x},\hat{u})\end{aligned} \tag{10.88}$$

$$\begin{aligned}\dot{\hat{s}} &= A\hat{s} + (\mathcal{E}^s)^{-1}K(u-\hat{u}) \\ \hat{u} &= C\hat{s},\end{aligned} \tag{10.89}$$

where $L = \mathit{block\text{-}diag}[L^1,\ldots,L^p], K = \mathit{block\text{-}diag}[K^1,\ldots,K^m]$, and L^i, K^j are Hurwitz vectors of dimension $k_i \times 1$ and $n_j \times 1$, respectively, for $i = 1,\ldots,p$, $j = 1,\ldots,m$. Analogously, we let

$$\begin{aligned}A &= \mathit{block\text{-}diag}[A^1,\ldots,A^m] \\ B &= \mathit{block\text{-}diag}[B^1,\ldots,B^m] \\ C &= \mathit{block\text{-}diag}[C^1,\ldots,C^m],\end{aligned}$$

where A^i, B^i,and C^i have dimensions $n_i \times n_i$, $n_i \times 1$, and $1 \times n_i$, respectively, and

$$A_i = \begin{bmatrix} 0 & & \\ \vdots & I & \\ 0 & \ldots & 0 \end{bmatrix}, B_i = \begin{bmatrix} 0 \\ \vdots \\ 0 \\ 1 \end{bmatrix}, C_i = [1 \quad 0 \quad \ldots \quad 0]. \tag{10.90}$$

Finally, $\mathcal{E}^x = \mathit{block\text{-}diag}[\mathcal{E}_1^x,\ldots,\mathcal{E}_p^x]$, where $\mathcal{E}_i^x = \mathrm{diag}[\eta_1,\eta_1^2,\ldots,\eta_1^{k_i}]$ and $\eta_1 \in \mathsf{R}$, and $\mathcal{E}^s = \mathit{block\text{-}diag}[\mathcal{E}_1^s,\ldots,\mathcal{E}_p^s]$, where $\mathcal{E}_j^s = \mathrm{diag}[\eta_2,\eta_2^2,\ldots,\eta_2^{n_j}]$ and $\eta_2 \in \mathsf{R}$. Let $U \triangleq [u_1^{(n_1)},\ldots,u_m^{(n_m)}]^\top$, then the vector $\dot{s}$ can be expressed as

$$\dot{s} = As + BU, \qquad u = Cs.$$

The following theorem replaces Theorem 10.1 and proves that the estimator above estimates x and s to an arbitrary degree of accuracy (but not asymptotically).

Theorem 10.3: *Assume that the plant (10.25) is uniformly completely observable, the state x of the plant and the vector s containing the derivatives of u are confined within a compact set Ω, i.e., $[x(t)^\top, s(t)^\top]^\top \in \Omega$, and $|U(t)| \leq M$, for all $t \geq 0$, with M a positive constant. If there exists a compact set $\Omega' \supset \Omega$ such that*

$$\hat{x}_a(t) \triangleq [\hat{x}(t)^\top, \hat{s}(t)^\top]^\top \in \Omega' \text{ for all } t \geq 0 \tag{10.91}$$

then the cascade (10.88) and (10.89) guarantee that the estimation error converges arbitrarily fast to an arbitrarily small neighborhood of the origin, that is, for all $\delta, T > 0$, and all $\hat{x}_a(0) \in \Omega'$ there exist $\bar{\eta}_1, \bar{\eta}_2$, $0 < \bar{\eta}_1, \bar{\eta}_2 \leq 1$, such that $|\hat{x} - x| \leq \delta$, $|\hat{s} - s| \leq \delta$, for all $t \geq T$, whenever $\eta_1 \in (0, \bar{\eta}_1), \eta_2 \in (0, \bar{\eta}_2)$.

While, on one hand, the cascade estimator (10.88), (10.89) eliminates the complication of adding integrators at the input side of the plant, on the other hand it introduces an asymptotic estimation error which can be made arbitrarily small but not identically zero. Thus, (10.88), (10.89) is *not* an observer for the plant (10.25) in the sense of Definition 10.1. Condition (10.91) plays a role similar to that of (10.52) in Theorem 10.1. In analogy to what was done before, we will show that such a condition is always guaranteed to hold when a dynamic projection is applied to the observer (10.88), (10.89).

Proof: The proof of this theorem is quite similar to that of Theorem 10.1. Define the following coordinate transformations: $y_e = \mathcal{H}(x, s)$ and $\hat{y}_e = \mathcal{H}(\hat{x}, \hat{s})$, where $x = \mathcal{H}^{-1}(y_e, s)$ and $\hat{x} = \mathcal{H}^{-1}(\hat{y}_e, \hat{s})$ are well-defined, unique, and smooth by the observability assumption. Next, let

$$\begin{aligned} A_c &= \textit{block-diag}[A_c^1, \ldots, A_c^\rho] \\ B_c &= \textit{block-diag}[B_c^1, \ldots, B_c^\rho] \\ C_c &= \textit{block-diag}[C_c^1, \ldots, C_c^\rho], \end{aligned}$$

where A_c^i, B_c^i, C_c^i have the form (10.90), and have dimension $k_i \times k_i$, $k_i \times 1$, and $1 \times k_i$, respectively, then in new coordinates one has

$$\begin{aligned} \dot{y}_e = & \frac{\partial \mathcal{H}}{\partial x} f(x, u) + \frac{\partial \mathcal{H}}{\partial s}[As + BU] \\ = & A_c y_e + B_c [\alpha(y_e, s, d) + \beta(y_e, s, U)], \end{aligned} \tag{10.92}$$

where $\alpha = [\alpha_1, \ldots, \alpha_\rho]^\top$, $\beta = [\beta_1, \ldots, \beta_\rho]^\top$,

$$\alpha_i(y_e, s) = \left[\frac{\partial \varphi_i^{k_i - 1}}{\partial x} f(x, u) + \sum_{k=0}^{k_i - 2} \frac{\partial \varphi_i^{k_i - 1}}{\partial u^{(k)}} u^{(k+1)} \right]_{x = \mathcal{H}^{-1}(y_e, s)}$$

$$\beta_i(y_e, s, U) = \left[\frac{\partial \varphi_i^{k_i - 1}}{\partial u^{(k_i - 1)}} u^{(k_i)} \right]_{x = \mathcal{H}^{-1}(y_e, s)}$$

for all $i = 1, \ldots, \rho$.

In a similar manner,

$$\begin{aligned} \dot{\hat{y}}_e = & \frac{\partial \mathcal{H}}{\partial \hat{x}} f(\hat{x}, \hat{u}) + \frac{\partial \mathcal{H}}{\partial \hat{s}} A\hat{s} + \left[\frac{\partial \mathcal{H}}{\partial \hat{x}}\right] \left[\frac{\partial \mathcal{H}}{\partial \hat{x}}\right]^{-1} \Big\{ (\mathcal{E}^x)^{-1} L C_c (y_e - \hat{y}_e) \\ & - \frac{\partial \mathcal{H}}{\partial \hat{s}} (\mathcal{E}^s)^{-1} K C (s - \hat{s}) \Big\} + \frac{\partial \mathcal{H}}{\partial \hat{s}} (\mathcal{E}^s)^{-1} K C (s - \hat{s}) \\ = & A_c \hat{y}_e + B_c \alpha(\hat{y}_e, \hat{s}) + (\mathcal{E}^x)^{-1} L C_c (y_e - \hat{y}_e). \end{aligned} \tag{10.93}$$

Define the estimation error in new coordinates, $\tilde{y}_e = \hat{y}_e - y_e$. Then, using (10.92) and (10.93), the estimator error dynamics are given by

$$\dot{\tilde{y}}_e = (A_c - (\mathcal{E}^x)^{-1} L C_c)\tilde{y}_e + B_c[\alpha(\hat{y}_e, \hat{s}) - \alpha(y_e, s) - \beta(y_e, s, U)]. \quad (10.94)$$

Introduce the scaling $\tilde{\nu} = \bar{\mathcal{E}}^x \tilde{y}_e$,

$$\bar{\mathcal{E}}^x \triangleq \textit{block-diag}[\bar{\mathcal{E}}_1^x, \ldots, \bar{\mathcal{E}}_\rho^x], \quad \bar{\mathcal{E}}_i^x \triangleq \operatorname{diag}\left[\frac{1}{\eta_1^{k_i - 1}}, \ldots, 1\right]$$

for $i = 1, \ldots, \rho$. After scaling, the error equation becomes

$$\dot{\tilde{\nu}} = \frac{1}{\eta_1}(A_c - LC_c)\tilde{\nu} + B_c\left[\alpha(\hat{y}_e, \hat{s}) - \alpha(y_e, s) - \beta(y_e, s, U)\right]. \quad (10.95)$$

Now, consider the estimator for the variable s. After defining the estimation error $\tilde{s} = \hat{s} - s$ and the scaling $\tilde{s}' = \bar{\mathcal{E}}^s \tilde{s}$,

$$\bar{\mathcal{E}}^s \triangleq \textit{block-diag}[\bar{\mathcal{E}}_1^s, \ldots, \bar{\mathcal{E}}_m^s], \quad \bar{\mathcal{E}}_i^s \triangleq \operatorname{diag}\left[\frac{1}{\eta_2^{n_i - 1}}, \ldots, 1\right]$$

for $i = 1, \ldots, m$, we obtain the error dynamics for the input derivatives estimator,

$$\dot{\tilde{s}}' = \frac{1}{\eta_2}(A - KC)\tilde{s}' - BU. \quad (10.96)$$

Recall that $[x^\top, s^\top]^\top \in \Omega$ and U is bounded for all $t \geq 0$, where Ω is a compact set. Next, consider the Lyapunov function candidate

$$V_o(\tilde{\nu}, \tilde{s}') = \tilde{\nu}^\top P_1 \tilde{\nu} + \tilde{s}'^\top P_2 \tilde{s}',$$

where P_1 and P_2 are solutions of the Lyapunov equations

$$P_1(A_c - LC_c) + (A_c - LC_c)^\top P_1 = -I \quad (10.97)$$
$$P_2(A - KC) + (A - KC)^\top P_2 = -I \quad (10.98)$$

(P_1 and P_2 exist and are positive definite because of the way L and K are defined). Recall that $[\hat{x}(0)^\top, \hat{s}(0)^\top]^\top \in \Omega$ and define the compact set

$$\tilde{\mathcal{K}} \triangleq \left\{\tilde{y}_e \in \mathbb{R}^n, \tilde{s} \in \mathbb{R}^{n_u} \mid [y_e^\top, s^\top]^\top \in \mathcal{F}^{-1}(\Omega), [\hat{y}_e^\top, \hat{s}^\top]^\top \in \mathcal{F}^{-1}(\Omega)\right\}.$$

By definition $\tilde{\mathcal{K}}$ contains the initial conditions $\tilde{y}_e(0), \tilde{s}'(0)$. Consider the bracketed term in (10.95) and notice that, due to the smoothness of α, β, condition (10.91), and the boundedness of U we have that there exists a positive scalar C such that

$$|\alpha(\hat{y}_e, \hat{s}) - \alpha(y_e, s) - \beta(y_e, s, U)| \leq C, \quad (10.99)$$

for all $[y_e^\top, s^\top]^\top \in \Omega$, $[\hat{y}_e^\top, \hat{s}^\top]^\top \in \Omega'$, and $|U| \leq M$. Calculate the time derivative of V_o along the $\tilde{\nu}, \tilde{s}'$ trajectories and use the fact that $|B_c| = |B| = 1$,

$$\begin{aligned}
\dot{V}_o &= -\frac{\tilde{\nu}^\top \tilde{\nu}}{\eta_1} - \frac{\tilde{s}'\tilde{s}'}{\eta_2} + 2\tilde{\nu}^\top P_1 B_c \left[\alpha(\hat{y}_e, \hat{s}, d^0) - \alpha(y_e, s, d) - \beta(y_e, s, U)\right] \\
&\quad + 2\tilde{s}' P_2 B U \\
&\leq -\frac{|\tilde{\nu}|^2}{\eta_1} - \frac{|\tilde{s}'|^2}{\eta_2} + 2|P_1|C|\tilde{\nu}| + 2|P_2|M|\tilde{s}'| \qquad (10.100)
\end{aligned}$$

for all $t \geq 0$. Let $\bar{C} = \max\{C, M\}$, by applying Young's inequality to the two positive terms in (10.100) we get

$$\begin{aligned}
\dot{V}_o &\leq -\frac{|\tilde{\nu}|^2}{\eta_1} - \frac{|\tilde{s}'|^2}{\eta_2} + |P_1|\bar{C}|\tilde{\nu}|^2 + |P_2|\bar{C}|\tilde{s}'|^2 + |P_1|\bar{C} + |P_2|\bar{C} \\
&\leq -\left(\frac{1}{\eta_1} - |P_1|\bar{C}\right)|\tilde{\nu}|^2 - \left(\frac{1}{\eta_2} - |P_2|\bar{C}\right)|\tilde{s}'|^2 + |P_1|\bar{C} + |P_2|\bar{C}.
\end{aligned}$$

Choose $\eta_1 < 1/(|P_1|\bar{C}), \eta_2 < 1/(|P_2|\bar{C})$, then it can be verified that when $V_o > 2\bar{V}(\eta_1, \eta_2)$, where

$$\bar{V}(\eta_1, \eta_2) = \frac{\max\{|P_1|, |P_2|\}}{\min\left\{\frac{1}{\eta_1} - |P_1|\bar{C}, \frac{1}{\eta_2} - |P_2|\bar{C}\right\}}(|P_1|\bar{C} + |P_2|\bar{C}),$$

we have that

$$\dot{V}_o < -\left(\frac{\min\left\{\frac{1}{\eta_1} - |P_1|\bar{C}, \frac{1}{\eta_2} - |P_2|\bar{C}\right\}}{\max\{|P_1|, |P_2|\}}\right)\frac{V_o}{2} < 0. \qquad (10.101)$$

Then, from (10.101) we have that the set $\{\tilde{\nu} \in \mathsf{R}^n, \tilde{s}' \in \mathsf{R}^{n_u} \mid V_o(\tilde{\nu}, \tilde{s}') \leq 2\bar{V}(\eta_1, \eta_2)\}$ is positively invariant, and it is reached in finite time. Moreover, the set $\{\tilde{\nu} \in \mathsf{R}^n, \tilde{s}' \in \mathsf{R}^{n_u} \mid V_o \leq 2\bar{V}\}$ can be made arbitrarily small by choosing sufficiently small η_1 and η_2. Finally, from (10.101) it follows that the rate of convergence of $\tilde{\nu}, \tilde{s}$ can be made as fast as desired by decreasing η_1 and η_2. From these observations, the fact that $\hat{x}_a - x_a = \mathcal{F}^{-1}(\hat{y}_e, \hat{s}) - \mathcal{F}^{-1}(y_e, s)$, and the continuity of $\mathcal{H}^{-1}$ with respect to both arguments, we conclude the proof. ∎

Next, we concentrate on employing the estimator (10.88), (10.89) to formulate a separation principle similar to that of Theorem 10.2. Instead of augmenting the system with a chain of integrators, directly consider the state-feedback controller $\bar{u}(x)$ which asymptotically stabilizes the origin $x = 0$ of (10.25) with, say, a domain of attraction $\mathcal{D}_x$. Applying the same converse Lyapunov theorem used earlier, we have that there exists a continuously differentiable Lyapunov function V_x defined on $\mathcal{D}_x$ satisfying, for all $x \in \mathcal{D}_x$,

$$\alpha_1(|x|) \leq V_x(x) \leq \alpha_2(|x|) \tag{10.102}$$

$$\lim_{x \to \partial \mathcal{D}_x} \alpha_1(|x|) = \infty \tag{10.103}$$

$$\frac{\partial V_x}{\partial x} f(x, \bar{u}(x)) \leq -\alpha_3(|x|), \tag{10.104}$$

where α_i, $i = 1, 2, 3$ are class $\mathcal{K}$ functions. Similarly to what we have done in the previous sections, given any positive scalar c, define the set

$$\Omega_c \triangleq \{x \in \mathrm{R}^n \mid V_x(x) \leq c\},$$

and, given two arbitrary positive scalars c_1 and c_2, with $c_1 < c_2$, choose a convex compact set $\mathcal{C}$ with smooth boundary such that

$$\mathcal{F}(\Omega_{c_2} \times \Omega^s) \subset \mathcal{C},$$

where Ω^s is the compact set containing $s(t) = [u(t), \dot{u}(t), \ldots, u^{(n_u-1)}(t)]^\top$ for all $t \geq 0$ when $x(t) \in \Omega_{c_2}$. The existence of Ω^s is guaranteed by the smoothness of $u(x)$ and the fact that Ω_{c_2} is compact. As before, let $\hat{\mathcal{Y}} = \mathcal{F}(\hat{x}_a)$, and denote by $N(\mathcal{Y})$ the normal vector to the boundary of $\mathcal{C}$ at $\mathcal{Y}$. Consider the following dynamic projection applied to the observer (10.88), (10.89)

$$\dot{\hat{x}}_a^P = \begin{cases} \dot{\hat{x}}_a - \left[\dfrac{\partial \mathcal{F}}{\partial \hat{x}_a}\right]^{-1} \Gamma \dfrac{N(\hat{\mathcal{Y}})N(\hat{\mathcal{Y}})^\top}{N(\hat{Y})^\top \Gamma N(\hat{\mathcal{Y}})} \dot{\hat{\mathcal{Y}}} & \text{if } N(\hat{\mathcal{Y}})^\top \dot{\hat{\mathcal{Y}}} \geq 0 \text{ and } \hat{\mathcal{Y}} \in \partial \mathcal{C} \\ \dot{\hat{x}}_a & \text{otherwise,} \end{cases} \tag{10.105}$$

where $\Gamma = \textit{block-diag}\{\Gamma_1, \Gamma_2\}$

$$\Gamma_1 = (S^1 \bar{\mathcal{E}}^x)^{-1} (S^1 \bar{\mathcal{E}}^x)^{-1} \qquad \Gamma_2 = (S^2 \bar{\mathcal{E}}^s)^{-1} (S^2 \bar{\mathcal{E}}^s)^{-1},$$

$S^1 = S^{1\top}$, $S^2 = S^{2\top}$ denote the matrix square roots of P_1, P_2 defined in (10.97), (10.98). Finally, $\dot{\hat{\mathcal{Y}}}$ is given by

$$\dot{\hat{\mathcal{Y}}} = \frac{\partial \mathcal{F}}{\partial \hat{x}_a} \dot{\hat{x}}_a = \begin{bmatrix} \frac{\partial \mathcal{H}}{\partial \hat{x}} \dot{\hat{x}} + \frac{\partial \mathcal{H}}{\partial \hat{s}} \dot{\hat{s}} \\ \dot{\hat{s}} \end{bmatrix}.$$

Lemma 10.4: *Let $\mathcal{C}$ be a convex compact set in $\mathcal{Y}$ coordinates with smooth boundary. The observer (10.88), (10.89) with dynamic projection (10.105) has the following properties*

(i) Boundedness: if $\hat{x}_a^P(0) \in \mathcal{F}^{-1}(\mathcal{C})$, then $\hat{x}^P(t) \in \mathcal{F}^{-1}(\mathcal{C})$ for all $t \geq 0$.

If, in addition, $x_a(t) \in \mathcal{F}^{-1}(\mathcal{C})$ for all $t \geq 0$ then the following holds

(ii) Preservation of original convergence characteristics: properties (i) and (ii) established by Theorem 10.3 remain valid for $\hat{x}_a^P$.

The proof of this lemma is essentially identical to that of Lemma 10.1 and is therefore omitted. Note that part (i) of the lemma implies that condition (10.91) in Theorem 10.3 is satisfied with $\hat{\Omega} = \mathcal{F}^{-1}(\mathcal{C})$.

Consider now the output-feedback controller

$$\hat{u} = \bar{u}(\hat{x}^P). \tag{10.106}$$

By the smoothness of f and the control law $\bar{u}(x)$, there exists a positive scalar $\bar{\gamma}_x$ such that

$$|f(x, \bar{u}(\hat{x}^P)) - f(x, \bar{u}(x))| \leq \bar{\gamma}_x |\hat{x}^P - x| \tag{10.107}$$

for all $x \in \Omega_{c_2}$, $\hat{x}^P \in \mathcal{F}^{-1}(\mathcal{C})$. Furthermore, from the fact that V_x is continuously differentiable, we have that there exists a positive scalar A_x such that $|\partial V_x/\partial x| \leq A_x$ for all x in Ω_{c_2}.

Notice now that, since $\tilde{x}$ does not vanish asymptotically, it follows that $x = 0$ is *not* an equilibrium point of the closed-loop system and, hence, it cannot be made asymptotically stable. The following theorem, however, shows that the output-feedback controller (10.106) enjoys desirable stability features.

Theorem 10.4: *Consider the closed-loop system formed by plant, estimator (10.88), (10.89), and controller (10.106). Then, given any triple (c_1, c_2, ϵ) such that $0 < c_1 < c_2$ and $0 < \epsilon < \alpha_3 \circ \alpha_2^{-1}(c_1)/\mu A_x \bar{\gamma}_x$, there exist scalars $\eta_1^*, 0 < \eta_1^* \leq 1$ and $\eta_2^*, 0 < \eta_2^* \leq 1$, such that, for all $\eta_1 \in (0, \eta_1^*]$ and $\eta_2 \in (0, \eta_2^*]$, every trajectory starting in Ω_{c_1} is bounded and enters the positively invariant set Ω_{d_ϵ}, where $d_\epsilon = \alpha_2 \circ \alpha_3^{-1}(\mu\, A_x\, \bar{\gamma}_x\, \epsilon)$, in finite time.*

Proof: See the proof of Lemma 10.2. ∎

Rather than stabilizing the origin of the closed-loop system, the proposed output-feedback controller stabilizes a neighborhood of the origin which can be made arbitrarily small by choosing sufficiently small η_1 and η_2 in the state estimator. We say in this case that this controller *practically stabilizes* the origin $x = 0$. Notice that, in analogy with the result in Theorem 10.2, if $\mathcal{D}_x = \mathbf{R}^n$ then Ω_{c_1} can be chosen to be any compact set in $\mathbf{R}^n$ and thus the controller (10.106) semiglobally practically stabilizes the origin $x = 0$. From a practical viewpoint this controller has more attractive features than that in (10.78) since, as discussed earlier, it provides a significant simplification in the control design.

10.7 Coping with Uncertainties

A drawback of the output-feedback controllers based on a separation principle introduced so far is that the plant is not allowed to be affected by any uncertainty. In this section we show that, under certain conditions, an output-feedback controller similar to that in (10.106) can be employed to recover the performance of robust state-feedback controllers such as those introduced in Chapter 6. Another limitation of the results derived in previous sections is that the state-feedback controller was restricted to be a static function of x. Since, in general, robust controllers may be dynamic, in this section we remove this restriction and show how to achieve separation between dynamic state-feedback control design and observer design.

Consider the MIMO system

$$\begin{aligned} \dot{x} &= f(x, u, \Delta(t)) \\ y &= h(x, u), \end{aligned} \tag{10.108}$$

where $\Delta(t)$ is a bounded disturbance, $|\Delta(t)| \leq \bar{\Delta}$. When no disturbance acts on the plant we replace $\Delta(t)$ by Δ^0, a known constant "nominal value" of $\Delta(t)$ (one can take $\Delta^0 = 0$). Our objective is to construct a robust output-feedback controller which makes a known compact set $\mathcal{N}$ uniformly asymptotically stable with a desired region of attraction.

Assume that, using the techniques introduced in Chapter 6, we have derived a state-feedback controller (in general, a dynamic controller) of the form

$$\begin{aligned} \dot{x}_c &= f_c(x_c, x) \\ u &= h_c(x_c, x), \end{aligned} \tag{10.109}$$

where $x_c \in \mathrm{R}^{n_c}$, f_c, h_c are sufficiently smooth, and such that $\mathcal{N} \subset \mathrm{R}^{n+n_c}$ is a positively invariant, uniformly asymptotically stable compact set for the closed-loop system

$$\begin{aligned} \dot{x} &= f(x, h_c(x_c, x), \Delta(t)) \\ \dot{x}_c &= f_c(x_c, x) \end{aligned} \tag{10.110}$$

for all $\Delta(t) \in \bar{\Delta}$, with domain of attraction $\mathcal{D}$. In other words, by letting $x_a = [x^\top, x_c^\top]^\top$ and

$$f_a(x_a, \Delta(t)) = [f(x, h_c(x_c, x), \Delta(t))^\top, f_c(x_c, x)^\top]^\top$$

then the closed-loop system

$$\dot{x}_a = f_a(x_a, \Delta(t)) \tag{10.111}$$

has the property that if $x_a(0) \subset \mathcal{D}$, the closed-loop state trajectory $x_a(t)$ is bounded, contained in $\mathcal{D}$, and asymptotically approaches the compact set

$\mathcal{N}$, for all $|\Delta(t)| \leq \bar{\Delta}$. Using a converse Lyapunov theorem it is possible to show (see [133]) that this implies the existence of a smooth function V defined in $\mathcal{D}$, two $\mathcal{K}_\infty$ functions α_1, α_2, and a positive function α_3 such that,

$$\alpha_1(\omega_{\mathcal{N}}(x_a)) \leq V(x_a) \leq \alpha_2(\omega_{\mathcal{N}}(x_a)), \quad \forall x_a \in \mathcal{D} \tag{10.112}$$

$$\dot{V}(x_a) \leq -\alpha_3(\omega_{\mathcal{N}}(x_a)), \quad \forall\{x_a \in \mathcal{D}, x_a \notin \mathcal{N}\}, \tag{10.113}$$

where $\omega_{\mathcal{N}}(x_a)$ is a positive definite function with respect to $\mathcal{N}$, continuous and proper in $\mathcal{D}$ (i.e., it approaches infinity on the boundary of $\mathcal{D}$ and it is identically zero in $\mathcal{N}$).

In order to introduce a nonlinear estimator for x, we form the observability mapping of the plant (10.108) which, with the usual notation, becomes

$$y_e \triangleq \begin{bmatrix} y_1 \\ \dot{y}_1 \\ \vdots \\ y_1^{(k_1-1)} \\ y_2 \\ \vdots \\ y_2^{(k_2-1)} \\ \vdots \\ y_p \\ \vdots \\ y_p^{(k_p-1)} \end{bmatrix} = \begin{bmatrix} h_1(x,u) \\ \varphi_1^1(x,u,u^{(1)},\Delta) \\ \vdots \\ \varphi_1^{k_1-1}\left(x,u,\ldots,u^{(k_1-1)},\Delta,\ldots,\Delta^{(k_1-2)}\right) \\ h_2(x,u) \\ \vdots \\ \varphi_2^{k_2-1}\left(x,u,\ldots,u^{(k_2-1)},\Delta,\ldots,\Delta^{(k_2-2)}\right) \\ \vdots \\ h_p(x,u) \\ \vdots \\ \varphi_p^{k_p-1}\left(x,u,\ldots,u^{(k_p-1)},\Delta,\ldots,\Delta^{(k_p-2)}\right) \end{bmatrix}$$

$$\triangleq \mathcal{H}\left(x,s,\Delta,\ldots,\Delta^{(n_\Delta-1)}\right), \tag{10.114}$$

where s is the vector containing the time derivatives of u that end up appearing in the observability mapping, φ_i^j, k_i and n_i have been defined in the previous section, and $n_\Delta - 1$ is the number of time derivatives of $\Delta(t)$ affecting the observability mapping. Since the unknown term $\Delta(t)$ in $\mathcal{H}$ does not allow us to define an observer for x, we cannot employ a separation principle to recover the performance of the robust state-feedback controller. To eliminate this obstacle, we need some simplifying assumption. In the following, we will assume that the plant has the property that

$$\frac{\partial \varphi_i^j}{\partial \Delta} = 0, \quad \text{for } i = 1,\ldots,p, \; j = 1,\ldots,k_i - 1, \tag{10.115}$$

on the domain of definition of φ_i^j. Therefore, (10.114) can be written as $y_e = \mathcal{H}(x,s)$. This assumption is always satisfied when, e.g., the disturbance has relative degree n with respect to the measurable output y. Notice that we do not require, however, that the plant be feedback linearizable.

Example 10.8 Consider the dynamical system

$$\begin{aligned} \dot{x}_1 &= x_2 \\ \dot{x}_2 &= x_1 u - x_2 + \Delta(t) x_1 \\ y &= x_1, \end{aligned} \tag{10.116}$$

where $\Delta(t)$ is an unknown bounded disturbance with bound $|\Delta(t)| \leq \bar{\Delta}$, for all $t \geq 0$. Note that this system is not feedback linearizable. Consider the Lyapunov function candidate

$$V(x) = \frac{3}{2}x_1^2 + x_1 x_2 + x_2^2$$

and its time derivative

$$\dot{V} = 2x_1x_2 - x_2^2 + (x_1^2 + 2x_1x_2)(u + \Delta(t)).$$

Using the nonlinear damping technique introduced in Chapter 6, given any positive scalar ϵ we choose

$$u = -1 - \frac{\bar{\Delta}^2}{4\epsilon}(2x_1x_2 + x_1^2) \triangleq \bar{u}(x)$$

so that

$$\begin{aligned} \dot{V} &= -x_1^2 - x_2^2 - \frac{\bar{\Delta}^2}{4\epsilon}(x_1^2 + 2x_1x_2)^2 + (x_1^2 + 2x_1x_2)\Delta(t) \\ &\leq -x_1^2 - x_2^2 + \epsilon, \end{aligned}$$

thus showing that the closed-loop system trajectories are UUB and the ultimate bound is of order ϵ. In other words, this static controller guarantees that $x(t)$ approaches a ball around the origin of radius proportional to ϵ (the set $\mathcal{N}$). Next, observe that the observability mapping of the system is given by

$$y_e = \begin{bmatrix} y \\ \dot{y} \end{bmatrix} = \mathcal{H}(x, s) = \begin{bmatrix} x_1 \\ x_2 \end{bmatrix},$$

which is independent of $\Delta(t)$. In conclusion, despite the fact that this system is not feedback linearizable, it satisfies our assumptions. $\triangle$

With the assumption above, defining an output-feedback controller that recovers the performance of the robust state-feedback controller (10.109) involves minor modifications to the theory presented in the previous sections.

Since the disturbance $\Delta(t)$ does not affect the observability mapping $\mathcal{H}$, we can employ a nonlinear estimator similar to (10.88), (10.89)

$$\begin{aligned}\dot{\hat{x}} &= f(\hat{x}, \hat{u}, \Delta^0) + \left[\frac{\partial \mathcal{H}(\hat{x}, \hat{s})}{\partial \hat{x}}\right]^{-1} \Big[(\mathcal{E}^x)^{-1} L\,(y - \hat{y}) \\ &\qquad - \frac{\partial \mathcal{H}(\hat{x}, \hat{s})}{\partial \hat{s}} (\mathcal{E}^s)^{-1} K C(u - \hat{u})\Big] \\ \hat{y} &= h(\hat{x}, \hat{u})\end{aligned} \tag{10.117}$$

$$\begin{aligned}\dot{\hat{s}} &= A\hat{s} + (\mathcal{E}^s)^{-1} K(u - \hat{u}) \\ \hat{u} &= C\hat{s},\end{aligned} \tag{10.118}$$

where $\Delta(t)$ in the vector field f has been replaced by its nominal value Δ^0 and other parameters and matrices have the the usual definitions. The analysis used in the proof of Lemma 10.2 shows that the estimator error converges to a desired small neighborhood of the origin in arbitrarily short time, provided η_1 and η_2 in the observer are chosen sufficiently small. This property allows us to define a robust output-feedback controller based on a separation principle. Once again, we have to apply the dynamic projection (10.105) to (10.117), (10.118) to eliminate the peaking phenomenon. Similarly to what done before, given any two positive scalars $c_1 < c_2$, the convex compact set $\mathcal{C}$ is found by making sure that

$$\Omega^x_{c_2} \times \Omega^s = \{x \in \mathbb{R}^n \mid V(x_a) \leq c_2, \text{ for all } x_c \in \mathbb{R}^{n_c}\} \times \Omega^s \subset \mathcal{F}^{-1}(\mathcal{C}),$$

where Ω^s was defined in Section 10.6. The output-feedback controller becomes

$$\begin{aligned}\dot{x}_c &= f_c(x_c, \hat{x}^P) \\ \hat{u} &= h_c(x_c, \hat{x}^P).\end{aligned} \tag{10.119}$$

By using essentially the same analysis as was carried out in the proof of Lemma 10.2 it is possible to show that, for all $x_a(0) \in \Omega_{c_1}$, the trajectories of the closed-loop system with the controller (10.119) are bounded, contained in Ω_{c_1}, and approach an arbitrarily small neighborhood of the set $\mathcal{N}$, provided η_1 and η_2 are chosen sufficiently small. The proof of this fact is left as an exercise to the reader: refer to Theorem 11.1 to understand the differences between this proof and that of Lemma 10.2.

10.8 Output-Feedback Tracking

In Chapters 6, 7, and 8 we have used a unified framework for stabilization and tracking by employing an error system $e = \chi(t, x)$ which quantifies the tracking performance. When available, the problem of finding a controller yielding boundedness of the internal dynamics and asymptotic tracking reduces to the problem of stabilizing the error system dynamics. Unfortunately, as seen in Section 10.2, assuming that e is measurable is, in some cases, too restrictive. When using a partial information approach, that is, when both the state of the plant and the tracking performance measure $e(t)$ are not directly measurable, tracking and stabilization become two different problems that must be analyzed separately.

Consider the problem of finding a controller that makes the output y of the plant (10.25) track a smooth reference trajectory $r(t) = [r_1(t), \ldots, r_p(t)]^\top$. We assume to work in a partial information framework, i.e., only the output y and the reference trajectory r are available for feedback. Recall from Section 6.2 the definition of a stable inverse of the plant, that is, two sufficiently smooth functions $x^r(t)$ and $c^r(t)$ satisfying

$$\begin{aligned} \dot{x}^r &= f(x^r, c^r) \\ r &= h(x^r, c^r). \end{aligned} \tag{10.120}$$

If one could solve (10.120), then the error system could be defined as

$$e(t) = x(t) - x^r(t) \stackrel{\triangle}{=} \chi(t, x). \tag{10.121}$$

Notice that there may be other choices of error systems satisfying Assumption 6.1 which are less general than (10.121) (e.g., using a stable manifold).

Consider the general error system (10.121) and its associated error dynamics

$$\dot{e} = f(x, u) - f(x^r, c^r) \stackrel{\triangle}{=} \alpha(t, x, u). \tag{10.122}$$

Since in a partial information framework the functions $x^r(t)$ and $c^r(t)$ are unknown, the function $\alpha(t, x, u)$ is partially unknown. Assume that, using the techniques developed in Chapter 6, one can find a *full information* controller (i.e., a controller which uses x and $x^r(t)$ as feedback) $\bar{u}(t, x, x^r, c^r)$ which has the property that

$$\bar{u}(t, x^r, x^r, c^r) = c^r \tag{10.123}$$

and it is such that the origin of the closed-loop system

$$\dot{e} = \alpha\left(t, x, \bar{u}(t, x, x^r, c^r)\right) \tag{10.124}$$

is uniformly asymptotically stable with a domain of attraction $\mathcal{D} \subset \mathsf{R}^n$. Condition (10.123) guarantees that $e = 0$ (i.e., $x = x^r$) is an equilibrium point of (10.124). The properties of the controller above imply the existence of a continuously differentiable Lyapunov function $V(t, e)$ defined for $e \in \mathcal{D}$ such that, for all $e \in \mathcal{D}$,

$$\alpha_1(|e|) \le V(t, e) \le \alpha_2(|e|) \tag{10.125}$$

$$\lim_{e \to \partial\mathcal{D}} \alpha_1(|e|) = \infty \tag{10.126}$$

$$\frac{\partial V}{\partial e}\alpha\left(t, x, \bar{u}(t, x, x^r, c^r)\right) \le -\alpha_3(|e|), \tag{10.127}$$

where α_i, $i = 1, 2, 3$ are class $\mathcal{K}$ functions.

Example 10.9 Consider the following input-output feedback linearizable system

$$\dot{x}_1 = x_2$$
$$\dot{x}_2 = x_1^3 + x_2^2 + u$$
$$y = x_1,$$

and the reference signal $r(t) = \sin t$. The stable inverse of the plant is found by solving (10.120)

$$x^r(t) = \begin{bmatrix} x^r{}_1(t) \\ x^r{}_2(t) \end{bmatrix} = \begin{bmatrix} \sin t \\ \cos t \end{bmatrix}, c^r(t) = -\sin t - \sin^3 t - \cos^2 t.$$

Define the error system as $e = x - x^r(t)$. Then, the error dynamics are given by

$$\begin{aligned} \dot{e}_1 &= x_2 - x^r{}_2(t) = x_2 - \cos t \\ \dot{e}_2 &= x_1^3 + x_2^2 + u - (x^r{}_1)^3(t) - (x^r{}_2)^2(t) - c^r(t) \\ &= x_1^3 + x_2^2 + \sin t + u. \end{aligned} \tag{10.128}$$

Letting

$$\alpha(t, x, u) = \begin{bmatrix} x_2 - \cos t \\ x_1^3 + x_2^2 + \sin t + u, \end{bmatrix} \tag{10.129}$$

the error dynamics are expressed in the form (10.122). Next, a *full information* stabilizing controller for (10.129) is given by

$$\begin{aligned} \bar{u}(t, x, x^r(t), c^r(t)) = \; & -x_1^3 - x_2^2 + (x^r{}_1)^2(t) + (x^r{}_2)^2(t) + c^r(t) \\ & - k_1(x_1 - x^r{}_1(t)) - k_2(x_2 - x^r{}_2(t)), \end{aligned}$$

where k_1 and k_2 are positive design parameters, so that the closed-loop system becomes

$$\begin{aligned} \dot{e}_1 &= e_2 \\ \dot{e}_2 &= -k_1 e_2 - k_2 e_2, \end{aligned} \tag{10.130}$$

and the origin $e = 0$ is globally uniformly asymptotically stable. Notice that the controller has the desired property that, when $x = x^r$, $\bar{u} = c^r(t)$. △

When dealing with special canonical forms for the system dynamics, such as strict-feedback systems, the tracking problem may be significantly simplified. If the derivatives of r are available for feedback (or if the plant output is to track q, which is a filtered version of r so that the derivatives of q are available), then one may use the techniques of Chapter 6 to solve for a stabilizing state-feedback controller without solving for x^r and c^r in (10.120). Since the stability of the observers developed earlier is independent of the reference, they work equally well for the tracking problem. Therefore the previous results based on the separation principle carry over to the tracking problem when a controller is defined for strict-feedback systems and the derivatives of r are available. The tracking problem may be extended to a more general class of nonlinear systems as will be shown next.

10.8.1 Practical Internal Models

Consider now the problem of stabilizing the origin $e = 0$ of the error dynamics (10.122) in a partial information framework, i.e., when x, $x^r(t)$, and $c^r(t)$ are not available for feedback. In previous sections we have seen how to build an estimator for x. If one could define an estimator for $x^r(t)$ and $c^r(t)$, then the tracking problem could be solved by using the separation principle developed earlier and replacing x, x^r, and c^r in $\bar{u}$ by their respective estimates (after appropriate projection). Unfortunately, the estimation of $x^r(t)$ and $c^r(t)$ may be, in general, a difficult or even impossible task. The estimation problem is not even clearly formulated. For instance, what kind of information can we use to estimate x^r and c^r? How do we assess whether such information is sufficient to carry out the estimation successfully? In order to answer these questions and to formulate the estimation problem in a precise way, consider once again (10.120) and think of it as a copy of the plant with an unknown state x^r, an unknown input c^r, but a measurable output $r(t)$ (see Figure 10.5). In the light of this observation, the problem can be posed as that of employing the output r to estimate the state x^r of the dynamical system (10.120) subject to an unknown input c^r. The presence of c^r, however, limits the applicability of the estimation techniques seen in previous sections because the observability mapping associated with (10.120) may in general depend on c^r and its time derivatives. We will see that the problem can be solved successfully if one can find a practical internal model. Before formally introducing this concept, we see its application in a simple example.

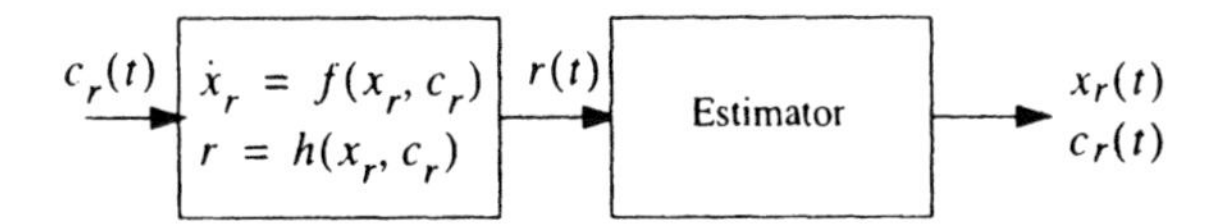

Figure 10.5. Posing the derivation of the stable inverse of the plant as an on-line estimation problem.

Example 10.10 Consider the following MIMO nonlinear system

$$\begin{aligned}
\dot{x}_1 &= x_2 \\
\dot{x}_2 &= x_1^2 + u_1 \\
\dot{x}_3 &= x_4 - u_1 - x_1^2 \\
\dot{x}_4 &= u_2,
\end{aligned} \tag{10.131}$$

with measurable output $y = [x_1, x_3]^\top$, and suppose we want to find a controller that achieves tracking of a smooth bounded reference trajectory $r(t) = [r_1(t), r_2(t)]^\top$. A stable inverse of (10.131) is given by

$$\begin{aligned}
x^r(t) &= [r_1(t), \dot{r}_1(t), r_2(t), \dot{r}_2(t) + \ddot{r}_1(t)]^\top \\
c^r(t) &= [\ddot{r}_1(t) - r_1^2(t), \ddot{r}_2(t) + r_1^{(3)}(t)]^\top .
\end{aligned} \tag{10.132}$$

Suppose now that $x^r(t)$ and $c^r(t)$ are not available for feedback and the only measurements are given by $y(t)$ and $r(t)$. This may happen because one may not know how to solve (10.120) or simply not know the time derivatives of $r(t)$. Choose $k_1 = 2$, $k_2 = 2$ and form the observability mapping associated with (10.131)

$$y_e = \mathcal{H}(x, u_1) = \left[x_1,\, x_2,\, x_3,\, x_4 - u_1 - x_1^2\right]^\top .$$

Now consider the copy of the plant with state x^r and input c^r

$$\begin{aligned}
\dot{x}^r{}_1 &= x^r{}_2 \\
\dot{x}^r{}_2 &= (x^r{}_1)^2 + c^r{}_1 \\
\dot{x}^r{}_3 &= x^r{}_4 - c^r{}_1 - (x^r{}_1)^2 \\
\dot{x}^r{}_4 &= c^r{}_2
\end{aligned} \qquad r = \begin{bmatrix} x^r{}_1 \\ x^r{}_3 \end{bmatrix}. \tag{10.133}$$

Clearly, the state and the input of (10.133) represent the stable inverse of the plant. Furthermore, it is clear that the observability mapping associated with (10.133) is given by

$$r_e = \mathcal{H}(x^r, c^r{}_1),$$

where $r_e \triangleq [r_1, \dot{r}_1, r_2, \dot{r}_2]^\top$. As mentioned earlier, since $c^r{}_1$ is unknown, one cannot simply apply the estimator developed in Section 10.6 to the output $r(t)$ and estimate $x^r(t)$.

Let $\zeta^r = [\zeta^r{}_1, \zeta^r{}_2, \zeta^r{}_3]^\top$, and augment the copy of the plant (10.133) with the the third-order compensator,

$$\begin{aligned} \dot{\zeta}^r{}_1 &= \zeta^r{}_2 + \zeta^r{}_3 \\ \dot{\zeta}^r{}_2 &= v^r{}_1 \\ \dot{\zeta}^r{}_3 &= v^r{}_2 \\ c^r &= [\zeta^r{}_1, \zeta^r{}_2]^\top, \end{aligned} \tag{10.134}$$

where $v^r = [v^r{}_1, v^r{}_2]^\top$ is the new *unknown* input and c^r is the output of the compensator. Notice that the compensator above is *regular*, that is, given any $c^r(t)$, one can find $\zeta^r(t)$ and $v^r(t)$ such that the output of the compensator is identically equal to $c^r(t)$, for all t. Choosing $k_1 = 4$ and $k_2 = 3$, the observability mapping associated with the composite system (10.133)-(10.134) is given by

$$r'_e \triangleq \begin{bmatrix} r_1 \\ \dot{r}_1 \\ r_1^{(2)} \\ r_1^{(3)} \\ r_2 \\ \dot{r}_2 \\ r_2^{(2)} \end{bmatrix} = \mathcal{H}'(x^r, \zeta^r) = \begin{bmatrix} x^r{}_1 \\ x^r{}_2 \\ (x^r{}_1)^2 + \zeta^r{}_1 \\ 2x^r{}_1 x^r{}_2 + \zeta^r{}_2 + \zeta^r{}_3 \\ x^r{}_3 \\ x^r{}_4 - \zeta^r{}_1 - (x^r{}_1)^2 \\ -\zeta^r{}_3 - 2x^r{}_1 x^r{}_2 \end{bmatrix}, \tag{10.135}$$

it does not depend on v^r, is invertible on $\mathsf{R}^4 \times \mathsf{R}^3$, and its inverse is given by

$$\begin{aligned} x^r &= \left[r_1, \dot{r}_1, r_2, \ddot{r}_1 + \dot{r}_2\right]^\top, \\ \zeta^r &= \left[\ddot{r}_1 - r_1^2, r_1^{(3)} + \ddot{r}_2, -2r_1\dot{r}_1 - \ddot{r}_2\right]^\top. \end{aligned} \tag{10.136}$$

Furthermore, once $x^r(t)$ and $\zeta^r(t)$ have been calculated, $c^r(t)$ is simply given by $c^r = [\zeta^r{}_1, \zeta^r{}_2]^\top = [\ddot{r}_1 - r_1^2, r_1^{(3)} + \ddot{r}_2]^\top$. Compare the expressions for x^r and c^r we just found with (10.132), and notice that indeed this is the stable inverse of the plant.

Now go back to the problem of estimating $x^r(t)$ and $c^r(t)$ and observe that, since the observability mapping of the composite system (10.133)-(10.134) is independent of the unknown function v^r, one can apply the estimator developed in Section 10.6 and approximate $[x^r(t)^\top, \zeta^r(t)^\top]^\top$ arbitrarily well without the knowledge of the time derivatives of r and without inverting the mapping $\mathcal{H}'$. From $[x^r(t)^\top, \zeta^r(t)^\top]^\top$ one also gets c^r.

Summarizing, we started from the copy of the plant (10.133) for which one cannot employ known estimation techniques, we introduced a regular compensator with the property that the observability mapping of the composite system does not depend on the new control input v^r, and we observed that this feature allows us to apply an estimator which employs r to approximate the stable inverse of the plant on-line. This compensator, which we call a *practical internal model*, plays a major role in the solution of the tracking problem. △

Definition 10.2: The compensator

$$\begin{aligned} \dot{\zeta}^r &= a(\zeta^r, x^r, v^r) \\ c^r &= b(\zeta^r, x^r), \end{aligned} \tag{10.137}$$

with state $\zeta^r \in \mathrm{R}^p$ (where $p \geq m$), input $v^r \in \mathrm{R}^m$, and sufficiently smooth functions a and b, is said to be a *practical internal model* for the copy of the plant (10.120) if

(i) The observability mapping of the composite system (10.120)-(10.137) $\mathcal{H}'$ does not depend on v^r and its derivatives, that is, $\mathcal{H}' = \mathcal{H}'(x^r, \zeta^r)$.

(ii) There exists a set of indices $\{\bar{k}_1, \ldots, \bar{k}_p\}$ such that the mapping $r_e' \triangleq [r_1, \ldots, r_1^{(\bar{k}_1-1)}, \ldots, r_p, \ldots, r_p^{(\bar{k}_p-1)}]^\top = \mathcal{H}'(x^r, \zeta^r)$ is invertible with respect to x^r and ζ^r, and its inverse is sufficiently smooth, for all $[x^{r\top}, \zeta^{r\top}]^\top \in \mathrm{R}^{n+p}$.

(iii) The compensator is regular, i.e., for each $x^r(0)$ and bounded $c^r(t)$ there exists $\zeta^r(0)$ and $v^r(t)$ such that, for all $t \geq 0$, $b(\zeta^r(t), x^r(t)) = c^r(t)$, and the functions $\zeta^r(t)$, $v^r(t)$ are bounded.

Note that, when the observability mapping of the plant does not depend on the control input, a practical internal model is simply given by m integrators

$$\dot{\zeta}^r{}_i = v^r{}_i, \qquad c^r{}_i = \zeta^r{}_i, \; i = 1, \ldots, m.$$

Example 10.11 We return to Example 10.9 and we show how to calculate the stable inverse of the plant using a practical internal model. Consider once again the input-output feedback linearizable system

$$\begin{aligned} \dot{x}_1 &= x_2 \\ \dot{x}_2 &= x_1^3 + x_2^2 + u \\ y &= x_1, \end{aligned}$$

and a generic reference signal $r(t)$. Notice that the observability mapping associated with this system does not depend on u

$$y_e = [y, \dot{y}]^\top = [x_1, x_2]^\top .$$

Hence, a practical internal model for the copy of the plant

$$\begin{aligned} \dot{x}^r{}_1 &= x^r{}_2 \\ \dot{x}^r{}_2 &= (x^r{}_1)^3 + (x^r{}_2)^2 + c^r \\ r &= x^r{}_1, \end{aligned} \tag{10.138}$$

is simply given by

$$\dot{\zeta}^r = v^r$$

The observability mapping of the composite system is given by

$$r_e' = [r, \dot{r}, \ddot{r}]^\top = \mathcal{H}'(x^r, \zeta^r) = [x^r{}_1, x^r{}_2, (x^r{}_1)^3 + (x^r{}_2)^2 + \zeta^r]^\top$$

which is invertible. By using an observer one can estimate x^r and ζ^r without inverting $\mathcal{H}'$ and without knowing the derivatives of r. Here, for the sake of illustration, we directly calculate the inverse of $\mathcal{H}'$

$$\begin{bmatrix} x^r \\ \zeta^r \end{bmatrix} = \begin{bmatrix} r \\ \dot{r} \\ \ddot{r} - r^3 - \dot{r}^3 \end{bmatrix}.$$

Recalling that $c^r = \zeta^r$, the stable inverse of the plant is found to be

$$x^r = [r, \dot{r}]^\top, \; c^r = \ddot{r} - r^3 - \dot{r}^3 .$$

Replace now $r(t)$ by the choice of reference trajectory used in Example 10.9, $r(t) = \sin t$, and notice that the stable inverse found using a practical internal model is identical to the one found solving (10.120). In conclusion, the integrator with state ζ^r allows us to estimate $c^r(t)$. If the full information controller $\bar{u}$ does not depend on c^r and hence its estimation is not needed, then the integrator can be removed.

△

10.8.2 Separation Principle for Tracking

We now have the elements to define a partial information controller solving the tracking problem. The control scheme we employ is illustrated in Figure 10.6. The scheme employs two estimators, the first one is used to estimate the functions x^r and ζ^r (and hence also c^r), while the second one estimates the state of the plant. A dynamic projection is employed to eliminate the peaking phenomenon and guarantee stability of the estimation errors, and the projected estimates $\hat{x}^P$, $\hat{x}^{rP}$, and $\hat{\zeta}^{rP}$ are used in the full information controller $\bar{u}$. The estimator for x is given by (10.88)-(10.89), and the corresponding projection is defined in (10.105). The estimator for $\hat{x}^r$ and $\hat{\zeta}^r$ is defined in what follows. Let $X^r \triangleq [x^{r\top}, \zeta^{r\top}]^\top$ and rewrite the copy of the plant (10.120) augmented with the practical internal model (10.137) as

$$\begin{aligned} \dot{X}^r &= F(X^r, v^r) \\ r &= H(X^r). \end{aligned} \tag{10.139}$$

Notice that, by the regularity of the practical internal model, the trajectory $X^r(t)$ is bounded for all $t \geq 0$ and, hence, there exists a compact set Ω^r such that $X^r(t) \in \Omega^r$, for all $t \geq 0$.

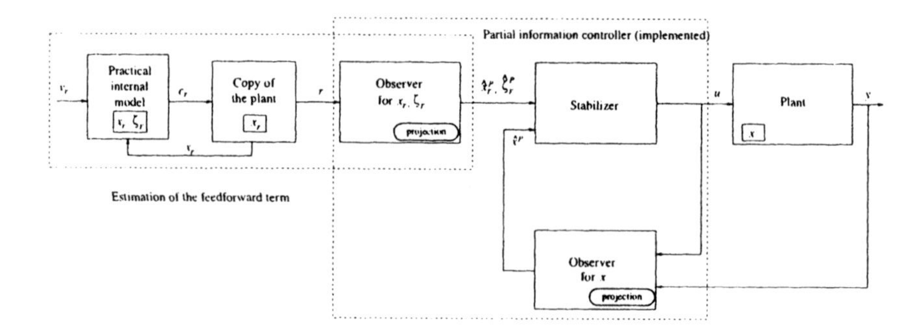

Figure 10.6. Partial information controller solving the tracking problem.

The observability mapping associated with (10.139) is trivially $r'_e = \mathcal{H}'(X^r)$. Since $\mathcal{H}'$ does not depend on the unknown v^r, we can think of (10.139) as a system without control input affected by an unknown disturbance $v^r(t)$, and apply the estimator (10.117)-(10.118) to (10.139). The nominal value of the "disturbance" input v^r can be set to zero. In this case the estimator is given by

$$\begin{aligned} \dot{\hat{X}}^r &= F(\hat{X}^r, 0) + \frac{\partial \mathcal{H}'(\hat{X}^r)}{\partial \hat{X}^r}\left[(\mathcal{E}^x)^{-1} L(r - \hat{r})\right] \\ \hat{r} &= H(\hat{X}^r), \end{aligned} \tag{10.140}$$

where $\mathcal{E}^x$ and L are defined in the usual way. Let $\mathcal{C}^r$ be a convex compact set with smooth boundary such that

$$\mathcal{H}'(\Omega^r) \subset \mathcal{C}^r,$$

let $N^r(r'_e)$ denote the normal vector to the boundary of $\mathcal{C}^r$ at r'_e and define the dynamic projection

$$\dot{\hat{X}}^{r^P} = \begin{cases} \dot{\hat{X}}^r - \left[\dfrac{\partial \mathcal{H}'}{\partial \hat{X}^r}\right]^{-1} \Gamma \dfrac{N^r(\hat{r}'_e) N^r(\hat{r}'_e)^\top}{N^r(\hat{r}'_e)^\top \Gamma N^r(\hat{r}'_e)} \dot{\hat{r}}'_e & \text{if } N^r(\hat{r}'_e)^\top \dot{\hat{r}}'_e \geq 0 \\ & \text{and } \hat{r}'_e \in \partial \mathcal{C}^r \\ \dot{\hat{X}}^r & \text{otherwise,} \end{cases} \tag{10.141}$$

where Γ is defined in the usual way, $\hat{r}'_e = \mathcal{H}'(\hat{X}^r)$ and $\dot{\hat{r}}'_e = \dfrac{\partial \mathcal{H}'}{\partial \hat{X}^r} \dot{\hat{X}}^r$.

Next, define

$$\Omega_c = \{x \in \mathsf{R}^n \mid V(t, e) \leq c, \text{ for all } t \geq 0\}$$

and notice that the boundedness of $x^r(t)$ implies the compactness of Ω_c. Pick any two positive scalars c_1 and c_2, choose a compact convex set $\mathcal{C}$ for the projection (10.105) such that

$$\mathcal{F}(\Omega_{c_2} \times \Omega^s) \subset \mathcal{C},$$

where Ω^s has been defined earlier, and consider the partial information tracking controller

$$\hat{u} = \bar{u}(t, \hat{x}^P, x^{rP}, b(\zeta^{rP}, x^{rP})). \tag{10.142}$$

Theorem 10.5: *Consider the closed-loop system formed by the error dynamics (10.122), the estimator for the state (10.88), (10.89) with projection (10.105), the estimator for the stable inverse (10.140) with projection (10.141), and the controller (10.142). Then, for any choice of c_1 and c_2, every trajectory starting in $\Omega_{c_1} \subset \mathcal{D}$ is contained in Ω_{c_2} and enters an arbitrarily small neighborhood of the origin $x = 0$ in finite time, provided the design parameters of the observers are chosen sufficiently small.*

Proof: The proof is a slight modification of that found in Lemma 10.2 and bears conceptual similarities to the proof of Theorem 11.1. We leave it as an exercise to the reader. ■

Observe that if the full information controller makes the origin of the error dynamics globally uniformly asymptotically stable (i.e., $\mathcal{D} = \mathsf{R}^n$), then the partial information controller makes $e = 0$ semiglobally practically uniformly stable.

10.9 Summary

In this chapter we have introduced tools for the design of output-feedback controllers achieving stabilization and tracking. When the plant can be transformed to output-feedback form, we have seen a systematic procedure which employs backstepping to achieve stabilization or tracking. In the case of tracking, the output-feedback backstepping controller we have found utilizes the first r time derivatives of the reference trajectory r and thus it cannot be considered a partial information controller.

For more general classes of nonlinear systems, we have seen that one can *separate* the control design into two parts. First, a state feedback controller (in the case of stabilization) or a full information controller (in the case of tracking) are designed to achieve desired control objectives. Next, a nonlinear high-gain type observer is developed. Finally, the output-feedback controller is found by replacing the unknown variables in the state-feedback (or full information) controller by their estimates and employing projection to guarantee the stability of the estimation error and avoid the destabilizing effects of the peaking phenomenon. This methodology, which is referred to as a *separation principle*, has two advantages. First, the control design is significantly simplified. Second, owing to the availability of more general control tools using state-feedback, a separation principle allows one to deal with more general classes of nonlinear systems, without restricting the attention to systems in output-feedback form.

We have also learned how to use the framework of a separation principle to solve the challenging tracking problem. Specifically, we have seen that if a *practical internal model* can be found, then the tracking problem can be converted into a standard stabilization problem relative to the error system $e = \chi(t,x) = x - x^r(t)$ by using one observer to estimate the state and another observer to estimate the stable inverse $(x^r(t), c^r(t))$ of the plant. We must stress, however, that there is no constructive procedure to find a practical internal model or to guarantee its existence.

10.10 Exercises and Design Problems

Exercise 10.1 (Output-Feedback Form) Use the backstepping method to develop a controller for the output-feedback system in Example 10.3 so that y tracks $r(t)$. Assuming that $|r| < b_1$, design the controller and observer when $|g(y) - \mathcal{F}(y,\theta)| < W$ only holds when $|y| \leq b_2$ with $b_2 > b_1$.

Exercise 10.2 (Properties of the Projection (10.105)) Prove Lemma 10.4.

Exercise 10.3 (Practical Stability of (10.106)) Prove Theorem 10.4.

Exercise 10.4 (Existence of a Practical Internal Model) Consider the system

$$\begin{aligned} \dot{x} &= f(x,u) \\ y &= h(x), \end{aligned} \tag{10.143}$$

and the associated observability mapping $y_e = \mathcal{H}(x, s)$. Assume that the system is UCO, that is, $\mathcal{H}$ is a global diffeomorphism and suppose that the plant is globally *dynamically feedback linearizable*, i.e., there exists a compensator

$$\begin{aligned} \dot{\xi} &= \gamma_1(\xi, x, v) \\ u &= \gamma_2(\xi, x) \end{aligned} \tag{10.144}$$

such that the augmented system

$$\begin{aligned} \dot{x} &= f(x, \gamma_2(\xi, x)) \\ \dot{\xi} &= \gamma_1(\xi, x) \\ y &= h(x) \end{aligned} \tag{10.145}$$

is globally feedback linearizable with output function h. Prove that the set of conditions above guarantees (sufficient condition) the existence of a practical internal model and that the output-feedback tracking problem can be solved semiglobally.

Exercise 10.5 (Ball and Beam) Consider the ball and beam system defined by

$$\begin{aligned} \dot{x} &= v \\ \dot{v} &= -g \sin\theta + x\omega^2 \\ \dot{\theta} &= \omega \\ \dot{\omega} &= -\frac{2mxv\omega}{J + mx^2} - \frac{mgx\cos\theta}{J + mx^2} + u, \end{aligned}$$

where the parameters are defined in Exercise 6.12. Design an observer for this system when x, θ, and u are measurable. Design an observer when only x and u are measurable.

Exercise 10.6 (Flexible Manipulator) The dynamics for a single-link manipulator with flexible joints is described by

$$\begin{aligned} I\ddot{q}_1 + MgL\sin(q_1) + k(q_1 - q_2) &= 0 \\ J\ddot{q}_2 - k(q_1 - q_2) &= u, \end{aligned} \tag{10.146}$$

where q_1 and q_2 are the angular positions, and u is the torque input [212]. Here I and J are the moments of inertia, k is the spring constant, M is the total mass, and $g = 9.81 m/s^2$ is the gravitational acceleration constant. Design an output-feedback controller when q_1 and u are measurable so that q_1 tracks $r(t)$.

Exercise 10.7 (Fuzzy/Neural Force Cancellation) Consider dynamics of the point mass

$$m\ddot{x} = f(x) + u, \tag{10.147}$$

where x is the position, m is the mass, and u is the force input. Assume that $f(x)$ may be approximated by the fuzzy/neural system $\mathcal{F}(x,\theta)$ so that $|f(x)-\mathcal{F}(x,\theta)| < W$ for all $x \in B_x$. Define an output-feedback controller so that $x \to 0$ and so that x never leaves the set $B_x = \{x \in \mathsf{R} : |x| < b\}$ when $\dot{x}$ is unknown.

Exercise 10.8 (Three-Phase Motor) Consider the three-phase motor defined in Exercise 7.13. Design an output-feedback controller so that θ tracks the reference $r(t)$ when the system parameters are known and only θ and u are measurable.

Exercise 10.9 (Steering Mirror) A steering mirror is used to direct a beam of light toward the center of a corner reflector target. The reflected beam travels back toward the mirror parallel to the original beam, but shifted in position when the beam does not hit the center of the corner reflector. The reflected beam is then detected by a quad-detector (a position-dependent optical sensor). A simplified model of a single axis of this system is described by

$$J\ddot{\theta} = u, \tag{10.148}$$

where θ is the pointing error, J is the moment of inertia and u is the control input to point the mirror. The output of the quad-detector is described by

$$y = k\frac{\theta}{|\theta| + \epsilon}, \tag{10.149}$$

where $k, \epsilon > 0$. Design an output-feedback controller that drives $\theta \to 0$ when $\dot{\theta}$ is not measurable.

Chapter 11

Adaptive Output Feedback Control

11.1 Overview

In Chapters 7 and 8 we saw how adaptive control may be used as a systematic design tool to develop dynamic controllers for systems with a great deal of uncertainty while still being able to achieve good closed-loop performance. The techniques presented there, however, assumed that full state information is available for feedback. In this chapter we will extend these tools to develop adaptive output-feedback controllers for both stabilization and tracking.

First, for systems with parametric uncertainties, we will provide a brief account of the adaptive output-feedback backstepping technique developed by Kanellakopoulos, Kokotović, and Morse (for more details, the reader is referred to [115]). Next, we will show how to employ the separation principle previously developed to extend the techniques seen in Chapters 7 and 8 to the output feedback framework. Recall that, in the spirit of a separation principle, one seeks to construct output-feedback controllers recovering the performance of given state-feedback controllers. This is advantageous in practice because the tools available to design output-feedback controllers without a separation principle are rather limited and work for restricted classes of nonlinear systems (see, e.g., Section 10.3) while, by separating the state estimation from the control design phase, one can exploit available state feedback design tools for quite general classes of systems. On the other hand, the implementation of a state estimator requires some knowledge about the plant and, in the adaptive control case, restricts the class of uncertainties affecting the system. These issues play a major role now and will somewhat limit the generality of the results illustrated in the previous chapter.

As we did earlier, we will separately study stabilization and tracking and, for the latter problem, we will once again make the distinction between a full information setting, where the state of the system and the error performance measure e are available for feedback, and a partial information setting, where both x and e are not directly measurable.

11.2 Control of Systems in Adaptive Tracking Form

In this section we consider systems in output-feedback form (10.5) affected by a parametric uncertainty. Specifically, we assume that the output-dependent nonlinearities $g_j(y)$ are unknown, but they can be exactly represented by a linear in the parameter approximator with no representation error. In other words, we consider the class of systems

$$
\begin{aligned}
\dot{x}_1 &= x_2 + g_{0,1}(y) + \sum_{j=1}^{p} \theta_j g_{j,1}(y) \\
\dot{x}_2 &= x_3 + g_{0,2}(y) + \sum_{j=1}^{p} \theta_j g_{j,2}(y) \\
&\vdots \\
\dot{x}_{r-1} &= x_r + g_{0,r-1}(y) + \sum_{j=1}^{p} \theta_j g_{j,r-1}(y) \\
\dot{x}_r &= x_{r+1} + g_{0,r}(y) + \sum_{j=1}^{p} \theta_j g_{j,r}(y) + d_m \sigma(y) u \\
&\vdots \\
\dot{x}_n &= g_{0,n}(y) + \sum_{j=1}^{p} \theta_j g_{j,n}(y) + d_0 \sigma(y) u \\
y &= x_1,
\end{aligned}
\tag{11.1}
$$

where each $g_{i,j}$ and $\sigma(y)$ are locally Lipschitz functions, $g_{i,j}(0) = 0$, $\sigma(y) \neq 0$ for all $y \in \mathbb{R}$, and $r = n - m$ is the relative degree of the system. The vector $\theta \in \mathbb{R}^p$ and the scalars d_i, $i = 1, \ldots m$ are unknown and such that the polynomial $p(s) = d_m s^m + \ldots + d_1 s + d_0$ is Hurwitz, thus implying that the zero dynamics of the system are globally exponentially stable. We assume that the sign of d_m is known and, without loss of generality, we assume that $d_m > 0$.

We seek to find an adaptive controller that, despite the presence of the unknown parameters θ_j and d_i, makes y asymptotically track a smooth reference trajectory $r(t)$ while guaranteeing boundedness of all internal variables. For $0 \le j \le p$ let

$$A = \begin{bmatrix} 0 & & \\ \vdots & I & \\ 0 & \dots & 0 \end{bmatrix}, d = \begin{bmatrix} 0 \\ \vdots \\ 0 \\ d_m \\ \vdots \\ d_0 \end{bmatrix}, g_j(y) = \begin{bmatrix} g_{j,1}(y) \\ \vdots \\ g_{j,n}(y) \end{bmatrix}, C^\top = \begin{bmatrix} 1 \\ 0 \\ \vdots \\ 0 \end{bmatrix}, \tag{11.2}$$

so that (11.1) can be rewritten in vector form as

$$\begin{aligned} \dot{x} &= Ax + g_0(y) + \sum_{j=1}^{p} \theta_j g_j(y) + d\sigma(y)u \\ y &= Cx. \end{aligned} \tag{11.3}$$

The class of systems (11.3), first considered by Marino and Tomei (see, e.g., [149] for geometric conditions that must hold in order for a system to be put in the form (11.3)), is said to be in *adaptive tracking form.* Systems in adaptive tracking form are particularly suitable for the design of adaptive output-feedback controllers. Using backstepping, one can devise a systematic design procedure to achieve global stabilization and tracking. For a more detailed discussion the reader is referred to [115] and [149]. Here, for the sake of illustration, we will provide a brief account, taken from [115], of one of the available techniques which employs the so-called K-filters to solve the tracking problem.

The main idea behind the design is to replace the state of the system, which is not measurable, by a θ-dependent estimate and employ the certainty equivalence principle to design an adaptive controller. If θ were known, then we could use the observer

$$\dot{\hat{x}} = A\hat{x} + g_0(y) + \sum_{j=1}^{p} \theta_j g_j(y) + d\sigma(y)u + L(y - C\hat{x}) \tag{11.4}$$

so that $\tilde{x} = x - \hat{x}$ decays as $\dot{\tilde{x}} = (A - LC)\tilde{x}$ where L is chosen so that $(A - LC)$ is Hurwitz. K-filters help defining a state estimate which, when θ is known, converges exponentially to the actual state. Consider the filters

$$\begin{aligned} \dot{\xi}_0 &= A\xi_0 + g_0(y) + L(y - C\xi_0) \\ \dot{\xi}_j &= (A - LC)\xi_j + g_j(y), \ 1 \le j \le p \\ \dot{v}_j &= (A - LC)v_j + \bar{e}_{n-j}\sigma(y)u, \ 0 \le j \le m, \end{aligned} \tag{11.5}$$

where $L = [l_1, \dots, l_n]^\top$ is a vector chosen so that $(A - LC)$ is Hurwitz (L exists because (A, C) is an observable pair), and $\bar{e}_i$ denotes the i-th basis vector for $\mathbf{R}^n$. When θ is known, the state of the system can be expressed as

$$x = \left(\xi_0 + \sum_{j=1}^{p} \theta_j \xi_j + \sum_{j=0}^{m} d_j v_j \right) + \tilde{x},$$

where $\tilde{x}$, the estimation error, has linear exponentially stable dynamics,

$$\dot{\tilde{x}} = (A - LC)\tilde{x}. \tag{11.6}$$

The filters in (11.5) are used in our analysis to replace the unknown state of the system by known quantities and using the certainty equivalence principle to replace the unknown θ in (11.5) by an estimate $\hat{\theta}$. Similarly to what we did in Section 10.3, nonlinear damping will be employed to account for the presence of the estimation error $\tilde{x}$.

e_1 subsystem Let $e_1 = x_1 - r$ and consider the subsystem defined by

$$\dot{e}_1 = x_2 + g_{0,1}(y) + \sum_{j=1}^{p} g_{j,1}(y) - \dot{r}.$$

Note that x_2 is not available for feedback and hence cannot be chosen as the virtual control input. In order to define a suitable virtual control input, express x_2 in terms of the states of the filters in (11.5) and rewrite $\dot{e}_1$ as

$$\dot{e}_1 = \xi_{0,2} + \sum_{j=1}^{p} \theta_j \left(\xi_{j,2} + g_{j,1}(y)\right) + g_{0,1}(y) + \sum_{j=0}^{m} d_j v_{j,2} + \tilde{x}_2 - \dot{r}. \tag{11.7}$$

Add and subtract the term $-\kappa e_1 - c_1 e_1$, where κ and c_1 are positive design scalars, and rewrite (11.7) as

$$\begin{aligned} \dot{e}_1 = \; & -\kappa e_1 - c_1 e_1 + \tilde{x}_2 + d_m \left[\frac{1}{d_m} \left(\kappa e_1 + c_1 e_1 + \xi_{0,2} + g_{0,1}(y) - \dot{r}\right) \right. \\ & \left. + \sum_{j=1}^{p} \frac{\theta_j}{d_m} \left(\xi_{j,2} + g_{j,1}(y)\right) + \sum_{j=0}^{m-1} \frac{d_j}{d_m} v_{j,2} + v_{m,2} \right]. \end{aligned} \tag{11.8}$$

The term $v_{m,2}$ has the lowest relative degree among all the variables in (11.8) and hence it is chosen to be our first virtual control input. Define the unknown parameter vector θ^1 and the first regressor ω_1 as

$$\begin{aligned} \theta^1 &= \left[\frac{1}{d_m}, \frac{\theta_1}{d_m}, \ldots, \frac{\theta_p}{d_m}, \frac{d_0}{d_m}, \ldots, \frac{d_{m-1}}{d_m} \right]^{\mathsf{T}} \\ \omega_1^{\mathsf{T}} &= [\kappa e_1 + c_1 e_1 + \xi_{0,2} + g_{0,1} - \dot{r}, \xi_{1,2} + g_{1,1}, \ldots, \xi_{p,2} + g_{p,1}, \\ & \quad v_{0,2}, \ldots, v_{m-1,2}] \end{aligned}$$

and rewrite (11.8) as

$$\dot{e}_1 = -\kappa e_1 - c_1 e_1 + \tilde{x}_2 + d_m \left[\omega_1^{\top}\theta^1 + v_{m,2}\right]. \tag{11.9}$$

Since θ^1 is not known, we use the certainty equivalence principle to define a stabilizing function for the virtual control $v_{m,2}$ which is used to define the second component of the error system

$$\begin{aligned} \nu_1(z_1) &= -\omega_1^{\top}\hat{\theta}^1, \; e_2 = v_{m,2} - \nu_1(z_1) \\ z_1 &= \left[y, \xi_0, \xi_1, \ldots, \xi_p, v_0, \ldots, v_m, \hat{\theta}^1, r, \dot{r}\right]. \end{aligned} \tag{11.10}$$

Recall that L has been chosen so that the matrix $A-LC$ is Hurwitz and let P be the positive definite and symmetric solution of the Lyapunov equation $P(A-LC)+(A-LC)^{\top}P = -I$. Consider now the function

$$V_1 = \frac{1}{2}e_1^2 + \frac{d_m}{2}(\theta^1 - \hat{\theta}^1)^{\top}\Gamma^{-1}(\theta^1 - \hat{\theta}^1) + \frac{1}{c_1}\tilde{x}^{\top}P\tilde{x}, \tag{11.11}$$

where $c_1 > 0$ and Γ is any positive definite symmetric matrix, and calculate its time derivative

$$\begin{aligned} \dot{V}_1 = \; & -\kappa e_1^2 - c_1 e_1^2 + e_1\tilde{x}_2 + d_m e_1\left[\omega_1^{\top}(\theta^1 - \hat{\theta}^1) + e_2\right] \\ & - d_m(\theta^1 - \hat{\theta}^1)^{\top}\Gamma^{-1}\dot{\hat{\theta}}^1 - \frac{1}{c_1}|\tilde{x}|^2. \end{aligned} \tag{11.12}$$

Choosing the update law

$$\dot{\hat{\theta}}^1 = \Gamma\omega_1 e_1, \tag{11.13}$$

and applying Young's inequality to $e_1\tilde{x}_2$,

$$e_1\tilde{x}_2 \le |e_1||\tilde{x}| \le c_1 e_1^2 + \frac{|\tilde{x}|}{4c_1}$$

we get

$$\dot{V}_1 \le -\kappa e_1^2 - \frac{3|\tilde{x}|}{4c_1} + d_m e_1 e_2. \tag{11.14}$$

e_2 subsystem Now calculate the time derivative of $e_2 = v_{m,2} - \nu_1(z_1)$,

$$\begin{aligned} \dot{e}_2 = \; & v_{m,3} - l_2 v_{m,1} - \frac{\partial \nu_1}{\partial y}\left[\xi_{0,2} + g_{0,1}(y) + \sum_{j=1}^{p}\theta_j\left(\xi_{j,2} + g_{j,1}(y)\right)\right. \\ & \left. + \sum_{j=0}^{m} d_j v_{j,2} + \tilde{x}_2\right] - \frac{\partial \nu_1}{\partial \xi_0}\left[A\xi_0 + g_0(y) + L(y - C\xi_0)\right] \\ & - \sum_{j=1}^{p}\frac{\partial \nu_1}{\partial \xi_j}\left[(A-LC)\xi_j + g_j(y)\right] \\ & - \sum_{j=0}^{m}\frac{\partial \nu_1}{\partial v_j}(A-LC)v_j - \frac{\partial \nu_1}{\partial \hat{\theta}^1}\dot{\hat{\theta}}^1 - \frac{\partial \nu_1}{\partial r}\dot{r} - \frac{\partial \nu_1}{\partial \dot{r}}\ddot{r}, \end{aligned} \tag{11.15}$$

and choose $v_{m,3}$ as the virtual control input. As we did earlier, in order to use the certainty equivalence principle we isolate the terms containing unknown parameters and we define a suitable regressor vector. To this end, let

$$\begin{aligned}\theta^2 &= [\theta_1, \ldots, \theta_p, d_0, \ldots, d_m]^\top \\ \omega_2^\top &= -\frac{\partial \nu_1}{\partial y}\left[\xi_{1,2} + g_{1,1}(y), \ldots, \xi_{p,2} + g_{p,1}(y), v_{0,2}, \ldots, v_{m,2}\right].\end{aligned}$$

and rewrite (11.15) as

$$\begin{aligned}\dot{e}_2 = \; & v_{m,3} - l_2 v_{m,1} - \frac{\partial \nu_1}{\partial y}\left[\xi_{0,2} + g_{0,1}(y) + \tilde{x}_2\right] + \omega_2^\top \theta^2 \\ & - \frac{\partial \nu_1}{\partial \xi_0}\left[A\xi_0 + g_0(y) + L(y - C\xi_0)\right] \\ & - \sum_{j=1}^{p} \frac{\partial \nu_1}{\partial \xi_j}\left[(A - LC)\xi_j + g_j(y)\right] \\ & - \sum_{j=0}^{m} \frac{\partial \nu_1}{\partial v_j}(A - LC)v_j - \frac{\partial \nu_1}{\partial \hat{\theta}^1}\dot{\hat{\theta}}^1 - \frac{\partial \nu_1}{\partial r}\dot{r} - \frac{\partial \nu_1}{\partial \dot{r}}\ddot{r}.\end{aligned} \tag{11.16}$$

Denote by $\hat{\theta}^2$ the estimate of the unknown θ^2 and define the function V_2 as

$$V_2 = V_1 + \frac{1}{2}e_2^2 + \frac{1}{2}(\theta^2 - \hat{\theta}^2)^\top \Gamma^{-1}(\theta^2 - \hat{\theta}^2) + \frac{1}{c_2}\tilde{x}^\top P \tilde{x}. \tag{11.17}$$

Using (11.14) and (11.16) and collecting the terms containing e_2 we have

$$\begin{aligned}\dot{V}_2 \leq \; & -\kappa e_1^2 - \frac{3|\tilde{x}|}{4c_1} + e_2 \Bigg\{ v_{m,3} + d_m e_1 - l_2 v_{m,1} - \frac{\partial \nu_1}{\partial y}\left[\xi_{0,2} + g_{0,1}(y)\right. \\ & \left. + \tilde{x}_2\right] + \omega_2^\top \theta^2 - \frac{\partial \nu_1}{\partial \xi_0}\left[A\xi_0 + g_0(y) + L(y - C\xi_0)\right] \\ & - \sum_{j=1}^{p} \frac{\partial \nu_1}{\partial \xi_j}\left[(A - LC)\xi_j + g_j(y)\right] \\ & - \sum_{j=0}^{m} \frac{\partial \nu_1}{\partial v_j}(A - LC)v_j - \frac{\partial \nu_1}{\partial \hat{\theta}^1}\dot{\hat{\theta}}^1 - \frac{\partial \nu_1}{\partial r}\dot{r} - \frac{\partial \nu_1}{\partial \dot{r}}\ddot{r} \Bigg\} \\ & + (\theta^2 - \hat{\theta}^2)^\top \Gamma^{-1} \dot{\hat{\theta}}^2 - \frac{1}{c_2}|\tilde{x}|^2.\end{aligned} \tag{11.18}$$

The choice of the stabilization function

$$
\begin{aligned}
\nu_2(z_2) = \quad & -\kappa e_2 - c_2 \left(\frac{\partial \nu_1}{\partial y}\right)^2 e_2 - d_m e_1 + l_2 v_{m,1} + \frac{\partial \nu_1}{\partial y}[\xi_{0,2} + g_{0,1}] \\
& - \omega_2^\top \hat{\theta}^2 + \frac{\partial \nu_1}{\partial \xi_0}[A\xi_0 + g_0 + L(y - C\xi_0)] + \sum_{j=1}^{p} \frac{\partial \nu_1}{\partial \xi_j}[(A - LC)\xi_j \\
& + g_j] + \sum_{j=0}^{m} \frac{\partial \nu_1}{\partial v_j}(A - LC)v_j + \frac{\partial \nu_1}{\partial \hat{\theta}^1}\dot{\hat{\theta}}^1 + \frac{\partial \nu_1}{\partial r}\dot{r} - \frac{\partial \nu_1}{\partial \dot{r}}\ddot{r},
\end{aligned}
\tag{11.19}
$$

where $c_2 > 0$ and $z_2 = [z_1, r^{(2)}, \hat{\theta}^2]$, yields

$$
\begin{aligned}
\dot{V}_2 \leq \quad & -\kappa e_1^2 - \frac{3|\tilde{x}|}{4c_1} - \kappa e_2^2 - c_2 \left(\frac{\partial \nu_1}{\partial y}\right)^2 e_2^2 - \frac{\partial \nu_1}{\partial y}\tilde{x}_2 e_2 + e_2 \omega_2^\top (\theta^2 - \hat{\theta}^2) \\
& + (\theta^2 - \hat{\theta}^2)^\top \Gamma^{-1} \dot{\hat{\theta}}^1 - \frac{1}{c_2}|\tilde{x}|^2 + e_2 e_3,
\end{aligned}
\tag{11.20}
$$

where $e_3 = v_{m,3} - \nu_2(z_2)$. In complete analogy with step 1, we now apply Young's inequality to the term containing $\tilde{x}_2$ and we choose an update law for $\hat{\theta}^2$ to cancel the term containing $\theta^2 - \hat{\theta}^2$,

$$
\begin{aligned}
-\frac{\partial \nu_1}{\partial y}\tilde{x}_2 e_2 &\leq c_2 \left(\frac{\partial \nu_1}{\partial y}\right)^2 e_2^2 + \frac{|\tilde{x}_2|^2}{4c_2} \\
\dot{\hat{\theta}}^2 &= \Gamma \omega_2 e_2,
\end{aligned}
\tag{11.21}
$$

so that

$$
\dot{V}_2 \leq -\kappa(e_1^2 + e_2^2) - \frac{3}{4}\left(\frac{1}{c_1} + \frac{1}{c_2}\right)|\tilde{x}|^2 + e_2 e_3. \tag{11.22}
$$

e_i subsystem ($1 \leq i < r$) The procedure above is iterated r times. At step i, we write the dynamics of $e_i = v_{m,i} - \nu_{i-1}(z_{i-1})$, where $z_{i-1} = [z_{i-2}, r^{(i-1)}, \hat{\theta}^{i-1}]$, as

$$
\begin{aligned}
\dot{e}_i = \quad & v_{m,i+1} - l_i v_{m,1} - \frac{\partial \nu_{i-1}}{\partial y}[\xi_{0,2} + g_{0,1}(y) + \tilde{x}_2] + \omega_i^\top \theta^2 \\
& - \frac{\partial \nu_{i-1}}{\partial \xi_0}[A\xi_0 + g_0(y) + L(y - C\xi_0)] \\
& - \sum_{j=1}^{p} \frac{\partial \nu_{i-1}}{\partial \xi_j}[(A - LC)\xi_j + g_j(y)] \\
& - \sum_{j=0}^{m} \frac{\partial \nu_{i-1}}{\partial v_j}(A - LC)v_j - \sum_{j=1}^{i-1} \frac{\partial \nu_{i-1}}{\partial \hat{\theta}^j}\dot{\hat{\theta}}^j - \sum_{j=0}^{i-1} \frac{\partial \nu_{i-1}}{\partial r^{(j)}} r^{(j+1)},
\end{aligned}
\tag{11.23}
$$

where $v_{m,i+1}$ is the virtual control input and

$$
\omega_i^\top = -\frac{\partial \nu_{i-1}}{\partial y}[\xi_{1,2} + g_{1,1}(y), \ldots, \xi_{p,2} + g_{p,1}(y), v_{0,2}, \ldots, v_{m,2}].
$$

Using the certainty equivalence principle, we approximate θ^2 by $\hat{\theta}^i$ and define the i-th stabilizing function

$$\begin{aligned}
\nu_i(z_i) = \quad & -\kappa e_i - c_i \left(\frac{\partial \nu_{i-1}}{\partial y}\right)^2 e_i - e_{i-1} + l_i v_{m,1} + \frac{\partial \nu_{i-1}}{\partial y}\left[\xi_{0,2} + g_{0,1}\right] \\
& - \omega_i^\top \hat{\theta}^i + \frac{\partial \nu_{i-1}}{\partial \xi_0}\left[A\xi_0 + g_0 + L(y - C\xi_0)\right] \\
& + \sum_{j=1}^{p} \frac{\partial \nu_{i-1}}{\partial \xi_j}\left[(A - LC)\xi_j + g_j\right] + \sum_{j=0}^{m} \frac{\partial \nu_{i-1}}{\partial v_j}(A - LC)v_j \\
& + \sum_{j=1}^{i-1} \frac{\partial \nu_{i-1}}{\partial \hat{\theta}^j}\dot{\hat{\theta}}^j + \sum_{j=0}^{i-1} \frac{\partial \nu_{i-1}}{\partial r^{(j)}} r^{(j+1)},
\end{aligned} \tag{11.24}$$

(where $z_i = [z_{i-1}, r^{(i)}, \hat{\theta}^i]$) and the continuous function

$$V_i = V_{i-1} + \frac{1}{2}e_i^2 + \frac{1}{2}(\theta^2 - \hat{\theta}^i)^\top \Gamma^{-1}(\theta^2 - \hat{\theta}^i) + \frac{1}{c_i}\tilde{x}^\top P\tilde{x}. \tag{11.25}$$

By letting $e_{i+1} = v_{m,i+1} - \nu_i(z_i)$ we get

$$\begin{aligned}
\dot{V}_i \leq \quad & -\kappa \sum_{j=1}^{i} e_j^2 - \frac{3}{4}\left(\sum_{j=1}^{i-1}\frac{1}{c_j}\right)|\tilde{x}|^2 - c_i\left(\frac{\partial \nu_{i-1}}{\partial y}\right)^2 e_i^2 \\
& - \frac{\partial \nu_i}{\partial y}\tilde{x}_2 e_i + e_i\omega_i^\top(\theta^2 - \hat{\theta}^i) + (\theta^2 - \hat{\theta}^i)^\top \Gamma^{-1}\dot{\hat{\theta}}^i \\
& - \frac{1}{c_i}|\tilde{x}|^2 + e_i e_{i+1}.
\end{aligned} \tag{11.26}$$

Using Young's inequality and a suitable update law we cancel the sign indefinite terms

$$-\frac{\partial \nu_{i-1}}{\partial y}\tilde{x}_2 e_i \leq c_i\left(\frac{\partial \nu_{i-1}}{\partial y}\right)^2 e_i^2 + \frac{|\tilde{x}_2|^2}{4c_i}$$

$$\dot{\hat{\theta}}^i = \Gamma\omega_i e_i \tag{11.27}$$

so that

$$\dot{V}_i \leq -\kappa\sum_{j=1}^{i} e_j^2 - \frac{3}{4}\left(\sum_{j=1}^{i}\frac{1}{c_j}\right)|\tilde{x}|^2 + e_i e_{i+1}. \tag{11.28}$$

e_r subsystem At the final step we use (11.24) and (11.27) to define the stabilizing function $\nu_r(z_r)$ and an adaptation law for $\hat{\theta}^r$. Then, the actual control input is found by defining the positive definite function

$$\begin{aligned}
V_r = \quad & \sum_{j=1}^{r}\left(\frac{1}{2}e_j^2 + \frac{1}{d_i}\tilde{x}^\top P\tilde{x}\right) + \frac{d_m}{2}(\theta^1 - \hat{\theta}^1)^\top \Gamma^{-1}(\theta^1 - \hat{\theta}^1) \\
& + \sum_{j=2}^{r}\frac{1}{2}(\theta^2 - \hat{\theta}^j)^\top \Gamma^{-1}(\theta^2 - \hat{\theta}^j),
\end{aligned} \tag{11.29}$$

whose time derivative contains now the control input u

$$\dot{V}_r \leq -\kappa \sum_{j=1}^{r} e_j^2 - \frac{3}{4}\left(\sum_{j=1}^{r} \frac{1}{c_j}\right) |\tilde{x}|^2 + e_r\left(\sigma(y)u - \nu_r(z_r) + v_{m,r+1}\right). \tag{11.30}$$

The control input

$$u = \frac{1}{\sigma(y)}\left(\nu_r(z_r) - v_{m,r+1}\right) \tag{11.31}$$

yields

$$\dot{V}_r \leq -\kappa \sum_{j=1}^{r} e_j^2 - \frac{3}{4}\left(\sum_{j=1}^{r} \frac{1}{c_j}\right) |\tilde{x}|^2 \leq 0, \tag{11.32}$$

thus proving the boundedness of $e_1, \ldots, e_r$ and their asymptotic convergence to zero (hence, $y \to r$), and the boundedness of $\tilde{x}$. From the boundedness of $e_1, \ldots, e_r$, $\tilde{x}$, and the smoothness of $r(t)$, we have that $x_1, \ldots, x_r$ are also bounded. Finally, from the exponential stability of the zero dynamics, the boundedness of $x_{r+1}, \ldots, x_n$ follows as well.

We conclude this section by pointing out that, despite its simplicity, this method suffers from the over-parameterization problem: at each step of the backstepping design, dynamics of order $p + m + 1$ for updating $\hat{\theta}^i$ have been introduced, and thus the total dynamic order of the controller is $r(p + m + 1)$, while the number of unknown parameters is only $p + m + 1$. To overcome this problem, a more advanced adaptive control design exists which employs the so-called "tuning functions." The reader is referred to [115] for a complete treatment of this topic.

11.3 Separation Principle for Adaptive Stabilization

While adaptive output-feedback backstepping provides a systematic tool to achieve global stabilization and tracking by output feedback, the class of systems for which the technique can be applied is quite restrictive since the nonlinearities and uncertainty may only depend upon the output y. On the other hand, an interesting feature of the output-feedback control techniques introduced in Chapter 10 is the generality of the class of systems taken in consideration. Here, we concentrate on the stabilization problem and we seek to exploit the generality of the tools in Chapter 10 to extend the results of Chapters 7 and 8 to the output-feedback framework.

Consider the nonlinear system

$$\begin{aligned} \dot{x} &= f(x, u, \Delta(t, x)) \\ y &= h(x, u), \end{aligned} \tag{11.33}$$

where $\Delta(t, x)$ represents either a modeling uncertainty or a time-varying disturbance. Suppose that, using the tools of Chapters 7 and 8, we have designed an adaptive controller for (11.33)

$$\begin{aligned}\dot{\hat{\theta}} &= f_\theta(\hat{\theta}, z) \\ u &= \nu_a(z, \hat{\theta}),\end{aligned} \tag{11.34}$$

where ν_a represents the adaptive control law and f_θ is the parameter update law. The controller may have been created using either a direct or indirect approach. It may also be defined using an approximator such as a fuzzy system or neural network. We will require, however, that the controller is Lipschitz continuous in z. As in the non-adaptive output-feedback case, this will allow us to study the stability of the closed-loop system when some of the states are replaced by their estimates. Since we are dealing with adaptive stabilization for the moment (rather than tracking), the vector z containing the measurable signals is simply given by x. Therefore, for the sake of clarity, throughout the rest of this section we will replace z by x.

Letting $\tilde{\theta} = \hat{\theta} - \theta$ represent, as usual, the parameter estimation error, we assume that (11.34) has the property that there exists a continuously differentiable Lyapunov-like function $V_a(x, \tilde{\theta})$ such that

$$\begin{aligned}\gamma_x^1(|x|) + \gamma_{\tilde{\theta}}^1(|\tilde{\theta}|) \le V_a(x, \tilde{\theta}) \le \gamma_x^2(|x|) + \gamma_{\tilde{\theta}}^2(|\tilde{\theta}|) \\ \dot{V}_a \le -k_1 V_a + k_2 \quad \text{for all } x \in B_r,\end{aligned} \tag{11.35}$$

where $\gamma_x^i(|x|)$, $\gamma_{\tilde{\theta}}^i(|\tilde{\theta}|)$, $i = 1, 2$, are class-$\mathcal{K}_\infty$, and B_r is the compact set where the approximator is assumed to be valid (see Chapters 7 and 8). The reader will notice that the adaptive controllers developed in Chapter 7 and Chapter 8 enjoy the properties in (11.35). Furthermore, as pointed out there, the second inequality in (11.35) implies that if k_2 is small enough the trajectories of the closed-loop system are uniformly ultimately bounded (UUB) and asymptotically approach the compact set

$$\mathcal{N} \triangleq \left\{ x \in \mathbb{R}^n, \tilde{\theta} \in \mathbb{R}^p \,:\, V_a(x, \tilde{\theta}) \le \frac{k_2}{k_1} \right\}. \tag{11.36}$$

To illustrate this concept, let $\alpha_1 = k_2/k_1$ and define

$$\alpha_2 = \min_{x \in \partial B_r,\, \tilde{\theta} \in \mathbb{R}^p} V_a,$$

(∂B_r denotes the boundary of the set B_r) so that $\{x \in \mathbb{R}^n, \tilde{\theta} \in \mathbb{R}^p : V_a \le \alpha_2\}$ is the largest level set of V_a guaranteeing, for all $\tilde{\theta}$, that $x \in B_r$. Notice that when $V_a \ge \alpha_1$, we find $\dot{V}_a \le 0$ by (11.35). Suppose k_2 is small enough that $\alpha_1 < \alpha_2$, and let c be any scalar satisfying $1 < c \le \alpha_2/\alpha_1$. From the definition of α_2 we have that $c\alpha_1 \leq V_a \leq \alpha_2$ implies $\dot{V}_a \leq -k_2(c-1) < 0$, thus implying that the closed-loop trajectories $(x(t), \tilde{\theta}(t))$ enter the set

$$\left\{x \in \mathbb{R}^n, \tilde{\theta} \in \mathbb{R}^p \,:\, V_a(x, \tilde{\theta}) \leq c\alpha_1\right\}$$

in finite time. Since this holds for any $1 < c \leq \alpha_2/\alpha_1$, we conclude that the closed-loop trajectories asymptotically approach the set $\Omega = \{V_a \leq \alpha_1\}$. Next, in order to extend the tools of Chapter 10 to the adaptive case, we need to modify V_a and construct a new Lyapunov-like function V_a' which is positive definite with respect to the set Ω, proper on the set $\{V_a \leq \alpha_2\}$, and such that its time derivative is negative on the set $\{\alpha_1 < V_a < \alpha_2\}$. In other words, we want to have

$$\begin{aligned} &\text{(i)} \quad V_a' = 0 \iff [x^\top, \tilde{\theta}^\top]^\top \in \mathcal{N} \\ &\text{(ii)} \quad V_a' = \infty \text{ on } \{V_a = \alpha_2\} \\ &\text{(iii)} \quad \dot{V}_a' < 0 \text{ on } \{\alpha_1 < V_a < \alpha_2\}. \end{aligned}$$

Choose

$$V_a' = \frac{\max\{0, V_a - \alpha_1\}}{\max\{\alpha_2 - V_a, 0\}}, \tag{11.37}$$

which clearly satisfies (i) and (ii) above. Now calculate the time derivative of V_a' on the set $\{\alpha_1 < V_a < \alpha_2\}$

$$\dot{V}_a' \;=\; \frac{\partial V_a'}{\partial V_a}\dot{V}_a = \frac{\alpha_2 - \alpha_1}{(\alpha_2 - V_a)^2}(-k_1 V_a + k_2) < 0,$$

and thus V_a' satisfies (iii) above. Compare now the properties of V_a' just defined to those of V in (10.112), (10.113) and observe that the only difference between the two is that V_a' is continuous but not continuously differentiable. It is easy to prove that the following function

$$V_a'' = \frac{(V_a - \alpha_1)^2}{(\alpha_2 - \alpha_1)^2 - (V_a - \alpha_1)^2}$$

has the same properties of V_a' and it is continuosly differentiable. Following the developments of Chapter 10, we have that there exist two $\mathcal{K}_\infty$ functions γ_1 and γ_2 and a positive function γ_3 satisfying

$$\begin{aligned} \gamma_1(\omega_{\mathcal{N}}(x, \tilde{\theta})) \leq V_a'(x, \tilde{\theta}) &\leq \gamma_2(\omega_{\mathcal{N}}(x, \tilde{\theta})) \\ \dot{V}_a' &\leq -\gamma_3(\omega_{\mathcal{N}}(x, \tilde{\theta})), \end{aligned} \tag{11.38}$$

where $\omega_{\mathcal{N}}(x, \tilde{\theta})$ is a positive definite function with respect to $\mathcal{N} = \{V_a \leq \alpha_1\}$ and proper on $\mathcal{D} = \{V_a < \alpha_2\}$.

In the spirit of the separation principle introduced in Chapter 10 we now try to find an adaptive output-feedback controller that relies on the estimation of x to recover the performance of (11.34). When using output feedback, the adaptive stabilization problem becomes more complex and more restrictive assumptions than the state feedback case are needed. In what follows we assume that (11.33) has special structures which will allow us to define a separation principle for adaptive systems.

11.3.1 Full State-Feedback Performance Recovery

Partition the state of the plant in two parts, $x = [{x_1}^\top, {x_2}^\top]^\top$, where $x_1 \in \mathsf{R}^{n_1}$, $x_2 \in \mathsf{R}^{n_2}$, $n_1 + n_2 = n$, and rewrite (11.33) as

$$\begin{aligned}\dot{x}_1 &= f^1(x_1, x_2, u, \Delta)\\ \dot{x}_2 &= f^2(x_1, x_2, u, \Delta).\end{aligned} \tag{11.39}$$

Consider now the case when the state of the upper subsystem is not directly measurable while the state of the lower subsystem is, and hence only x_1 needs to be estimated. In other words, assume that the output of the system is given by

$$y = \begin{bmatrix} y_1 \\ y_2 \end{bmatrix} = \begin{bmatrix} h^1(x_1, x_2, u) \\ x_2 \end{bmatrix}, \tag{11.40}$$

where $y_1 \in \mathsf{R}^{\rho_1}$, so that $y \in \mathsf{R}^{\rho}$, and $\rho = \rho_1 + n_2$. The components of the vector y_1 will be denoted by $y_{1,i}$, $i = 1, \ldots, \rho_1$, and a similar notation will be used, when needed, for other vectors. Assume further that the observability mapping associated with the upper subsystem is a diffeomorphism and does not depend on the uncertainty Δ, that is,

$$y_{e,1} \triangleq \begin{bmatrix} y_{1,1} \\ \vdots \\ y_{1,1}^{(k_1-1)} \\ \vdots \\ y_{1,\rho_1} \\ \vdots \\ y_{1,\rho_1}^{(k_{\rho_1}-1)} \end{bmatrix} = \begin{bmatrix} h_1^1(x_1, x_2, u) \\ \varphi_1^1(x_1, x_2, u, u^{(1)}) \\ \vdots \\ \varphi_1^{k_1-1}\left(x_1, x_2, u, \ldots, u^{(k_1-1)}\right) \\ \vdots \\ h_{\rho_1}^1(x_1, u) \\ \vdots \\ \varphi_{\rho_1}^{k_{\rho_1}-1}\left(x_1, x_2, u, \ldots, u^{(k_\rho-1)}\right) \end{bmatrix} \triangleq \mathcal{H}_1(x_1, x_2, s), \tag{11.41}$$

where, as usual, $s \triangleq [u_1, \ldots, u_1^{(\bar{n}_1-1)}, \ldots, u_m, \ldots, u_m^{(\bar{n}_m-1)}]^\top \in \mathsf{R}^{n_u}$, $\sum_{i=1}^{\rho_1} k_i = n_1$, $n_u \triangleq \bar{n}_1 + \ldots + \bar{n}_m$, $0 \le \bar{n}_i \le \max\{k_1, \ldots, k_\rho\}$, and the functions φ_i^j are defined as in (10.81).

Example 11.1 For single-input single-output systems, the situation described above arises when, for instance, the plant is described by a differential equation in input-output form of the type

$$y_1^{(n_1)} = \bar{f}\left(y_1, \dot{y}_1, \dots, y_1^{(n_1-1)}, v, \dot{v}, \dots, v^{(n_2)}, \Delta\right), \tag{11.42}$$

where v is the plant input, y_1 is the measurable output, and Δ is the disturbance or modeling uncertainty affecting the plant. With the definitions

$$x_1 = \begin{bmatrix} y \\ \dot{y} \\ \vdots \\ y^{(n_1-1)} \end{bmatrix}, \quad x_2 = \begin{bmatrix} v \\ \dot{v} \\ \vdots \\ v^{(n_2-1)} \end{bmatrix}, \quad v^{(n_2)} = u \tag{11.43}$$

the input-output system is transformed into the state-space representation

$$\begin{aligned} \dot{x}_{1,1} &= x_{1,2} \\ &\vdots \\ \dot{x}_{1,n_1} &= \bar{f}(x_1, x_2, u, \Delta) \\ \dot{x}_{2,1} &= x_{2.2} \\ &\vdots \\ \dot{x}_{2,n_2} &= u, \end{aligned} \tag{11.44}$$

which has the form (11.39). Notice that since the control input u is known, the state x_2 of the lower subsystem is available for feedback as it can be obtained by using a chain of $n_2 - 1$ integrators. Moreover, the observability mapping of the upper subsystem is simply given by the identity mapping

$$y_{e,1} = \mathcal{H}_1(x_1) = [x_{1,1}, \dots, x_{1,n_1}]^\top,$$

which is independent of the disturbance Δ and is clearly a diffeomorphism. Thus, the input-output system satisfies the assumptions above. $\triangle$

With the assumptions above and the definition of V_a' we can use the ideas developed in Section 10.7 to estimate x_1 from y_1 and hence recover the performance of full state feedback controllers (recall that we assumed x_2 to be available for feedback). Let Δ^0 denote the known nominal value of the vector Δ, possibly dependent on x_1 and x_2 (if we have no *a priori* knowledge about Δ, then we can set $\Delta^0 = 0$), and given any scalar $c > 0$, let Ω_c denote the level set $\{V_a' \leq c\}$ and notice that Ω_c is compact. In analogy to what was done in Section 10.7, let $\Omega_c^{x_1}$ denote the projection of the set Ω_c onto R^{n_1} defined as

$$\Omega_c^{x_1} = \left\{x_1 \in \mathsf{R}^{n_1} : [x_1^\top, x_2^\top, \tilde{\theta}^\top]^\top \in \Omega_c, \text{ for all } x_2 \in \mathsf{R}^{n_2} \text{ and all } \tilde{\theta} \in \mathsf{R}^p\right\} \tag{11.45}$$

Using a separation principle and the fact that x_2 is directly available for feedback, we can now employ the estimate of x_1 in the state-feedback adaptive controller (11.34) to define an adaptive output-feedback controller. To this end, recall the definition of Ω^s in Section 10.6, pick any two positive scalars $c_1 < c_2$ and a convex compact set $\mathcal{C}$ satisfying

$$\mathcal{F}(\Omega^{x_1}_{c_2} \times \Omega^s) \subset \mathcal{C},$$

where

$$\mathcal{F} = \begin{bmatrix} \mathcal{H}(x_1, s) \\ s \end{bmatrix}.$$

Thus $\mathcal{C}$ is a compact space that contains all $\mathcal{F}$ when $V_a' < c$. Now consider the adaptive output-feedback controller

$$\begin{aligned}
\dot{\hat{x}}_1 &= f^1(\hat{x}_1, x_2, \hat{u}, \Delta^0) + \left[\frac{\partial \mathcal{H}_1(\hat{x}_1, x_2, \hat{s})}{\partial \hat{x}_1}\right]^{-1} \Bigg[(\mathcal{E}^{x_1})^{-1} L\,(y_1 - \hat{y}_1) \\
&\qquad - \frac{\partial \mathcal{H}_1(\hat{x}_1, x_2, \hat{s})}{\partial \hat{s}} (\mathcal{E}^s)^{-1} K C(u - \hat{u})\Bigg] \\
\hat{y}_1 &= h^1(\hat{x}_1, x_2, \hat{u}) \\
\dot{\hat{s}} &= A\hat{s} + (\mathcal{E}^s)^{-1} K(u - \hat{u}) \\
\hat{u} &= C\hat{s}, \\
\begin{bmatrix} \dot{\hat{x}}_1^P \\ \dot{\hat{s}}^P \end{bmatrix} &= \mathcal{P}_{\mathcal{C}}\,(\dot{\hat{x}}_1^P, \dot{\hat{s}}^P, \hat{x}_1^P, \hat{s}^P)
\end{aligned} \tag{11.46}$$

$$\begin{aligned}
\dot{\hat{\theta}} &= f_\theta(\hat{\theta}, \hat{x}_1^{P\top}, x_2^\top]^\top) \\
u &= \nu_a\left([\hat{x}_1^{P\top}, x_2^\top]^\top, \hat{\theta}\right).
\end{aligned} \tag{11.47}$$

The map $\mathcal{P}$ represents the dynamic projection operation (10.105) performed over the set $\mathcal{C}$. The estimator for x_1 used here is structurally identical to the one in (10.117), (10.118) (refer to Section 10.6 for a definition of $L, K, A, C, \mathcal{E}^{x_1}, \mathcal{E}^s$), just recall that its performance is governed by the choice of two design parameters, η_1 and η_2. In preparation for the stability analysis, let $x_a = [x_1, x_2, \tilde{\theta}]^\top$, $\hat{x}_a^P = [\hat{x}_1^P, x_2, \tilde{\theta}]^\top$, and

$$\begin{aligned}
F(x_a, u, \Delta) &= \begin{bmatrix} f^1(x_1, x_2, u, \Delta) \\ f^2(x_1, x_2, u, \Delta) \\ f_\theta(\tilde{\theta} + \theta, x_1, x_2) \end{bmatrix} \\
\hat{F}(x_a, \hat{x}_1^P, u, \Delta) &= \begin{bmatrix} f^1(x_1, x_2, u, \Delta) \\ f^2(x_1, x_2, u, \Delta) \\ f_\theta(\tilde{\theta} + \theta, \hat{x}_1^P, x_2) \end{bmatrix}
\end{aligned}$$

so that the closed-loop system using state feedback can be rewritten as

$$\dot{x}_a = F(x_a, \nu_a(x_a), \Delta), \tag{11.48}$$

and the closed-loop system using the output-feedback controller (11.47) reads as

$$\dot{x}_a = \hat{F}(x_a, \hat{x}_1^P, \nu_a(\hat{x}_a^P), \Delta). \tag{11.49}$$

Observe now that $\hat{F}(x_a, x_1, \nu_a(x_a), \Delta) = F(x_a, \nu_a(x_a), \Delta)$ and that, by the Lipschitz continuity of f^1, f^2, f_θ, and ν_a, there exists a positive scalar $\bar{\gamma}$ such that, for all x_a in Ω_{c_2} and $\hat{x}_1^P \in \mathcal{H}_1^{-1}(\mathcal{C})$,

$$\left|\hat{F}(x_a, \hat{x}_1^P, \nu_a(\hat{x}_a^P), \Delta) - F(x_a, \nu_a(x_a), \Delta)\right| \leq \bar{\gamma}|\hat{x}_1^P - x_1|.$$

Furthermore, note that the gradient of V_a' is bounded on any compact set and let A denote the upper bound on its norm over the compact set Ω_{c_2}, that is, for all x_a in Ω_{c_2},

$$\left|\frac{\partial V_a'}{\partial x_a}\right| \leq A|x_a|.$$

We are now ready to state the closed-loop stability theorem.

Theorem 11.1: *Consider the closed-loop system formed by the plant (11.39) and the adaptive output-feedback controller (11.46), (11.47). Then, given any triple (c_1, c_2, ϵ) such that $0 < c_1 < c_2$ and $0 < \epsilon < \gamma_3 \circ \gamma_2^{-1}(c_1)/A\bar{\gamma}$, there exist scalars $\eta_1^* \in (0,1]$ and $\eta_2^* \in (0,1]$ such that, for all $\eta_1 \in (0, \eta_1^*]$ and $\eta_2 \in (0, \eta_2^*]$, every trajectory starting in $\Omega_{c_1} = \{V_a' \leq c_1\}$ is bounded within $\Omega_{c_2} = \{V_a' \leq c_2\}$ and asymptotically approaches the positively invariant set Ω_{d_ϵ}, where $d_\epsilon = \gamma_2 \circ \gamma_3^{-1}(A\,\bar{\gamma}\,\epsilon)$.*

Proof: The logic of this proof is the same of that in Lemma 10.2. From the assumptions made on the observability mapping $\mathcal{H}_1$ associated with the x_1 subsystem, we have that (11.46) is an estimator for x_1 enjoying the properties of Theorem 10.3 and Lemma 10.4. As in the proof of Lemma 10.2, notice that $V_a'(x_a(0)) \leq c_1 < c_2$ and there exists a positive exit time from the set Ω_{c_2}. Since $\hat{x}_1^P$ is constrained to lie inside $\mathcal{H}_1^{-1}(\mathcal{C})$, $|\hat{x}_1^P|$ has an upper bound which does not depend on η_1 and η_2, and hence there exists a uniform upper bound T_1 to the exit time which is independent of η_1 and η_2, and such that $V_a'(x_a(t)) \leq c_2$, for all $t \in [0, T_1)$ and for all $\eta_1, \eta_2 \in (0,1]$. Choose T_0 such that $0 < T_0 < T_1$. Then, in the time interval $t \in [0, T_1)$, we can apply Theorem 10.3, part (ii), and conclude that for any positive ϵ there exist $\eta_1^*, \eta_2^* \in (0,1]$ such that, for all $\eta_1 \in (0, \eta_1^*]$ and all $\eta_2 \in (0, \eta_2^*]$, $|\hat{x}_1^P - x_1| \leq \epsilon, \forall t \in [T_0, T_1)$. Hence, for all $t \in [T_0, T_1)$, we have that $V_a'(x_a(t)) \leq c_2$ and $|\hat{x}_1^P(t) - x_1(t)| \leq \epsilon$. It now remains to show that this latter fact guarantees that the domain Ω_{c_2} is positively invariant and, hence, $T_1 = \infty$. To this end, take the time derivative of $V_a'(x_a)$ along the trajectories of the closed-loop system for $t \in [T_0, T_1)$

$$\begin{aligned}
\dot{V}_a' &= \frac{\partial V_a'}{\partial x_a} F\left(x_a, \nu_a(x_a), \Delta\right) + \frac{\partial V_a'}{\partial x_a}\left[\hat{F}(x_a, \hat{x}_1^P, \nu_a(\hat{x}_a^P), \Delta)\right. \\
&\qquad \left. - F\left(x_a, \nu_a(x_a), \Delta\right)\right] \\
&\le -\gamma_3(\omega_{\mathcal{N}}(x, \tilde{\theta})) + \left|\frac{\partial V_a'}{\partial x_a}\right| \left|\hat{F}(x_a, \hat{x}_1^P, \nu_a(\hat{x}_a^P), \Delta) - F\left(x_a, \nu_a(x_a), \Delta\right)\right| \\
&\le -\gamma_3(\omega_{\mathcal{N}}(x, \tilde{\theta})) + A\bar{\gamma}|\hat{x}_1^P - x_1| \\
&\le -\gamma_3 \circ \gamma_2^{-1}(V_a') + A\bar{\gamma}\epsilon.
\end{aligned} \tag{11.50}$$

Since, on the boundary of the set $\{V_a' \le c_2\}$, $\dot{V}_a' \le -\gamma_3 \circ \gamma_2^{-1}(c_2) + A\bar{\gamma}\epsilon$ which is strictly negative, we have that the set Ω_{c_2} is positively invariant, and hence $T_1 = \infty$ and the closed-loop trajectories $x_a(t)$ are bounded. Furthermore, from the last inequality in (11.50) we conclude that the trajectories $x_a(t)$ asymptotically approach the set $\{V_a' \le d_\epsilon\}$, thus concluding the proof of the theorem. ■

From the fact that the constant d_ϵ can be made arbitrarily small, we have that Theorem 11.1 proves practical stability of the set $\mathcal{N}$ (defined by (11.36)) under output-feedback control. Recall that, when $c_1 \to \infty$, we find Ω_{c_1}, the estimate of the domain of attraction for Ω_{d_ϵ}, coincides with the set $\{V_a \le \alpha_2\}$, the estimate of the domain of attraction for $\{V_a \le \alpha_1\}$ using state feedback. Hence, by choosing the estimator design parameters η_1 and η_2 sufficiently small, the adaptive output-feedback controller (11.47) recovers the performance of the adaptive state-feedback controller.

Example 11.2 An important class of nonlinear systems satisfying the assumptions of Theorem 11.1 is given by systems in adaptive tracking form (11.1). In order to see that, let $u' = \sigma(y)u$ and note that (11.1) can be rewritten in input-output form as

$$y^{(n)} = \psi_0(y, \ldots, y^{(n-1)}) + \sum_{i=1}^{p} \psi_i(y, \ldots, y^{(n-1)})\theta_i + \sum_{i=0}^{m} d_i u'^{(i)}, \tag{11.51}$$

where ψ_i, $0 \le i \le p$, are smooth functions. The proof of this fact is rather straightforward and is left as an exercise. Note now that (11.51) is a particular type of system in input-output form (11.42) and therefore it can be represented in the form (11.39), where Δ is simply a vector of unknown constant parameters,

$$\Delta = [\theta_1, \ldots, \theta_p, d_0, \ldots, d_m]^\top,$$

and

$$\begin{aligned}
\dot{x}_{1,1} &= x_{1,2} \\
&\vdots \\
\dot{x}_{1,n} &= \psi_0(x_1) + \sum_{i=1}^{p} \psi_i(x_1)\theta_i + \sum_{i=0}^{m-1} d_i x_{2,i+1} + d_m U \\
\dot{x}_{2,1} &= x_{2,2} \\
&\vdots \\
\dot{x}_{2,m} &= U,
\end{aligned} \tag{11.52}$$

where U denotes the m-th time derivative of u'. By adding a chain of m integrators at the input side we have that U is our new control input and x_2 is available for feedback. In order to solve the adaptive state-feedback stabilization problem, it is sufficient to drive the state of the upper subsystem to the origin. The exponential stability of the zero dynamics of (11.52) guarantees then that the state of the lower subsystem is bounded and converges to the origin as well.

Next, we will show that we can use the tools of Chapter 7 to develop an adaptive controller for the upper subsystem in (11.52). Since we are considering a stabilization problem, we take $e = x_1$ and we rewrite the upper subsystem as

$$\dot{e} = \alpha(t, x_1) + \beta U, \tag{11.53}$$

where the time dependence in α comes from the presence of $x_2(t)$, viewed as an external measurable signal. Assumption 7.1 is satisfied with $k_2 = 0$ since, if Δ was known, a static stabilizing controller would be given by

$$\nu_s(x, \Delta) \triangleq -\frac{1}{d_m}\left[\psi_0(x_1) + \sum_{i=1}^{p} \psi_i(x_1)\theta_i + \sum_{i=0}^{m-1} d_i x_{2,i+1} + k^\top x_1\right],$$

where $k \in \mathbb{R}^n$ is any vector such that the polynomial $s^n + k_n s^{n-1} + \ldots + k_1$ is Hurwitz. Notice that, when $U = \nu_s$, the upper subsystem becomes linear

$$\dot{x}_1 = A_0 x_1,$$

and A_0 is a Hurwitz matrix with eigenvalues given by the roots of $s^n + k_n s^{n-1} + \ldots + k_1$. Hence, $V_s = x_1^\top P x_1$, where P is the solution of

$$A_0^\top P + P A_0 = -I,$$

is a Lyapunov function for the closed-loop system. Further, Assumption 7.2 is satisfied with $c = 1/d_m$ and $\bar{\beta} = [0, \ldots, 0, 1]^\top$. Next, we apply Theorem 7.2 to find an adaptive controller of the type

$$\begin{aligned}\dot{\hat{\theta}} &= f_\theta(\hat{\theta}, x_1, x_2)\\ U &= \mathcal{F}(x_1, x_2, \hat{\theta}),\end{aligned} \tag{11.54}$$

where $\hat{\theta} \in \mathsf{R}^{p+m+1}$, such that the origin of the upper subsystem is globally asymptotically stable. To this end, we employ the following approximator

$$\mathcal{F} = \hat{\theta}^\top \Psi(x_1, x_2) - \eta \left(\frac{\partial V_s}{\partial e}\bar{\beta}\right)^\top \tag{11.55}$$

where $\eta > 0$,

$$\Psi(x_1, x_2) = -\left[\psi_0(x_1) + k^\top x_1, \psi_1(x_1), \ldots, \psi_p(x_1), x_{2,1}, \ldots, x_{2,m}\right]^\top,$$

and $\hat{\theta}$ is the estimate of the vector

$$\Theta = \left[\frac{1}{d_m}, \frac{\theta_1}{d_m}, \ldots, \frac{\theta_p}{d_m}, \frac{d_0}{d_m}, \ldots, \frac{d_{m-1}}{d_m}\right]^\top.$$

The particular approximation structure (11.55) has the property that, when $\hat{\theta} = \Theta$, we have $\mathcal{F} = \nu_s - \eta[(\partial V_s/\partial e)\bar{\beta}]^\top$ with zero representation error over the entire state space. Thus, in Theorem 7.2, we can set $W = 0$ and $B_r = \mathsf{R}^{n+m+1}$ and conclude that the update law (7.15), with $\sigma = 0$, guarantees global asymptotic stability of the origin of the upper subsystem.

Now defining an adaptive output-feedback controller using the results of this section is a rather straightforward task and is left as an exercise. Here, we just highlight the fact that the output-feedback controller has the property the the origin of the closed-loop system is semiglobally practically stable.

$\triangle$

The example above shows that the tools developed in this section and Chapter 7 can be applied instead of the adaptive output-feedback backstepping technique seen earlier for the stabilization of systems in adaptive tracking form. The control design here is simpler than the recursive backstepping design, but the stability result is slightly weaker (semiglobal practical stability rather than global asymptotic stability). An advantage of Theorem 11.1, besides the design simplicity, is that it deals with more general classes of nonlinear systems.

11.3.2 Partial State-Feedback Performance Recovery

In this section we consider the situation, which frequently arises in nonlinear control, when the adaptive state-feedback controller uses partial state feedback to achieve a given control objective. Suppose that the plant can be partitioned as the interconnection of two subsystems

$$\begin{aligned}\dot{x}_1 &= f^1(x_1, x_2, u, \Delta)\\ \dot{x}_2 &= f^2(x_1, x_2, \Delta)\\ y &= h(x_1, u),\end{aligned} \tag{11.56}$$

where $x_1 \in \mathsf{R}^{n_1}$, $x_2 \in \mathsf{R}^{n_2}$, and $n_1 + n_2 = n$. In contrast to the class of systems considered in the previous section, here the lower subsystem does not depend on the control input u and x_2 is not available for feedback (thus, for instance, x_2 may be thought of as a dynamic uncertainty). Assume that, for the state-feedback adaptive controller (11.34), the variable z be given by x_1 only, and that the properties in (11.35) still hold for the entire state $x = [x_1^\top x_2^\top]^\top$ of the system. Further, assume that the observability mapping associated with the upper subsystem is a diffeomorphism and does not depend on x_2 and the uncertainty Δ, that is,

$$y_{e,1} \triangleq \left[y_1, \ldots, y_1^{(k_1-1)}, \ldots, y_p, \ldots, y_p^{(k_p-1)}\right]^\top = \mathcal{H}_1(x_1, s), \tag{11.57}$$

where k_i and s are defined in the usual way.

Example 11.3 For single-input single-output systems, the situation described above arises when, for instance, the system is described by differential equations with a part in lower-triangular form

$$\begin{aligned}\dot{\xi}_1 &= f_1(\xi_1) + g_1(\xi_1)\xi_2\\ &\;\;\vdots\\ \dot{\xi}_{n_1-1} &= f_{n_1-1}(\xi_1, \ldots, \xi_{n_1-1}) + g_{n_1-1}(\xi_1, \ldots, \xi_{n_1-1})\,\xi_n\\ \dot{\xi}_{n_1} &= f_{n_1}(\xi) + g_{n_1-1}(\xi)\,(\Delta_1(\xi) + \Pi_1(\xi)u)\\ \dot{x}_2 &= f^2(\xi, x_2, \Delta_2(\xi))\\ y = \xi_1,&\end{aligned} \tag{11.58}$$

which, letting $x_1 = [\xi_1, \ldots, \xi_{n_1}]^\top$, has the form (11.56). The functions g_i are bounded away from zero, the unknown functions $\Delta_1(x_1)$, $\Delta_2(x_1)$, $\Pi_1(x_1)$ are assumed to be smooth, and the x_2-subsystem is assumed to be input-to-state practically stable with respect to x_1. The class of systems represented by (11.58) is similar to that considered in Section 8.3.2 (there, however, the x_2 subsystem was not included), but here the uncertainty is not allowed to affect the functions f_i and g_i for $i = 1, n_1 - 1$. The lower subsystem may represent unmodeled dynamics whose state x_2 is not available for feedback.

The observability mapping associated with the upper subsystem is given by

$$\mathcal{H}_1(x_1) = \begin{bmatrix} \xi_1 \\ f_1(\xi_1) + g_1(\xi_1)\xi_2 \\ \vdots \\ f_{n_1-1}(\xi_1, \ldots, \xi_{n_1-1}) + g_{n_1-1}(\xi_1, \ldots, \xi_{n_1-1})\,\xi_n \end{bmatrix},$$

which, from the smoothness of f_i, g_i and the fact that all g_i are bounded away from zero, is a diffeomorphism. Notice also that in this particular case $\mathcal{H}_1$ does not depend on u. Next, by applying the stabilization technique illustrated in Section 8.3.2, we get an adaptive state-feedback controller for the x_1-subsystem

$$\begin{aligned} \dot{\hat{\theta}} &= f_\theta(\hat{\theta}, x_1) \\ u &= \nu_a(x_1, \hat{\theta}), \end{aligned} \tag{11.59}$$

and a Lyapunov-like function $V_1(x_1, \tilde{\theta})$ such that

$$\begin{aligned} \gamma^1_{x_1}(|x_1|) + \gamma^1_{\tilde{\theta}}(|\tilde{\theta}|) \le V_1(x_1, \tilde{\theta}) &\le \gamma^2_{x_1}(|x_1|) + \gamma^2_{\tilde{\theta}}(|\tilde{\theta}|) \\ \dot{V}_1 \le -k_1 V_1 + k_2 \quad &\text{on } \{V_1 \le \alpha_2\}. \end{aligned} \tag{11.60}$$

In other words, the adaptive controller (11.59) guarantees the uniform ultimate boundedness of x_1. In order to show that the class of systems in lower-triangular form (11.58) with the controller (11.59) fulfill our assumptions, we have to find a Lyapunov-like function V_a which depends on both x_1 and x_2 and enjoys the properties in (11.35). To this end, observe that since the function Δ_2 is smooth, if x_1 is bounded, $\Delta_2(x_1)$ is bounded. Then, since the x_2-subsystem is input-to-state practically stable, we know that there exists a continuously differentiable function $V_2(x_2)$ and constants $c > 0$, $d \ge 0$ such that

$$\begin{aligned} \gamma^1_{x_2}(|x_2|) \le V_2(x_2) &\le \gamma^2_{x_2}(|x_2|) \\ \dot{V}_2(x_2) &\le -cV_2 + \psi(|x_1|) + d, \end{aligned} \tag{11.61}$$

where $\gamma^1_{x_2}$, $\gamma^2_{x_2}$, and ψ are class $\mathcal{K}_\infty$. Considering now the whole system (11.58), choose $V_a(x, \tilde{\theta}) = V_1(x_1, \tilde{\theta}) + V_2(x_2)$ and notice that

$$\gamma^1_x(|x|) + \gamma^1_{\tilde{\theta}}(|\tilde{\theta}|) \le V_a(x, \tilde{\theta}) \le \gamma^2_x(|x|) + \gamma^2_{\tilde{\theta}}(|\tilde{\theta}|), \tag{11.62}$$

where

$$\gamma_x^1(|x|) = \gamma_{x_1}^1(|x_1|) + \gamma_{x_2}^1(|x_2|), \ \gamma_x^2(|x|) = \gamma_{x_1}^2(|x_1|) + \gamma_{x_2}^2(|x_2|).$$

Furthermore, on the set $\{V_1 \leq \alpha_2\}$,

$$\begin{aligned} \dot{V}_a & \leq -k_1 V_1 + k_2 - cV_2 + \psi(|x_1|) + d \\ & \leq -k_1 V_1 - cV_2 + \psi \circ \left(\gamma_{x_1}^1\right)^{-1}(\alpha_2) + k_2 + d \\ & \leq -\min\{k_1, c\} V_a + \psi \circ \left(\gamma_{x_1}^1\right)^{-1}(\alpha_2) + k_2 + d. \end{aligned} \tag{11.63}$$

Set $\bar{k}_1 = \min\{k_1, c\}$, $\bar{k}_2 = \psi \circ \left(\gamma_{x_1}^1\right)^{-1}(\alpha_2) + k_2 + d$, and note that $V_a \leq \alpha_2$ implies $V_1 \leq \alpha_2$ and thus the inequality above reads as

$$\dot{V}_a \leq -\bar{k}_1 V_a + \bar{k}_2 \text{ on } \{V_a \leq \alpha_2\}. \tag{11.64}$$

From (11.62) and (11.64) we conclude that systems in lower-triangular form (11.58) satisfy our assumptions. △

For a system with the structure (11.56), satisfying the assumptions made above, it is easy to define an adaptive output feedback controller using, as we did in Section 11.3.1, a separation principle. We start by using V_a to construct a continuous function $V_a'(x_1, x_2, \tilde{\theta})$ as in (11.37), defining its generic level set $\Omega_c = \{V_a' \leq c\}$ (where c is any positive scalar) and its projection $\Omega_c^{x_1}$ onto R^{n_1} (see (11.45)). Given any pair of scalars c_1 and c_2 such that $0 < c_1 < c_2$, choose $\mathcal{C}$ satisfying

$$\mathcal{F}(\Omega_{c_2}^{x_1} \times \Omega^s) \subset \mathcal{C},$$

and employ the adaptive state-feedback controller (11.34) (where $z = x_1$), together with an estimator for x_1 and a dynamic projection, to define the following adaptive output-feedback controller

$$\begin{aligned} \dot{\hat{x}}_1 &= f^1(\hat{x}_1, 0, \hat{u}, \Delta^0) + \left[\frac{\partial \mathcal{H}_1(\hat{x}_1, \hat{s})}{\partial \hat{x}_1}\right]^{-1} \left[(\mathcal{E}^{x_1})^{-1} L\,(y - \hat{y}) \right. \\ & \qquad \left. - \frac{\partial \mathcal{H}_1(\hat{x}_1, \hat{s})}{\partial \hat{s}} (\mathcal{E}^s)^{-1} K C(u - \hat{u})\right] \\ \hat{y} &= h(\hat{x}_1, \hat{u}) \\ \dot{\hat{s}} &= A\hat{s} + (\mathcal{E}^s)^{-1} K(u - \hat{u}) \\ \hat{u} &= C\hat{s}, \\ \begin{bmatrix} \dot{\hat{x}}_1^P \\ \dot{\hat{s}}^P \end{bmatrix} &= \mathcal{P}_\mathcal{C}(\dot{\hat{x}}_1^P, \dot{\hat{s}}^P, \hat{x}_1^P, \hat{s}^P) \end{aligned} \tag{11.65}$$

$$\begin{aligned} \dot{\hat{\theta}} &= f_\theta(\hat{\theta}, \hat{x}_1^P) \\ u &= \nu_a\left(\hat{x}_1^P, \hat{\theta}\right), \end{aligned} \tag{11.66}$$

where $\mathcal{P}_C$ is defined as in (10.105), $L, K, A, C, \mathcal{E}^{x_1}, \mathcal{E}^s$ are defined as in Section 10.6. The unknown state x_2 is replaced by 0 in the estimator (11.65), and the uncertainty Δ is replaced by its nominal value Δ^0. The assumption we made that the observability mapping $\mathcal{H}_1$ does not depend on x_2 turns out to be crucial for the solution of the adaptive stabilization problem by output feedback, as it allows us to treat x_2 as an *unknown disturbance* and use Theorem 10.3 to guarantee that, provided η_1 and η_2 are chosen sufficiently small, (11.65) achieves an arbitrarily good estimation of x_2 with arbitrarily fast convergence. This property in turn allows us to employ the same analysis carried out in the proof of Theorem 11.1 to prove the following result.

Theorem 11.2: *Consider the closed-loop system formed by the plant (11.56) and the adaptive output-feedback controller (11.65), (11.66). Then, given any triple (c_1, c_2, ϵ) such that $0 < c_1 < c_2$, and $\epsilon > 0$ is sufficiently small, there exist scalars $\eta_1^* \in (0,1]$, $\eta_2^* \in (0,1]$, and a class-$\mathcal{K}_\infty$ function γ such that, for all $\eta_1 \in (0, \eta_1^*]$ and $\eta_2 \in (0, \eta_2^*]$, every trajectory starting in Ω_{c_1} is bounded within Ω_{c_2} and asymptotically approaches the positively invariant set $\Omega_{\gamma(\epsilon)}$.*

This result (whose proof is essentially identical to that of Theorem 11.1 and hence is omitted) shows that the adaptive output-feedback controller (11.65), (11.66) recovers the properties of the state-feedback controller (11.34).

Example 11.4 Consider again the problem, analyzed in Example 8.4, of designing a controller to accurately point an antenna driven by a permanent magnet DC motor and assume that an optical encoder is available to measure θ, but we want to avoid using the noisy signal coming from a tachometer to measure ω. Recall that the system is defined by

$$\begin{aligned}
\dot{\theta} &= \omega \\
\dot{\omega} &= \left(\frac{T_c \cos\phi}{J}\right)\sin(N\theta) + \left(\frac{T_c \sin\phi}{J}\right)\cos(N\theta) - \frac{q}{J} + \frac{u}{J} \\
\dot{q} &= \sigma\omega\left[1 - \frac{q}{T_f}\operatorname{sgn}(\omega)\right] \\
y &= \theta,
\end{aligned} \tag{11.67}$$

where the dynamic uncertainty Δ is given by

$$\Delta(t,\theta,\omega) = \left(\frac{T_c \cos\phi}{J}\right)\sin(N\theta) + \left(\frac{T_c \sin\phi}{J}\right)\cos(N\theta) - \frac{q(t)}{J},$$

and q is not available for feedback. Let $x_1 = [\theta, \omega]^\top$ and $x_2 = q$. In Example 8.4 we have designed a partial state-feedback controller achieving the property in (11.35). Here we are concerned with stabilization and hence we assume that r is a constant reference and we seek to drive the angle θ of the antenna to the desired value r while keeping all other variables bounded. The state-feedback controller developed in Example 8.4 is now given by

$$\begin{aligned} \dot{\hat{\theta}} &= -\Gamma\left[\left(\frac{\partial V_s}{\partial e}\beta\left(\frac{\partial \mathcal{F}_\Delta}{\partial \theta} - A_{\mathcal{F}}\right)\right)^\top + \sigma(\hat{\theta} - \theta^0)\right] \triangleq f_\theta(\hat{\theta}, x_1) \\ u &= \mathcal{F}_\Pi^{-1}(x_1, \hat{\theta})\left(\mathcal{F}_\Delta(x_1, \hat{\theta}) - \eta\left(\frac{\partial V_s}{\partial e}\beta\right)^\top + \nu_s(x_1)\right) \triangleq \nu_a(x_1, \hat{\theta}), \end{aligned} \tag{11.68}$$

where $e_1 = x_{1,1} - r$, $e_2 = x_{1,2} + \kappa(x_{1,1} - r)$, $\hat{\theta} = [\hat{\theta}_1, \hat{\theta}_2, \hat{\theta}_3, \hat{\bar{\theta}}^\top]^\top$,

$$\begin{aligned} &\nu_s(x_1) = -\kappa(e_2 - \kappa e_1) - \kappa e_2 - e_1, \quad V_s = \tfrac{1}{2}e_1^2 + \tfrac{1}{2}e_2^2, \\ &\frac{\partial \mathcal{F}_\Delta}{\partial \theta} = \left[\sin(N x_1), \cos(N x_1), 0, -\eta_\psi \frac{\partial \mathcal{F}_{f_s}}{\partial \bar{\theta}}\right]^\top \\ &A_{\mathcal{F}} = [0, 0, 1, 0, \ldots, 0]^\top, \ \beta = [0, 1]^\top, \end{aligned}$$

and $\Gamma, \sigma, \theta^0, \eta, \eta_\psi$ are design parameters. The function $\mathcal{F}_\Delta$ is used to approximate unknown nonlinearities and is given by

$$\mathcal{F}_\Delta = \hat{\theta}_1 \sin(N x_{1,1}) + \hat{\theta}_2 \cos(N x_{1,1}) - \eta_\psi \mathcal{F}_{fs}(V_s, \hat{\bar{\theta}})\left(\frac{\partial V_s}{\partial e}\beta\right)^\top,$$

and $\mathcal{F}_{f_s}(V_s, \hat{\theta})$ is a fuzzy system with input V_s and 10 fixed output membership functions with centers defined by $\bar{\theta}$.

The observability mapping of the upper subsystem is simply given by $\mathcal{H}_1(x_1) = [x_{1,1}, x_{1,2}]^\top$ which is the identity map and does not depend on x_2 and Δ. Hence, all our assumptions are satisfied and we can apply the controller (11.65), (11.66) to recover the performance of the state-feedback controller. The estimator (11.65) is in this case a standard high-gain observer

$$\begin{aligned} \dot{\hat{x}}_{1,1} &= \hat{x}_{1,2} + \frac{l_1}{\eta_1}(y - \hat{x}_{1,1}) \\ \dot{\hat{x}}_{1,2} &= \frac{u}{J} + \frac{l_2}{\eta_1^2}(y - \hat{x}_{1,1}), \end{aligned} \tag{11.69}$$

where the dynamics for $\hat{s}$ are not included because $\mathcal{H}_1$ does not depend on s. Recall from Chapter 10 that since $\mathcal{H}_1$ is the identity map, the projection operation in (11.65) can be replaced by a saturation. Specifically the observer (11.69) remains unchanged (there are no terms that need to be saturated), whereas $\hat{x}_1^P$ is replaced by $\hat{x}_1^s = \mathcal{V}\,\mathrm{sat}(\mathcal{V}^{-1}\hat{x}_1)$, with $\mathcal{V}_i = \sup_{x_1 \in \Omega_{r_2}^{x_1}}\{x_{1,i}\}$.

△

An interesting feature of the plant in Example 11.4 is the fact that the partial state-feedback controller (11.68) is able to stabilize the dynamics of the upper subsystem despite the presence of the unknown state x_2. Note that the plant in Example 11.3 does not have this feature, as the x_1 dynamics do not depend on x_2. The plant in Example 11.4 is in fact representative of an important class of dynamical systems, namely, input-output feedback linearizable systems with stable zero dynamics,

$$\begin{aligned} \dot{x}_{1,i} &= x_{1,i+1}, \; i = 1, \ldots, n_1 - 1 \\ \dot{x}_{1,n_1} &= \bar{f}(x_1, x_2, \Delta) + \bar{g}(x_1, x_2, \Delta) u \\ \dot{x}_2 &= f^2(x_1, x_2). \end{aligned} \tag{11.70}$$

If all the nonlinearities are smooth, $\bar{g}$ is bounded away from zero, and the origin of

$$\dot{x}_2 = f^2(0, x_2)$$

is globally asymptotically stable, then one can find a partial state-feedback controller u which depends only on x_1 and guarantees that all the state trajectories all uniformly ultimately bounded with an arbitrarily small ultimate bound and arbitrarily large domain of attraction. This can be achieved, for instance, using linear high-gain controllers of the type

$$u = -\left[a_1 K^{n_1 - 1} x_{1,1} + a_2 K^{n_1 - 2} x_{1,2} + \ldots, a_{n_1 - 1} K x_{1,n_1 - 1} + x_{1,n_1}\right],$$

where K is a sufficiently high gain, and $a_1, \ldots, a_{n_1 - 1}$ are such that the polynomial

$$\lambda^{n_1} + a_{n_1 - 1}\lambda^{n_1 - 1} + \ldots + a_2 \lambda + a_1$$

has roots with negative real part (refer to [81]). Unfortunately, despite its simplicity, this controller has the disadvantage of requiring a large control action to overcome the uncertainties and, in some practical applications, may have the undesirable effect of exciting unmodeled dynamics or damaging the actuators. To avoid this problem one may develop an adaptive controller using partial state feedback to achieve the same control objective with a milder control action. Following the methodology developed in Chapter 7, we first assume that Δ is known and we find a partial state feedback controller which stabilizes a neighborhood of the origin of (11.70) over a given compact set (this is easily achieved by using the fact that $\bar{g}$ is bounded away from zero and that $\bar{f}$ is bounded on any compact set). Next, when Δ is unknown, we use a tunable nonlinearity $u = \mathcal{F}(x_1, \hat{\theta})$ to approximate the ideal controller. The controller so obtained has the property that the closed-loop trajectories are uniformly ultimately bounded over a predetermined compact set. Once such a controller is found, noticing that (11.70) satisfies all our assumptions, we can use the techniques illustrated in this section to implement an adaptive output-feedback controller yielding essentially the same stability properties of the adaptive state-feedback controller.

11.4 Separation Principle for Adaptive Tracking

We now turn our attention to the adaptive tracking problem. In Chapter 10 we have introduced the concept of a practical internal model to perform output-feedback tracking in a partial information framework (that is, when only u, the input of the plant, y, the output of the plant, and r, the reference trajectory, are known at each instant of time). Here, we consider the extension of this theory to the adaptive framework.

Consider an uncertain plant

$$\begin{aligned}\dot{x} &= f(x, u, \Delta(t, x)) \\ y &= h(x, u),\end{aligned} \tag{11.71}$$

where Δ represents a modeling uncertainty or a time varying disturbance, and suppose that, in a *partial information* framework, we want to find an adaptive controller that makes the output y track a smooth reference trajectory $r(t) = [r_1(t), \ldots, r_\rho(t)]^\top$. In order to have a well-defined tracking problem we must characterize a stable inverse of the plant. Suppose there exist functions $x^r(t)$ and $c^r(t)$ such that

$$\begin{aligned}\dot{x}^r &= f(x^r, c^r, \Delta(t, x^r)) \\ r &= h(x^r, c^r).\end{aligned} \tag{11.72}$$

If the pair $(x^r(t), c^r(t))$, together with the state x of the plant, were available for feedback, then one could employ the tools illustrated in previous chapters to find a *full information* adaptive controller. In order to do that, we first define an error system $e = \chi(t, x)$ and the associated error dynamics

$$\dot{e} = \alpha(t, x) + \beta(x) u, \tag{11.73}$$

and then, using the tools of Chapters 7 and 8, we find an adaptive controller

$$\begin{aligned}\dot{\hat{\theta}} &= f_\theta(\hat{\theta}, z) \\ u &= \nu_a(z, \hat{\theta}),\end{aligned} \tag{11.74}$$

with the property that there exists a continuously differentiable function $V_a(t, e, \tilde{\theta})$ satisfying

$$\begin{aligned}\gamma_e^1(|e|) + \gamma_{\tilde{\theta}}^1(|\tilde{\theta}|) \leq V_a(t, e, \tilde{\theta}) &\leq \gamma_e^2(|e|) + \gamma_{\tilde{\theta}}^2(|\tilde{\theta}|) \\ \dot{V}_a \leq -k_1 V_a + k_2 \quad &\text{for all } e \in B_r,\end{aligned} \tag{11.75}$$

where B_r is a fixed compact set. Using V_a, define a new continuous function V_a' in an identical manner to (11.37), let

$$\Omega_c = \{e \in \mathbb{R}^n, \tilde{\theta} \in \mathbb{R}^p \,:\, V_a'(t, e, \tilde{\theta}) \leq c, \text{ for all } t \geq 0\}$$

denote the time-varying level set of the function V'_a and let

$$\Omega_c^x = \{x \in \mathrm{R}^n \; : \; [e^\top, \tilde{\theta}^\top]^\top \in \Omega_c, \text{ for all } t \geq 0\}$$

denote its projection on the x domain. From the properties of V'_a, we find Ω_c is uniformly compact for all $c > 0$ and Ω_c^x is uniformly compact because Ω^r is compact.

In order to use (11.74) to define a controller yielding similar properties in a partial information framework, we first need to understand what signals are included in the vector z and whether they can be estimated using available information. In general, given an error system $e = \chi(t, x)$ satisfying Assumption 6.1, the vector z may contain

$$z = [x^\top, e^\top, x^{r\top}, c^{r\top}]^\top.$$

However, recalling from Chapter 10 that $e(t)$ can be expressed as a smooth function of $x(t)$ and $x - x^r(t)$, that is, $e(t) = \pi(x(t), x - x^r(t))$, we have that z reduces to $z = [x^\top, x^{r\top}, c^{r\top}]^\top$. In some cases, when dealing with particular classes of nonlinear systems such as systems in strict feedback form or input-output feedback linearizable systems, z is simply given by x, r, and a finite number of the time derivatives of r. In these cases the analysis is significantly simplified as one does not need to solve (11.72) to get $x^r(t)$ and $c^r(t)$. However, even in this situation, the information contained in r and its time derivatives is *equivalent* to the information contained in x^r and c^r (see example below). In conclusion, in order to develop a unified approach to deal with adaptive tracking of general nonlinear systems, without loss of generality we will say that in a full information framework z is given by

$$z = [x^\top, x^{r\top}, c^{r\top}].$$

Example 11.5 Consider the strict feedback system in Example 6.9 and let, for simplicity, $n = 2$

$$\begin{aligned} \dot{x}_1 &= f_1(x_1) + g_1(x_1)x_2 \\ \dot{x}_2 &= f_2(x_1, x_2) + g_2(x_1, x_2)u \\ y &= x_1. \end{aligned} \tag{11.76}$$

Assume that the functions g_1 and g_2 are bounded away from zero and let $r(t)$ be a smooth reference trajectory. The functions $x^r(t)$ and $c^r(t)$ satisfying (11.72) are found by noticing that the mapping $\mathcal{H}$ defined by

$$r_e = \begin{bmatrix} r \\ \dot{r} \end{bmatrix} = \mathcal{H}(x^r) = \begin{bmatrix} x^r{}_1 \\ f_1(x^r{}_1) + g_1(x^r{}_1)x^r{}_2 \end{bmatrix} \tag{11.77}$$

is invertible and $x^r = \mathcal{H}^{-1}(r_e)$. From x^r we calculate c^r as follows:

$$c^r = \frac{1}{g_2({x^r}_1, {x^r}_2)} ({\dot{x}^r}_2 - f_2({x^r}_1, {x^r}_2)) \triangleq a_1(x^r) + a_2(x^r)\ddot{r}, \quad (11.78)$$

where a_1 and a_2 are two smooth functions well-defined everywhere. From (11.77) and (11.78) we conclude that x^r and c^r can be expressed as functions of $r, \dot{r}$, and $\ddot{r}$, and a general error system can be defined as $e = \chi(t, x) = x - x^r(t)$. Consider instead the error system defined in Example 6.9, which represents a more natural choice for the special class of systems at hand

$$\begin{aligned} e' &= \begin{bmatrix} x_1 - r \\ x_2 - \dfrac{\dot{r} - f_1(x_1) - \kappa e_1}{g_1(x_1)} \end{bmatrix} \\ &= \begin{bmatrix} e_1 \\ e_2 + \dfrac{\dot{r} - f_1(r)}{g_1(r)} - \dfrac{\dot{r} - f_1(x_1) - \kappa e_1}{g_1(x_1)} \end{bmatrix} \triangleq \pi(x, e), \end{aligned}$$

and observe that $\pi(x^r, 0) = 0$. Furthermore, when $e'(t) = 0$, we have

$$e_1 = e_1' = 0, \; e_2 = e_2' - \frac{\dot{r} - f_1(r)}{g_1(r)} + \frac{\dot{r} - f_1(r) - \kappa 0}{g_1(r)} = 0.$$

Recall that these two properties, namely the fact that

- $e = 0$ implies $e' = 0$
- $e' = 0$ implies $e \to 0$ (in this case, $e = 0$)

are general features which have been discussed in Chapter 10.

Now, recall from Example 6.9 that the vector z containing the feedback signals to be used to find a tracking controller (adaptive as well as nonadaptive) is $z = [x, r, \dot{r}, \ddot{r}]^\top$. Using the fact that $[r, \dot{r}]^\top = \mathcal{H}^{-1}(x^r)$ and $\ddot{r} = 1/a_2(x^r)(c^r - a_1(x^r))$, we conclude that, without loss of generality, z above can be rewritten as

$$z = [x^\top, {x^r}^\top, c^r]^\top.$$

This discussion can be easily extended to strict feedback systems of arbitrary degree. $\triangle$

We now turn our attention to the problem of estimating the vector z in a partial information framework when only u, y, and r are available for feedback. The effect of the uncertainty Δ on the solution of the adaptive tracking problem is twofold. First, as seen in Section 10.7, in order to estimate x we need to assume that Δ does not affect the observability mapping associated with the plant, that is,

$$y_e \triangleq \left[y_1, \ldots, y_1^{(k_1-1)}, \ldots, y_p, \ldots, y_p^{(k_p-1)}\right]^\top = \mathcal{H}(x, s), \tag{11.79}$$

where k_i and s are defined in the usual way. Second, since Δ affects (11.72), one cannot apply the ideas found in Chapter 10 to estimate $(x^r(t), c^r(t))$ from r. To see that, view (as we did in Section 10.8) (11.72) as a copy of the plant with unknown input c^r, unknown state x^r, and a measurable output r. Recall that, if a practical internal model with state ζ^r exists, the observability mapping associated with the copy of the plant (11.72) augmented with the practical internal model depends only on the state of the augmented system $[x^{r\top}, \zeta^{r\top}]^\top$, and hence one can employ r to estimate x^r and ζ^r. Now, however, such an observability mapping also depends on Δ, and hence $[x^{r\top}, \zeta^{r\top}]^\top$ cannot be estimated from r. In what follows, we will illustrate a way to circumvent this problem by modifying the practical internal model and requiring an additional assumption on the uncertainty. We will see that the assumption we will introduce is always satisfied in the particular case when Δ satisfies the matching condition.

11.4.1 Practical Internal Models for Adaptive Tracking

Assume that there exists a smooth function $m(x, u, \Delta(t, x))$ such that the plant (11.33) can be rewritten as

$$\begin{aligned} \dot{x} &= f\left(x, m(x, u, \Delta(t, x))\right) \\ y &= h(x, u), \end{aligned} \tag{11.80}$$

and that the adaptive controller (11.74) is such that

$$z = [x^\top, x^{r\top}, m(x^r, c^r, \Delta(t, x^r))^\top].$$

Furthermore, consider the copy of the plant

$$\begin{aligned} \dot{x}^r &= f\left(x^r, m(x^r, c^r, \Delta(t, x^r))\right) \\ r &= h(x^r, c^r), \end{aligned} \tag{11.81}$$

and assume there exists a practical internal model

$$\begin{aligned} \dot{\zeta}^r &= a(\zeta^r, x^r, v^r) \\ m(x^r, c^r, \Delta(t, x^r)) &= b(\zeta^r, x^r) \end{aligned} \tag{11.82}$$

which enjoys the properties in Definition 10.2 and, in particular, such that the observability mapping $\mathcal{H}'$ of the composite system (11.81)-(11.82) does not depend on v^r and its time derivatives, that is, $\mathcal{H}'(x^r, \zeta^r)$.

A sufficient condition for this assumption to hold true is that the uncertainty Δ satisfies a matching condition, as seen in the following example.

Example 11.6 Consider the following MIMO nonlinear system, slightly modified from Example 10.10

$$\begin{aligned}\dot{x}_1 &= x_2\\ \dot{x}_2 &= x_1^2 + u_1\\ \dot{x}_3 &= x_4 - u_1 - x_1^2\\ \dot{x}_4 &= \Delta_1(x) + \Delta_2(x)u_2,\end{aligned} \tag{11.83}$$

where Δ_1 and Δ_2 are unknown smooth functions and Δ_2 is bounded away from zero. Let $\Delta(x)$ denote the vector with components Δ_1 and Δ_2, then one can see that the uncertainty Δ satisfies the matching condition. Suppose for now that we know the stable inverse of the plant $(x^r(t), c^r(t))$ and define the error system $e = x - x^r(t)$. Its dynamics are then given by

$$\begin{aligned}\dot{e}_1 &= e_2\\ \dot{e}_2 &= x_1^2 + u_1 - (x^r{}_1)^2 - c^r{}_1\\ \dot{e}_3 &= e_4 - x_1^2 - u_1 + (x^r{}_1)^2 + c^r{}_1\\ \dot{e}_4 &= \Delta_1(x) + \Delta_2(x)u_2 - \Delta_1(x^r) - \Delta_2(x^r)c^r{}_2.\end{aligned} \tag{11.84}$$

If Δ was known, a full information controller would be given by

$$\begin{aligned}u_1 &= c^r{}_1 + (x^r{}_1)^2 - x_1^2 - k_1^\top e\\ u_2 &= -\frac{\Delta_1(x) + k_2^\top e - m(\Delta(x^r), c^r{}_2)}{\Delta_2(x)},\end{aligned} \tag{11.85}$$

where $m(\Delta(x^r), c^r{}_2) = \Delta_1(x^r) + \Delta_2(x^r)c^r{}_2$, $k_1 = [8,6,3,3]^\top$, and $k_2 = [3,5,7,3]^\top$. This controller makes the closed-loop system linear and globally exponentially stable. Since Δ is unknown, one can use the tools of Chapter 7 to find a direct adaptive controller which directly estimates the ideal controller (11.85), or the tools of Chapter 8 to find an indirect adaptive controller which estimates Δ to implement (11.85). In either case, the full information adaptive controller would depend on the vector of feedback signals

$$z = [x, x^r, c^r{}_1, m(\Delta(x^r), c^r{}_2)]^\top.$$

Hence, there is no need to estimate $c^r{}_2(t)$, only an indirect measure of it, namely, $m(\Delta(x^r), c^r{}_2)$, needs to be estimated.

In order to do that, consider the copy of the plant

$$\begin{aligned}\dot{x}^r{}_1 &= x^r{}_2\\ \dot{x}^r{}_2 &= x^{r2}_1 + c^r{}_1\\ \dot{x}^r{}_3 &= x^r{}_4 - c^r{}_1 - x^{r2}_1\\ \dot{x}^r{}_4 &= \Delta_1(x^r) + \Delta_2(x^r)c^r{}_2 = m(\Delta(x^r), c^r{}_2),\end{aligned} \tag{11.86}$$

and define the following practical internal model, a slight modification of that used in Example 10.10

$$
\begin{aligned}
\dot{\zeta}^r{}_1 &= \zeta^r{}_2 + \zeta^r{}_3 \\
\dot{\zeta}^r{}_2 &= v^r{}_1 \\
\dot{\zeta}^r{}_3 &= v^r{}_2 \\
[c^r{}_1, m(\Delta(x^r), c^r{}_2)]^\top &= [\zeta^r{}_1, \zeta^r{}_2]^\top .
\end{aligned}
\tag{11.87}
$$

The only modification we made to this compensator, which we have employed earlier to solve the tracking problem in the absence of uncertainties, is to replace one of its outputs, $c^r{}_2$, by $m((\Delta(x^r), c^r{}_2))$. With that in mind, we find that the observability mapping associated with the composite system (11.86)-(11.87) is identical to that in Example 10.10, despite the presence of the uncertainty Δ

$$
r'_e \triangleq \begin{bmatrix} r_1 \\ \dot{r}_1 \\ r_1^{(2)} \\ r_1^{(3)} \\ r_2 \\ \dot{r}_2 \\ r_2^{(2)} \end{bmatrix} = \mathcal{H}'(x^r, \zeta^r) = \begin{bmatrix} x^r{}_1 \\ x^r{}_2 \\ (x^r{}_1)^2 + \zeta^r{}_1 \\ 2x^r{}_1 x^r{}_2 + \zeta^r{}_2 + \zeta^r{}_3 \\ x^r{}_3 \\ x^r{}_4 - \zeta^r{}_1 - (x^r{}_1)^2 \\ -\zeta^r{}_3 - 2x^r{}_1 x^r{}_2 \end{bmatrix}, \tag{11.88}
$$

from which one can immediately see that x^r and ζ^r can be estimated from r by defining a suitable estimator. Once x^r and ζ^r have been estimated, $c^r{}_1$ and $m(\Delta(x^r), c^r{}_2)$ are simply the output of the practical internal model,

$$
[c^r{}_1, m(\Delta(x^r), c^r{}_2)]^\top = [\zeta^r{}_1, \zeta^r{}_2]^\top .
$$

Finally, we conclude this example by noting that the observability mapping associated with the system is given by

$$
y_e = \mathcal{H}(x, u_1) = \left[x_1,\, x_2,\, x_3,\, x_4 - u_1 - x_1^2\right]^\top .
$$

It depends on u_1 but it is independent of the uncertainty Δ. Summarizing, all our assumptions are satisfied and one can use a separation principle to define a partial information controller recovering the performance of (11.85). $\triangle$

We are now ready to define an adaptive controller solving the tracking problem in a partial information framework. In order to do that, we introduce the same notations used in Chapter 10 to solve the nonadaptive tracking problem. Let $X^r \stackrel{\Delta}{=} [x^{r\,\mathsf{T}}, \zeta^{r\,\mathsf{T}}]^{\mathsf{T}}$ and rewrite the copy of the plant (11.81) augmented with the practical internal model (11.82) as

$$\begin{aligned} \dot{X}^r &= F(X^r, v^r) \\ r &= H(X^r). \end{aligned} \tag{11.89}$$

Notice that the uncertainty Δ is embedded in the definition of the practical internal model. Further, observe that, by the regularity of the practical internal model, the trajectory $X^r(t)$ is bounded for all $t \geq 0$ and, hence, there exists a compact set Ω^r such that $X^r(t) \in \Omega^r$, for all $t \geq 0$. Let $\mathcal{C}^r$ be a convex compact set with smooth boundary such that

$$\mathcal{H}'(\Omega^r) \subset \mathcal{C}^r.$$

Further, given any two positive scalars $c_1 < c_2$, choose another convex compact set $\mathcal{C}$ satisfying

$$\mathcal{F}(\Omega^x_{c_2} \times \Omega^s) \subset \mathcal{C}.$$

Consider the following partial information adaptive controller

$$\begin{aligned} \dot{\hat{X}}^r &= F(\hat{X}^r, 0) + \frac{\partial \mathcal{H}'(\hat{X}^r)}{\partial \hat{X}^r}\left[\left(\mathcal{E}^{x^r}\right)^{-1} L(r - \hat{r})\right] \\ \hat{r} &= H(\hat{X}^r) \\ \dot{\hat{x}}^{rP} &= \mathcal{P}_{\mathcal{C}^r}(\dot{\hat{x}}^r, \hat{x}^r) \end{aligned} \tag{11.90}$$

$$\begin{aligned} \dot{\hat{x}} &= f\left(\hat{x}, m(\hat{x}, \hat{u}, \Delta^0(t, \hat{x}))\right) + \left[\frac{\partial \mathcal{H}(\hat{x}, \hat{s})}{\partial \hat{x}}\right]^{-1}\Big[(\mathcal{E}^x)^{-1} L\,(y - \hat{y}) \\ &\qquad - \frac{\partial \mathcal{H}(\hat{x}, \hat{s})}{\partial \hat{s}}(\mathcal{E}^s)^{-1} KC(u - \hat{u})\Big] \\ \hat{y} &= h(\hat{x}, \hat{u}) \\ \dot{\hat{s}} &= A\hat{s} + (\mathcal{E}^s)^{-1} K(u - \hat{u}) \\ \hat{u} &= C\hat{s}, \\ \begin{bmatrix} \dot{\hat{x}}^P \\ \dot{\hat{s}}^P \end{bmatrix} &= \mathcal{P}_{\mathcal{C}}\,(\dot{\hat{x}}^P, \dot{\hat{s}}^P, \hat{x}^P, \dot{\hat{s}}^P) \end{aligned} \tag{11.91}$$

$$\begin{aligned} \dot{\hat{\theta}} &= f_\theta\left(\hat{\theta}, \hat{z}^P\right) \\ u &= \nu_a\left(\hat{z}^P, \hat{\theta}\right), \end{aligned} \tag{11.92}$$

where $\mathcal{P}_{\mathcal{C}}$ and $\mathcal{P}_{\mathcal{C}^r}$ are defined in (10.105) and (10.141), respectively, $\hat{z}^P = [\hat{x}^{P\mathsf{T}}, \hat{x}^{rP\mathsf{T}}, b(\hat{\zeta}^{rP}, \hat{x}^{rP})^{\mathsf{T}}]^{\mathsf{T}}$, $\Delta^0(t, \hat{x}^P)$ is the nominal value of the uncertainty Δ (if no a priori information about Δ is available, one can set

$\Delta^0 = 0$), and L, K, A, C, $\mathcal{E}^{x^r}, \mathcal{E}^x, \mathcal{E}^s$ are defined as in Section 10.6. It is useful to recall the the performance of the estimators (11.90) and (11.91) is controlled by the choice of three design parameters, η_1, (relative to (11.90)), η_2, and η_3 (relative to (11.91)).

The following theorem states that the proposed partial information controller recovers the performance of the original full information controller (11.74).

Theorem 11.3: *Consider the closed-loop system formed by the plant (11.80) and the adaptive output-feedback controller (11.90)-(11.92). Then, given any triple (c_1, c_2, ϵ) such that $0 < c_1 < c_2$, and $\epsilon > 0$ is sufficiently small, there exist scalars $\eta_1^* \in (0,1]$, $\eta_2^* \in (0,1]$, $\eta_3^* \in (0,1]$, and a class-$\mathcal{K}_\infty$ function γ such that, for all $\eta_1 \in (0,\eta_1^*]$, $\eta_2 \in (0,\eta_2^*]$, and $\eta_3 \in (0,\eta_3^*]$, every $(x,\tilde{\theta})$ trajectory starting in Ω_{c_1} is bounded within Ω_{c_2} and asymptotically approaches the positively invariant set $\Omega_{\gamma(\epsilon)}$.*

The proof of this theorem exploits the separation principle proved in Theorem 11.1 and is left as an exercise.

11.4.2 Partial State-Feedback Performance Recovery

We now apply the results of the previous section to the particular case when the full information controller utilizes partial state feedback to accomplish the control task, and we show that in this case our assumptions can be relaxed. Consider again the class of systems investigated in Section 11.3.2

$$\begin{aligned} \dot{x}_1 &= f^1(x_1, x_2, m(x_1, x_2, u, \Delta(t, x_1, x_2))) \\ \dot{x}_2 &= f^2(x_1, x_2, \Delta(t, x_1, x_2)) \\ y &= h(x_1, u). \end{aligned} \tag{11.93}$$

Suppose the full information controller yields desired stability properties (see (11.75)) by employing the feedback signals

$$z = \left[x_1^\top, {x^r}_1^\top, m\left(x^r{}_1, x^r{}_2, c^r, \Delta(t, x^r{}_1, x^r{}_2)\right)^\top \right]^\top$$

and there exists a practical internal model for the upper subsystem

$$\begin{aligned} \dot{\zeta}^r &= a(\zeta^r, x^r{}_1, v^r) \\ m\left(x^r{}_1, x^r{}_2, c^r, \Delta(t, x^r{}_1, x^r{}_2)\right) &= b(\zeta^r, x^r{}_1) \end{aligned} \tag{11.94}$$

satisfying all the assumptions made in the previous section. Suppose further, as we did in Section 11.3.2, that the observability mapping associated with the upper subsystem is a diffeomorphism and does not depend on x_2 and the uncertainty Δ, that is,

$$y_{e,1} \triangleq \left[y_1, \ldots, y_1^{(k_1-1)}, \ldots, y_p, \ldots, y_p^{(k_p-1)} \right]^\top = \mathcal{H}_1(x_1, s), \tag{11.95}$$

where k_i and s are defined in the usual way. Then, the methodology illustrated in the previous section can be applied to solve this problem with a partial information adaptive controller which does not need to estimate x_2 and ${x^r}_2$. Notice that the particular structure of this problem allows us to weaken the assumptions on the uncertainty Δ (no assumption is made on how Δ affects the lower subsystem).

Example 11.7 An important classe of nonlinear systems fitting the structure in (11.93) is that described by differential equations of the form

$$\begin{aligned} \dot{x}_{1,i} &= x_{1,i+1}, \; i = 1, \ldots, n_1 - 1 \\ \dot{x}_{1,n_1} &= \Delta_1(x_1) + \Pi_1(x_1)u \\ \dot{x}_2 &= f^2\left(x_1, x_2, \Delta_2(x_1, x_2)\right). \end{aligned} \tag{11.96}$$

Assume that, if x_1 is bounded for all $t \geq 0$, $x_2(t)$ is also bounded. This is achieved, for instance, when the lower subsystem, viewed as a system with state x_2 and input x_1, is input-to-state practically stable and notice that, for any initial condition $x_2(0)$, the solution $({x^r}_1(t), x_2(t))$ of

$$\begin{aligned} \dot{x}^r{}_{1,i} &= {x^r}_{1,i+1}, \; i = 1, \ldots, n_1 - 1 \\ \dot{x}^r{}_{1,n_1} &= \Delta_1({x^r}_1) + \Pi_1({x^r}_1)c^r \\ \dot{x}_2 &= f^2\left({x^r}_1, x_2, \Delta_2({x^r}_1, x_2)\right) \\ r(t) &= {x^r}_{1,1}(t) \end{aligned} \tag{11.97}$$

is bounded and hence it is a stable inverse of the plant, that is, we can redefine $x_2(t)$ to be ${x^r}_2(t)$ (notice that ${x^r}_1$ is simply given by $n_1 - 1$ time derivatives of $r(t)$).

Next, in order to find a full information controller, consider the error system $e = x_1 - {x^r}_1$ which does not include the state of the lower subsystem. In the new coordinates, the dynamics read as

$$\begin{aligned} \dot{e}_i &= e_{i+1}, \; i = 1, \ldots, n_1 - 1 \\ \dot{e}_{n_1} &= \Delta_1(x_1) + \Pi_1(x_1)u - m\left(\Delta_1({x^r}_1), \Pi_1({x^r}_1)c^r\right) \\ \dot{x}_2 &= f^2\left(e_1 + {x^r}_1, x_2, \Delta_2(e_1 + {x^r}_1, x_2)\right). \end{aligned} \tag{11.98}$$

If the uncertainty Δ_1 and the function m were known, a tracking controller could be defined as $u = \nu_s(z)$, where

$$z = [x_1^\top, {x^r}_1^\top, m(\Delta_1({x^r}_1), \Pi_1({x^r}_1), c^r)^\top]^\top,$$

and ν_s is defined as

$$\nu_s = \frac{1}{\Delta_1(x_1)}\left(-\Pi_1(x_1) + m(\Delta_1({x^r}_1), \Pi_1({x^r}_1), c^r) - k^\top e\right).$$

In the definition above, the vector $k \in \mathrm{R}^{n_1}$ can be chosen so that the origin of the error dynamics is globally exponentially stable. By assumption, the boundedness of $e(t)$ and $x^r(t)$ guarantee that the state of the lower subsystem is bounded. Since Δ_1 and Π_1 are unknown, one can employ the tools developed in Chapter 7 to find a direct adaptive controller

$$\nu_a(z) = \mathcal{F}(z, \hat{\theta})$$

and a continuously differentiable function V_a satisying (11.75) or, equivalently, use the tools of Chapter 8 to find an indirect adaptive controller

$$\nu_a(z) = \frac{1}{\mathcal{F}_{\Pi_1}(x_1, \hat{\theta})} \left(-\mathcal{F}_{\Delta_1}(x_1, \hat{\theta}) + m(\Delta_1(x^r{}_1), \Pi_1(x^r{}_1), c^r) - k^\top e \right),$$

yielding similar properties.

In order to recover the performance of the full information adaptive controller $\nu_a(z)$, we need to estimate z. To this end, notice that the observability mapping associated with the upper subsystem is simply the identity mapping $\mathcal{H}(x_1) = x_1$ (i.e., x_1 is given by y and its first $n_1 - 1$ derivatives) and hence we can estimate x_1 from y despite the presence of Δ_1. Next, let

$$m(\Delta^1(x_1), \Pi_1(x_1), u) = \Delta_1(x_1) + \Pi_1(x_1)u$$

and consider the following practical internal model (a simple integrator)

$$\begin{aligned} \dot{\zeta}^r &= v^r \\ m\left(\Delta_1(x^r{}_1), \Pi_1(x^r{}_1), c^r\right) &= \zeta^r. \end{aligned} \tag{11.99}$$

Hence, in this case, $X^r = [x^r{}_1^\top, \zeta^{r\top}]^\top$ is given by the first n_1 derivatives of $r(t)$ which can be approximated using the estimator (11.90), where $\mathcal{H}'$ is the identity mapping $\mathcal{H}'(X^r) = X^r$. In a similar manner, the state x_1 can be approximated by using the estimator (11.91) with $\mathcal{H}$ the identity mapping and with no $\hat{s}$ dynamics. In conclusion, a partial information controller is found replacing z in $\nu_a(z)$ by the estimates of $n_1 - 1$ derivatives of y (corresponding to the state x_1) and n_1 derivatives of r (corresponding to x^r and $\zeta^r = m$), and saturating z on a suitable compact set. This result, which is not surprising, is found as a particular case of the theory developed in the previous section.

△

In the example above the requirement that Δ_1 and Π_1 depend only on x_1 can be relaxed provided the full information controller does not employ x_2 to solve the tracking problem. This is seen in the next example.

Example 11.8 Go back to the problem, considered in Example 8.4, of designing a controller to accurately point an antenna driven by a permanent magnet DC motor. Recall the system model

$$\begin{aligned} \dot{\theta} &= \omega \\ \dot{\omega} &= \left(\frac{T_c \cos\phi}{J}\right)\sin(N\theta) + \left(\frac{T_c \sin\phi}{J}\right)\cos(N\theta) - \frac{q}{J} + \frac{u}{J} \\ \dot{q} &= \sigma\omega\left[1 - \frac{q}{T_f}\mathrm{sgn}(\omega)\right] \\ y &= \theta, \end{aligned} \tag{11.100}$$

where the dynamic uncertainty Δ is given by

$$\Delta(\theta,\omega,q) = \left(\frac{T_c \cos\phi}{J}\right)\sin(N\theta) + \left(\frac{T_c \sin\phi}{J}\right)\cos(N\theta) - \frac{q(t)}{J}.$$

In Example 11.4 we assumed the reference angle r to be constant (which corresponds to considering a stabilization problem) and we extended the results of Example 8.4 to the case when only θ is available for feedback. Now we seek to find a partial information controller which solves the more general tracking problem (i.e., r is time-varying) by using only θ and r as feedback signals. This is easily accomplished by letting $x_1 = [\theta,\omega]^\top$ and $x_2 = q$, and noticing that the plant has the form (11.96) and $q(t)$ is bounded provided $\omega(t)$ is bounded, for all $t \geq 0$. Furthermore,

$$m(\Delta(x_1,x_2),u) = \Delta(\theta,\omega,q) + \frac{u}{J}.$$

Notice that, unlike in (11.96), Δ here depends on both x_1 and x_2, but the full information controller developed in Example 8.4 does not employ x_2. Thus, the performance of the full information controller can be recovered by using (11.69) to estimate x_1 and defining the simple practical internal model

$$\begin{aligned} \dot{\zeta}^r &= v^r \\ m(\Delta({x^r}_1,{x^r}_2),c^r) &= \zeta^r, \end{aligned}$$

to estimate $[x^r(t)^\top,\zeta^r(t)]^\top = [r,\dot{r},\ddot{r}]^\top$ from $r(t)$. The estimator in this case reduces to

$$\begin{aligned} \dot{\hat{x}}^r_1 &= {\hat{x}^r}_2 + \frac{l'_1}{\eta_2}(r-\hat{r}) \\ \dot{\hat{x}}^r_2 &= \hat{\zeta}^r + \frac{l'_2}{\eta_2^2}(r-\hat{r}) \\ \dot{\hat{\zeta}}^r &= \frac{l'_3}{\eta_2^3}(r-\hat{r}) \\ \hat{r} &= {\hat{x}^r}_1, \end{aligned} \tag{11.101}$$

where l_1', l_2', l_3', and ρ_2 are design parameters chosen as in (10.39).

△

11.5 Summary

In this chapter we have seen how to design adaptive systems coping with partial information. When the plant can be transformed into adaptive tracking form, we have seen that there exists a systematic design procedure to build adaptive controllers solving the stabilization and tracking problems. For more general classes of nonlinear systems, we have learned how to design adaptive output-feedback controllers for stabilization and tracking based on a separation principle. It is worth mentioning that, when using backstepping to achieve tracking, one needs to know the reference trajectory and its first r time derivatives, and hence the resulting controller does not work in a partial information framework. On the other hand, using a separation principle and practical internal models, the controllers we found use only u, y, and r as feedback signals.

When designing an adaptive output-feedback controller using the separation principle, one must first place the dynamics in a form so that an observer may be defined to estimate the unmeasurable states of the system. Once in this form, a state-feedback adaptive controller is defined using either the direct approach of Chapter 7 or the indirect approach of Chapter 8. Using the Lyapunov function associated with the controller and system dynamics, we then proceed to design an observer that may then be used to estimate the unmeasred states. As long as the gain of the observer is chosen to be high enough, we find that the previously designed controller may be used with the true state replaced by its estimate.

11.6 Exercises and Design Problems

Exercise 11.1 (Input-Output Forms) Prove that systems is adaptive tracking form (11.1) can be represented in input-output form as in (11.51), and find the functions ψ_i, $0 \leq i \leq p$. Read Example 11.2 and use Theorem 11.1 to define an output-feedback controller which semiglobally practically stabilizes the origin of (11.1).

Exercise 11.2 (Performance Recovery) Prove Theorem 11.2.

Exercise 11.3 (Input-Output Feedback Linearizable Systems) Following the indications given at the end of Section 11.3.2, develop an output-feedback controller for the class of systems (11.70) assuming input-to-state stable zero dynamics.

Exercise 11.4 (Adaptive Tracking) Prove Theorem 11.3.

Exercise 11.5 (Orientation Control of an Antenna) Simulate the adaptive tracking controller in Example 11.8 for different choices of the design parameters η_1 and η_2.

Exercise 11.6 (Satellite Control) Consider the satellite system in Example 6.3 whose dynamics may be described by

$$\begin{aligned}\dot{z} &= v \\ \dot{v} &= z\omega^2 - \frac{k_g}{mz^2} + \frac{u_1}{m} \\ \dot{\omega} &= -\frac{2v\omega}{z} + \frac{u_2}{mz}.\end{aligned}$$

Use both the backstepping approach described in Section 11.2 and the the separation principle to design adaptive output-feedback controllers so that $z \to r_1$ and $\omega \to r_2$ when v is not measurable and k_g is an unknown constant. How do the two designs compare in terms of performance and ease of implementation.

Exercise 11.7 (Structural Dynamics) Consider the system defined by

$$\begin{aligned}\dot{q}_1 &= q_2 \\ \dot{q}_2 &= -\omega^2 q_1 - 2\zeta\omega q_2 + b_1 u \\ \dot{q}_3 &= q_4 \\ m\dot{q}_4 &= f(q_3) + b_2 u,\end{aligned}$$

where $y = q_3 + cq_1$ is the output and u is the force input. Here q_3, q_4 represent the rigid body motion of the system, while q_1, q_2 include motion due to structural dynamics. The parameters are defined as follows: ω is the modal frequency, ζ is the modal damping, c defines the influence of the elastic state on the output (based on the mode shape), and m is the system mass. Design a state-feedback controller for this system so $y \to 0$ when the states are measurable and $f(q_3)$ is known. Then design an observer and adaptive output-feedback controller for this system when $f(q_3)$ is unknown, but may be approximated by a neural network. Test your controller when $f(q_3) = q_3 - q_3^2$.

Exercise 11.8 (Output-Feedback Fuzzy Control Design) Consider the system in adaptive tracking form

$$\begin{aligned}\dot{\xi}_1 &= f(\xi_1) + \xi_2 \\ \dot{\xi}_2 &= u,\end{aligned} \tag{11.102}$$

where $y = \xi_1$. Show that this may be equivalently expressed as

$$\begin{aligned} \dot{x}_1 &= x_2 \\ \dot{x}_2 &= \frac{\partial f}{\partial x_1} x_2 + u \end{aligned} \tag{11.103}$$

with $y = x_1$. Assuing that x_1 and x_2 are available for feedback, design an adpative controller when $(\partial f / \partial x_1) x_2$ is unknown, but may be represented by a fuzzy system. Design the controller to work when

- $f(x_1) = \sin(x_1)$
- $f(x_1) = x_1^2$.

Then design an observer for the system when x_2 is not measurable. Show that when the observer gain is increased, the performance of the state-feedback controller is recovered.

Exercise 11.9 (Adaptive Observer) Consider the SISO system

$$\begin{aligned} \dot{x} &= Ax + g_0(y) + b\theta^\top g(y) + d\sigma(y)u \\ y &= cx, \end{aligned} \tag{11.104}$$

where (A, b, c) forms a strictly positive real (SPR) system with $\theta \in R^p$ unknown. A property of a SPR system [191] is that there exists positive definite matricies P, Q such that

$$\begin{aligned} PA + A^\top P &= -Q \\ Pb &= c^\top. \end{aligned}$$

Given the observer

$$\dot{\hat{x}} = A\hat{x} + g_0(y) + b\hat{\theta}^\top g(y) + d\sigma(y)u + L(y - c\hat{x}), \tag{11.105}$$

use the SPR property to design an update law for $\hat{\theta}$ so that $\hat{x} \to x$.

Chapter 12

Applications

12.1 Overview

In Chapters 10 and 11 we discussed techniques for output-feedback control of nonlinear systems both via non-adaptive and adaptive methods. In this chapter we apply these techniques to three illustrative examples.

In the first example we consider the problem of controlling stall and surge in a jet engine compressor and we seek to find a controller which only employs the measurement of the differential pressure across the compressor to reject stall and surge while regulating the pressure at a desired value. This design presents an interesting challenge in that, when there is no mass flow through the compressor, the system looses observability. Thus, the system is not uniformly completely observable and the tools presented in Chapter 10 cannot be straightforwardly applied. We show, however, that the dynamic projection used in Chapter 10 to eliminate the peaking phenomenon can be used here to bound the observer states away from the unobservable region of the state space.

The second example considers the problem of controlling the horizontal position of a beam by using two electromagnets at its sides in the presence of an unknown force representing, for example, friction acting on the beam. We first solve the problem assuming that the unknown force is zero and then develop an adaptive controller to recover the closed-loop performance when the unknown force is not zero. Finally, using the tools of Chapter 11, we show that a simple linear high-gain observer and control saturation allow us to define an adaptive output feedback controller.

The last example of the chapter focuses on the tracking problem for a simple vertical take-off and landing (VTOL) aircraft model. The main challenge in solving the tracking problem is to find a practical internal model, as defined in Definition 10.2, and a full information controller (i.e., a controller which exploits the knowledge of the state of the plant and its stable inverse). Once this is done, one can use the tools of Chapter 10 to define estimators for the state of the plant and the stable inverse, and invoke a separation principle to find a partial information controller. Some design details in this example are skipped and are left to the reader as an exercise.

12.2 Nonadaptive Stabilization: Jet Engine

The problem of controlling surge and stall in jet engine compressors is of fundamental importance in preventing damage and lengthening the life of these components. Rotating stall develops when there is a region of stagnant flow rotating around the circumference of the compressor causing undesired vibrations in the blades and reduced pressure rise of the compressor. Surge is an axisymmetric oscillation of the flow through the compressor that can cause undesired vibrations in other components of the compression system and damage to the engine. In [158], Moore and Greitzer developed a three-state finite dimensional Galerkin approximation of a nonlinear PDE model describing the compression system. This model is described by (see [115] for an analogous exposition)

$$\begin{aligned} \dot{\Phi} &= -\Psi + \Psi_C(\Phi) - 3\Phi R \\ \dot{\Psi} &= \tfrac{1}{\beta^2}(\Phi - \Phi_T) \\ \dot{R} &= \sigma R(1 - \Phi^2 - R), \;\; R(0) \geq 0, \end{aligned} \tag{12.1}$$

where Φ represents the mass flow, Ψ is the plenum pressure rise, $R \geq 0$ is the normalized stall cell squared amplitude, Φ_T is the mass flow through the throttle, $\sigma = 7$, and $\beta = 1/\sqrt{2}$. The functions $\Psi_c(\Phi)$ and $\Phi_T(\Psi)$ are the compressor and throttle characteristics, respectively, and are defined as

$$\begin{aligned} \Psi_C(\Phi) &= \Psi_{C_0} + 1 + 3/2\Phi - 1/2\Phi^3, \\ \Psi &= \tfrac{1}{\gamma}(1 + \Phi_T(\Psi))^2, \end{aligned}$$

where Ψ_{C_0} is a constant and γ is the throttle opening (the control input). Given the static relationship existing between Φ_T and γ, without loss of generality, in what follows we will design a controller assuming that Φ_T is our control input and use

$$\gamma = \frac{(1 + \Phi_T(\Psi))^2}{\Psi}$$

to define γ. Our control objective is to stabilize system (12.1) around the critical equilibrium $R^e = 0, \Phi^e = 1, \Psi^e = \Psi_C(\Phi^e) = \Psi_{C_0}+2$, which achieves the peak operation on the compressor characteristic, by only measuring the plenum pressure rise Ψ. We shift the origin to the desired equilibrium with the change of variables $\phi = \Phi - 1, \psi = \Psi - \Psi_{C_0} - 2$. System (12.1) then becomes

$$\begin{aligned}\dot{R} &= -\sigma R^2 - \sigma R(2\phi + \phi^2)\\ \dot{\phi} &= -\psi - 3/2\phi^2 - 1/2\phi^3 - 3R\phi - 3R\\ \dot{\psi} &= -\tfrac{1}{\beta^2}(\Phi_T - 1 - \phi)\end{aligned} \tag{12.2}$$

The pressure rise (and hence ψ) is the only measurable state variable.

12.2.1 State-Feedback Design

For convenience, in the remainder of the section we will redefine the control input to be $u = \Phi_T - 1$ and we will let $x = [R, \phi, \psi]^\top$.

Lemma 12.1: *For system (12.2), with the choice of the control law*

$$\bar{u}(x) = (1 - \beta^2 k_1 k_2)\phi + \beta^2 k_2 \psi + 3\beta^2 k_1 R\phi, \tag{12.3}$$

where k_1 and k_2 are positive scalars satisfying the inequalities,

$$k_1 > \frac{17}{8} + \frac{(2C\sigma + 3)^2}{2} \tag{12.4}$$

$$\left(C\sigma - \frac{105}{64}\right)k_1^2 + \frac{3}{4}\left(-\frac{1}{2}C\sigma + \frac{21}{4}\right)k_1 - (C\sigma + 3)^2 > 0 \tag{12.5}$$

$$k_2 > k_1 + \frac{9}{4}k_1^2 + \frac{9k_1}{4k_1 - 9/2} + \frac{(k_1^2 - 1)^2}{4} \tag{12.6}$$

$$C > \frac{3}{2\sigma} \tag{12.7}$$

the origin is an asymptotically stable equilibrium point with domain of attraction $\mathcal{D} = \{x \in \mathrm{R}^3 | R \geq 0\}$.

Proof: For the sake of simplicity, redefine the control input to be $u' = -\frac{1}{\beta^2}(u - \phi)$, so that the last equation in (12.2) becomes $\dot{\psi} = u'$. Next, notice that system (12.2) can be viewed as the interconnection of two subsystems:

$$S_1 : \dot{R} = -\sigma R^2$$

$$S_2 : \begin{cases} \dot{\phi} = -\psi - \dfrac{3}{2}\phi^2 - \dfrac{1}{2}\phi^3 \\ \dot{\psi} = -u', \end{cases} \tag{12.8}$$

where we are ignoring the interconnection term for now. A Lyapunov function for S_1, defined on the domain $\{R \in \mathrm{R} : R \geq 0\}$, is $V_1 = R$, and its time derivative is readily found to be $\dot{V}_1 = -\sigma R^2$ thus showing that the origin of S_1 is an asymptotically stable equilibrium point of S_1, and its domain of attraction is $\{R \in \mathrm{R} : R \geq 0\}$. As for subsystem S_2 the analysis found in Section 2.4.3 in [115] suggests using $V_2 = \frac{1}{2}\phi^2 + \frac{k_1}{8}\phi^4 + \frac{1}{2}(\phi - k_1\psi)^2$, where k_1 is a positive design constant. Furthermore, in [115], a stabilizing control law for S_2 is found to be $u' = -c_1\phi + c_2\psi$, where c_1 and c_2 are two appropriate positive constants.

In the following we will show that, in order to stabilize the interconnection of systems S_1 and S_2, one needs to add to $u' = -c_1\phi + c_2\psi$ a term which is proportional to the product $R\phi$. Based on these considerations, consider the following Lyapunov function candidate for system (12.2),

$$V = CV_1 + V_2 = CR + \frac{1}{2}\phi^2 + \frac{k_1}{8}\phi^4 + \frac{1}{2}(\psi - k_1\phi)^2, \tag{12.9}$$

where $C > 0$ is a scalar. After noticing that V is positive definite on the domain $\mathcal{D}$, and letting $\tilde{\psi} = \psi - k_1\phi$, we calculate the time derivative of V as follows,

$$\begin{aligned} \dot{V} = \; & -C\sigma R^2 - C\sigma R(2\phi + \phi^2) + \left(\phi + \tfrac{k_1}{2}\phi^3\right)\left(-\psi - \tfrac{3}{2}\phi^2 - \tfrac{1}{2}\phi^3\right. \\ & \left. - 3R\phi - 3R\right) + \tilde{\psi}\left(u' + k_1\psi + \frac{3}{2}k_1\phi^2 + \frac{1}{2}k_1\phi^3 + 3k_1R\phi + 3k_1R\right). \end{aligned} \tag{12.10}$$

Here, as in [115], we use the identity $-\frac{3}{2}\phi^2 - \frac{1}{2}\phi^3 = -\frac{1}{2}\left(\phi + \frac{3}{2}\right)^2\phi + \frac{9}{8}\phi$ to eliminate the potentially destabilizing term $-\left(\phi + k_1/2\phi^3\right)3/2\phi^2$. Next, substituting (12.3) into (12.10) (after taking in account the definition of u'), letting $\bar{k}_1 = k_1 - 9/8$, and using the definition of $\tilde{\psi}$, we get

$$\begin{aligned} \dot{V} = & -C\sigma R^2 - C\sigma R(2\phi + \phi^2) \\ & + \left(\phi + \tfrac{k_1}{2}\phi^3\right)\left(-\tilde{\psi} - \bar{k}_1\phi - \tfrac{1}{2}\left(\phi + \tfrac{3}{2}\right)^2\phi - 3R\phi - 3R\right) \\ & + \tilde{\psi}\left(-(k_2 - k_1)\tilde{\psi} + k_1^2\phi + \frac{3}{2}k_1\phi^2 + \frac{1}{2}k_1\phi^3 + 3k_1R\right). \end{aligned} \tag{12.11}$$

Now notice that the expression $-\left(\phi + \frac{k_1}{2}\phi^3\right)\frac{1}{2}\left(\phi + \frac{3}{2}\right)^2\phi$ can be discarded since it is negative definite, and that the term $\frac{k_1}{2}\phi^3\tilde{\psi}$ cancels out. After collecting the remaining terms, we get

$$\begin{aligned} \dot{V} \leq \; & -C\sigma R^2 - (2C\sigma + 3)R\phi - (C\sigma + 3)R\phi^2 - \bar{k}_1\phi^2 \\ & - \left(\frac{k_1\bar{k}_1}{2} + \frac{3k_1}{2}R\right)\phi^4 - \frac{3k_1}{2}R\phi^3 + \\ & + \tilde{\psi}\left(-(k_2 - k_1)\tilde{\psi} + (k_1^2 - 1)\phi + \frac{3}{2}k_1\phi^2 + 3k_1R\right). \end{aligned} \tag{12.12}$$

By using Young's inequality five times we have

$$
\begin{aligned}
-(2C\sigma+3)R\phi &\le \frac{1}{2}R^2 + \frac{(2C\sigma+3)^2}{2}\phi^2 \\
-\frac{3k_1}{2}R\phi^3 &\le \frac{3k_1}{2}\left(\frac{R\phi^2}{4} + R\phi^4\right) \\
(k_1^2-1)\phi\tilde{\psi} &\le \phi^2 + \frac{(k_1^2-1)^2}{4}\tilde{\psi}^2 \\
3k_1 R\tilde{\psi} &\le R^2 + \frac{9}{4}k_1^2\tilde{\psi}^2 \\
\frac{3}{2}k_1\phi^2\tilde{\psi} &\le \frac{k_1\bar{k}_1}{4}\phi^4 + \frac{9k_1}{4\bar{k}_1}\tilde{\psi}^2
\end{aligned}
$$

Applying the inequalities above to (12.12) we get

$$
\begin{aligned}
\dot{V} \;&\le -\left(C\sigma - \frac{3}{2}\right)R^2 - \left(\bar{k}_1 - \frac{(2C\sigma+3)^2}{2} - 1\right)\phi^2 - \left(k_2 - k_1 - \frac{9}{4}k_1^2\right. \\
&\quad \left. - \frac{9k_1}{4\bar{k}_1} - \frac{(k_1^2-1)^2}{4}\right)\tilde{\psi}^2 - \left(C\sigma + 3 - \frac{3}{8}k_1\right)R\phi^2 - \frac{k_1\bar{k}_1}{4}\phi^4 \\
&\le -\begin{bmatrix} R \\ \phi^2 \end{bmatrix}^{\mathsf{T}} \begin{bmatrix} C\sigma - \frac{3}{2} & \frac{1}{2}\left(C\sigma + 3 - \frac{3}{8}k_1\right) \\ \frac{1}{2}\left(C\sigma + 3 - \frac{3}{8}k_1\right) & \frac{1}{4}k_1\bar{k}_1 \end{bmatrix} \begin{bmatrix} R \\ \phi^2 \end{bmatrix} \\
&\quad - \left(\bar{k}_1 - \frac{(2C\sigma+3)^2}{2} - 1\right)\phi^2 - \left(k_2 - k_1 - \frac{9}{4}k_1^2 - \frac{9k_1}{4\bar{k}_1}\right. \\
&\quad \left. - \frac{(k_1^2-1)^2}{4}\right)\tilde{\psi}^2 .
\end{aligned}
\tag{12.13}
$$

Hence, $\dot{V}$ is negative definite on the domain $\mathcal{D}$, provided that the quadratic form above is positive definite and that the coefficients multiplying ϕ^2 and $\tilde{\psi}^2$ be positive. By imposing the positive definiteness of the quadratic form we obtain $C\sigma - \frac{3}{2} > 0$, $\left(C\sigma - \frac{3}{2}\right)\frac{1}{4}k_1\bar{k}_1 - \frac{1}{4}\left(C\sigma + 3 - \frac{3}{8}k_1\right)^2 > 0$, while by imposing the positivity of the coefficients of the remaining two terms we get $\bar{k}_1 > \frac{(2C\sigma+3)^2}{2} + 1$, $k_2 > k_1 + \frac{9}{4}k_1^2 + \frac{9k_1}{4\bar{k}_1} + \frac{(k_1^2-1)^2}{4}$. By using the definition of $\bar{k}_1$, inequalities (12.4), (12.5), (12.6), and (12.7) follow. In conclusion, if k_1, k_2, and C are chosen so that (12.4)-(12.7) hold, we have that $\dot{V}$ is negative definite on $\mathcal{D}$ which contains the origin. This leads to the conclusion that $\{R = 0, \phi = 0, \tilde{\psi} = 0\}$ is an asymptotically stable equilibrium point, which in turn implies that $\{R = 0, \phi = 0, \psi = 0\}$ is an asymptotically stable equilibrium point.

Our next objective is to show that $\mathcal{D}$ is a region of attraction for the origin. This, however, is not immediately evident from our result, since the set $\{x \in \mathsf{R}^3 : V \le K, K > 0\}$ is unbounded and, due to the presence of the term CR in V, it is not completely contained in $\mathcal{D}$. In other words, it may happen that, while the Lyapunov function is decreasing, R becomes negative, and thus the state trajectory exits the set $\mathcal{D}$, where $\dot{V}$ is guaranteed to be negative definite. Therefore, in order to complete our analysis, we need to show that $\mathcal{D}$ is invariant, which, together with $\dot{V} < 0$, implies that the set $\{x \in \mathrm{R}^3 : V \leq K, K > 0\} \cap \mathcal{D}$ is a region of attraction of the origin for any $K > 0$. This is readily seen by noticing that, on the boundary of $\mathcal{D}$, $R = 0$. From (12.2), $R = 0$ implies $\dot{R} = 0$, thus proving that no trajectory of the system can cross the boundary of $\mathcal{D}$, and therefore $\mathcal{D}$ is invariant. In conclusion, given any initial condition $x(0)$ in $\mathcal{D}$, there exists a constant $K > 0$ such that the initial condition is contained in the set $\{x \in \mathrm{R}^3 : V \leq K, K > 0\} \cap \mathcal{D}$, thus proving that the origin of system (12.2) is an asymptotically stable equilibrium point with domain of attraction $\mathcal{D}$. ■

We conclude this section by remarking that inequalities (12.4)-(12.7) represent conservative bounds on k_1 and k_2. In practical implementation, these parameters may be chosen significantly smaller after some tuning.

12.2.2 Output-Feedback Design

The control design carried out in the previous section relies on the knowledge of the mass flow Φ and the normalized stall cell squared amplitude R. In this section we extend our design to the more realistic situation when only the plenum pressure rise Ψ (and hence ψ) is available for feedback. To this end, we will seek to employ the separation principle found in Chapter 10 to define an observer-based output-feedback controller.

Recalling that the plant (12.2) is SISO with input $u = \Phi_T - 1$ and output ψ, the observability mapping is given by

$$y_e = \mathcal{H}(x_a) = \begin{bmatrix} \psi \\ -\dfrac{1}{\beta^2}(u - \phi) \\ \dfrac{1}{\beta^2}\left(-\dot{u} - \psi - \dfrac{3}{2}\phi^2 - \dfrac{1}{2}\phi^3 - 3R\phi - 3R\right) \end{bmatrix}, \tag{12.14}$$

where $y_e = [y, \dot{y}, \ddot{y}]^\top$ and $x_a = [x^\top, u, \dot{u}]^\top$. In what follows, we will implement the output feedback control scheme illustrated in Section 10.4.4. Specifically, since $\mathcal{H}$ depends on u and $\dot{u}$, we augment the system with two integrators at the input side

$$\dot{s}_1 = s_2, \quad \dot{s}_2 = v \tag{12.15}$$

and, using backstepping, we redesign the control input for the augmented system with state $x_a = [x^\top, s_1, s_2]^\top$ as follows:

$$v = \dot{\alpha} - \tilde{s}_1 - k_4\tilde{s}_2 \stackrel{\triangle}{=} \bar{v}(x_a), \tag{12.16}$$

where

$$\begin{aligned} \tilde{s}_1 &= s_1 - \bar{u} \\ \alpha &= -k_3\tilde{s}_1 - \frac{\partial V}{\partial x}g(x) + \frac{\partial \bar{u}}{\partial x}[f(x) + g(x)\,s_1] \\ \tilde{s}_2 &= s_2 - \alpha, \end{aligned} \tag{12.17}$$

and k_3, k_4 are arbitrary positive constants. The Lyapunov function of the closed-loop augmented system is

$$\bar{V}(x_a) = V(x) + \frac{1}{2}\bar{s}_1^2 + \frac{1}{2}\bar{s}_2^2.$$

Notice that, following the same reasoning as in the proof of Lemma 12.1 the set $\{[R, \phi, \psi, s_1, s_2]^\top \in \mathbb{R}^5 \; : \; R \geq 0\}$ is invariant; hence by employing backstepping we guarantee that the origin of the augmented system is asymptotically stable with domain of attraction $\bar{\mathcal{D}} = \mathcal{D} \times \mathbb{R}^2$. Note that, by using the design outlined in Section 10.6, one could avoid the dynamic extension (12.15) and hence significantly simplify the state-feedback control design, at the expense of loosing the asymptotic stability of the closed-loop system formed by the plant and the output-feedback controller. We leave the implementation of the control scheme of Section 10.6 as an exercise to the reader.

We now seek to design a nonlinear observer to estimate the state of the system. Before doing that we have to check whether the uniform complete observability assumption which is key to the results of Theorem 10.2 is satisfied for system (12.2). To this end, we are interested in checking whether the mapping $\mathcal{H}$ in (12.14) is invertible with respect to x, for any $x \in \mathbb{R}^3$, and for all $[u, \dot{u}]^\top \in \mathbb{R}^2$. Noting that when $\phi = -1$ $\mathcal{H}$ does not depend on R, we have that R cannot be calculated as a function of y_e. In other words, when $\phi = -1$ $\mathcal{H}$ is *not* a diffeomorphism and hence the system *is not* uniformly completely observable. Note however that $\mathcal{H}$ is a diffeomorphism on $\mathcal{X} = \{x \in \mathbb{R}^3 \,|\, \phi \neq -1\}$ and that the operating point $\phi = -1$ at which the observability mapping is singular corresponds to the situation when $\Phi = 0$, i.e., there is no mass flow through the compressor, which is a condition one would like to avoid during normal engine operation. Note also that the singularity in the observability mapping poses a theoretical problem in that the separation principle illustrated in Theorem 10.2 requires the plant to be uniformly completely observable. When the UCO assumption is violated, one may have that during the transient the observer-generated estimate crosses the region where the observability mapping has a singularity, i.e., in this case, the domain $\partial\mathcal{X} = \{x \in \mathbb{R}^3 \; : \; \phi = -1\}$. When this happens, the observer (10.39) becomes undefined because $[\partial\mathcal{H}(\hat{x}, s)/\partial\hat{x}]^{-1}$ does not exist whenever $x \in \partial\mathcal{X}$.

In conclusion, in order to achieve separation between the controller and observer designs when the plant in not UCO, one must make sure that the observer estimate does not cross the unobservable boundary $\partial\mathcal{X}$ during transient or, in other words, one must constrain the observer estimate to lie inside the observable set $\mathcal{X}$. Keeping this in mind, recall that the dynamic projection (10.63), when applied to the nonlinear observer (10.39), constrains its state in within the compact set $\mathcal{H}^{-1}(\mathcal{C})$, where $\mathcal{C}$ is a convex compact set such that $\mathcal{F}(\Omega_{c_2}) \subset \mathcal{C}$, where in this case $\mathcal{F} : \mathbb{R}^5 \to \mathbb{R}^5$ and

Ω_{c2} are defined by

$$\mathcal{F}(x_a) = \begin{bmatrix} \mathcal{H}(x, s_1, s_2) \\ s_1 \\ s_2 \end{bmatrix},$$

$$\Omega_{c_2} = \{x_a \in \mathsf{R}^5 \ : \ \bar{V}(x_a) \leq c_2, R \geq 0\}.$$

Notice that the set Ω_{c_2} is compact. If we choose $\mathcal{C}$ such that $\mathcal{F}^{-1}(\mathcal{C}) \subset \mathcal{X} \times \mathsf{R}^2$ then the dynamic projection (10.63), besides eliminating the peaking phenomenon of the observer, guarantees that the observer estimates are contained in within the observable region $\mathcal{X}$. In conclusion, we seek to find a compact convex set $\mathcal{C}$ satisfying the following condition

$$\mathcal{F}(\Omega_{c_2}) \subset \mathcal{C} \subset \mathcal{F}(\mathcal{X} \times \mathsf{R}^2). \tag{12.18}$$

Given a pair of scalars a_i, b_i, $i = 1, \ldots, 5$, it is easy to see that the set

$$\begin{aligned} \mathcal{C} = \Bigg\{ & [y_e^\top, s^\top]^\top \in \mathsf{R}^5 \ : \ y_{e,1} \in [a_1, b_1], y_{e,2} \in \left[\frac{a_2 - s_1}{\beta^2}, \frac{b_2 - s_1}{\beta^2}\right], \\ & y_{e,3} \in \left[\frac{1}{\beta^2}(-s_2 + a_3), \frac{1}{\beta^2}(-s_2 + b_3)\right], s_1 \in [a_4, b_4], s_2 \in [a_5, b_5] \Bigg\} \end{aligned}$$

is contained in $\mathcal{F}(\mathcal{X} \times \mathsf{R}^2)$ for all $a_2 < 1$ and it is compact. Furthermore, the set $\mathcal{C}$ is convex. In order to see that, note that $\mathcal{C}$ can be written as the cross product of three sets $\mathcal{C}_1$, $\mathcal{C}_2$, and $\mathcal{C}_3$, i.e., $\mathcal{C} = \mathcal{C}_1 \times \mathcal{C}_2 \times \mathcal{C}_3$, where

$$\begin{aligned} \mathcal{C}_1 &= \left\{ [y_{e,2}, s_1]^\top \in \mathsf{R}^2 \ : \ y_{e,2} \in \left[\frac{a_2 - s_1}{\beta^2}, \frac{b_2 - s_1}{\beta^2}\right], s_1 \in [a_4, b_4] \right\}, \\ \mathcal{C}_2 &= \left\{ [y_{e,3}, s_2] \in \mathsf{R}^2 \ : \ y_{e,3} \in \left[\frac{-s_2 + a_3}{\beta^2}, \frac{-s_2 + b_3}{\beta^2}\right], s_2 \in [a_5, b_5] \right\}, \\ \mathcal{C}_3 &= \{y_{e,1} \in \mathsf{R} \ : \ y_{e,1} \in [a_1, b_1]\}. \end{aligned}$$

The sets $\mathcal{C}_1$, $\mathcal{C}_2$, $\mathcal{C}_3$ are immediately seen to be convex and, hence, the resulting product set $\mathcal{C}$ is convex as well (it is easy to show that the product set of any finite number of convex sets is a convex set). Hence, in order to satisfy the condition in (12.18), it remains to use the Lyapunov function $\bar{V}$ to find the largest value of c_2 such that $\Omega_{c_2} \subset \mathcal{X} \times \mathsf{R}^2$ and subsequently pick values for the scalars a_i, b_i, $i = 1, \ldots, 5$ such that $a_2 < 1$ and $\mathcal{F}(\Omega_{c_2}) \subset \mathcal{C}$. A more practical way to address the design of $\mathcal{C}$ entails running a number of simulations for the closed-loop system under state feedback corresponding to several initial conditions $x_a(0)$ and calculating upper and lower bounds of $\psi(t)$, $-\phi(t)$, $-\psi(t) - 3/2\phi^2(t) - 1/2\phi^3(t) - 3R(t)\phi(t) - 3R(t)$, $s_1(t)$, $s_2(t)$. These will provide the values of a_i, b_i, $i = 1, \ldots, 5$, respectively. By doing that, we found that whenever $x_a(0) \in \Omega_0 \triangleq \{x_a \in \mathsf{R}^5 \ : \ R \in [0, 0.1], \phi \in [-0.1, 0.1], \psi \in [-0.5, 0.5], s_1 \in [-0.1, 0.1], s_2 \in [-0.1, 0.1]\}$, we have that

$a_1 = -1.15$, $b_1 = 0.5$, $a_2 = 0.3$, $b_2 = -0.1$, $a_3 = -0.75$, $b_3 = 0.4$, $a_4 = -2$, $b_4 = 7$, $a_5 = -70$, $b_5 = 250$. Notice that the set $\mathcal{C}$ defined above does not have a smooth boundary. Its boundary is continuous but not differentiable at some corners. This, in general, may generate some numerical problems in the projection, since at the points when the boundary is not differentiable the normal vector to $\mathcal{C}$ is not uniquely defined. Should numerical problems arise, one may slightly modify the definition of $\mathcal{C}$ by smoothing out the corners.

Having chosen the set $\mathcal{C}$, we define the nonlinear observer

$$\begin{aligned}
\dot{\hat{R}} = & \; -\sigma \hat{R}^2 - \sigma R(2\hat{\phi} + \hat{\phi}^2) \\
& - \frac{(l_1/\eta) + \beta^2(3\hat{\phi} + 3\hat{R} + (3/2)\hat{\phi}^2)(l_2/\eta^2) + \beta^2(l_3/\eta^3)}{3(1+\hat{\phi})}(\psi - \hat{\psi}) \\
\dot{\hat{\phi}} = & \; -\hat{\psi} - 3/2\,\hat{\phi}^2 - 1/2\,\hat{\phi}^3 - 3\hat{R}\hat{\phi} - 3\hat{R} + \beta^2(l_2/\eta^2)(\psi - \hat{\psi}) \\
\dot{\hat{\psi}} = & \; -\frac{z_1 - \hat{\phi}}{\beta^2} + (l_1/\eta)(\psi - \hat{\psi})
\end{aligned}$$

where η is a positive design parameter and the vector $L = [l_1, l_2, l_3]^\top \in \mathrm{R}^3$ is chosen to be $[3, 3, 1]$, which is Hurwitz. Next, we calculate the solution P of the Lyapunov equation $P(A_c - LC_c) + (A_c - LC_c)^\top P = -I$, where (A_c, C_c) is a canonical observable pair. The dynamic projection confining the observer state in within the compact set $\mathcal{F}^{-1}(\mathcal{C})$ is given by

$$\dot{\hat{x}}^P = \left[\frac{\partial \mathcal{H}}{\partial \hat{x}}\right]^{-1} \left\{\mathcal{P}\left(\hat{y}_e, \dot{\hat{y}}_e, s, \dot{s}\right) - \frac{\partial \mathcal{H}}{\partial s}\dot{s}\right\}$$

$$\mathcal{P}(\hat{y}_e, \dot{\hat{y}}_e, s, \dot{s}) = \begin{cases} \dot{\hat{y}}_e - \Gamma \dfrac{N_{y_e}(\hat{y}_e, s)\left(N_{y_e}(\hat{y}_e, s)^\top \dot{\hat{y}}_e + N_s(\hat{y}_e, s)^\top \dot{s}\right)}{N(\hat{y}_e, s)^\top \Gamma N(\hat{y}_e, s)} \\ \qquad \text{if } N(\hat{y}_e, s)^\top \dot{\hat{y}}_e + N_s(\hat{y}_e, s)^\top \dot{s} \geq 0 \text{ and } \hat{y}_e \in \partial\mathcal{C} \\ \dot{\hat{y}}_e \quad \text{otherwise,} \end{cases}$$

where $\Gamma = (S\mathcal{E}')^{-1}(S\mathcal{E}')^{-1}$, $S = S^\top$ denotes the matrix square root of P, $\hat{y}_e = \mathcal{H}(\hat{x}, s)$, $\dot{\hat{y}}_e = \left\{\frac{\partial \mathcal{H}}{\partial \hat{x}}\dot{\hat{x}} + \frac{\partial \mathcal{H}}{\partial s}\dot{s}\right\}$ and $N_{y_e}(\hat{y}_e, s)$, $N_s(\hat{y}_e, s)$ denote the y_e and s components of the normal vector to the set $\mathcal{C}$, which are given by

$$N_{y_e}(\hat{y}_e, s) = \begin{cases} [1, 0, 0]^\top & \text{if } \hat{y}_{e,1} = b_1 \\ [-1, 0, 0]^\top & \text{if } \hat{y}_{e,1} = a_1 \\ [0, 1, 0]^\top & \text{if } \hat{y}_{e,2} = -\frac{1}{\beta^2}(s_1 - b_2) \\ [0, -1, 0]^\top & \text{if } \hat{y}_{e,2} = -\frac{1}{\beta^2}(s_1 - a_2) \\ [0, 0, 1]^\top & \text{if } \hat{y}_{e,3} = \frac{1}{\beta^2}(-s_2 + b_3) \\ [0, 0, -1]^\top & \text{if } \hat{y}_{e,3} = \frac{1}{\beta^2}(-s_2 + a_3) \\ [0, 0, 0]^\top & \begin{cases} \text{if } s_1 = a_4, \text{ or } s_1 = b_4, \\ \text{or } s_2 = a_5, \text{ or } s_2 = b_5 \end{cases} \end{cases}$$

and

$$N_s(\hat{y}_e, s) = \begin{cases} [0,0]^\top & \text{if } \hat{y}_{e,1} = b_1 \text{ or } \hat{y}_{e,1} = a_1 \\ \left[\frac{1}{\beta^2}, 0\right]^\top & \text{if } \hat{y}_{e,2} = -\frac{1}{\beta^2}(s_1 - b_2) \\ \left[-\frac{1}{\beta^2}, 0\right]^\top & \text{if } \hat{y}_{e,2} = -\frac{1}{\beta^2}(s_1 - a_2) \\ \left[0, \frac{1}{\beta^2}\right]^\top & \text{if } \hat{y}_{e,3} = \frac{1}{\beta^2}(-s_2 + b_3) \\ \left[0, -\frac{1}{\beta^2}\right]^\top & \text{if } \hat{y}_{e,3} = \frac{1}{\beta^2}(-s_2 + a_3) \\ [1,0]^\top & \text{if } s_1 = b_4 \\ [-1,0]^\top & \text{if } s_1 = a_4 \\ [0,1]^\top & \text{if } s_2 = b_5 \\ [0,-1]^\top & \text{if } s_2 = a_5. \end{cases}$$

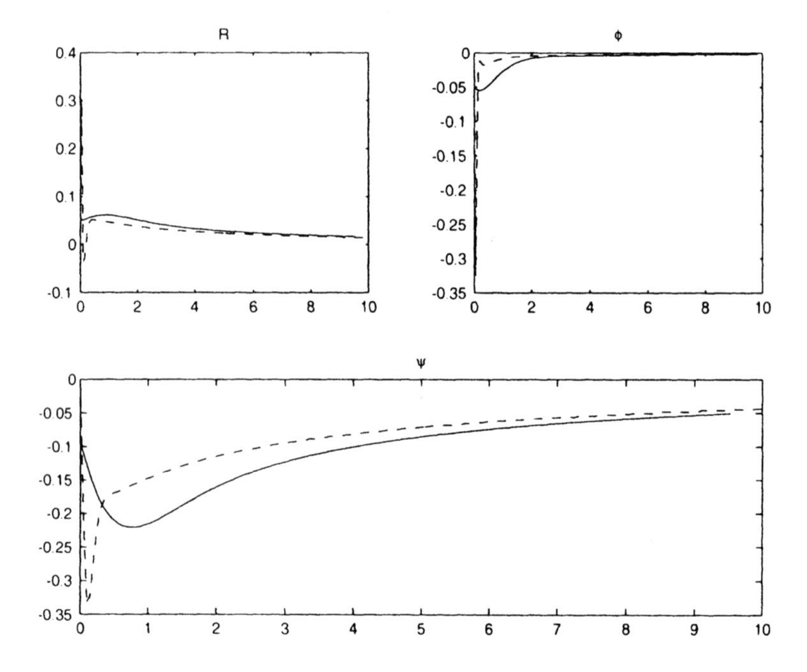

Figure 12.1. Time evolutions of the states and the observer estimates for $\eta = 1/10$.

Having designed a nonlinear observer and a dynamic projection, we are ready to define the output-feedback controller based on a separation principle

$$u = (1 - \beta^2 k_1 k_2)\hat{\phi}^P + \hat{\beta}^{P2} k_2 \hat{\psi}^P + 3\beta^2 k_1 \hat{R}^P \hat{\phi}^P,$$

where the choice of $k_1 = 25$ and $k_2 = 1.1 \cdot 10^5$ fulfills inequalities (12.4)-(12.7) in Lemma 12.1.

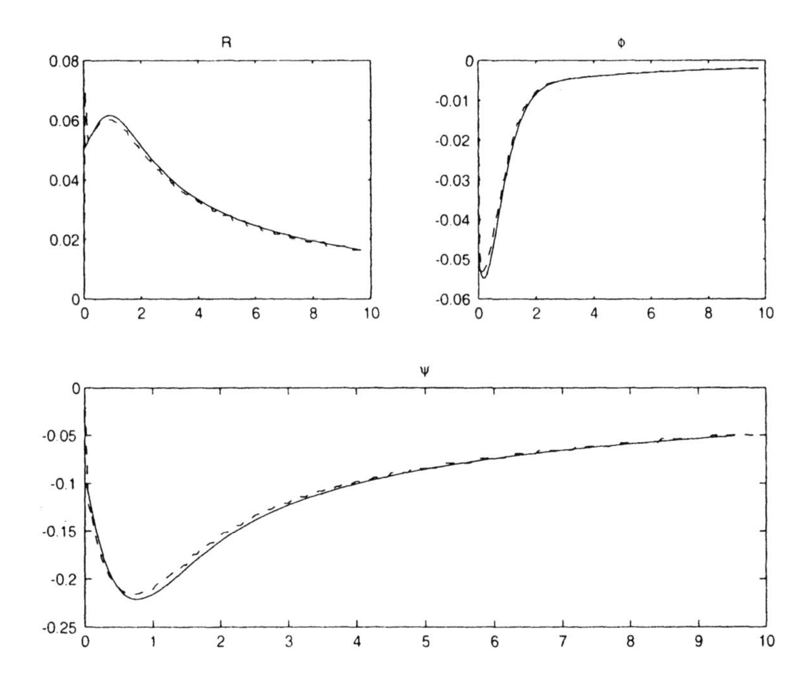

Figure 12.2. Time evolutions of the states and the observer estimates for $\eta = 1/50$.

Figures 12.1 and 12.2 depict the time evolutions of the states and the observer estimates for $\eta = 1/10$ and $\eta = 1/50$, respectively (where the performance obtained using state feedback is shown by the solid line). Notice that the observer convergence when $\eta = 1/50$ is significantly faster, as predicted by the theory in Chapter 10. In Figure 12.3 the trajectories of the system under state feedback are compared to the ones under output feedback for decreasing values of η. The smaller η is, the closer the output-feedback trajectories are to the state feedback ones.

12.3 Adaptive Stabilization: Electromagnet Control

In this section we consider the problem of using a pair of electromagnets to position a beam of mass m at a desired position (see Figure 12.4). Each electromagnet is independently driven by a linear power amplifier providing a voltage u_a and u_b across its windings to electromagnet a and b, respectively. In order to model the dynamics of the plant, we start by defining x as the beam position with $x = 0$ when the beam is centered between the electromagnets. Let G denote the nominal gaps of the electromagnets (i.e., G is the distance between each electromagnet and the beam when

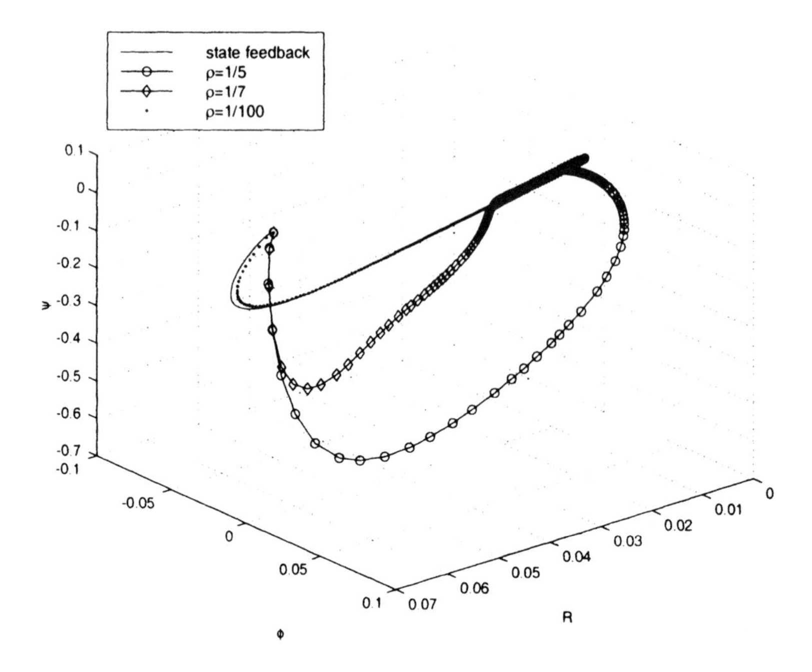

Figure 12.3. State trajectory of the closed-loop system under state feedback and output feedback for decreasing values of η.

$x = 0$). Next, the force exerted by each electromagnet is assumed to be proportional to the square of the corresponding phase current and inversely proportional to the squared distance to the beam.

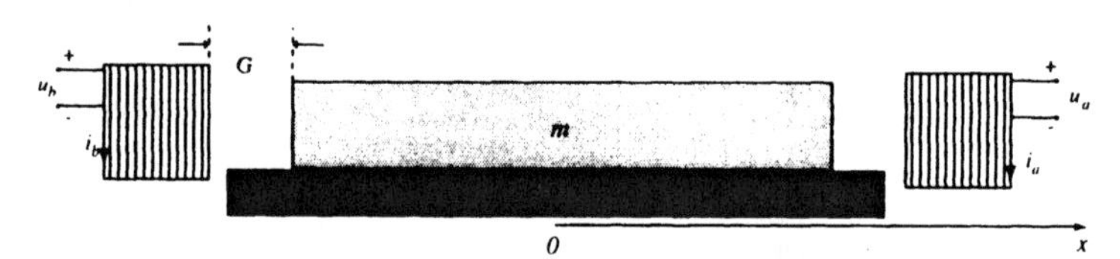

Figure 12.4. Electromagnet control of a beam.

With this framework we obtain the system model defined by

$$\begin{aligned}
\dot{x} &= v \\
m\dot{v} &= \frac{k_a}{2}\frac{i_a^2}{g_a^2(x)} - \frac{k_b}{2}\frac{i_b^2}{g_b^2(x)} + \Delta(x) \\
L_a\dot{i}_a &= -R_a i_a + u_a \\
L_b\dot{i}_b &= -R_b i_b + u_b,
\end{aligned} \tag{12.19}$$

where v is the beam velocity, and i_a and i_b are the currents for phase a and

b, respectively. Here we also have the gaps $g_a(x) = G - x$, $g_b(x) = G + x$, actuator constants k_a and k_b, inductance L_a and L_b, and resistance R_a and R_b. According to our model, the phase-a electromagnet pulls in the positive x direction, while phase-b pulls in the negative x direction. Here $\Delta(x)$ is a position-dependent force created by unmodeled components such as, e.g., friction. Our control objective is to regulate the position of the beam at some desired location r (where $|r| < G$) while keeping the state of the system bounded. Notice that if either $x = G$ or $x = -G$, that is, when there is physical contact between the beam and one of the electromagnets, then the system equation (12.19) has a singularity and hence its solution is not well-defined. In order to avoid this situation, the controller must be designed in such a way that the singularities are avoided.

We will start our control design by assuming that the unknown friction force $\Delta(x) = 0$ and we will find an *ideal* state-feedback controller fulfilling our design objectives. After that, we will use state-feedback adaptive control techniques of Chapters 7 and 8 to find a controller when $\Delta \neq 0$. Our design will be completed by replacing the adaptive state-feedback controller by an output-feedback controller which estimates the beam velocity and invokes the separation principle developed in Chapter 11 to recover the performance of the adaptive state-feedback controller.

12.3.1 Ideal Controller Design

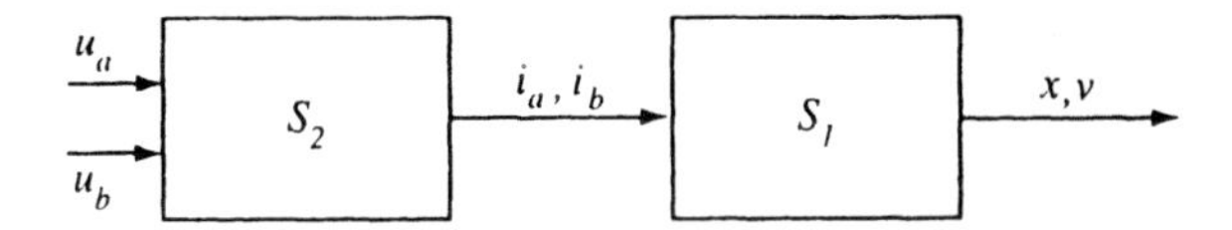

Figure 12.5. Block diagram of the magnetic levitation system.

Rewrite (12.19) as the interconnection of two subsystems with the topology depicted in Figure 12.5, where S_1, S_2 are given by

$$
\begin{aligned}
S_1 : &\begin{cases} \dot{x} = v \\ m\dot{v} = \dfrac{k_a}{2}\dfrac{i_a^2}{g_a^2(x)} - \dfrac{k_b}{2}\dfrac{i_b^2}{g_b^2(x)} + \Delta \end{cases} \\
S_2 : &\begin{cases} L_a \dot{i}_a = -R_a i_a + u_a \\ L_b \dot{i}_b = -R_b i_b + u_b . \end{cases}
\end{aligned}
\tag{12.20}
$$

Notice that the pair (S_1, S_2) is cascade-connected and hence has a triangular structure: we will utilize this structure to simplify the control design procedure. Specifically, referring to Figure 12.5, we will first design a controller for S_1 assuming that i_a and i_b are control inputs. Then, applying the backstepping methodology to S_2, we will derive stabilizing voltage inputs u_a and u_b.

Lemma 12.2: *Consider subsystem S_1 with $\Delta = 0$, viewed as a system with state $[x, v]^\top$ and input (i_a, i_b). Define a controller*

$$\begin{aligned} i_a &= \sqrt{\frac{2m\eta_a(z)g_a^2}{k_a}} \\ i_b &= \sqrt{\frac{2m\eta_b(z)g_b^2}{k_b}}, \end{aligned} \tag{12.21}$$

where $\eta_a(z)$ and $\eta_b(z)$ are chosen such that $\eta_a(z) - \eta_b(z) = z$ and such that $\eta_a(z), \eta_b(z) \geq 0$ for all z where z is defined by (12.25). Then the closed-loop system has an equilibrium at $(x, v) = (r, 0)$ which is exponentially stable.

Proof: Define the first error term as $e_1 = x - r$ so that

$$\dot{e}_1 = v. \tag{12.22}$$

Ideally we could choose $v = -\kappa_1 e_1$ with $\kappa_1 > 0$ so that $\dot{e}_1 = -\kappa_1 e_1$ has $e_1 = 0$ as an exponentially stable equilibrium point. But v is not an input (it is the velocity), so instead we choose $e_2 = v + \kappa_1 e_1$ (so if $e_2 \to 0$, then the e_1 dynamics have the desired form). Taking the derivative of e_2, we find

$$\dot{e}_2 = \frac{k_a i_a^2}{2mg_a^2} - \frac{k_b i_b^2}{2mg_b^2} + \kappa_1 \dot{e}_1, \tag{12.23}$$

where $\dot{e}_1 = -\kappa_1 e_1 + e_2$.

Now consider the Lyapunov candidate $V_1 = \frac{1}{2}e_1^2 + \frac{1}{2}e_2^2$ so that

$$\dot{V}_1 = e_1(-\kappa_1 e_1 + e_2) + e_2\left(\frac{k_a i_a^2}{2mg_a^2} - \frac{k_b i_b^2}{2mg_b^2} + \kappa_1 \dot{e}_1\right). \tag{12.24}$$

If we can choose

$$\frac{k_a i_a^2}{2mg_a^2} - \frac{k_b i_b^2}{2mg_b^2} = z,$$

where

$$z = -\kappa_1 \dot{e}_1 - e_1 - \kappa_1 e_2, \tag{12.25}$$

then $\dot{V}_1 = -2\kappa_1 V_1$, so $V_1 = 0$ is an exponentially stable equilibrium point. Notice that this is achieved by letting

$$\begin{aligned} i_a &= \nu_a = \sqrt{\frac{2m\eta_a(z)g_a^2}{k_a}} \\ i_b &= \nu_b = \sqrt{\frac{2m\eta_b(z)g_b^2}{k_b}} \end{aligned} \tag{12.26}$$

where $\eta_a(z)$ and $\eta_b(z)$ are chosen such that $\eta_a(z)-\eta_b(z) = z$ for all z. Keeping $\eta_a, \eta_b \geq 0$ ensures that the arguments to the square-root function are positive as long as $m, k_a, k_b > 0$, which must be the case for any physically realizable system. ■

All that is left to do for the S_1 controller is to find some η_a and η_b satisfying the assumptions of Lemma 12.2. One choice is to simply let

$$\eta_a = \begin{cases} z+b & \text{if } z \geq 0 \\ b & \text{otherwise} \end{cases} \qquad \eta_b = \begin{cases} -z+b & \text{if } z \leq 0 \\ b & \text{otherwise,} \end{cases} \tag{12.27}$$

where $b \geq 0$ is a bias current applied to each electromagnet. Notice that the $\eta_a(z)$ term is used when $z \geq 0$ and the $\eta_b(z)$ term is used when z is negative. The combination then works together to create the desired forces in the system.

This above choice for η_a, η_b is Lipschitz continuous so that the solution of the differential equations will be well defined and unique, but it is not smooth. Since we plan to use the backstepping approach to define the controller for the cascaded system, we will want to define a controller that is smooth (or at least once differentiable). A smooth approximation to (12.27) may be defined by

$$\eta_a = \frac{z+\sqrt{z^2+\epsilon}}{2} + b \qquad \eta_b = \frac{-z+\sqrt{z^2+\epsilon}}{2} + b \tag{12.28}$$

with $\epsilon, b > 0$. Notice that as $\epsilon \to 0$ we find (12.28) becomes equivalent to (12.27). Notice that even though this is an approximation to our original choice of η_a, η_b, the smooth version does satisfy both requirements of Lemma 12.2 (i.e., $\eta_a(z) - \eta_b(z) = z$ and $\eta_a, \eta_b \geq 0$ for all z) so that we still achieve exponential stability of the closed-loop system.

Having found a controller for subsystem S_1, we proceed to the derivation of a controller for the cascade (S_1, S_2) when $\Delta = 0$ (see Figure 12.5).

Lemma 12.3: *Consider the cascade connection of system S_2 and S_1 and let $e_3 = i_a - \nu_a$, $e_4 = i_b - \nu_b$ with ν_a and ν_b defined by (12.26). The controller (12.33) guarantees that $e = 0$ is an exponentially stable equilibrium point and the set*

$$\mathcal{D} = \left\{ e \in \mathbb{R}^4 : |e|^2 \leq |e(0)|^2 \right\},$$

is positively invariant when $\Delta = 0$.

Notice that Lemma 12.3 implies that all trajectories of the cascade connection (S_1, S_2) originating in $\mathcal{D}$ at $t = 0$ remain in $\mathcal{D}$ for all future time. Furthermore, noting that $e \in \mathcal{D}$ implies that

$$|x - r|^2 = e_1^2 \leq |e(0)|^2,$$

we conclude that the initial conditions may be chosen such that the closed-loop system trajectories do not cross the singularity at $x = \pm G$. The proof of Lemma 12.3 is a straightforward application of the backstepping methodology illustrated in Chapter 6.

Proof: Consider the cascade interconnection of S_1 and S_2 and the following Lyapunov function candidate

$$V = V_1 + \frac{1}{2}e_3^2 + \frac{1}{2}e_4^2, \tag{12.29}$$

where $e_3 = i_a - \nu_a$ and $e_4 = i_b - \nu_b$ and V_1 is defined in Lemma 12.2. Taking the derivative of this Lyapunov candidate we find

$$\dot{V} = \dot{V}_1 + (i_a - \nu_a)(\dot{i}_a - \dot{\nu}_a) + (i_b - \nu_b)(\dot{i}_b - \dot{\nu}_b). \tag{12.30}$$

Using the cascaded system definition we find

$$\begin{aligned}\dot{V} &= e_1(-\kappa_1 e_1 + e_2) + e_2\left(\frac{k_a i_a^2}{2mg_a^2} - \frac{k_b i_b^2}{2mg_b^2} + \kappa_1 \dot{e}_1\right) \\ &\quad + (i_a - \nu_a)(\dot{i}_a - \dot{\nu}_a) + (i_b - \nu_b)(\dot{i}_b - \dot{\nu}_b).\end{aligned} \tag{12.31}$$

Adding and subtracting the ideal control laws ν_a and ν_b for the first subsystem

$$\begin{aligned}\dot{V} &= e_2\left(\frac{k_a(\nu_a^2 + i_a^2 - \nu_a^2)}{2mg_a^2} - \frac{k_b(\nu_b^2 + i_b^2 - \nu_b^2)}{2mg_b^2} + \kappa_1 \dot{e}_1\right) \\ &\quad + e_1(-\kappa_1 e_1 + e_2) + (i_a - \nu_a)(\dot{i}_a - \dot{\nu}_a) + (i_b - \nu_b)(\dot{i}_b - \dot{\nu}_b)\end{aligned}$$

so that

$$\begin{aligned}\dot{V} &= -2\kappa_1 V_1 + e_2\left(\frac{k_a(i_a^2 - \nu_a^2)}{2mg_a^2} - \frac{k_b(i_b^2 - \nu_b^2)}{2mg_b^2}\right) \\ &\quad + (i_a - \nu_a)(\dot{i}_a - \dot{\nu}_a) + (i_b - \nu_b)(\dot{i}_b - \dot{\nu}_b).\end{aligned} \tag{12.32}$$

Using the electrical dynamics for each phase, we then choose the controllers

$$\begin{bmatrix} u_a \\ u_b \end{bmatrix} = \begin{bmatrix} R_a i_a + L_a\left(-e_2\left(\frac{k_a(i_a + \nu_a)}{2mg_a^2}\right) + \dot{\nu}_a - \kappa_2 e_3\right) \\ R_b i_b + L_b\left(e_2\left(\frac{k_b(i_b + \nu_b)}{2mg_b^2}\right) + \dot{\nu}_b - \kappa_2 e_4\right) \end{bmatrix}, \tag{12.33}$$

where $\kappa_2 > 0$ so that $\dot{V} \leq -2\kappa_m V$ with $\kappa_m = \min(\kappa_1, \kappa_2)$. With this inequality we find that $V(t) \leq V(0)$ for all t. ■

12.3.2 Adaptive Controller Design

In the previous section we developed a controller which yields desired stability properties of the closed-loop system when the uncertainty $\Delta(x) = 0$. Here we focus on the development of an adaptive controller which employs on-line approximation to recover the properties of (12.33) when $\Delta(x)$ is not necessarily zero. We will assume that a linear in the parameter approximator has been defined such that there exists an ideal parameter set θ such that $|\Delta(x) + \mathcal{F}(x, \theta)| \leq W$ for all $x \in [-G, G]$.

Lemma 12.4: *Consider the cascade connection of system S_2 and S_1 and let $e_3 = i_a - \nu_a(\hat{z})$, $e_4 = i_b - \nu_b(\hat{z})$ with ν_a and ν_b defined by (12.26) with $\hat{z} = -\kappa_1 \dot{e}_1 - e_1 - (\kappa_1 + \eta)e_2 + \mathcal{F}(x, \hat{\theta})/m$.*

The controller

$$\begin{bmatrix} u_a \\ u_b \end{bmatrix} = \begin{bmatrix} R_a i_a + L_a \left(q_a - e_2 \left(\frac{k_a (i_a + \nu_a(\hat{z}))}{2 m g_a^2} \right) - \zeta_a e_3 - \frac{\partial \nu_a}{\partial v} \frac{\mathcal{F}(x, \hat{\theta})}{m} \right) \\ R_b i_b + L_b \left(q_b + e_2 \left(\frac{k_b (i_b + \nu_b(\hat{z}))}{2 m g_b^2} \right) - \zeta_b e_4 - \frac{\partial \nu_b}{\partial v} \frac{\mathcal{F}(x, \hat{\theta})}{m} \right) \end{bmatrix}$$
$$\zeta_a = \kappa_2 + \eta \left(\frac{\partial \nu_a}{\partial v} \right)^2, \qquad \zeta_b = \kappa_2 + \eta \left(\frac{\partial \nu_b}{\partial v} \right)^2 \tag{12.34}$$

with $\kappa_2, \eta > 0$ and parameter update law

$$\dot{\hat{\theta}} = -\Gamma \left[\left(\frac{\partial \mathcal{F}}{\partial \theta} \right)^{\top} \left(e_2 - \frac{e_3}{m} \frac{\partial \nu_a(\hat{z})}{\partial v} - \frac{e_4}{m} \frac{\partial \nu_b(\hat{z})}{\partial v} \right) + \sigma(\hat{\theta} - \theta^0) \right], \tag{12.35}$$

is such that, if the following condition is satisfied

$$\frac{1}{2} \sum_{i=1}^{4} e_i(0)^2 + \tilde{\theta}(0)^{\top} \Gamma^{-1} \tilde{\theta}(0) \leq \frac{(G - |r|)^2}{2},$$

the closed-loop system trajectories are uniformly ultimately bounded with arbitrarily small ultimate bound. The values of q_a and q_b will be defined in the following proof.

Proof: Assume that for a given approximator $\mathcal{F}(x, \hat{\theta})$ there exists some θ such that $|\Delta(x) + \mathcal{F}(x, \theta)| \leq W$ for all $x \in [-G, G]$. Then one may want to define a stabilizing signal for the S_1 subsystem using

$$\bar{z} = -\kappa_1 \dot{e}_1 - e_1 - (\kappa_1 + \eta)e_2 + \frac{\mathcal{F}(x, \theta)}{m}, \tag{12.36}$$

based on (12.25) where the η term has been included to account for the ideal approximation error. The S_2 controller could then be defined based on this stabilizing signal as was done for the static case using backstepping. Since θ is unknown, however, we will instead consider the signal

$$\hat{z} = -\kappa_1 \dot{e}_1 - e_1 - (\kappa_1 + \eta)e_2 + \frac{\mathcal{F}(x, \hat{\theta})}{m}, \tag{12.37}$$

which is defined using the parameter estimate. With this, consider the cascade interconnection of S_1 and S_2 and the Lyapunov-like function

$$V_s = V_1 + \frac{1}{2}e_3^2 + \frac{1}{2}e_4^2, \tag{12.38}$$

where $e_3 = i_a - \nu_a(\hat{z})$ and $e_4 = i_b - \nu_b(\hat{z})$ and V_1 is defined in Lemma 12.2. Notice that V_s does not contain the parameters $\hat{\theta}$ so it can not be used to assess the stability properties of the entire system. Taking the derivative of this function we find

$$\begin{aligned}\dot{V}_s &= e_1(-\kappa_1 e_1 + e_2) + e_2\left(\frac{k_a i_a^2}{2mg_a^2} - \frac{k_b i_b^2}{2mg_b^2} + \kappa_1 \dot{e}_1 + \frac{\Delta(x)}{m}\right) \\ &\quad + (i_a - \nu_a(\hat{z}))(\dot{i}_a - \dot{\nu}_a(\hat{z})) + (i_b - \nu_b(\hat{z}))(\dot{i}_b - \dot{\nu}_b(\hat{z})).\end{aligned}$$

Adding and subtracting the ideal control laws $\nu_a^2(\bar{z})$ and $\nu_b^2(\bar{z})$ for the first subsystem

$$\begin{aligned}\dot{V}_s &= e_1(-\kappa_1 e_1 + e_2) + e_2\left(\kappa_1 \dot{e}_1 + \frac{\Delta(x)}{m}\right) \\ &\quad + e_2\left(\frac{k_a(\nu_a^2(\bar{z}) + i_a^2 - \nu_a^2(\bar{z}))}{2mg_a^2} - \frac{k_b(\nu_b^2(\bar{z}) + i_b^2 - \nu_b^2(\bar{z}))}{2mg_b^2}\right) \\ &\quad + (i_a - \nu_a(\hat{z}))(\dot{i}_a - \dot{\nu}_a(\hat{z})) + (i_b - \nu_b(\hat{z}))(\dot{i}_b - \dot{\nu}_b(\hat{z})),\end{aligned}$$

so that

$$\begin{aligned}\dot{V}_s &= -2\kappa_1 V_1 + e_2\frac{w(x)}{m} - \eta e_2^2 + e_2\left(\frac{k_a(i_a^2 - \nu_a^2(\bar{z}))}{2mg_a^2} - \frac{k_b(i_b^2 - \nu_b^2(\bar{z}))}{2mg_b^2}\right) \\ &\quad + (i_a - \nu_a(\hat{z}))(\dot{i}_a - \dot{\nu}_a(\hat{z})) + (i_b - \nu_b(\hat{z}))(\dot{i}_b - \dot{\nu}_b(\hat{z})),\end{aligned}$$

where $w(x) = \Delta(x) + \mathcal{F}(x, \theta)$ is the ideal approximation error. Now adding and subtracting $\nu_a^2(\hat{z})$ and $\nu_b^2(\hat{z})$

$$\begin{aligned}\dot{V}_s &= -2\kappa_1 V_1 + e_2\frac{w(x)}{m} - \eta e_2^2 + e_2\left(\frac{k_a(i_a^2 - \nu_a^2(\hat{z}))}{2mg_a^2} - \frac{k_b(i_b^2 - \nu_b^2(\hat{z}))}{2mg_b^2}\right) \\ &\quad + e_2\left(\frac{k_a(\nu_a^2(\hat{z}) - \nu_a^2(\bar{z}))}{2mg_a^2} - \frac{k_b(\nu_b^2(\hat{z}) - \nu_b^2(\bar{z}))}{2mg_b^2}\right) \\ &\quad + (i_a - \nu_a(\hat{z}))(\dot{i}_a - \dot{\nu}_a(\hat{z})) + (i_b - \nu_b(\hat{z}))(\dot{i}_b - \dot{\nu}_b(\hat{z})).\end{aligned} \tag{12.39}$$

But

$$\begin{aligned}\frac{k_a(\nu_a^2(\hat{z}) - \nu_a^2(\bar{z}))}{2mg_a^2} - \frac{k_b(\nu_b^2(\hat{z}) - \nu_b^2(\bar{z}))}{2mg_b^2} &= \eta_a(\hat{z}) - \eta_b(\hat{z}) - \eta_a(\bar{z}) + \eta_b(\bar{z}) \\ &= \hat{z} - \bar{z} = \frac{\partial \mathcal{F}}{\partial \theta}\tilde{\theta}\end{aligned}$$

with $\tilde{\theta} = \hat{\theta} - \theta$ so that

$$\begin{aligned} \dot{V}_s &= -2\kappa_1 V_1 + e_2 \frac{w(x)}{m} - \eta e_2^2 + e_2 \frac{\partial \mathcal{F}}{\partial \theta} \tilde{\theta} \\ &+ e_2 \left(\frac{k_a (i_a^2 - \nu_a^2(\hat{z}))}{2 m g_a^2} - \frac{k_b (i_b^2 - \nu_b^2(\hat{z}))}{2 m g_b^2} \right) \\ &+ (i_a - \nu_a(\hat{z}))(\dot{i}_a - \dot{\nu}_a(\hat{z})) + (i_b - \nu_b(\hat{z}))(\dot{i}_b - \dot{\nu}_b(\hat{z})). \end{aligned} \tag{12.40}$$

Notice that

$$\begin{aligned} \dot{\nu}_a(\hat{z}) &= \frac{\partial \nu_a}{\partial x} \dot{x} + \frac{\partial \nu_a}{\partial v} \dot{v} + \frac{\partial \nu_a}{\partial \theta} \dot{\hat{\theta}} \\ &= \frac{\partial \nu_a}{\partial x} v + \frac{\partial \nu_a}{\partial v} \left(\frac{k_a i_a^2}{2 m g_a^2} - \frac{k_b i_b^2}{2 m g_b^2} + \frac{\Delta(x)}{m} \right) + \frac{\partial \nu_a}{\partial \theta} \dot{\hat{\theta}} \end{aligned}$$

contains the unknown $\Delta(x)$ term. Substituting this into (12.40) we now find

$$\begin{aligned} \dot{V}_s &= -2\kappa_1 V_1 + e_2 \frac{w(x)}{m} - \eta e_2^2 + e_2 \left(\frac{k_a (i_a^2 - \nu_a^2(\hat{z}))}{2 m g_a^2} - \frac{k_b (i_b^2 - \nu_b^2(\hat{z}))}{2 m g_b^2} \right) \\ &+ (i_a - \nu_a(\hat{z})) \left(\dot{i}_a - q_a - \frac{\partial \nu_a(\hat{z})}{\partial v} \frac{\Delta(x)}{m} \right) \\ &+ (i_b - \nu_b(\hat{z})) \left(\dot{i}_b - q_b - \frac{\partial \nu_b(\hat{z})}{\partial v} \frac{\Delta(x)}{m} \right) + e_2 \frac{\partial \mathcal{F}}{\partial \theta} \tilde{\theta}, \end{aligned}$$

where

$$\begin{aligned} q_a &= \frac{\partial \nu_a(\hat{z})}{\partial x} v + \frac{\partial \nu_a(\hat{z})}{\partial v} \left(\frac{k_a i_a^2}{2 m g_a^2} - \frac{k_b i_b^2}{2 m g_b^2} \right) + \frac{\partial \nu_a(\hat{z})}{\partial \theta} \dot{\hat{\theta}} \\ q_b &= \frac{\partial \nu_b(\hat{z})}{\partial x} v + \frac{\partial \nu_b(\hat{z})}{\partial v} \left(\frac{k_a i_a^2}{2 m g_a^2} - \frac{k_b i_b^2}{2 m g_b^2} \right) + \frac{\partial \nu_b(\hat{z})}{\partial \theta} \dot{\hat{\theta}}. \end{aligned}$$

Using the electrical dynamics for each electromagnet and the controller (12.34) we get

$$\begin{aligned} \dot{V}_s &= -2\kappa_1 V_1 + e_2 \frac{w(x)}{m} - \eta e_2^2 + e_2 \frac{\partial \mathcal{F}}{\partial \theta} \tilde{\theta} \\ &- e_3 \frac{\partial \nu_a(\hat{z})}{\partial v} \frac{w(x)}{m} - \zeta_a e_3^2 - \frac{e_3}{m} \frac{\partial \nu_a(\hat{z})}{\partial v} \frac{\partial \mathcal{F}}{\partial \theta} \tilde{\theta} \\ &- e_4 \frac{\partial \nu_b(\hat{z})}{\partial v} \frac{w(x)}{m} - \zeta_b e_4^2 - \frac{e_4}{m} \frac{\partial \nu_b(\hat{z})}{\partial v} \frac{\partial \mathcal{F}}{\partial \theta} \tilde{\theta}. \end{aligned}$$

Using the definition of ζ_a and ζ_b and the inequality $-x^2 \pm 2xy \leq y^2$ three times we have

$$\dot{V}_s \leq -2\kappa_m V_s + \frac{3W^2}{4\eta m^2} + \left(e_2 - \frac{e_3}{m} \frac{\partial \nu_a}{\partial v} - \frac{e_4}{m} \frac{\partial \nu_b}{\partial v} \right) \frac{\partial \mathcal{F}}{\partial \theta} \tilde{\theta} \tag{12.41}$$

when $x \in [-G, G]$ (so that $|w(x)| \leq W$), where $\kappa_m = \min(\kappa_1, \kappa_2)$.

Now consider a Lyapunov candidate for the adaptive system defined as $V_a = V_s + \frac{1}{2}\tilde{\theta}^\top \Gamma^{-1}\tilde{\theta}$, where Γ^{-1} is a positive definite symmetric matrix. Then, when $x \in [-G, G]$,

$$\dot{V}_a = -2\kappa_m V_s + \frac{3W^2}{4\eta m^2} + \left(e_2 - \frac{e_3}{m}\frac{\partial \nu_a}{\partial v} - \frac{e_4}{m}\frac{\partial \nu_b}{\partial v}\right)\frac{\partial \mathcal{F}}{\partial \theta}\tilde{\theta} + \tilde{\theta}^\top \Gamma^{-1}\dot{\tilde{\theta}}. \quad (12.42)$$

With the choice of parameter update law (12.35) we get

$$\dot{V}_a = -2\kappa_m V_s + \frac{3W^2}{4\eta m^2} - \sigma\tilde{\theta}^\top\left(\hat{\theta} - \theta^0\right). \quad (12.43)$$

Since $-2\tilde{\theta}^\top\left(\hat{\theta} - \theta^0\right) \leq -|\tilde{\theta}|^2 + |\theta - \theta^0|^2$ we find

$$\begin{aligned}
\dot{V}_a &\leq -2\kappa_m V_s - \frac{\sigma}{2}|\tilde{\theta}|^2 + \frac{3W^2}{4\eta m^2} + \frac{\sigma}{2}|\theta - \theta^0|^2 \\
&\leq -2\kappa_m V_s - \frac{\sigma}{2\lambda_{\max}(\Gamma^{-1})}\tilde{\theta}^\top\Gamma^{-1}\tilde{\theta} + \frac{3W^2}{4\eta m^2} + \frac{\sigma}{2}|\theta - \theta^0|^2.
\end{aligned}$$

Thus, we have shown that on the set $\{[x, v, i_a, i_b, \tilde{\theta}]^\top \in \mathrm{R}^{p+5} : x \in [-G, G]\}$ $\dot{V}_a \leq -k_1 V_a + k_2$, where

$$k_1 = \min\left(2\kappa_m, \frac{\sigma}{\lambda_{\max}(\Gamma^{-1})}\right)$$

and

$$k_2 = \frac{3W^2}{4\eta m^2} + \frac{\sigma}{2}|\theta - \theta^0|^2.$$

Note now that, by proper choice of the controller parameters, k_1 may be made arbitrarily large and thus such that

$$\frac{k_2}{k_1} \leq \frac{(G - |r|)^2}{2}.$$

From this fact, if the initial conditions of the adaptive system are chosen so that

$$V_a(0) \leq \frac{(G - |r|)^2}{2},$$

then it is readily seen that

$$V_a(t) = \frac{1}{2}\sum_{i=1}^{4} e_i^2(t) + \frac{1}{2}\tilde{\theta}(t)^\top\Gamma^{-1}\tilde{\theta}(t) \leq \frac{(G - |r|)^2}{2}, \text{ for all } t \geq 0,$$

and hence, in particular,

$$\frac{1}{2}e_1^2(t) \le V_a(t) \le \frac{(G - |r|)^2}{2}, \text{ for all } t \ge 0,$$

or $|x| \le G$ for all $t \ge 0$. In other words, if the initial conditions and the controller parameters are chosen appropriately, one can guarantee that the closed-loop trajectories are confined inside the set where the approximation holds, and the solution $(e(t), \tilde{\theta}(t))$ is uniformly ultimately bounded with arbitrarily small ultimate bound. ■

The performance of the closed-loop system using the above adaptive state-feedback controller was tested before moving to the output-feedback case to ensure that proper performance may be met when full state information is available. In this study, we will assume that $m = 1kg$, $G = 0.02m$, $k_a, k_b = 0.1Nm^2/A^2$, $R_a, R_b = 2\Omega$, and $L_a, L_b = 0.001H$.

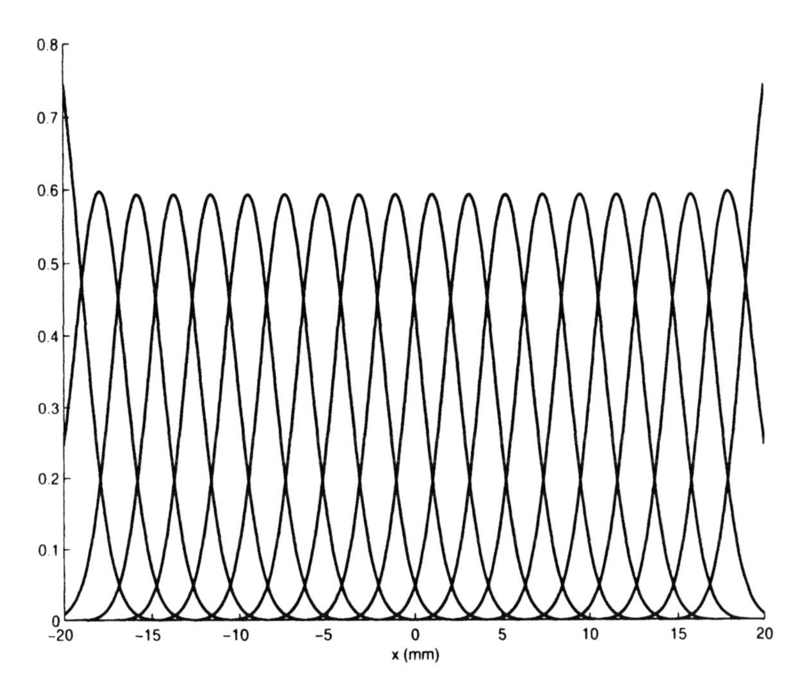

Figure 12.6. Radial basis neural network activation functions.

The parameters of the controller were chosen by first simulating the system with the static state-feedback controller with $\Delta = 0$. It was found that choosing $\kappa_1 = 50$, $\kappa_2 = 400$, and $b = 1$ results in desirable performance. The approximator was then designed using a radial basis neural network with $p = 20$ activation functions as shown in Figure 12.6. The parameters for the adaptive controller were chosen by first simulating the case when the adaptive terms were set to zero (i.e., $\Gamma = 0$). The rate of adaptation was then increased until desired performance was obtained. We chose to use values of $\Gamma = 5000I$, $\sigma = 0.0001$, $\theta^0 = 0$, and $\eta = 1$.

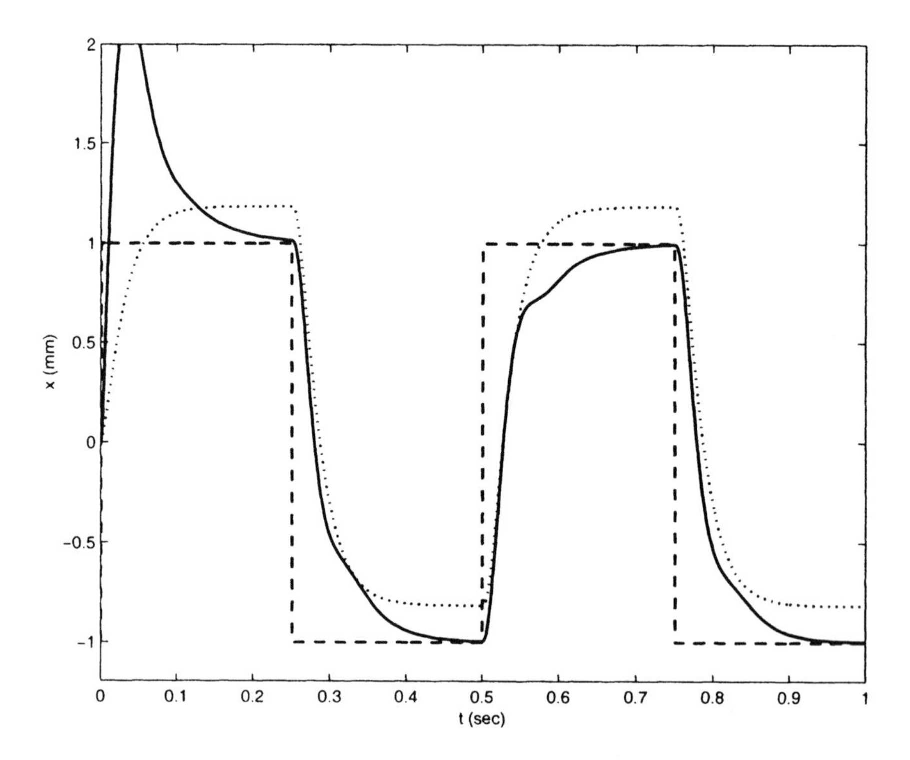

Figure 12.7. Position x when using static feedback control (–) and adaptive state-feedback control ($\cdots$). Here the dashed line (- -) is the reference signal r.

To test the controller, we set $r = \pm 1mm$ through a series of steps. Both the static and adaptive controller were tested with $\Delta(x) = 0.5N$ for all x. The beam position is shown in Figure 12.7. We see that the static case has a constant offset from the ideal position due to the disturbance force, while the adaptive controller is able to cancel its effects. Figure 12.8 shows the current supplied to the two electromagnets when using the static state-feedback controller (the currents for the adaptive case look similar).

12.3.3 Output-Feedback Extension

In the previous section we have defined an adaptive controller which employs state-feedback to achieve the objective of positioning the beam at a desired location. In this section we use the tools developed in Chapter 11 to extend these results to the case when the beam velocity is not available

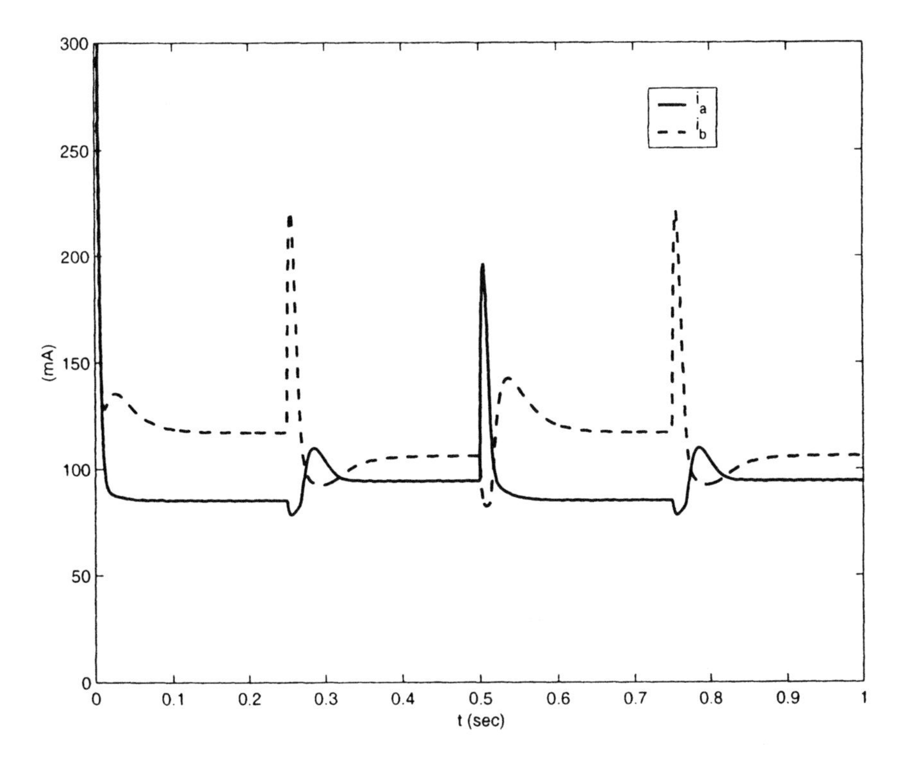

Figure 12.8. Actuator current using state-feedback control.

for feedback, while the position and currents are measurable.

We will begin the output-feedback design by first partitioning the system dynamics of (12.19). let $x_1 = [x, v]^\top$ and $x_2 = [i_a, i_b]$ and view the system dynamics (12.19) as the connection of two subsystems with states x_1 and x_2 and outputs $y_1 = x$ and $y_2 = x_2$. In other words, the state of the first subsystem must be estimated from its output y_1 (notice that the output of the first subsystem is in this case its first state variable), while the state of the second subsystem is directly available for feedback. Hence, the problem of controlling the magnetic levitation system by output feedback is identical to that stated in Section 11.3.1. Following the procedure outlined there, we write the subsystem dynamics as

$$\begin{bmatrix} \dot{x}_1 \\ \hline \dot{x}_2 \end{bmatrix} = \begin{bmatrix} v \\ \frac{k_a}{2m}\frac{i_a^2}{g_a^2(x)} - \frac{k_b}{2m}\frac{i_b^2}{g_b^2(x)} + \frac{\Delta(x)}{m} \\ \hline -\frac{R_a}{L_a} i_a + \frac{1}{L_a} u_a \\ -\frac{R_b}{L_b} i_b + \frac{1}{L_b} u_b, \end{bmatrix}. \tag{12.44}$$

The observability mapping for the x_1 dynamics is given as

$$y_{e,1} = \mathcal{H}_1(x_1, x_2, u_a, u_b, \Delta) = \begin{bmatrix} y_1 \\ \dot{y}_1 \end{bmatrix} = \begin{bmatrix} x \\ v \end{bmatrix}, \tag{12.45}$$

so that

$$\frac{\partial \mathcal{H}_1(\hat{x}_1, x_2, u_a, u_b, \Delta)}{\partial \hat{x}_1} = I. \tag{12.46}$$

Using (11.46) with $\Delta = 0$, the high gain observer becomes

$$\begin{bmatrix} \dot{\hat{x}}_{1,1} \\ \dot{\hat{x}}_{1,2} \end{bmatrix} = \begin{bmatrix} \hat{x}_{1,2} \\ \frac{k_a}{2m}\frac{i_a^2}{g_a^2(x)} - \frac{k_b}{2m}\frac{i_b^2}{g_b^2(x)} \end{bmatrix} + \mathcal{E}^{-1}L(y_1 - \hat{x}_{1,1}), \tag{12.47}$$

where $\mathcal{E}^{-1} = \mathrm{diag}(1/\eta, 1/\eta^2)$ with $\eta > 0$, and $L = [l_1, l_2]^\top$ with $s^2 + l_1 s + l_2$ Hurwitz. Even though we set $\Delta = 0$ to define the observer, it is possible to make the estimation error for the velocity signal arbitrarily small by choosing the observer gain, η, large enough.

In order to avoid the peaking phenomenon one can choose to replace the dynamic projection with a saturation applied to $\hat{x}_1$ so that, for instance, the σ-modified adaptive output-feedback controller is defined by (12.34) and (12.35) with the velocity v replaced with

$$\hat{v}^s = \mathcal{V}\mathrm{sat}(\mathcal{V}^{-1}\hat{x}_{1,2}), \tag{12.48}$$

where $\mathcal{V}$ is the largest velocity magnitude obtainable using state-feedback. Notice that there is no need here to replace x by $\hat{x}^s$, since x in this case is the output of the system. In conclusion, not surprisingly, the design of an output-feedback controller based on a separation principle reduces simply to the employment of a high-gain observer to estimate the beam velocity and estimate saturation to eliminate the destabilizing effect of the peaking phenomenon.

A value of $L = [10, 25]^\top$ was chosen for L so that $s^2 + 10s + 25$ is Hurwitz with both poles at $s = -5$. The performance of the adaptive output-feedback controller is shown in Figure 12.9 when η was chosen to be 0.1, 0.01, and 0.001. Notice that as η is decreased, the performance of the state-feedback controller is recovered.

Two alternative approaches to this stabilization problem are presented in exercises at the end of this chapter. The first one (see Exercise 12.2) is a variation of the cascaded design presented here, while the second one (see Exercise 12.3) entails designing a controller for the desired actuator force rather than desired currents in S_1.

12.4 Tracking: VTOL Aircraft

In this section we apply the theory of output-feedback tracking developed in Chapter 10 to the VTOL (vertical take off and landing) aircraft depicted

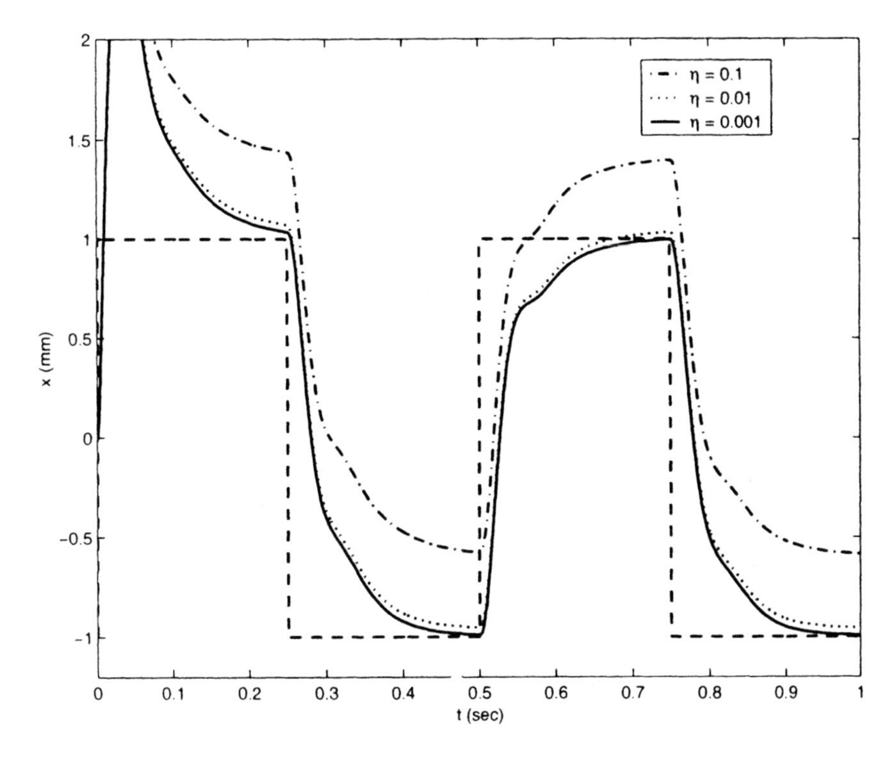

Figure 12.9. Position x when using output-feedback control with different values of η.

in Figure 12.10 whose simplified mathematical model, taken from [150], is

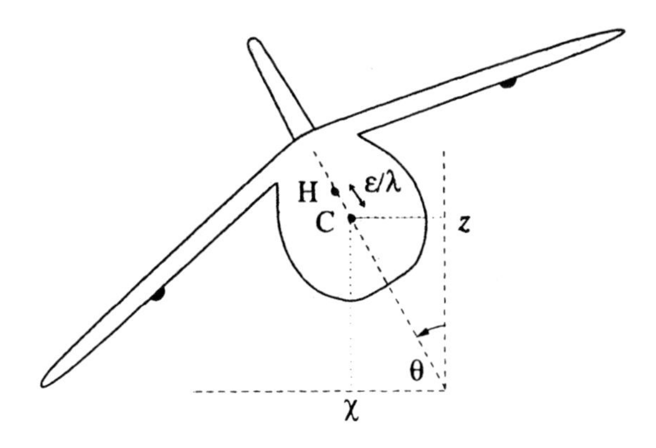

Figure 12.10. VTOL aircraft.

given by

$$\begin{aligned}\ddot{\chi} &= -u_1 \sin\theta + \epsilon u_2 \cos\theta,\\ \ddot{z} &= u_1 \cos\theta + \epsilon u_2 \sin\theta - g,\\ \ddot{\theta} &= \lambda u_2,\end{aligned} \tag{12.49}$$

where χ and z are the coordinates of the center of mass of the aircraft on a fixed inertial frame, and θ indicates its inclination with respect to the vertical axis. The state of this system is $x = [\chi, \dot{\chi}, z, \dot{z}, \theta, \dot{\theta}]^\top$, and we assume the measurable output is given by the *Huygens center of oscillation* of the aircraft (denoted by H in Figure 12.10),

$$y = \left[\chi - \frac{\epsilon}{\lambda}\sin\theta,\ \ z + \frac{\epsilon}{\lambda}\cos\theta\right]^\top. \tag{12.50}$$

The problem we want to solve entails finding a partial information controller (i.e., a controller using only the information given by the plant output, the reference trajectory, and the control input) making the output track a smooth reference trajectory.

12.4.1 Finding the Practical Internal Model

Following the development in Section 10.8, we seek to find a practical internal model satisfying Definition 10.2. We will start by considering a copy of the plant (12.49) with state x^r, input c^r, and output r

$$\begin{aligned}\dot{x}^r{}_1 &= x^r{}_2\\ \dot{x}^r{}_2 &= -c^r{}_1 \sin x^r{}_5 + \epsilon c^r{}_2 \cos x^r{}_5\\ \dot{x}^r{}_3 &= x^r{}_4\\ \dot{x}^r{}_4 &= c^r{}_1 \cos x^r{}_5 + \epsilon c^r{}_2 \sin x^r{}_5 - g\\ \dot{x}^r{}_5 &= x^r{}_6\\ \dot{x}^r{}_6 &= \lambda c^r{}_2\\ r &= \left[x^r{}_1 - \frac{\epsilon}{\lambda}\sin x^r{}_5,\ \ x^r{}_3 + \frac{\epsilon}{\lambda}\cos x^r{}_5\right]\end{aligned} \tag{12.51}$$

and forming the associated observability mapping. To do that, following the procedure outlined in Section 10.5, select two indices k_1 and k_2 corresponding to the number of times each of the two outputs of the system needs to be differentiated to yield a diffeomorphism. Selecting $k_1 = k_2 = 3$

we get

$$r_e = \begin{bmatrix} r_1 \\ \dot{r}_1 \\ \ddot{r}_1 \\ r_2 \\ \dot{r}_2 \\ \ddot{r}_2 \end{bmatrix} = \mathcal{H}(x^r, c^r) = \begin{bmatrix} x^r{}_1 - \frac{\epsilon}{\lambda}\sin(x^r{}_5) \\ x^r{}_2 - \frac{\epsilon}{\lambda}\cos(x^r{}_5)x^r{}_6 \\ \sin(x^r{}_5)\left(\frac{\epsilon}{\lambda}x^{r2}_6 - c^r{}_1\right) \\ x^r{}_3 + \frac{\epsilon}{\lambda}\cos(x^r{}_5) \\ x^r{}_4 - \frac{\epsilon}{\lambda}\sin(x^r{}_5)x^r{}_6 \\ -\cos(x^r{}_5)\left(\frac{\epsilon}{\lambda}x^{r2}_6 - c^r{}_1\right) - g \end{bmatrix}. \tag{12.52}$$

Note now that the observability mapping depends on the unknown inputs $c^r{}_1$, $c^r{}_2$ and thus we have to find a compensator which, besides being regular, is such that the observability mapping of the plant augmented with it is a diffeomorphism independent of the control input. Consider, to this end, the compensator

$$\begin{aligned} \dot{\zeta}^r{}_1 &= \zeta^r{}_2 \\ \dot{\zeta}^r{}_2 &= -\bar{v}^r{}_1 \sin x^r{}_5 + \bar{v}^r{}_2 \cos x^r{}_5 + \zeta^r{}_2 (x^r{}_6)^2 \\ c^r &= \left[\zeta^r{}_1 + \frac{\epsilon}{\lambda}(x^r{}_6)^2,\ \frac{1}{\lambda\zeta^r{}_1}\left(-\bar{v}^r{}_1\cos x^r{}_5 - \bar{v}^r{}_2 \sin x^r{}_5 - 2\zeta^r{}_2 x^r{}_6\right)\right]^{\mathsf{T}}, \end{aligned} \tag{12.53}$$

which, in [150], was shown to make the augmented system (12.51), (12.53) state-feedback linearizable with respect to the output r.

Indeed the change of coordinates $\xi_r = \mathcal{T}(x^r, \zeta^r{}_1, \zeta^r{}_2)$ given by

$$\begin{aligned} \xi_{r1} &= r_1 = x^r{}_1 - \frac{\epsilon}{\lambda}\sin x^r{}_5 \\ \xi_{r2} &= \dot{r}_1 = x^r{}_2 - \frac{\epsilon}{\lambda} \\ \xi_{r3} &= \ddot{r}_1 = -\zeta^r{}_1 \sin x^r{}_5 \\ \xi_4 &= r_1^{(3)} = -\dot{\zeta}^r{}_2 \sin x^r{}_5 - \zeta^r{}_1 \dot{x}^r{}_6 \cos x^r{}_5 \\ \xi_{r5} &= r_2 = x^r{}_3 + \frac{\epsilon}{\lambda}\cos x^r{}_5 \\ \xi_{r6} &= \dot{r}_2 = x^r{}_4 - \frac{\epsilon}{\lambda}\dot{x}^r{}_6 \sin x^r{}_5 \\ \xi_7 &= \ddot{r}_2 = \zeta^r{}_1 \cos x^r{}_5 - g \\ \xi_{r8} &= r_2^{(3)} = \dot{\zeta}^r{}_2 \cos x^r{}_5 - \zeta^r{}_1 \dot{x}^r{}_6 \sin x^r{}_5 \end{aligned} \tag{12.54}$$

is a diffeomorphism on $\{[x^{r\mathsf{T}}, \zeta^r{}_1, \zeta^r{}_2]^{\mathsf{T}} \in \mathsf{R}^6 \times \mathsf{R}^2 : \zeta^r{}_1 \neq 0\}$, and transforms the composite system (12.51), (12.53) into the linear system

$$\dot{\xi}_r = A_c \xi_r + B_c \bar{v}^r,$$

where

$$A_c = \mathrm{diag}[a_c, a_c], \qquad B_c = \mathrm{diag}[b_c, b_c],$$

and

$$a_c = \begin{bmatrix} 0 & 1 & 0 & 0 \\ 0 & 0 & 1 & 0 \\ 0 & 0 & 0 & 1 \\ 0 & 0 & 0 & 0 \end{bmatrix}, \qquad b_c = \begin{bmatrix} 0 \\ 0 \\ 0 \\ 1 \end{bmatrix}.$$

A closer look at the mapping at (12.54) shows that ξ_r is the vector

$$[r_1, \dot{r}_1, r_1^{(2)}, r_1^{(3)}, r_2, \dot{r}_2, r_2^{(2)}, r_2^{(3)}]^{\mathsf{T}} \in \mathsf{R}^8$$

and hence T is the observability mapping associated with the copy of the plant (12.51) augmented with the compensator (12.53) and the choice of indices $k_1 = k_2 = 4$. Since T does not depend on $\bar{v}^r$ and it is a diffeomorphism on $\{[x^{r\mathsf{T}}, \zeta^r{}_1, \zeta^r{}_2]^{\mathsf{T}} \in \mathsf{R}^6 \times \mathsf{R}^2 \,:\, \zeta^r{}_1 \neq 0\}$, we conclude that the compensator (12.53) is a good candidate for a practical internal model. Note, however, that the output c^r of the compensator is a function of both the state of the augmented system and its input $\bar{v}^r$, while the output of a practical internal model should only be a function of the state so that the estimation of the state of the augmented systems allows for the estimation of $c^r(t)$. In order to remove this obstacle one can further extend the compensator dynamics by adding integrators at the input side,

$$\begin{aligned} \dot{\zeta}^r{}_3 &= v^r{}_1 \\ \dot{\zeta}^r{}_4 &= v^r{}_2 \\ \zeta^r{}_3 &= v^{r\prime}{}_1, \qquad \zeta^r{}_4 = v^{r\prime}{}_2 \end{aligned}$$

therefore obtaining the compensator

$$\begin{aligned} \dot{\zeta}^r{}_1 &= \zeta^r{}_2 \\ \dot{\zeta}^r{}_2 &= -\zeta^r{}_3 \sin x^r{}_5 + \zeta^r{}_4 \cos x^r{}_5 + \zeta^r{}_2 (x^r{}_6)^2 \\ \dot{\zeta}^r{}_3 &= v^r{}_1 \\ \dot{\zeta}^r{}_4 &= v^r{}_2 \\ c^r &= \left[\zeta^r{}_1 + \frac{\epsilon}{\lambda}(x^r{}_6)^2,\ \frac{1}{\lambda \zeta^r{}_1}(-\zeta^r{}_3 \cos x^r{}_5 - \zeta^r{}_4 \sin x^r{}_5 - 2\zeta^r{}_2 x^r{}_6)\right]^{\mathsf{T}}, \end{aligned} \tag{12.55}$$

with state $\zeta^r = [\zeta^r{}_1, \zeta^r{}_2, \zeta^r{}_3, \zeta^r{}_4]^{\mathsf{T}}$. It is now easy to see that (12.55) is a practical internal model in the sense of Definition 10.2:

(i) The observability mapping of the composite system (12.51), (12.55) is given by

$$r_e = \begin{bmatrix} r_1 \\ \vdots \\ r_1^{(5)} \\ r_2 \\ \vdots \\ r_2^{(5)} \end{bmatrix} = \mathcal{H}'(x^r, \zeta^r) = \begin{bmatrix} \xi_{r1} \\ \vdots \\ \xi_{r4} \\ \zeta^r{}_3 \\ \xi_{r5} \\ \vdots \\ \xi_{r8} \\ \zeta^r{}_4 \end{bmatrix} \tag{12.56}$$

and thus does not depend on the control input v^r.

(ii) The mapping $\mathcal{H}'(x^r, \zeta^r)$, with the choice of indices $\bar{k}_1 = 5$, $\bar{k}_2 = 5$ (corresponding to four differentiations of r_1 and r_2), is a diffeomorphism on $\mathcal{X} = \{[x^{r\,\mathsf{T}}, \zeta^r]^\mathsf{T} \in \mathsf{R}^6 \times \mathsf{R}^4 : \zeta^r{}_1 \neq 0\}$

(iii) The compensator (12.55), viewed as a system with input v^r, output c^r, and state x^r, is regular on $\mathcal{X}$. This is seen by calculating $c^r{}_1^{(3)}$ and $\dot{c}^r{}_2$,

$$\begin{aligned} c^r{}_1^{(3)} &= -v^r{}_1 \sin x^r{}_5 + v^r{}_2 \cos x^r{}_5 + f_1(\zeta^r, x^r) \\ \dot{c}^r{}_2 &= \frac{1}{\lambda \zeta^r{}_1} (-v^r{}_1 \cos x^r{}_5 - v^r{}_2 \sin x^r{}_5) + f_2(\zeta^r, x^r) \end{aligned} \tag{12.57}$$

for some functions f_1 and f_2, and noticing that, on $\mathcal{X}$,

$$\begin{aligned} v^r{}_1 = {} & -c^r{}_1^{(3)} \sin x^r{}_5 - \lambda \zeta^r{}_1 \cos x^r{}_5 \dot{c}^r{}_2 + f_1(\zeta^r, x^r) \sin x^r{}_5 \\ & + \lambda f_2(\zeta^r, x^r) \zeta^r{}_1 \cos x^r{}_5 \\ v^r{}_2 = {} & c^r{}_1^{(3)} \cos x^r{}_5 - \lambda \zeta^r{}_1 \sin x^r{}_5 \dot{c}^r{}_2 - f_1(\zeta^r, x^r) \cos x^r{}_5 \\ & + \lambda f_2(\zeta^r, x^r) \zeta^r{}_1 \sin x^r{}_5 . \end{aligned} \tag{12.58}$$

Similarly, on $\mathcal{X}$, $\zeta^r{}_1$ and $\zeta^r{}_2$ can be expressed as functions of $(x^r, c^r{}_1)$ and $(x^r, \dot{c}^r{}_1)$, respectively, while $\zeta^r{}_3$ and $\zeta^r{}_4$ can be expressed as functions of $(x^r, \ddot{c}^r{}_1, c^r{}_2)$.

Notice, from the discussion above, that the practical internal model (12.55) is well-defined only when $[x^{r\,\mathsf{T}}, \zeta^{r\,\mathsf{T}}]^\mathsf{T} \in \mathcal{X}$. Thus, in order to apply the theory illustrated in Section 10.8 and recover the performance of a full information controller, the ideal controller has to be designed so that the closed-loop trajectory avoids the singularity in $\zeta^r{}_1 = 0$, while the dynamic projection has to bound the observers estimates away from it.

12.4.2 Full Information Controller

We now turn our attention to the problem of finding a full information *ideal* controller which solves the tracking problem assuming that x, x^r, and c^r were known. In order to do that, augment the plant with a copy of the compensator (12.53)

$$\begin{aligned} \dot{\zeta}_1 &= \zeta_2 \\ \dot{\zeta}_2 &= -v_1 \sin x_5 + v_2 \cos x_5 + \zeta_2 x_6^2 \\ u &= \left[\zeta_1 + \frac{\epsilon}{\lambda} x_6^2, \ \frac{1}{\lambda \zeta_1} \left(-v_1 \cos x_5 - v_2 \sin x_5 - 2\zeta_2 x_6 \right) \right]^{\mathsf{T}} \end{aligned} \tag{12.59}$$

and notice from the previous section that the transformation

$$\xi = T(x, \zeta_1, \zeta_2)$$

is a diffeomorphism for all $\zeta_1 \neq 0$ and transforms the augmented system (12.49), (12.59) into the linear system

$$\dot{\xi} = A_c \xi + B_c v,$$

where A_c and B_c were defined earlier. It is now easy to solve the tracking problem: consider the error system $e = T(x, \zeta_1, \zeta_2) - T(x^r, {\zeta^r}_1, {\zeta^r}_2)$, which is well-defined when $\zeta_1 \neq 0$, ${\zeta^r}_1 \neq 0$. The error dynamics are linear and read as

$$\dot{e} = A_c e + B_c \left(v - [{\zeta^r}_3, {\zeta^r}_4]^{\mathsf{T}} \right). \tag{12.60}$$

A globally exponentially stabilizing controller is given by

$$v = \begin{bmatrix} {\zeta^r}_3 \\ {\zeta^r}_4 \end{bmatrix} - Ke, \tag{12.61}$$

where K is chosen so that the matrix $A_c - B_c K$ is Hurwitz (this can be done since the pair (A_c, B_c) is controllable). In conclusion, in original state coordinates the full information controller reads as

$$\begin{aligned} \dot{\zeta}_1 &= \zeta_2 \\ \dot{\zeta}_2 &= -v_1 \sin x_5 + v_2 \cos x_5 + \zeta_2 x_6^2 \\ v &= \begin{bmatrix} {\zeta^r}_3 \\ {\zeta^r}_4 \end{bmatrix} - K \left(T(x, \zeta_1, \zeta_2) - T(x^r, {\zeta^r}_1, {\zeta^r}_2) \right) \\ u &= \left[\zeta_1 + \frac{\epsilon}{\lambda} x_6^2, \ \frac{1}{\lambda \zeta_1} \left(-v_1 \cos x_5 - v_2 \sin x_5 - 2\zeta_2 x_6 \right) \right]^{\mathsf{T}}. \end{aligned} \tag{12.62}$$

If the reference trajectory to be tracked is such that ${\zeta^r}_1(t) \neq 0$ for all $t \geq 0$, it is not difficult to find the set of initial conditions such that the singularity in $\zeta_1 = 0$ is not crossed.

12.4.3 Partial Information Controller

Having found a full information controller and a practical internal model, we now turn our attention to the design of a partial information controller based on a separation principle. We follow the developments in Section 10.8 and design first an estimator for $x^r(t)$ and $\zeta^r(t)$. This is easily done by applying an estimator of the form (10.39) to the system formed by the copy of the plant (12.51) and the practical internal model (12.55)

$$
\begin{aligned}
\dot{x}^r{}_1 &= x^r{}_2 \\
\dot{x}^r{}_2 &= -c^r{}_1 \sin x^r{}_5 + \frac{\epsilon}{\lambda\zeta^r{}_1}\left(-\zeta^r{}_3 \cos x^r{}_5 - \zeta^r{}_4 \sin x^r{}_5 - 2\zeta^r{}_2 x^r{}_6\right)\cos x^r{}_5 \\
\dot{x}^r{}_3 &= x^r{}_4 \\
\dot{x}^r{}_4 &= \left(\zeta^r{}_1 + \frac{\epsilon}{\lambda}(x^r{}_6)^2\right)\cos x^r{}_5 + \frac{\epsilon}{\lambda\zeta^r{}_1}\left(-\zeta^r{}_3 \cos x^r{}_5 - \zeta^r{}_4 \sin x^r{}_5 \right. \\
&\qquad \left. -2\zeta^r{}_2 x^r{}_6\right)\sin x^r{}_5 - g \\
\dot{x}^r{}_5 &= x^r{}_6 \\
\dot{x}^r{}_6 &= \frac{1}{\zeta^r{}_1}\left(-\zeta^r{}_3 \cos x^r{}_5 - \zeta^r{}_4 \sin x^r{}_5 - 2\zeta^r{}_2 x^r{}_6\right) \\
\dot{\zeta}^r{}_1 &= \zeta^r{}_2 \\
\dot{\zeta}^r{}_2 &= -\zeta^r{}_3 \sin x^r{}_5 + \zeta^r{}_4 \cos x^r{}_5 + \zeta^r{}_2 (x^r{}_6)^2 \\
\dot{\zeta}^r{}_3 &= v^r{}_1 \\
\dot{\zeta}^r{}_4 &= v^r{}_2 \\
r &= \left[x^r{}_1 - \frac{\epsilon}{\lambda}\sin x^r{}_5, x^r{}_3 + \frac{\epsilon}{\lambda}\cos x^r{}_5\right],
\end{aligned}
\tag{12.63}
$$

and recalling that the observability mapping associated with (12.63) is given by $\mathcal{H}'(x^r, \zeta^r) = [\xi_{r1}, \ldots, \xi_{r4}, \zeta^r{}_3, \xi_{r5}, \ldots, \xi_{r8}, \zeta^r{}_4]^\top$, with ξ_r defined in (12.54).

Next, we design a state observer using the fact that the full information controller includes the compensator (12.59). This is useful because by choosing indices $k_1 = 2$, $k_2 = 4$ and forming the observability mapping of the plant (12.49) we get the observability mapping

$$
y_e = \begin{bmatrix} y_1 \\ \dot{y}_1 \\ y_2 \\ \dot{y}_2 \\ \ddot{y}_2 \\ y_2^{(3)} \end{bmatrix} = \begin{bmatrix} x_1 - \frac{\epsilon}{\lambda}\sin x_5 \\ x_2 - \frac{\epsilon}{\lambda}\cos x_5\, x_6 \\ x_3 + \frac{\epsilon}{\lambda}\cos x_5 \\ x_4 - \frac{\epsilon}{\lambda}\sin x_5\, x_6 \\ \zeta_1 \cos x_5 - g \\ \zeta_2 \cos x_5 - \zeta_1 \sin x_5\, z_6 \end{bmatrix},
\tag{12.64}
$$

which does not depend on the control input of the augmented system, v, and is a diffeomorphism on $\{x \in \mathsf{R}^6 : x_5 \in (-\pi/2, \pi/2), z_1 \neq 0\}$. Hence, an asymptotic estimate of the state of the plant is found by applying the observer in (10.39) to the system

$$\begin{aligned}
\dot{x}_1 &= x_2 \\
\dot{x}_2 &= -\left(\zeta_1 + \frac{\epsilon}{\lambda}x_6^2\right)\sin x_5 + \epsilon\left(\frac{1}{\lambda\zeta_1}(-v_1\cos x_5 - v_2 \sin x_5\right. \\
&\quad \left. - 2\zeta_2 x_6)\right)\cos x_5 \\
\dot{x}_3 &= x_4 \\
\dot{x}_4 &= c^r{}_1 \cos x_5 + \epsilon c^r{}_2 \sin x_5 - g \\
\dot{x}_5 &= x_6 \\
\dot{x}_6 &= \lambda c^r{}_2 \\
\dot{\zeta}_1 &= \zeta_2 \\
\dot{\zeta}_2 &= -v_1 \sin x_5 + v_2 \cos x_5 + \zeta_2 (x_6)^2 \\
y &= \left[x_1 - \frac{\epsilon}{\lambda}\sin x_5, x_3 + \frac{\epsilon}{\lambda}\cos x_5\right],
\end{aligned} \tag{12.65}$$

where ζ is a measurable vector (it is part of the state of the controller) and the observability mapping if given by (12.64).

The design of the partial information controller is concluded by designing two projections, one for the stable inverse estimator and one for the state estimator. As mentioned earlier, the projections, besides eliminating the peaking phenomenon, should prevent the estimators states from crossing the regions where the mappings $\mathcal{H}$ in (12.64) and $\mathcal{H}'$ in (12.56) are not invertible. To do that, one can follow a procedure analogous to that illustrated in the stall and surge control problem: we leave this part, as well as the simulation of the partial information controller, as an exercise to the reader. In conclusion, the partial information controller solving the tracking problem is implemented by replacing the unknown values of x, x^r, and ζ^r in the full information controller (12.62) by their projected estimates.

12.5 Summary

In this chapter we have considered three design examples. In the first example, the control of stall and surge in jet engine compressors, we have seen how to apply the idea of dynamic projection introduced in Chapter 10 to achieve a separation principle in the presence of singularities in the observability mapping, that is, when the system is not uniformly completely observable. In the second design example, we studied the problem of controlling the horizontal position of a beam by using two electromagnets. We first designed a state-feedback adaptive controller and then we easily derived an adaptive output-feedback controller by using a linear high-gain observer and control saturation. Finally, in the third example, we applied the tracking theory of Chapter 10 to the control of the VTOL aircraft. This was done by first finding a practical internal model and a full information controller, and then invoking a separation principle to find a partial information controller.

12.6 Exercises and Design Problems

Exercise 12.1 (Surge and Stall without Integrators) For the stall and surge control problem, implement an output-feedback controller without using dynamic extension, that is, without augmenting the system with two integrators at the input side. Specifically, use the tools of Section 10.6 to define an appropriate state estimator.

Exercise 12.2 (Electromagnet Control) Consider the electromagnet control problem and prove that the controller for subsystem S_1 defined by

$$\begin{aligned} i_a &= g_a(x)\sqrt{\tfrac{2}{k_a\gamma}} \\ i_b &= g_b(x)\sqrt{\tfrac{2\gamma}{k_b}}(x+v+\tfrac{1}{\gamma}), \end{aligned}$$

where $\gamma > 0$ is a design parameter, is such that the origin of S_1 is exponentially stable provided γ is sufficiently small and, furthermore, its domain of attraction becomes larger as γ is made smaller. Use this result to define a state-feedback controller for the cascaded system (S_1, S_2), its output feedback counterpart, and simulate the closed-loop system under state and output feedback.

Exercise 12.3 (Electromagnet Control) Rewrite the dynamics of the electromagnet control application as

$$\begin{aligned} S_1 &: \begin{cases} \dot{x} &= v \\ m\dot{v} &= F(x, i_a, i_b) + \Delta \end{cases} \\ S_2 &: \begin{cases} L_a\dot{i}_a &= -R_a i_a + u_a \\ L_b\dot{i}_b &= -R_b i_b + u_b, \end{cases} \end{aligned} \tag{12.66}$$

where

$$F(x, i_a, i_b) = \frac{k_a}{2}\frac{i_a^2}{g_a^2(x)} - \frac{k_b}{2}\frac{i_b^2}{g_b^2(x)}.$$

Design a smooth state-feedback controller $F = \nu_1$ for S_1 using F as an input when $\Delta = 0$. Then define the augmented error $e_a = F - \nu_1$ when F is not an input, and use the backstepping approach to define a controller for the composite system. Finally, design adaptive state-feedback and output-feedback controllers for the system when $\Delta \neq 0$.

Exercise 12.4 (VTOL Aircraft) Consider the VTOL aircraft design problem and design two suitable projections to bound the stable inverse and state estimators away from regions where the associated observability mappings are singular.

Exercise 12.5 (VTOL Aircraft: Adaptive Extension) Suppose now that some uncertainty affects the system. Develop an adaptive output-feedback tracker. To do that, develop first a full information adaptive controller (i.e., an adaptive controller which exploits the knowledge of the system state and the stable inverse), and then, using the ideas introduced in Chapter 11, modify the practical internal model introduced in this chapter to deal with the presence of the uncertainty and define appropriate state and stable inverse estimators (in order to do that, you may have to introduce some restriction on the uncertainty).

Part IV

Extensions

Chapter 13

Discrete-Time Systems

13.1 Overview

Thus far we have discussed how to control continuous-time systems for both state-feedback and output-feedback problems. With today's high performance real-time systems, the control algorithms developed in the continuous-time framework are typically implemented on a sampled-data system. As long as the sampling rate of the controller is high with respect to the system dynamics, there is typically no problem with this approach. When the delays associated with the discrete-time nature of the implementation are large, however, it is possible that closed-loop system performance will become poor or even unstable. In this chapter, we will study how discrete-time designs may be used to improve performance when dealing with discrete-time systems.

Developing a controller in the continuous-time framework (even if the controller is to be implemented in a sampled-data system) allows the designer to take advantage of a number of intuitive nonlinear design techniques such as nonlinear damping. There may be times, however, when the sampled-data nature of the implementation forces us to consider the system delays. In these cases, it may be possible to develop a discrete-time model of the plant and directly develop a controller in the discrete-time framework. These two approaches (continuous and discrete) to the design of a controller for a sampled-data system are shown in Figure 13.1.

We will be interested in the control of discrete-time systems which may be transformed into

$$\begin{aligned} x(k+1) &= f_d(x(k)) + g_d(x(k))u(k) \\ y(k) &= h_d(x(k)), \end{aligned} \tag{13.1}$$

where k is the sequence index, $x \in \mathrm{R}^n$ is the state vector, $u \in \mathrm{R}^m$ is the system input and $y \in \mathrm{R}^p$ is the output. The analysis of discrete-time systems typically follows the same general approach as when dealing with continuous-time systems. One usually defines an error system that quantifies the closed-loop system performance. Once an error system is defined,

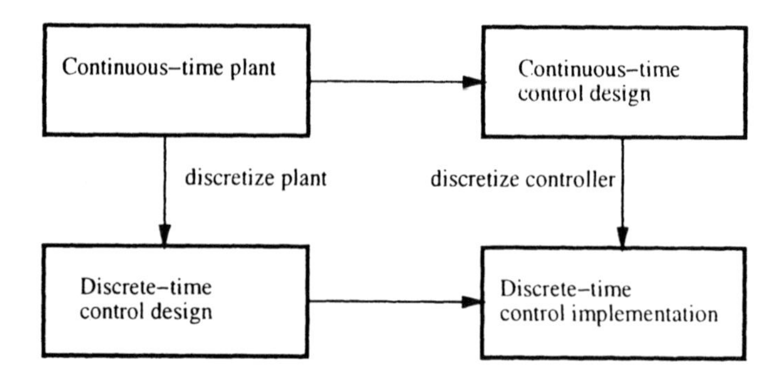

Figure 13.1. Common approaches used to design controllers in sampled-data systems.

a Lyapunov candidate is used to reduce the dimension of the problem. A controller is then chosen which forces the Lyapunov candidate to decrease with time.

Even though the methods used to analyze a discrete-time system closely match that of a continuous-time framework, the delay associated with the feedback forces more restrictive designs. Adding to the complexity of the analysis, a relatively simple continuous-time plant may become rather complex when moving to a discrete-time framework. For example, a nonlinear continuous-time system which contains no uncertainty may have a discrete-time counterpart that does contain uncertainties not satisfying matching conditions, arising from approximating the continuous-time dynamics.

In this chapter, we will provide a few basic tools that may be used to design both static and adaptive controllers in the discrete-time framework. Since the design of discrete-time controllers has been the subject of research for many years, a rather large volume of techniques exist, each possessing their own particular advantages. The approach we have chosen in this chapter tends to show many of the similarities in the design/analysis of continuous and discrete-time systems.

13.2 Discrete-Time Systems

Before learning how to design static and adaptive controllers for discrete-time systems, we will consider how to convert from a continuous-time system to canonical discrete-time representations.

13.2.1 Converting from Continuous-Time Representations

Consider the continuous-time system defined by

$$\dot{\xi} = f(\xi) + g(\xi)u \tag{13.2}$$

with state $\xi \in R^n$ and input $u \in R$. For now we will assume that the functions $f, g : R^n \to R^n$ are known. Later we will see that an adaptive control approach may be used when the exact forms of f and g are unknown. If the controller is implemented in a digital system with sampling period T, then a zero-order hold is typically used to model the digital output. In particular, we will let

$$u(kT + \tau) = v(k)$$

for all $\tau \in [0, T)$ with $v(k)$ the discrete-time controller output sequence. The integer k will be used as the discrete-time sequence index.

Integrating the state trajectory over a sampling period one obtains

$$\xi(kT + T) - \xi(kT) = \int_{\tau=kT}^{kT+T} [f(\xi(\tau)) + g(\xi(\tau))v(k)]d\tau.$$

Since $v(k)$ is constant over the sampling period we find

$$\xi(kT + T) = \left(\xi(kT) + \int_{\tau=kT}^{kT+T} f(\xi(\tau))d\tau \right) + \left(\int_{\tau=kT}^{kT+T} g(\xi(\tau))d\tau \right) v(k).$$

Define the sequence $x(k) = \xi(kT)$ as the sampled value of the continuous-time state vector. Also, assume that there exist some known functions $f_d(\xi)$ and $g_d(\xi)$ such that

$$f_d(x(k)) \approx \xi(kT) + \int_{\tau=kT}^{kT+T} f(\xi(\tau))d\tau, \tag{13.3}$$

and

$$g_d(x(k)) \approx \int_{\tau=kT}^{kT+T} g(\xi(\tau))d\tau. \tag{13.4}$$

We will show how to obtain f_d and g_d later in this section.

Then we may express the discrete-time state sequence as

$$x(k + 1) = f_d(x(k)) + g_d(x(k))v(k) + \omega(k), \tag{13.5}$$

where $\omega(k)$ is an integration approximation error due to the intersample behavior of the system (it accounts for the approximation used to define f_d and g_d). Assuming that the continuous-time dynamics are not "too nonlinear" and that the approximations (13.3) and (13.4) are reasonably accurate, then we can assume that $\omega(k)$ is a bounded sequence. Even though this is often a reasonable assumption, there are continuous-time systems (such as ones with finite escape times) where it may be difficult or impossible to choose f_d and g_d such that $\omega(k)$ is bounded.

There are several techniques available for the conversion of a continuous-time plant representation to a discrete-time representation (that is finding some f_d and g_d in (13.5)). If the plant dynamics are defined by the linear representation

$$\dot{\xi} = A\xi + Bu, \tag{13.6}$$

then it is possible to develop an exact representation assuming that the intersample character of the controller output may be defined by a zero-order hold. This is demonstrated in the following example:

Example 13.1 Consider the scalar system defined by

$$\dot{\xi} = a\xi + bu(t), \tag{13.7}$$

with $b \neq 0$. Assume that u is the output of a zero-order hold so that

$$u(kT + \tau) = v(k)$$

for any $\tau \in [0, T)$ with $v(k)$ the output sequence of the discrete-time controller. Suppose that we wish to determine the values of f_d and g_d in (13.5) so that $\omega(k) \equiv 0$ (the output of the discrete-time representation exactly matches that of the continuous-time representation at the sample times).

The solution of (13.7) from $t = kT$ to $t = kT + T$ is given by

$$\xi(t) = -\frac{bv(k)}{a} + \left(\xi(kT) + \frac{bv(k)}{a}\right) e^{a(t-kT)}.$$

Thus with $x(k) = \xi(kT)$ and $t = kT + T$, we obtain

$$\begin{aligned} x(k+1) &= -\frac{bv(k)}{a} + \left(x(k) + \frac{bv(k)}{a}\right) e^{aT} \\ &= e^{aT} x(k) + \left(e^{aT} - \frac{b}{a}\right) v(k). \end{aligned}$$

Matching the coefficients to (13.5), we find

$$f_d = e^{aT} \qquad g_d = \left(e^{aT} - \frac{b}{a}\right)$$

for all k. $\triangle$

This same procedure may be applied to general linear systems. In fact there exist routines in software packages such as MATLAB which will automatically perform this calculation.

When the plant is nonlinear, it is not possible in general to define an exact discrete-time representation since the solution to the differential equation may be unknown. Also, if there is uncertainty in the system, it may not be possible to exactly solve the differential equations. In these cases, we will use an approximation to represent the effect of the integration. Given a function $f(\xi(t))$, Euler's method may be used to obtain

$$\int_{\tau=kT}^{kT+T} f(\xi(\tau))d\tau \approx Tf(x(k)). \tag{13.8}$$

When $\xi(\tau)$ is a constant over $\tau \in [0,T)$, then (13.8) becomes an equality. Thus (13.8) generally becomes more accurate as the sampling period T decreases ensuring that $\xi(\tau)$ is nearly constant on $\tau \in [0,T)$. In general, the magnitude of the error due to intersample behavior ω will decrease on the order of T^2.

Using the trapezoidal rule for integration, we find

$$\int_{\tau=kT}^{kT+T} f(\xi(\tau))d\tau \approx \frac{T}{2}\left(f(x(k)) + f(x(k+1))\right). \tag{13.9}$$

Since the trapezoid method is not causal, it is typically only used when the dynamics are linear. The following example demonstrates how one may convert from a continuous-time to discrete-time representation.

Example 13.2 Suppose that a discrete-model is to be developed for the system

$$\begin{aligned} \dot{\xi}_1 &= \xi_2 \\ \dot{\xi}_2 &= 2\xi_1 - \xi_1^3/10. \end{aligned} \tag{13.10}$$

Using trapezoidal integration on the ξ_1 term, we may use (13.9) with (13.3) to obtain

$$\xi_1(kT+T) \approx \xi_1(kT) + \frac{T}{2}\left(\xi_2(kT) + \xi_2(kT+T)\right).$$

Similarly, if Euler's method is used for ξ_2 we obtain

$$\xi_2(kT+T) \approx \xi_2(kT) + T(2\xi_1(kT) - \xi_1^3(kT)/10).$$

Letting $x_1(k) = \xi_1(kT)$ and $x_2(k) = \xi_2(kT)$ we find

$$\begin{aligned} x_1(k+1) &= x_1(k) + \tfrac{T}{2}\left(2x_2(k) + T(2x_1(k) - \tfrac{x_1^3(k)}{10})\right) + \omega_1(k) \\ x_2(k+2) &= x_2(k) + T(2x_1(k) - \tfrac{x_1^3(k)}{10}) + \omega_2(k), \end{aligned}$$

where ω_1 and ω_2 represent the error due to intersample behavior.

Choosing $T = 0.1$, we obtain the discrete state trajectory shown in Figure 13.2. As stated earlier, the intersample error will tend to decrease on the order of T^2 when using Euler's method to approximate the integration. This is demonstrated in Figure 13.3 where the sequence $\omega_2(k)$ is shown when $T - 0.1$, 0.05, and 0.01. $\triangle$

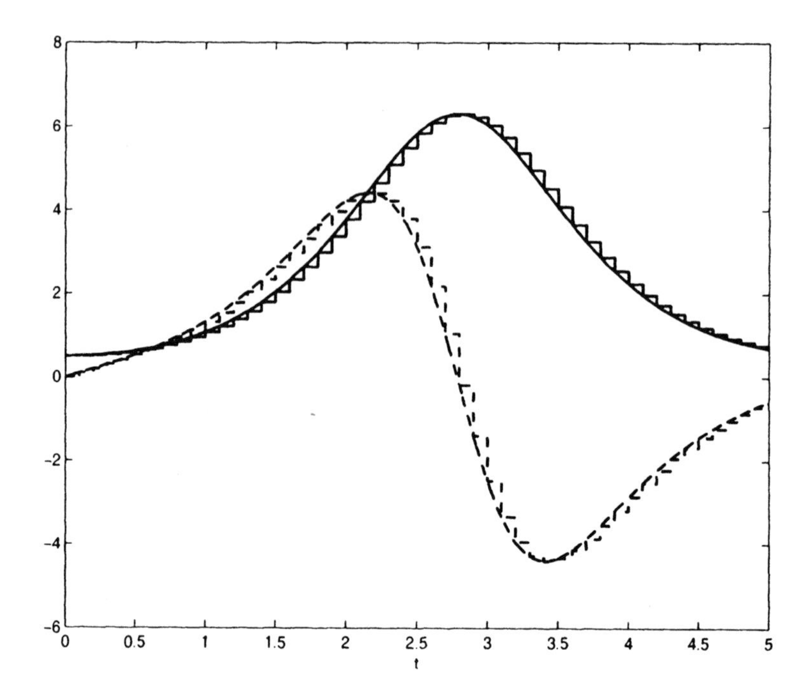

Figure 13.2. State trajectory and sequence for ξ_1 (–) and ξ_2 (- -).

13.2.2 Canonical Forms

To help generate widely applicable control methodologies, we will consider systems transformed into special canonical forms. The discrete-time feedback linearizable canonical form is defined by

$$\begin{aligned} x_1(k+1) &= x_2(k) \\ &\vdots \\ x_{n-1}(k+1) &= x_n(k) \\ x_n(k+1) &= f_d(q(k), x(k)) + g_d(q(k), x(k))u(k), \end{aligned} \tag{13.11}$$

where $y = x_1$, and g_d is bounded away from zero. As with continuous-time systems, the q states represent the zero dynamics of the system. When designing controllers for (13.11) we will typically assume that f_d and g_d are known. Since the states in (13.11) are defined using a chain of delays, it is often easy to place an arbitrary discrete-time system into the feedback linearizable canonical form.

Example 13.3 Consider the discrete-time system defined by

$$\begin{aligned} \xi_1(k+1) &= \xi_1(k) + T\xi_2(k) \\ \xi_2(k+1) &= \xi_2(k) + T\left(\xi_1^2(k) + \xi_2^2(k) + u(k)\right), \end{aligned}$$

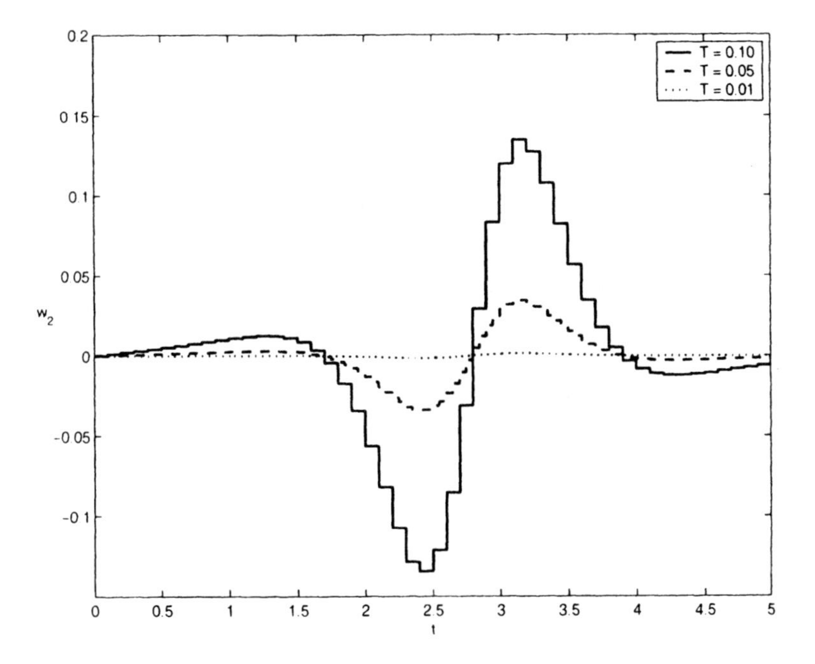

Figure 13.3. Intersample error $\omega_2(k)$ with $T = 0.1$ (–), $T = 0.05$ (- -), and $T = 0.01$ ($\cdots$).

with $y = \xi_1$. Letting $x_1 = \xi_1$ and $x_1(k+1) = x_2(k)$ we find

$$\begin{aligned} x_2(k+1) &= x_1(k+2) = \xi_1(k+1) + T\xi_2(k+1) \\ &= x_2(k) + T\left[\xi_2(k) + T\left(x_1^2(k) + \xi_2^2(k) + u(k)\right)\right]. \end{aligned}$$

Since $x_2(k) = x_1(k+1) = x_1(k) + T\xi_2(k)$ we find $\xi_2(k) = (x_2(k) - x_1(k))/T$. Thus

$$\begin{aligned} x_1(k+1) &= x_2 \\ x_2(k+1) &= 2x_2 - x_1 + (x_2 - x_1^2)^2 + T^2x_1^2 + T^2u, \end{aligned}$$

where it is assumed that each variable is defined for index k unless otherwise specified. $\triangle$

In many cases the intersample character may not be ignored. In such cases a perturbed feedback linearizable form may be used and is defined as

$$\begin{aligned} x_1(k+1) &= x_2(k) + \omega_1(k) \\ &\vdots \\ x_{n-1}(k+1) &= x_n(k) + \omega_{n-1}(k) \\ x_n(k+1) &= f_d(q(k), x(k)) + g_d(q(k), x(k))u(k) + \omega_n(k), \end{aligned} \tag{13.12}$$

where $\omega = [\omega_1, \ldots, \omega_n]^\top$ is a vector of unknown signals. For a general nonlinear system, there is no guarantee that each ω_i is a bounded sequence. In practice, however, it is often reasonable to assume that the sampling rate is chosen high enough so that the integration error due to the intersample behavior is relatively small.

The following example shows how the intersample behavior may be propagated when trying to convert to the canonical form:

Example 13.4 Consider the discrete-time system defined by

$$\begin{aligned}\xi_1(k+1) &= \xi_1(k) + T\xi_2(k) + \omega_1(k)\\ \xi_2(k+1) &= \xi_2(k) + T\left(\xi_1^2(k) + \xi_2^2(k) + u(k)\right) + \omega_2(k),\end{aligned}$$

where $\omega(k) = [\omega_1(k), \omega_2(k)]^\top$ is an unknown bounded sequence. Since $\omega(k)$ is unknown, we may not let $x_2(k) = x_1(k+1) = \xi_1(k) + T\xi_2(k) + \omega_1(k)$ and proceed along the same steps as in the previous example. Instead, we will ignore the effects of $\omega_1(k)$ and define

$$\begin{aligned}x_1(k) &= \xi_1(k)\\ x_2(k) &= \xi_1(k) + T\xi_2(k).\end{aligned}$$

With this transformation, exploiting the additive nature of the approximation error we find

$$\begin{aligned}x_1(k+1) &= x_2 + \omega_1\\ x_2(k+1) &= 2x_2 - x_1 + (x_2 - x_1^2)^2 + T^2x_1^2 + T^2u + \omega_1 + T\omega_2\end{aligned}$$

△

13.3 Static Controller Design

As with the case in the design of controllers for continuous-time systems, we will specify an error system to quantify the performance of the controller. The goal of the controller design will then be to reduce the magnitude of the values in the error system over time.

13.3.1 The Error System and Lyapunov Candidate

We will once again assume that the error system is designed to quantify the closed-loop system performance and satisfies the following property:

Assumption 13.1: The error system $e = \chi(k, x)$ is defined so that $|x| \leq \psi_x(k, |e|)$ where $\psi_x(k, s)$ is nondecreasing with respect to $s \in \mathsf{R}^+$ for all k.

Thus if we are able to force the error system to be bounded, then the plant states will also be bounded. As with the continuous-time case, we will use a positive definite Lyapunov candidate to help study the trajectory of $e(k)$. Most often the choice $V(k) = e^\top(k)Pe(k)$ will be used, with P a positive definite matrix. This way if $V(k)$ is forced to zero, then $e(k) \to 0$. The following lemma is the discrete-time counter part of Lemma 2.1 and will be helpful in studying the sequence $V(k)$.

Lemma 13.1: *Given the nonnegative sequence $V(k)$, if*

$$V(k+1) - V(k) \leq -k_1 V(k) + k_2$$

with $0 < k_1 \leq 1$ and $k_2 \geq 0$, then $V(k)$ is bounded by

$$V(k) \leq \alpha^k V(0) + \left(\frac{1-\alpha^k}{k_1}\right) k_2, \tag{13.13}$$

where $\alpha = 1 - k_1$.

Proof: Define the signal $\eta(k+1) = \eta(k) - k_1\eta(k) + k_2$, with $\eta(0) = V(0)$. If $\eta(k) \geq V(k)$ then

$$\eta(k+1) - V(k+1) \geq (1-k_1)\left[\eta(k) - V(k)\right] \geq 0$$

so $\eta(k) \geq V(k)$ for all k by induction provided that $k_1 \leq 1$.

Notice that

$$\begin{aligned}
\eta(1) &= (1-k_1)\,\eta(0) + k_2 \\
\eta(2) &= (1-k_1)^2\,\eta(0) + (1-k_1)\,k_2 + k_2 \\
\eta(3) &= (1-k_1)^3\,\eta(0) + (1-k_1)^2\,k_2 + (1-k_1)\,k_2 + k_2.
\end{aligned}$$

This may be continued to obtain the sequence

$$\eta(k) = (1-k_1)^k\,\eta(0) + k_2 \sum_{i=0}^{k-1} (1-k_1)^i. \tag{13.14}$$

Since

$$\sum_{i=0}^{k-1} a^i = \frac{1-a^k}{1-a} \tag{13.15}$$

for any $a \neq 1$ we obtain

$$\eta(k) = (1-k_1)^k\,\eta(0) + \left(\frac{1-(1-k_1)^k}{k_1}\right) k_2. \tag{13.16}$$

With $|1-k_1| \leq 1$ we find the sequence $\eta(k)$ is bounded, and thus by the comparison principle conclude (13.13). ■

When $\alpha < 1$ and $k_2 \equiv 0$ in Lemma 13.1, then $V(k) = 0$ is an exponentially stable equilibrium point. If $\alpha < 1$ and $k_2 \neq 0$, then $V(k)$ is a UUB sequence with ultimate bound defined by

$$\lim_{k \to \infty} V(k) \leq \frac{k_2}{k_1}. \tag{13.17}$$

Recall that for continuous-time systems if a Lyapunov candidate satisfies the property $\dot{V} \leq -k_1 V + k_2$, then the ultimate bound is again defined by k_2/k_1. In the continuous-time framework, however, it is often possible to define the feedback controller so that k_1 is arbitrarily large so that the ultimate bound may be made arbitrarily small. When working in the discrete-time framework, k_1 must be less than one in general. This can be shown as follows: Let $k_2 = 0$ and assume $k_1 > 1$. Then $V(k+1) = V(k) - k_1 V(k) < 0$. But $V(k)$ is a positive definite signal which cannot become negative. Thus $k_1 > 1$ cannot occur due to the positive definiteness of V. Since k_1 may not be made arbitrarily large in the discrete-time case, it is often not possible to make the ultimate bound arbitrarily small when $k_2 \neq 0$.

13.3.2 State Feedback Design

In this section we will design controllers for the discrete-time feedback linearizable system (13.11) where the intersample behavior $\omega(k)$ is ignored for now. We will be interested in the tracking problem where a controller is to be designed so that $x_1(k) \to r(k)$ when f_d and g_d in (13.11) are known.

In the continuous-time state-feedback case, we assumed that the derivatives of the reference signal were available to the controller. In general, this could be done by forcing the plant output to track the output of a reference model. Since the states of the reference model are available, it was possible to obtain the higher-order derivatives of the output. In the discrete-time case we will assume knowledge of $r(k), \ldots, r(k+n)$. Once again a reference model may be used. As a simple example, it is possible to define the reference model as $r(k) = v(k-n)$ where it is desired that the plant output $y(k)$ track $v(k)$, and $r(k)$ is the output of the reference model. Since the reference model output is a delayed version of $v(k)$, the signals $r(k), \ldots, r(k+n)$ are available for use in the controller definition.

When full state feedback is available, the traditional approach to developing a suitable controller for the linear discrete-time system

$$x(k+1) = A_x x(k) + bu(k), \tag{13.18}$$

where

$$A_x = \begin{bmatrix} 0 & & \\ \vdots & & I \\ 0 & & \\ -c_1 & \cdots & -c_{n-1} \end{bmatrix} \tag{13.19}$$

and $b = [0, \ldots, 0, 1]^\top$ is via the solution of a discrete-time Lyapunov matrix equation. In particular, we choose $u(k) = Kx(k)$ so that the eigenvalues of $A = A_x + bK$ all lie within the unit circle. The following discrete-time equivalent to the continuous-time Lyapunov matrix equation may then be used with the Lyapunov candidate $V(k) = x^\top(x)Px(x)$ to establish that $V(k)$ decreases with time.

Lemma 13.2: *If all the eigenvalues of A have magnitude less than 1, then for any positive definite symmetric Q, there exists a unique positive definite symmetric P satisfying*

$$A^\top PA - P = -Q \tag{13.20}$$

with $\lambda_{\min}(Q) \leq \lambda_{\max}(P)$.

Proof: Positive definiteness of P may be shown as done in [97]. Since $V(k+1) - V(k) = -x^\top(k)Qx(k)$ and $V(k) \leq \lambda_{\max}(P)|x(k)|^2$, we find

$$V(k+1) \leq \left(1 - \frac{\lambda_{\min}(Q)}{\lambda_{\max}(P)}\right) V(k). \tag{13.21}$$

Assume for a moment that $\lambda_{\min}(Q) > \lambda_{\max}(P)$. Then $V(k+1) < 0$ for any $V(k) > 0$. But $V(k+1)$ may not become negative by definition, thus $\lambda_{\min}(Q) > \lambda_{\max}(P)$ is not possible. This completes the proof. ∎

Using Lemma 13.2 we are now ready to design static controllers for (13.11).

Scalar Error System Approach

Since the system dynamics (13.11) are defined by a chain of delays, it is possible to just consider the transient response of $e(k) = x_1(k) - r(k)$, so $e \in \mathrm{R}$ is a scalar. Notice that

$$e(k+n) = -r(k+n) + f_d(x(k)) + g_d(x(k))u(k),$$

or

$$e(k+n) = \alpha(z(k)) + \beta(z(k))u(k), \tag{13.22}$$

where $\alpha = -r(k+n) + f_d(x(k))$ and $\beta = g_d(x(k))$. As was done for the continuous-time case, the term $z(k)$ is shorthand used to represent all the known signals for a given representation. In (13.22), for example, let $z(k) = [r(k+n), x^\top(k)]^\top$.

Now consider the controller defined by

$$\nu_s(k) = \frac{r(k+n) - f_d(k, x(k)) + \kappa e(k)}{g_d(x(k))} \tag{13.23}$$

with $|\kappa| < 1$. Using this control law, we find

$$e(k+n) = \kappa e(k). \tag{13.24}$$

Theorem 13.1: *The state-feedback control law (13.23) ensures that the error system (13.22) is exponentially stable in the large.*

Proof: Consider the Lyapunov candidate $V_s(k) = e^2(k)$. Using the controller defined by (13.23), we find

$$V_s(k+n) - V_s(k) = (\kappa^2 - 1)e^2(k). \tag{13.25}$$

Letting $v_i(k) = V_s(kn+i)$ for $i = 1, \ldots, n$ we find

$$\begin{aligned} v_i(k+1) - v_i(k) &= V_s((k+1)n+i) - V_s(kn+i) \\ &= (\kappa^2 - 1)V_s(kn+i) \\ &= (\kappa^2 - 1)v_i(k) \end{aligned} \tag{13.26}$$

(13.27)

for each sequence $v_i(k)$, $i = 1, \ldots, n$. Since $\kappa^2 < 1$, we may use Lemma 13.1 to conclude that each sequence $v_i(k)$ is exponentially stable so $e \to 0$. ■

When $\kappa = 0$, then the error decays to 0 in n steps. This type of controller is often referred to as a dead-beat controller.

Vector Error System Approach

We will now design a control law using a multi-dimensional error system. Since we want $x_1(k) \to r(k)$, assign the first error variable as $e_1(k) = x_1(k) - r(k)$. Defining $e_j(k) = x_j(k) - r(k+j-1)$ for $j = 2, \ldots, n$ we obtain

$$\begin{aligned} e_1(k+1) &= e_2(k) \\ &\vdots \\ e_{n-1}(k+1) &= e_n(k+1) \\ e_n(k+1) &= -r(k+n) + f_d(x) + g_d(x)u(k). \end{aligned} \tag{13.28}$$

Thus

$$e(k+1) = \alpha(z(k)) + \beta(z(k))u(k), \tag{13.29}$$

where

$$\alpha = \begin{bmatrix} e_2(k) \\ \vdots \\ e_n(k+1) \\ -r(k+n) + f_d(x(k)) \end{bmatrix} \qquad \beta = \begin{bmatrix} 0 \\ \vdots \\ 0 \\ g_d(x) \end{bmatrix}$$

with $z(k)$ a vector of measurable signals.

Now consider the control law

$$\nu_s(k) = \frac{r(k+n) - f_d(k) - c_1 e_1(k) - \ldots - c_{n-1} e_{n-1}(k)}{g_d(k)} \tag{13.30}$$

so that $e_n(k+1) = -c_1 e_1(k) - \ldots - c_{n-1} e_{n-1}(k)$. Notice that the error dynamics may be expressed as

$$e(k+1) = Ae(k), \tag{13.31}$$

where

$$A = \begin{bmatrix} 0 & & \\ \vdots & & I \\ 0 & & \\ -c_1 & \cdots & -c_{n-1} \end{bmatrix}. \tag{13.32}$$

It is possible to choose the coefficients $c_1, \ldots, c_{n-1}$ so that the eigenvalues of A are all contained within the unit circle. Placing the poles within the unit circle results in a stable closed-loop system as shown in the following theorem:

Theorem 13.2: *The state-feedback control law (13.30) with the eigenvalues of A defined with magnitude less than 1 ensures that the error system (13.29) is exponentially stable in the large.*

Proof: Consider the Lyapunov candidate

$$V_s(k) = e^\top(k) P e(k),$$

where P is a positive definite matrix to be chosen shortly. The difference becomes

$$\begin{aligned} V_s(k+1) - V_s(k) &= e^\top(k+1) P e(k+1) - e^\top(k) P e(k) \quad (13.33) \\ &= e^\top(k) \left[A^\top P A - P \right] e(k). \end{aligned}$$

Choosing P to satisfy the discrete-time Lyapunov matrix equation $A^\top PA - P = -Q$ with Q positive definite, we find

$$\begin{aligned} V_s(k+1) - V_s(k) &= -e^\top(k) Q e(k) \quad (13.34) \\ &\leq -\lambda_{\min}(Q) e^\top(k) e(k) \leq -\frac{\lambda_{\min}(Q)}{\lambda_{\max}(P)} V. \end{aligned}$$

Let $\bar{k}_1 = \lambda_{\min}(Q)/\lambda_{\max}(P)$. Using Lemma 13.1 we obtain

$$V_s(k) \leq \left(1 - \bar{k}_1\right)^k V_s(0), \tag{13.35}$$

for all k. Thus $V \to 0$ so $e \to 0$ as $k \to \infty$. ∎

We will now see how the scalar and vector error approaches may be used to develop a discrete-time control law.

Example 13.5 Consider the discrete-time plant defined by

$$\begin{aligned} x_1(k+1) &= x_2(k) \\ x_2(k+1) &= x_3(k) \\ x_3(k+1) &= x_1^2(k) - 2x_3(k) + 2u(k), \end{aligned} \tag{13.36}$$

with output $y(k) = x_1(k)$.

Scalar error system: Using the scalar error system approach, we first define the error system $e(k) = y(k) - r(k)$, where $r(k)$ is the desired plant output. Advancing the error by three time steps we find

$$e(k+3) = -r(k+3) + x_1^2(k) - 2x_3(k) + 2u(k).$$

Choosing the control law $u = \nu_s$ defined by (13.23), we obtain

$$\nu_s = \frac{r(k+3) - x_1^2(k) + 2x_3(k) + \kappa e(k)}{2} \tag{13.37}$$

with $|\kappa| < 1$.

Vector error system: Using the vector approach, we first define the error system

$$e = \begin{bmatrix} x_1(k) - r(k) \\ x_2(k) - r(k+1) \\ x_3(k) - r(k+2) \end{bmatrix}, \tag{13.38}$$

where $r(k)$ is again the desired output. Defining $u = \nu_s$ with ν_s defined by (13.30), we obtain

$$\nu_s = \frac{r(k+3) - x_1^2(k) + 2x_3(k) - c_1 e_1(k) - c_2 e_2(k) - c_3 e_3(k)}{2}, \tag{13.39}$$

where we choose the controller coefficients such that

$$A = \begin{bmatrix} 0 & 1 & 0 \\ 0 & 0 & 1 \\ -c_1 & -c_2 & -c_3 \end{bmatrix}$$

has all its eigenvalues within the unit circle.

The performance of the two approaches is shown in Figure 13.4. In these simulations, the initial conditions were set to $x(0) = [0, 0, 0]^\top$ and the reference was chosen to be $r(k) = 1$. For the scalar case we chose $\kappa = 0.8$, while for the vector case we chose $c_1 = -0.512$, $c_2 = 1.92$, and $c_3 = -2.4$ so that the eigenvalues of A were all 0.8. △

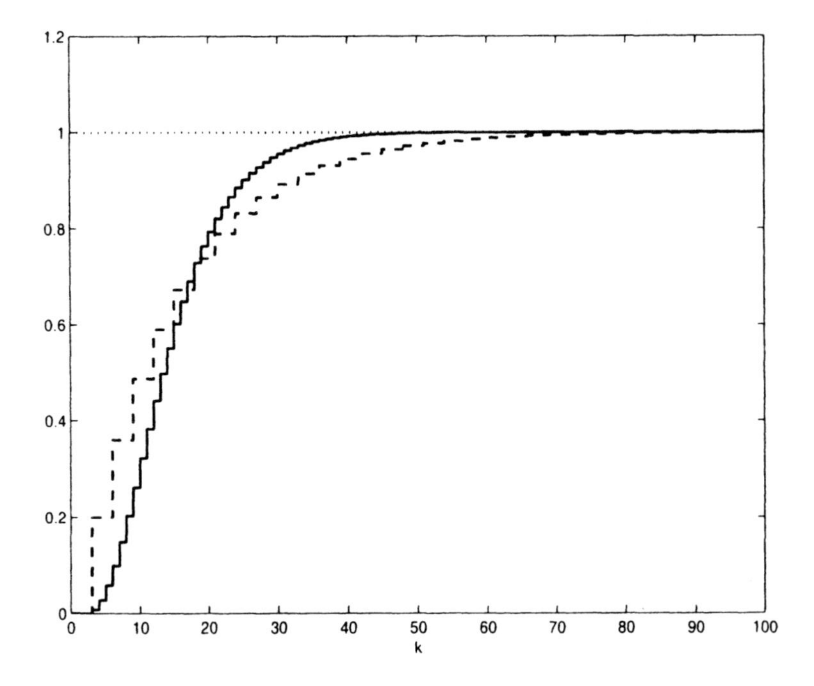

Figure 13.4. Response of the closed-loop system for both the vector-based control law (–), and scalar-based control law (- -).

13.3.3 Zero Dynamics

We will now consider the discrete-time system defined by

$$\begin{aligned} q(k+1) &= \phi(q(k), x(k)) \\ x_1(k+1) &= x_2(k) \\ &\vdots \\ x_{n-1}(k+1) &= x_n(k+1) \\ x_n(k+1) &= f_d(q(k), x(k)) + g_d(q(k), x(k))u(k), \end{aligned} \tag{13.40}$$

where the $q(k)$ dynamics have been included to represent unmodeled or zero dynamics. Since we assume full state feedback (we assume that q is measurable), it is still possible to define control laws that cancel the effects of f_d and g_d. Thus the closed-loop error dynamics for the scalar error system case (13.24) and the vector error system case (13.31) still hold. Since the error dynamics are independent of q, we have a triangular system so that Theorems (13.1) and (13.2) still hold which guarantee that x is bounded.

We will assume the q-subsystem, with x as an input, is input-to-state stable, so that there exists some positive definite V_q such that

$$\gamma_{q1}(|q|) \le V_q(q) \le \gamma_{q2}(|q|) \tag{13.41}$$

$$V_q(k+1) - V_q(k) \le -k_3 V_q(k) + \psi(x(k)), \tag{13.42}$$

where γ_{q1} and γ_{q2} are class $\mathcal{K}_\infty$, $0 < k_3 < 1$, and $\psi : \mathrm{R}^n \to \mathrm{R}$ is nonnegative and bounded for any bounded x. When $\psi(x) = 0$ the input-to-state stability assumption (13.42) becomes $V_q(k+1) - V_q(k) \le -k_3 V_q$, thus implying global asymptotic stability of the origin.

When the x-dynamics are bounded, then $\psi(x)$ is bounded. Let

$$k_4 = \sup_{x \in S_x} \psi(x),$$

where S_x is the space to which x is confined. Then $V_q(k+1) - V_q(k) \le -k_3 V_q(k) + k_4$ so that V_q is a bounded sequence which ensures that $q(k)$ is also bounded.

13.3.4 State Trajectory Bounds

As with the continuous-time case, we will modify the standard Lyapunov approach so that the Lyapunov candidate only needs to decay on some region. By considering only regional results we may consider the use of controllers defined with finite approximators. If, for example, a feedback linearizing control law is used where $f_d(x)$ is only (approximately) known over some region S_x, then we want to define the controller so that x never leaves the region S_x. The following lemma is the extension of Lemma 13.1 to the regional case.

Lemma 13.3: *Given the nonnegative sequence $V_s(k)$, if*

$$V_s(k+1) - V_s(k) \le -k_1 V_s(k) + k_2,$$

with $0 < k_1 \le 1$ and $k_2 \ge 0$ when $V_s \le V_r$ with

$$V_r = \max(V_s(0), k_2/k_1), \tag{13.43}$$

then $V_s(k)$ is UUB with ultimate bound k_2/k_1.

Proof: From Lemma 13.1 we know that

$$V_s(k) \le \left(V_s(0) - \frac{k_2}{k_1}\right)\alpha^k + \frac{k_2}{k_1}, \tag{13.44}$$

where $\alpha = 1 - k_1$ when $V \le V_r$. Since α^k is monotonically decreasing, we find $V_s(k) \le \max(V_s(0), k_2/k_1)$. Thus $V_s(k)$ never becomes larger than V_r so that assumptions of the theorem hold for all k. The ultimate bound is then found using (13.44) with $k \to \infty$. ■

Consider the case where $e(k+1) = Ae(k)$ with all the eigenvalue of A contained in the unit circle. If $V_s = e^\top Pe$, then $V_s(k+1) - V_s(k) = e^\top(A^\top PA - P)e = -e^\top Qe$ when P is chosen to satisfy the discrete-time Lyapunov matrix equation and Q is chosen to be positive definite. Now if uncertainty is included in the system dynamics, it may only be possible to establish that $V_s(k+1) - V_s(k) \leq -e^\top(k)Qe(k) + k_2$, where $k_2 > 0$ is a constant resulting from the uncertainty (this type of term will be studied in more detail when we consider robust control of discrete-time systems). Letting $k_1 = \lambda_{\min}(Q)/\lambda_{\max}(P)$, we find $V_s(k+1) - V_s(k) \leq -k_1 V_s(k) + k_2$ which is in the form used in Lemma 13.3. Comparing terms, we find that k_1 is dependent upon P and Q (and thus upon the eigenvalues of the system matrix A). The term k_2 tends to result from uncertainties in the system.

If $V_s(k) = e^\top(k)Pe(k)$, then Lemma 13.3 may be used to establish maximum and ultimate bounds on $e(k)$. In particular, $|e(k)|^2 \leq V_s(k)/\lambda_{\min}(P)$ so $e \in B_e$ for all k, where

$$B_e = \left\{ e \in \mathsf{R}^p : |e| \leq \sqrt{\max\left(\frac{V_s(0)}{\lambda_{\min}(P)}, \frac{k_2}{k_1 \lambda_{\min}(P)} \right)} \right\}. \tag{13.45}$$

The ultimate bound may also be found using Lemma 13.3 since

$$\lim_{k\to\infty} |e(k)| \leq \lim_{k\to\infty} \sqrt{\frac{V_s(k)}{\lambda_{\min}(P)}} = \sqrt{\frac{k_2}{k_1 \lambda_{\min}(P)}}. \tag{13.46}$$

If e may be bounded, then it is possible to use Assumption 13.1 to place bounds on the state vector x. Here $x \in B_x$ for all k where

$$B_x = \left\{ x \in \mathsf{R}^n : \sup_k \psi_x \left(k, \sqrt{\max\left(\frac{V_s(0)}{\lambda_{\min}(P)}, \frac{k_2}{k_1 \lambda_{\min}(P)} \right)} \right) \right\}. \tag{13.47}$$

This may seem like an odd bound since at first glance it appears that we may simply choose P so that $\lambda_{\min}(P)$ is made arbitrarily large (which does not change the control law). As discussed above, k_1 may depend upon both Q and A (and thus P). Therefore changing the value of Q (or equivalently P) will change the value of k_1. Therefore it is not possible to simply choose some parameters independent of the control law that will force B_x to become arbitrarily small. If a fuzzy system with valid inputs $x \in S_x$ is used in the definition of a control law, then we need to choose the controller parameters such that $B_x \subset S_x$. This way the states x will never leave S_x.

The reader should keep in mind that the results here are "regional" and not necessarily local. With a local stability result, it is possible to define $x(0)$ so that the state trajectory remains bounded for all k within a set whose size cannot be fixed, but is rather determined by the characteristics of the problem at hand. With a regional result, we not only place requirements on $x(0)$, but we also require that the controller is defined so that $V_s(k+1) - V_s(k) \le -k_1 V_s(k) + k_2$ when $V \le V_r$. Thus when $V \le V_r$, the closed-loop system exhibits stability which will ensure that $x \in S_x$ for all k, and the size of S_x can be fixed by the designer. The requirement of regional stability is particularly important when dealing with controllers designed with approximators defined only over some compact space.

13.4 Robust Control of Discrete-Time Systems

We will now be interested in the control of systems whose error dynamics are defined by

$$e(k+d) = \alpha(z(k)) + \beta(z(k))u(k) + \omega(k), \tag{13.48}$$

where $\omega(k)$ is bounded such that $|\omega| \le \Omega$ for all k with $\Omega \in \mathbb{R}$. Notice that (13.48) may be used to represent either the scalar error system (13.22) with $d = n$ or the vector error system (13.29) with $d = 1$. It will be assumed that β is bounded away from zero for all k.

13.4.1 Inherent Robustness

The following theorem shows that if a stabilizing discrete-time control law is defined ignoring $\omega(k)$ in (13.48), it may also stabilize the error system when $|\omega| \le \Omega$ (though possibly with degraded performance).

Theorem 13.3: *If the static discrete-time control law $u(k) = \nu_s(k)$ is defined so that the Lyapunov function $V_s(k) = e^\top P e$ decreases according to $V_s(k+d) - V_s(k) \le -k_1 V_s(k) + k_2$ along the solutions of (13.48) when $w \equiv 0$, then the solutions of (13.48) are UUB when $|\omega| \le \Omega$ for all k with ultimate bound (13.57).*

Proof: Using the Lyapunov candidate $V_s(k) = e^\top(k) P e(k)$ with P a symmetric positive definite matrix, we find

$$V_s(k+d) - V_s(k) = [\alpha + \beta\nu_s + \omega]^\top P [\alpha + \beta\nu_s + \omega] - V_s(k). \tag{13.49}$$

Since

$$(x+y)^\top P (x+y) = x^\top P x + 2x^\top P y + y^\top P y, \tag{13.50}$$

and

$$\begin{aligned} 0 &\le \left(\sqrt{\epsilon}x - \frac{y}{\sqrt{\epsilon}}\right)^\top P \left(\sqrt{\epsilon}x - \frac{y}{\sqrt{\epsilon}}\right) \\ &= \epsilon x^\top P x - 2x^\top P y + \frac{1}{\epsilon} y^\top P y, \end{aligned} \tag{13.51}$$

we may add (13.50) and (13.51) to obtain

$$(x+y)^\top P(x+y) \le (1+\epsilon)x^\top Px + \left(1+\frac{1}{\epsilon}\right)y^\top Py, \tag{13.52}$$

for any $\epsilon > 0$. Using (13.52) with $x = \alpha + \beta\nu_s$ and $y = \omega$, we find

$$\begin{aligned} V_s(k+d) - V_s(k) &\le (1+\epsilon)(\alpha+\beta\nu_s)^\top P(\alpha+\beta\nu_s) \\ &\quad + \left(1+\frac{1}{\epsilon}\right)\omega^\top(k)P\omega(k) - V_s(k). \end{aligned}$$

Rearranging terms we obtain

$$\begin{aligned} V_s(k+d) - V_s(k) &\le (1+\epsilon)\left[(\alpha+\beta\nu_s)^\top P(\alpha+\beta\nu_s) - V_s(k)\right] \\ &\quad + \left(1+\frac{1}{\epsilon}\right)\lambda_{\max}(P)\Omega^2 + \epsilon V_s(k). \end{aligned}$$

Since $V_s(k+d) - V_s(k) \le -k_1V_s(k) + k_2$ along the solutions of (13.48) when $u(k) = \nu_s(k)$ and $\omega \equiv 0$, we know that

$$(\alpha+\beta\nu_s)^\top P(\alpha+\beta\nu_s) - V_s(k) \le -k_1V_s(k) + k_2$$

since $V_s(k+d) = (\alpha+\beta\nu_s)^\top P(\alpha+\beta\nu_s)$ when $\omega \equiv 0$. Thus

$$\begin{aligned} V_s(k+d) - V_s(k) &\le -(k_1 + \epsilon k_1 - \epsilon)V_s(k) \\ &\quad + (1+\epsilon)k_2 + \left(1+\frac{1}{\epsilon}\right)\lambda_{\max}(P)\Omega^2. \end{aligned} \tag{13.53}$$

Now define the sequence $v_i(k) = V_s(dk+i)$ for $i = 1,\ldots,d$. Then

$$\begin{aligned} v_i(k+1) - v_i(k) &\le -(k_1 + \epsilon k_1 - \epsilon)v_i(k) \\ &\quad + (1+\epsilon)k_2 + \left(1+\frac{1}{\epsilon}\right)\lambda_{\max}(P)\Omega^2. \end{aligned} \tag{13.54}$$

for $i = 1,\ldots,d$.

If ϵ is chosen such that

$$0 < k_1 + \epsilon k_1 - \epsilon \le 1, \tag{13.55}$$

then according to Lemma 13.1, the solutions of each $v_i(k)$ are UUB. The range on ϵ may be bound using (13.55) so

$$\begin{aligned} 0 > -k_1 + (1-k_1)\epsilon &\ge -1 \\ \tfrac{k_1}{1-k_1} > \epsilon &\ge \tfrac{-1}{1-k_1}. \end{aligned} \tag{13.56}$$

But we already require that $\epsilon > 0$, so $\epsilon \in S_\epsilon$ with $S_\epsilon = (0, \frac{k_1}{1-k_1})$. Using Lemma 13.1 the ultimate bound on the sequence $v_i(k)$ is found to be

$$\lim_{k\to\infty} |v_i(k)| \leq \inf_{\epsilon\in S_\epsilon} \frac{(1+\epsilon)k_2 + \left(1+\frac{1}{\epsilon}\right)\lambda_{\max}(P)\Omega^2}{k_1 + \epsilon k_1 - \epsilon}, \tag{13.57}$$

which is also the ultimate bound for V_s. ∎

Theorem 13.3 shows that the design of a state feedback controller tends to be robust with respect to bounded uncertainties that do not necessarily satisfy matching conditions. Unfortunately, the ultimate bound specified by (13.57) may be rather conservative. Assume for now that the integration error may be ignored so that $\Omega = 0$. In this case the ultimate bound becomes

$$\lim_{k\to\infty} |V_s(k)| \leq \inf_{\epsilon\in S_\epsilon} \frac{(1+\epsilon)k_2}{k_1 + \epsilon k_1 - \epsilon}.$$

Then according to (13.55) we may choose $\epsilon \approx 0$ so that the ultimate bound is defined by k_2/k_1 which according to Lemma 13.1 is the value obtained when $\omega \equiv 0$. When Ω may not be chosen to be zero, the size of the ultimate bound defined by (13.57) grows.

13.4.2 A Dead-Zone Modification

If the bound on ω is known, it is possible to use this information in the design of the control law. To do this we will consider the use of a dead-zone nonlinearity. A continuous dead-zone is defined as

$$D_c(x,b) = \begin{cases} x-b & \text{if } x \geq b \\ 0 & \text{if } -b < x < b \\ x+b & \text{if } x < -b \end{cases}, \tag{13.58}$$

where $b > 0$ is the size of the dead-zone. A special property of the continuous dead-zone is described in the following lemma:

Lemma 13.4: *If $y = x + \omega$ where $|\omega| < \Omega$ for some $\Omega > 0$, then $D_c(y, \Omega) = \pi x$, with $0 \leq \pi < 1$.*

Proof: Consider the following three cases:

1. If $|y| \leq \Omega$ then $D_c(y,\Omega) = 0$. That is, $D_c(y,\Omega) = \pi x$ with $\pi = 0$.
2. If $\Omega < y$ then $D_c(y,\Omega) = x + \omega - \Omega < x$. Also, $0 = \Omega - \Omega < y - \Omega = D_c(y,\Omega)$, so that $D_c(y,\Omega) = \pi x$ where $0 < \pi < 1$.
3. If $y < -\Omega$ then $D_c(y,\Omega) = x+\omega+\Omega > x$. Also, $0 = \Omega - \Omega > \Omega + y = D_c(y,\Omega)$, so that $D_c(y,\Omega) = \pi x$ where $0 < \pi < 1$.

Thus $D_c(y,\Omega) = \pi x$, where $0 \leq \pi < 1$ for all y. ∎

We will now consider (13.48) when $e \in \mathbb{R}$ is a scalar error system with $|\omega(k)| \leq \Omega$ for all k. Define a new sequence $q(k) = D_c(e(k), \Omega)$ so that the error variable is passed through a dead-zone of size Ω. Using Lemma 13.4 we find

$$q(k+d) = \pi(k)\left(\alpha(k) + \beta(k)u(k)\right) \tag{13.59}$$

with $0 \leq \pi(k) < 1$. We will now consider the control law $u(k) = \nu_s(k)$ where

$$\nu_s(k) = \frac{-\alpha(z(k)) + \kappa q(k)}{\beta(z(k))} \tag{13.60}$$

with $|\kappa| < 1$. With this control law we now find

$$q(k+d) = \kappa\pi(k)q(k) \tag{13.61}$$

for all k. This results in a stable closed-loop system as summarized by the following theorem:

Theorem 13.4: *The control law (13.60) ensures that the solutions of (13.48) are UUB with ultimate bound*

$$\lim_{k\to\infty} |e(k)| \leq \Omega, \tag{13.62}$$

where $|\omega| \leq \Omega$ *for all* k.

Proof: Consider the Lyapunov candidate $V_s(k) = q^2(k)$. Notice that when using the control law (13.60), we find

$$V_s(k+d) - V_s(k) = \kappa^2\pi^2(k)V_s(k) - V_s(k). \tag{13.63}$$

Since $\kappa^2\pi^2(k) - 1 < \kappa^2 - 1$, we may let $k_1 = \kappa^2 - 1$ so that

$$V_s(k+d) - V_s(k) \leq -k_1 V_s(k) \tag{13.64}$$

with $0 < k_1 < 1$. Therefore each sequence $v_i = V_s(kd + i)$ is exponentially stable for $i = 1, \ldots, d$ so $q(k) \to 0$. Since $|e(k)| = |q(k) + e(k) - q(k)|$ with $|e(k) - q(k)| \leq \Omega$, we may conclude that

$$\begin{aligned} \lim_{k\to\infty} |e(k)| &\leq \lim_{k\to\infty} |q(k)| + \lim_{k\to\infty} |e(k) - q(k)| \\ &= \lim_{k\to\infty} |e(k) - q(k)|. \end{aligned} \tag{13.65}$$

But from the definition of the dead-zone we have $|e(k) - q(k)| \leq \Omega$. Thus $\lim_{k\to\infty} |e(k)| \leq \Omega$. ■

Even though the ultimate bound obtained in Theorem 13.4 tends to be less conservative than the one of Theorem 13.3, the control law (13.60) does require that the bound Ω is known.

13.5 Adaptive Control

We have seen that it is possible to define controllers for discrete-time systems which may contain bounded uncertainties. In the static case, however, these uncertainties may cause the tracking errors to become large. In this section we will investigate the use of an adaptive approach which attempts to cancel the effects of an uncertainty.

Here we will be interested in developing an adaptive controller for the error dynamics

$$e(k+d) = \alpha(z(k)) + \beta(x)[\Delta(x(k)) + u(k)] + \omega(k), \tag{13.66}$$

where $e \in \mathbb{R}^q$, and the scalar function $\Delta(x)$ is an uncertainty. Here z is again a vector of known signals that may be used to define α. We will assume that β is finite and bounded away from zero for all x.

Notice that when $\Delta \equiv 0$, a static controller for the error system (13.66) may be defined using the techniques presented earlier. The purpose of the adaptive terms will thus be to cancel the effects of $\Delta(x)$. We will assume that a linear in the parameter approximator, $\mathcal{F}(x, \hat{\theta})$, is defined so that there exists some $\theta \in \mathbb{R}^p$ such that $|\Delta(x) + \mathcal{F}(z, \theta)| \leq W$ for all $x \in S_x$. We will also assume that the error system is defined such that if $e \in S_e$ (where it is assumed that S_e contains a ball centered at the origin), then $x \in S_x$. Our goal will thus be to define an adaptive controller which ensures that $e \in B_e$ where it is possible to make $B_e \subset S_e$ by proper choice of the controller parameters.

13.5.1 Adaptive Control Preliminaries

There have been a few different approaches used to solve the discrete-time adaptive control problem. The traditional approach is to define an identifier which is used to obtain an estimate of an ideal parameter set for the approximator. Using the Key Technical Lemma (see, e.g., Goodwin and Sin [63]), it is possible to show that the error system converges to zero when the parameter estimate remains bounded. The Key Technical Lemma, however, does not directly provide bounds on the state trajectory. Since we wish to use regional approximators in the control law, a different approach is used here.

To help develop adaptive controllers for discrete-time systems, we will make use of the following theorem:

Theorem 13.5: *Given some nonnegative sequences $V_s(k)$ and $V_{\tilde{\theta}}(k)$, if $V(k) = V_s(k) + V_{\tilde{\theta}}(k)$ with $V_{\tilde{\theta}}(k+1) - V_{\tilde{\theta}}(k) \leq 0$ and $V(k+1) - V(k) \leq -k_1 V_s(k) + k_2$, then $V_{\tilde{\theta}}(k)$ is bounded and $V_s(k)$ is UUB with maximum bound (13.69) and ultimate bound (13.72)*

Proof: Since $V_{\tilde{\theta}}(k)$ is a nonincreasing sequence and is bounded from below (it is nonnegative) we may conclude that $V_{\tilde{\theta}}(k) \le V_{\tilde{\theta}}(0)$ for all k. Additionally, since $V_s = V - V_{\tilde{\theta}}$ we find

$$V_s(k+1) - V_s(k) \le -k_1 V_s(k) + k_2 + V_{\tilde{\theta}}(k) - V_{\tilde{\theta}}(k+1). \tag{13.67}$$

But $V_{\tilde{\theta}}(k) - V_{\tilde{\theta}}(k+1) \le V_{\tilde{\theta}}(0)$ since $V_{\tilde{\theta}}$ may not decrease by more than $V_{\tilde{\theta}}(0)$. Since

$$V_s(k+1) - V_s(k) \le -k_1 V_s(k) + k_2 + V_{\tilde{\theta}}(0) \tag{13.68}$$

we may conclude that V_s is bounded by use of Lemma 13.1 with maximum bound $V_s(k) \le V_r$ where

$$V_r = \max\left(V_s(0), \frac{k_2 + V_{\tilde{\theta}}(0)}{k_1}\right). \tag{13.69}$$

Additionally, since $V_{\tilde{\theta}}(k)$ is a nonincreasing nonnegative sequence, given any $\delta > 0$ there exists some K so that $V_{\tilde{\theta}}(k+1) - V_{\tilde{\theta}}(k) \le \delta$ for all $k \ge K$. Thus

$$V_s(k+1) - V_s(k) \le -k_1 V_s(k) + k_2 + \delta \tag{13.70}$$

for all $k > K$. Using the steps in Lemma 13.1 we find

$$V_s(k) \le \alpha^{(k-K)} V_s(K) + \left(\frac{1 - \alpha^{(k-K)}}{k_1}\right)(k_2 + \delta) \tag{13.71}$$

for all $k \ge K$, where $\alpha = 1 - k_1$. Thus

$$\lim_{k\to\infty} V_s(k) \le \frac{k_2 + \delta}{k_1}.$$

But since δ may be chosen arbitrarily small we may ignore its contribution and conclude that

$$\lim_{k\to\infty} V_s(k) \le \frac{k_2}{k_1}, \tag{13.72}$$

which is independent of any initial conditions. ■

Using Theorem 13.5 we may find maximum and ultimate bounds on e. If $V_s(k) = e^\top(k)Pe(k)$ with P positive definite, then

$$|e(k)| \le \sqrt{\frac{V_s(k)}{\lambda_{\min}(P)}}. \tag{13.73}$$

Notice that since $V_s = e^\top Pe$, we cannot make $|e(k)|$ in (13.73) smaller by simply increasing $\lambda_{\min}(P)$. If the assumptions of Theorem 13.5 are satisfied, then using Lemma 13.3 with (13.68) we find

$$|e(k)| \le \sqrt{\max\left(\frac{V_s(0)}{\lambda_{\min}(P)}, \frac{k_2 + V_{\tilde{\theta}}(0)}{k_1 \lambda_{\min}(P)}\right)} \tag{13.74}$$

for all k. Similarly using (13.72) the ultimate bound on e is found to be

$$\lim_{k\to\infty} |e(k)| \leq \sqrt{\frac{k_2}{k_1 \lambda_{\min}(P)}}. \tag{13.75}$$

With these tools, we are now ready to design an adaptive controller. As for the non-adaptive case, the value of k_1 will be influenced by our choice of P. Thus simply choosing $\lambda_{\min}(P)$ to be some large number will not necessarily decrease the ultimate bound since this in turn may decrease k_1. For this reason, and also because the choice of P does not affect the control law, one should not treat the bounds (13.73) and (13.75) as design parameters that one can directly manipulate to improve performance. Rather, the bound (13.73) is simply an upper bound (although not necessarily a *least* upper bound) that will be useful to determine a region within which the approximator used in the adaptive controller should be defined.

13.5.2 The Adaptive Controller

If the uncertainty in (13.66) may be ignored, then a static controller may be defined that stabilizes the system using the control techniques presented earlier. Based on this, we make the following assumption:

Assumption 13.2: The error system $e = \chi(k, x)$ satisfies Assumption 13.1 and a known Lyapunov function $V_s = e^\top P e$ exists such that $V_s(k+d) - V_s(k) \leq -k_1 V_s(k) + k_2$ along the solutions of (13.66) when $u = \nu_s$ and $\Delta \equiv 0$.

The adaptive control law is now defined by $u(k) = \nu_a(k)$ with

$$\nu_a = \nu_s(k) + \mathcal{F}(x(k), \hat{\theta}(k)), \tag{13.76}$$

where $\hat{\theta}$ is an estimate of the ideal parameter vector θ. Thus we are using certainty equivalence to try to cancel the effects of the uncertainty Δ.

We will now pick a variable which may be used to measure the mismatch between $\mathcal{F}(k, \hat{\theta})$ and $\mathcal{F}(k, \theta)$. Consider the scalar signal

$$q(k) = \beta^\top(k-d)\left[e(k) - \alpha(k-d) - \beta(k-d)\nu_s(k-d)\right].$$

Notice that q is measurable since α and β are known. Using the definition of the error system

$$q(k+d) = \beta^\top(k)\left[\beta(k)(\Delta(k) + \mathcal{F}(k, \hat{\theta})) + \omega(k)\right].$$

Also

$$\begin{aligned} \Delta(k) + \mathcal{F}(k, \hat{\theta}) &= \Delta(k) + \frac{\partial \mathcal{F}}{\partial \hat{\theta}}\left(\tilde{\theta}(k) + \theta\right) \\ &= w(k) + \frac{\partial \mathcal{F}}{\partial \hat{\theta}}\tilde{\theta}(k), \end{aligned} \tag{13.77}$$

where $w(k) = \Delta(x(k)) + \mathcal{F}(x(k), \theta)$ and $|w| \le W$ when $x \in S_x$. We thus find

$$q(k+d) = \beta^\top \beta w(k) + \beta^\top \omega + \beta^\top \beta \frac{\partial \mathcal{F}}{\partial \hat{\theta}} \tilde{\theta}. \tag{13.78}$$

Choose some $\rho \in \mathbb{R}$ such that

$$\rho \ge \beta^\top \beta W + \beta^\top \omega, \tag{13.79}$$

for all k. Since β and ω are assumed to be bounded, some finite ρ exists. A dead-zone nonlinearity will now be used to define

$$\bar{q}(k) = D_c(q(k), \rho), \tag{13.80}$$

so that

$$\bar{q}(k+d) = \pi(k+d)\beta^\top \beta \frac{\partial \mathcal{F}}{\partial \hat{\theta}} \tilde{\theta}, \tag{13.81}$$

when $x \in S_x$. Similar to the static controller design case, we will find that using a dead-zone in the adaptive control law will allow for the design of a robust adaptive controller.

Here we will study the closed-loop system when using the normalized gradient update law defined as

$$\hat{\theta}(k) = \hat{\theta}(k-d) - \frac{\eta \zeta(k-d, \hat{\theta}(k-d))}{|\beta(k-d)|^2 \left(1 + \gamma \left|\zeta(k-d, \hat{\theta}(k-d))\right|^2\right)} \bar{q}(k), \tag{13.82}$$

where $\gamma > 0$ and $0 < \eta < 2\gamma$ and we have defined

$$\zeta(k, s) = \frac{\partial \mathcal{F}(k, s)}{\partial s}$$

for notational convenience. If the parameter error is defined as $\tilde{\theta} = \hat{\theta} - \theta$, then

$$\tilde{\theta}(k) = \tilde{\theta}(k-d) - \frac{\eta \zeta(k-d, \hat{\theta}(k-d))}{|\beta(k-d)|^2 \left(1 + \gamma \left|\zeta(k-d, \hat{\theta}(k-d))\right|^2\right)} \bar{q}(k).$$

The stability results may be summarized as follows:

Theorem 13.6: *Let Assumption 13.2 hold with $V_s(e) = e^\top P e$. Assume that for a given linear in the parameter approximator $\mathcal{F}(x, \hat{\theta})$ there exists some θ such that $|\Delta(x) + \mathcal{F}(x, \theta)| \le W$ for all $x \in S_x$. Let $b_{\mathcal{F}} > 0$ such that $|\frac{\partial \mathcal{F}(x,\hat{\theta})}{\partial \hat{\theta}}| \le b_{\mathcal{F}}$ for all $x \in S_x$. Assume $e \in S_e$ contains the origin and implies $x \in S_x$. If the control parameters are chosen such that $B_e \subseteq S_e$ with B_e defined by (13.94), then the parameter update law (13.82)* *with adaptive controller (13.76) guarantee that the solutions of (13.66) are UUB with ultimate bound (13.95).*

Proof: We will start by studying the behavior of $\tilde{\theta}(k)$. Let $V_{\tilde{\theta}} = \tilde{\theta}^{\top}(k)\tilde{\theta}(k)$. We will consider the case where q is within the dead zone separate from the case where q is outside the dead zone. We may express these two sets as

$$\begin{aligned}\mathcal{I}_0 &= \{k : |q(k)| \leq \rho\} \\ \mathcal{I}_1 &= \{k : \text{otherwise}\}.\end{aligned}$$

That is, $\mathcal{I}_0$ are the times that q is within the dead zone, while $\mathcal{I}_1$ are the times when q is outside the dead zone. First consider the case that $k + d \in \mathcal{I}_0$. Then $q(k+d) = 0$ so $\tilde{\theta}(k+d) = \tilde{\theta}(k)$. Since the parameter error does not change, we have $V_{\tilde{\theta}}(k+d) - V_{\tilde{\theta}}(k) = 0$.

If $k + d \in \mathcal{I}_1$ ($\bar{q}$ is nonzero), then

$$\begin{aligned}V_{\tilde{\theta}}(k+d) - V_{\tilde{\theta}}(k) &= \left|\tilde{\theta}(k) - \frac{\eta\zeta(k,\hat{\theta})\bar{q}(k+d)}{\beta^{\top}\beta\left(1+\gamma\left|\zeta(k,\hat{\theta})\right|^2\right)}\right|^2 - |\tilde{\theta}(k)|^2 \\ &= \frac{-2\eta\zeta^{\top}(k,\hat{\theta})\tilde{\theta}(k)\bar{q}(k+d)}{\beta^{\top}\beta\left(1+\gamma\left|\zeta(k,\hat{\theta})\right|^2\right)} \\ &\quad + \left|\frac{\eta\zeta(k,\hat{\theta})\bar{q}(k+d)}{\beta^{\top}\beta\left(1+\gamma\left|\zeta(k,\hat{\theta})\right|^2\right)}\right|^2.\end{aligned}$$

Since we are outside the deadzone, $\pi(k+d) > 0$ so

$$\zeta(k,\hat{\theta}(k))\tilde{\theta}(k) = \frac{\bar{q}(k+d)}{\pi(k+d)\beta^{\top}(k)\beta(k)}.$$

Thus

$$V_{\tilde{\theta}}(k+d) - V_{\tilde{\theta}}(k) = \frac{\eta}{(\beta^{\top}\beta)^2}\left[\frac{-2}{\pi(k+d)} + \frac{\eta|\zeta(k)|^2}{1+\gamma|\zeta(k)|^2}\right]\frac{\bar{q}^2(k+d)}{1+\gamma|\zeta(k)|^2}.$$

Since $0 < \pi < 1$ we find

$$V_{\tilde{\theta}}(k+d) - V_{\tilde{\theta}}(k) \leq \frac{\eta}{(\beta^{\top}\beta)^2}\left[-2 + \frac{\eta|\zeta(k)|^2}{1+\gamma|\zeta(k)|^2}\right]\frac{\bar{q}^2(k+d)}{1+\gamma|\zeta(k)|^2}.$$

The choice for η and γ will ensure that

$$2 - \frac{\eta|\zeta(k)|^2}{1+\gamma|\zeta(k)|^2} \geq 2 - \frac{\eta}{\gamma} > 0. \tag{13.83}$$

Let $k_3 = 2 - \eta/\gamma$. Now

$$V_{\tilde{\theta}}(k+d) - V_{\tilde{\theta}}(k) \leq -\frac{\eta k_3}{(\beta^\top \beta)^2} \frac{\bar{q}^2(k+d)}{(1+\gamma|\zeta(k)|^2)}$$

for all $k \in \mathcal{I}_0 \cap \mathcal{I}_1$. This ensures that $V_{\tilde{\theta}}(k)$ is a monotonically nonincreasing positive sequence so that $V_{\tilde{\theta}}(k)$ is bounded.

We will now study the trajectory of $e(k)$ using the positive definite function $V_s(k) = e^\top(k) P e(k)$. In order to use Theorem 13.5 we must now show given $V = V_s + k_4 V_{\tilde{\theta}}$ we have $V(k+d) - V(k) \leq -\bar{k}_1 V_s(k) + \bar{k}_2$ where $\bar{k}_1, k_4 > 0$ and $\bar{k}_2 \geq 0$.

Notice that the error dynamics may be expressed as

$$e(k+d) = \alpha(k,x) + \beta(x)\left(\Delta(x) + \mathcal{F}(x,\theta) + \nu_s(k) + \frac{\partial \mathcal{F}}{\partial \hat{\theta}}\tilde{\theta}\right) + \omega(k), \quad (13.84)$$

where we have substituted the definition of the control law (13.76) and have added and subtracted $\mathcal{F}(x,\theta)$. Using (13.84) and (13.52) we obtain

$$\begin{aligned} V_s(k+d) - V_s(k) \leq\ & (1+\epsilon)\left[\alpha + \beta\nu_s + \omega\right]^\top P \left[\alpha + \beta\nu_s + \omega\right] - V_s(k) \\ & + \left(1 + \frac{1}{\epsilon}\right)\beta^\top P \beta \left(\Delta(x) + \mathcal{F}(x,\theta) + \frac{\partial \mathcal{F}}{\partial \hat{\theta}}\tilde{\theta}\right)^2 \end{aligned}$$

for any $\epsilon > 0$. Using Assumption 13.2 we are guaranteed that

$$\begin{aligned} V_s(k+d) - V_s(k) &= \left[\alpha + \beta\nu_s + \omega\right]^\top P \left[\alpha + \beta\nu_s + \omega\right] - V_s(k) \\ &\leq -k_1 V_s(k) + k_2, \end{aligned} \quad (13.85)$$

where we have used the error dynamics (13.66) with $\Delta \equiv 0$ and $u(k) = \nu_s(k)$. Therefore when $x \in S_x$ we find

$$\begin{aligned} V_s(k+d) - V_s(k) \leq\ & (1+\epsilon)\left(-k_1 V_s(k) + k_2\right) + \epsilon V_s(k) \quad (13.86) \\ & + \left(1 + \frac{1}{\epsilon}\right)\beta^\top(k) P \beta(k) \left(W + \left|\frac{\partial \mathcal{F}}{\partial \hat{\theta}}\tilde{\theta}\right|\right)^2. \end{aligned}$$

Now consider the positive definite function $V = V_s + k_4 V_{\tilde{\theta}}$ where $k_4 > 0$. Taking the difference

$$\begin{aligned} V(k+d) - V(k) =\ & V_s(k+d) - V_s(k) + k_4\left(V_{\tilde{\theta}}(k+d) - V_{\tilde{\theta}}(k)\right) \\ \leq\ & (1+\epsilon)\left(-k_1 V_s(k) + k_2\right) + \epsilon V_s(k) \\ & + \left(1 + \frac{1}{\epsilon}\right)\beta^\top(k) P \beta(k) \left(W + \left|\frac{\partial \mathcal{F}}{\partial \hat{\theta}}\tilde{\theta}\right|\right)^2 \\ & - \frac{\eta k_3 k_4}{(\beta^\top \beta)^2} \frac{\bar{q}^2(k+d)}{(1+\gamma|\zeta(k)|^2)}. \end{aligned}$$

Letting $\bar{k}_1 = k_1 + \epsilon k_1 - \epsilon$ we find

$$V(k+d) - V(k) \leq -\bar{k}_1 V_s + (1+\epsilon)k_2 - \frac{\eta k_3 k_4}{(\beta^\top \beta)^2} \frac{\bar{q}^2(k+d)}{1+\gamma|\zeta(k)|^2} + \left(1+\frac{1}{\epsilon}\right) \beta^\top(k) P \beta(k) \left(W + \left|\frac{\partial \mathcal{F}}{\partial \hat{\theta}} \tilde{\theta}\right|\right)^2. \tag{13.87}$$

Notice from (13.78) that

$$\begin{aligned} \beta^\top \beta \frac{\partial \mathcal{F}}{\partial \hat{\theta}} \tilde{\theta} &= q(k+d) - \beta^\top \beta w(k) - \beta^\top \omega \\ &= \bar{q}(k+d) + z(k+d), \end{aligned} \tag{13.88}$$

where

$$z(k+d) = q(k+d) - \bar{q}(k+d) - \beta^\top \beta w(k) - \beta^\top \omega.$$

Notice that $|z(k)| \leq 2\rho$ when $x \in S_x$ since ρ was defined such that $\rho \geq \beta^\top \beta W + \beta^\top \omega$ and $|q(k) - \bar{q}(k)| \leq \rho$ by the definition of the dead zone. Thus

$$\begin{aligned} \left|\frac{\partial \mathcal{F}}{\partial \hat{\theta}} \tilde{\theta}\right| &= \left|\frac{\bar{q}(k+d)}{\beta^\top \beta} + \frac{z(k+d)}{\beta^\top \beta}\right| \\ &\leq \left|\frac{\bar{q}(k+d)}{\beta^\top \beta}\right| + \frac{2\rho}{\beta^\top \beta} \end{aligned} \tag{13.89}$$

when $x \in S_x$.

Since $(|x| + |y|)^2 \leq 2x^2 + 2y^2$ we find

$$\begin{aligned} \left(W + \frac{\partial \mathcal{F}}{\partial \hat{\theta}} \tilde{\theta}\right)^2 &\leq \left(W + \frac{2\rho}{\beta^\top \beta} + \frac{|\bar{q}(k+d)|}{\beta^\top \beta}\right)^2 \\ &\leq 2\left(W + \frac{2\rho}{\beta^\top \beta}\right)^2 + 2\left(\frac{\bar{q}(k+d)}{\beta^\top \beta}\right)^2. \end{aligned}$$

Since $|\frac{\partial \mathcal{F}}{\partial \hat{\theta}}| \leq b_{\mathcal{F}}$ when $x \in S_x$ we find

$$\begin{aligned} \left(W + \frac{\partial \mathcal{F}}{\partial \hat{\theta}} \tilde{\theta}\right)^2 &\leq 2\left(W + \frac{2\rho}{\beta^\top \beta}\right)^2 + \frac{2(1+\gamma b_{\mathcal{F}}^2)}{(1+\gamma b_{\mathcal{F}}^2)} \left(\frac{\bar{q}(k+d)}{\beta^\top \beta}\right)^2 \\ &\leq 2\left(W + \frac{2\rho}{\beta^\top \beta}\right)^2 + \frac{2(1+\gamma b_{\mathcal{F}}^2)}{(1+\gamma|\zeta|^2)} \frac{\bar{q}^2(k+d)}{(\beta^\top \beta)^2}. \end{aligned} \tag{13.90}$$

Combining (13.87) and (13.90) we find

$$V(k+d) - V(k) \leq -\bar{k}_1 V_s + \bar{k}_2 - \frac{p(k)}{(\beta^\top \beta)^2} \frac{\bar{q}^2(k+d)}{(1+\gamma|\zeta(k)|^2)}, \tag{13.91}$$

where

$$p(k) = \eta k_3 k_4 - 2\left(1+\frac{1}{\epsilon}\right)\beta^\top(k)P\beta(k)(1+\gamma b_{\mathcal{F}}^2),$$

and we choose

$$\bar{k}_2 \geq (1+\epsilon)k_2 + 2\left(1+\frac{1}{\epsilon}\right)\beta^\top(k)P\beta(k)\left(W+\frac{2\rho}{\beta^\top\beta}\right)^2.$$

Now choosing

$$k_4 \geq \frac{2\left(1+\frac{1}{\epsilon}\right)\beta^\top(k)P\beta(k)(1+\gamma b_{\mathcal{F}}^2)}{\eta k_3} \tag{13.92}$$

ensures that $p(k) \geq 0$ so so $V(k+d) - V(k) \leq -\bar{k}_1 V_s(k) + \bar{k}_2$. Since we additionally have $V_{\tilde{\theta}}(k+d) - V_{\tilde{\theta}}(k) \leq 0$, it is now possible to use Theorem 13.5 to place bounds on $e(k)$.

Using Theorem 13.5 we may conclude that $V_s(k) \leq V_r$ for all k where

$$V_r = \max\left(V_s(0), \frac{\bar{k}_2 + k_4 V_{\tilde{\theta}}(0)}{\bar{k}_1}\right). \tag{13.93}$$

Thus $e \in B_e$ for all k where

$$B_e = \left\{e : |e| \leq \sqrt{V_r/\lambda_{\min}(P)}\right\}. \tag{13.94}$$

If the control parameters are chosen such that $B_e \subseteq S_e$ then the inputs to the approximator remain properly bounded. From Theorem 13.5, the ultimate bound on V_s is $\bar{k}_2/\bar{k}_1$ so

$$\lim_{k\to\infty} |e(k)| \leq \sqrt{\frac{\bar{k}_2}{\bar{k}_1 \lambda_{\min}(P)}}, \tag{13.95}$$

which completes the proof. ∎

As with the design of continuous-time adaptive controllers, one may pick the space over which a reasonable approximation may be obtained. Based on this region, then the controller is designed so that the states never leave this region. To design a discrete-time adaptive controller, we typically take the following steps:

1. Place the plant in a canonical representation so that an error system may be defined (e.g., as defined by (13.12)).

2. Define an error system and Lyapunov candidate V_s for the static problem.

3. Define a static control law $u = \nu_s$ which ensures that $V_s(k+d) - V_s(k) \leq -k_1 V_s(k) + k_2$ when $\Delta = 0$ in (13.66).

4. Choose an approximator $\mathcal{F}(x, \hat{\theta})$ such that there exists some θ where $|\mathcal{F}(x, \theta) + \Delta(x)| \leq W$ for all $x \in S_x$. Estimate an upper bound for W.

5. Find some B_r such that $e \in B_r$ implies $x \in S_x$.

6. Choose the initial conditions, control parameters, and update law parameters such that $B_e \subseteq B_r$ with B_e defined by (13.94).

In the continuous-time case, we could always define gains for the feedback law and update law so that the states remained in B_x. For the discrete-time case, it may be difficult (or impossible) to find gains which satisfy the above requirements for a given sampling period T. By making T smaller, the choice of the controller parameters typically becomes less restrictive.

As with the continuous-time approaches, the above closed-loop system using an adaptive controller is stable without persistency of excitation requirements. Since we are using approximators in the feedback law that may contain a large number of adjustable parameters, this is an important feature. Also the ultimate bound is independent of the system initial conditions.

The following example demonstrates how to design a discrete-time adaptive controller:

Example 13.6 Here we will design a discrete-time controller for the system

$$x(k+1) = x(k) + T\Delta(x) + Tu(k), \tag{13.96}$$

where $\Delta(x)$ represents unknown dynamics and T is the sample period. Assume that for a particular application we have

$$\Delta = c_1 + c_2 \sin(2\pi Tk + \phi), \tag{13.97}$$

when $|x| \leq 10$. Additionally, $T = 0.01$ and $c_1 = 10, c_2 = 0.1$, and $\phi = 0.1$ are unknown constants. Once $|x| > 10$, the form of the uncertainty is no longer valid. For this example it will be desirable for the plant output to track $r(k) = 1$.

Since the plant dynamics are already in a canonical form, the first step in developing the adaptive controller will be to define an error system and Lyapunov candidate for the nominal plant using a static control law (i.e., when $\Delta = 0$). We will consider $e(k) = x(k) - r(k)$ and $V_s(k) = Pe^2(k)$ with $P > 0$. Notice that this choice of the error dynamics fits the form (13.66) with $\alpha = -r(k+1) + x(k)$ and $\beta = T$. We will now study the properties of the closed-loop system with a static control law define by $u = \nu_s$, where

$$\nu_s = \frac{r(k+1) - x(k) + \kappa e}{T} \tag{13.98}$$

with $0 < |\kappa| < 1$. Notice that when $\Delta = 0$, we find

$$V_s(k+1) - V_s(k) = (\kappa^2 - 1)Pe^2(k) = -k_1 V_s(k),$$

where $k_1 = 1 - \kappa^2$. Since $|\kappa| < 1$, we find $0 < k_1 < 1$. Thus $V_s(k+1) - V_s(k) \leq -k_1 V_s(k) + k_2$ with $k_2 = 0$.

The next step to define the discrete-time adaptive controller is to define an appropriate approximator. Due to the form of the uncertainty, we will consider the approximator

$$\mathcal{F}(\hat{\theta}) = \hat{\theta}, \tag{13.99}$$

so that

$$\begin{aligned} |\mathcal{F}(\theta) + \Delta| &= |\theta + c_1 + c_2 \sin(2\pi T k + \phi)| \\ &= |c_2 \sin(2\pi T k + \phi)| \leq |c_2| \end{aligned}$$

when $\theta = -c_1$ and $|x| \leq 10$. The bound in the representation error, W, may thus be chosen such that $W \geq |c_2|$ (an inequality is used here since $|c_2|$ may not be known, but we assume its upper bound is known). We also know from the choice of the system error and reference trajectory that $|x| < |e| + |r|$. Since $r = 1$, if $|e| \leq 9$, then $|x| \leq 10$ so that the plant dynamics and approximator remain valid.

We are now ready to define the parameters of the control system to ensure that $|e(k)| \leq 9$ for all k. Recall that the ultimate bound is given by $\lim_{k\to\infty} |e(k)| \leq b(\epsilon)$, where

$$b(\epsilon) = \sqrt{\frac{\bar{k}_2}{\bar{k}_1 P}}, \tag{13.100}$$

with

$$\bar{k}_1 = k_1 + \epsilon k_1 - \epsilon \qquad \bar{k}_2 = 2\left(1 + \tfrac{1}{\epsilon}\right) P T^2 \left(W + 2\rho/T^2\right)^2 .$$

Here we have used the values $k_2 = 0$ and $\beta = T$. Notice that the P in $\bar{k}_2$ and the denominator of (13.100) will cancel one another so that the choice of P has no effect upon the ultimate bound of the error. Recall from (13.79) that

$$\rho \geq T^2 W + T\omega$$

represents the total adaptive controller uncertainty when the ideal parameters are known (here we have again substituted $\beta = T$). Since we are assuming a true discrete-time system (rather than a sampled-data system), we may take $\omega = 0$. A plot of $b(\epsilon)$ is shown in Figure 13.5

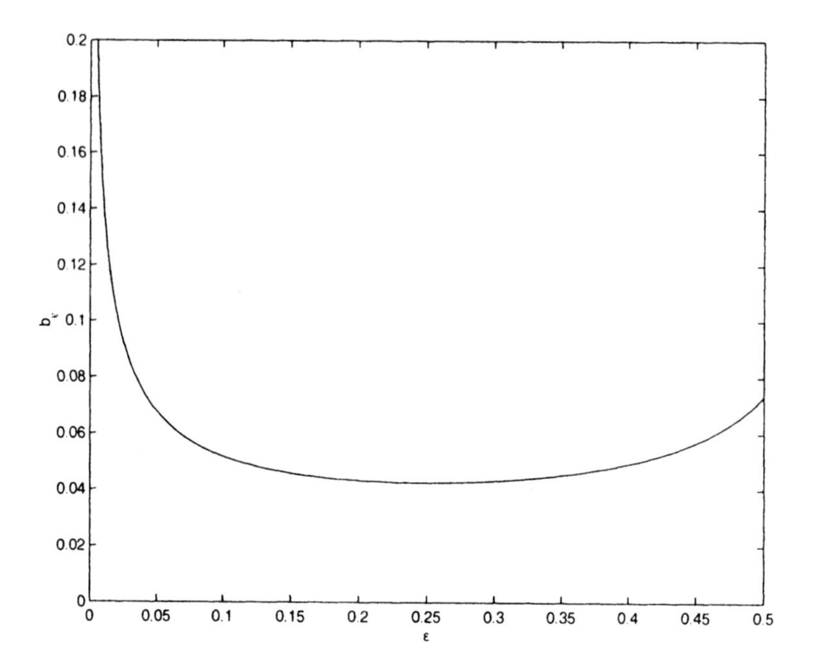

Figure 13.5. Plot of the ultimate bound function.

where we have chosen $\kappa = 0.8$ and $W = 0.2$ (i.e., $|W| = 2|c_2|$). The minimum value for $b(\epsilon)$ is 0.0424 and occurs when $\epsilon = 0.25$. For this value of ϵ, we also find that $\bar{k}_1 = 0.2$ and $\bar{k}_2 = 0.00036$. Even though the choice for ϵ is not used in the control law (and thus will not affect the true ultimate bound), it will help us determine a bound on the error transients when an adaptive controller is used, as will be shown next.

Recall that $|e| \le \sqrt{V_r/\lambda_{\min}(P)}$ where

$$V_r = \max\left(V_s(0), \frac{\bar{k}_2 + k_4 V_{\tilde{\theta}}(0)}{\bar{k}_1}\right). \tag{13.101}$$

Here

$$k_4 = \frac{2\left(1 + \frac{1}{\epsilon}\right) PT^2(1 + \gamma b_{\mathcal{F}}^2)}{\eta k_3},$$

and $k_3 = 2 - \eta/\gamma$, where $\gamma > 0$ and $0 < \eta < 2\gamma$ are the parameters used in the update law. Also by the definition of the approximator, we find $|\frac{\partial \mathcal{F}(x,\hat{\theta})}{\partial \hat{\theta}}| = 1$ so choosing $b_{\mathcal{F}} = T$ satisfies the requirements of Theorem 13.6. Using $\gamma = 1$ and $\eta = 0.2$, we find $k_3 = 1.8$ and $k_4 = 0.0056$.

All that is left to do is to place bounds on the initial conditions so that the error is properly bounded. To do this we need to ensure that both elements in the right-hand side of (13.101) are defined such that $|e| \leq 9$. The first term is easily satisfied by requiring that $|e(0)| \leq 9$. The second term requires that

$$\sqrt{\frac{\bar{k}_2 + k_4(\hat{\theta}(0) - \theta)^2}{k_1 \lambda_{\min}(P)}} \leq 9, \tag{13.102}$$

or that $|\hat{\theta}(0) - \theta| \leq 53.9$, which is acheived with the initial choice of $\hat{\theta}(0) = 0$ since $\theta = -10$.

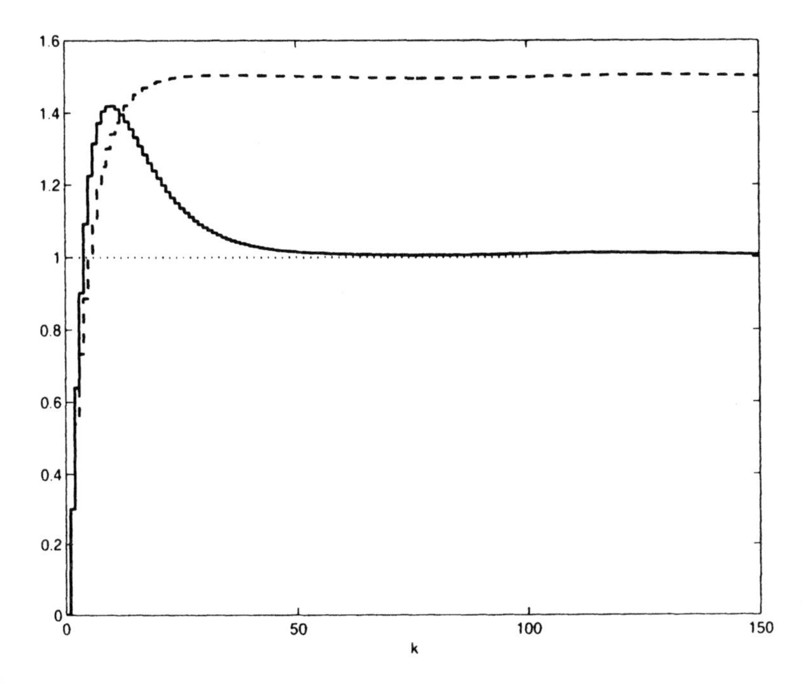

Figure 13.6. Closed-loop system response when using the adaptive controller (--), and when the adaptive portion is turned off (- -).

Figure 13.6 shows the results of the closed-loop system using the parameters chosen above and $x(0) = 0$ (shown by the solid line). As a comparison, the simulation was also run when we set $u(k) = \nu_s(k)$ so that only the static control portion was used (effectively removing the adaptive control portion). As might be expected, the adaptive controller improves the steady state performance. $\triangle$

13.6 Summary

In this chapter we learned how to design controllers for nonlinear discrete-time systems. Since today's microprocessors are so efficient, the continuous-time approaches studied in the previous chapters may be used for most real-world applications. There may be cases, however, when the delay associated with the sample rate is significant with respect to the desired closed-loop system bandwidth. In these cases it may be necessary to design the controller using a discrete-time framework. The disadvantage with going to a discrete-time framework is that the mathematics may become more involved, limiting the number of design techniques available for robust controller design.

13.7 Exercises and Design Problems

Exercise 13.1 (Discrete-Time Systems, Uniform Boundedness) Suppose you are given the scalar difference equation

$$x(k+1) = a\,\mathrm{sgn}(x(k))x(k) + b,$$

where $a, b > 0$ and $\mathrm{sgn}(x(k)) = 1$ if $x(k) \geq 0$ and $\mathrm{sgn}(x(k)) = -1$ if $x(k) < 0$. Are the solutions to this difference equation uniformly bounded? If so, prove it.

Exercise 13.2 (Sample Rate and Stability) Consider the continuous-time system

$$\dot{x} = u$$

and control law $u = -\kappa x$, where $\kappa > 0$ so that $x = 0$ is an exponentially stable equilibrium point. Use Euler's method to convert this to a discrete-time system and determine the range that κ may take on so that the resulting discrete-time system is stable given a sample period T.

Exercise 13.3 (Prediction) Given the system

$$\begin{aligned} x_1(k+1) &= x_2(k) \\ &\vdots \\ x_{n-1}(k+1) &= x_n(k) \\ x_n(k+1) &= f_d(x(k)) + g_d(x(k))u(k) \end{aligned}$$

with output $y = x_1$, we find

$$y(k+n) = f_d(x(k)) + g_d(x(k))u(k).$$

Use this form to show that

$$\hat{y}(k+n) = f_d(x(k)) + g_d(x(k))u(k) + \kappa\,(y(k) - \hat{y}(k)) \qquad (13.103)$$

forms a stable predictor so that $\hat{y}(k+n) \to y(k+n)$ when $|\kappa| < 1$.

Exercise 13.4 (Output-Feedback Control) Consider the system defined by

$$\begin{aligned} x_1(k+1) &= x_2(k) \\ &\vdots \\ x_{n-1}(k+1) &= x_n(k) \\ x_n(k+1) &= f_d(y(k)) + g_d(y(k))u(k), \end{aligned}$$

where $y(k) = x(k)$. Define a state estimator and control law so that $x \to 0$ when only $y(k)$ is measurable.

Exercise 13.5 (Surge Tank) Use Euler's method to discretize the surge tank whose dynamics are given by

$$A(x)\dot{x} = -c\sqrt{2gx} + u, \qquad (13.104)$$

given a sample period T. See Example 7.11 for a description of the variables. Design a static discrete-time controller so that $x \to r$ assuming that the plant dynamics are known. Simulate the performance of your controller design 1) using your discretized plant dynamics and no intersample error, and 2) using a continuous-time simulation where the controller output is only changed at the sample rate. How do the two simulations compare as T is increased?

Exercise 13.6 (Fuzzy Control of a Point Mass) Consider the discrete-time model of a point mass

$$\begin{aligned} x(k+1) &= x(k) + Tv(k) \\ v(k+1) &= v(k) + \tfrac{T}{m}(u(k) + \Delta(k)), \end{aligned} \qquad (13.105)$$

where x is the position, v is the velocity, u is the force input, and m is the mass. Given the sample period T, design a static controller so that $x \to 0$ when the disturbance force satisfies $\Delta = 0$. Then design a single adaptive controller using a fuzzy system for the cases when

- $\Delta(x) = \cos(x)$
- $\Delta(x) = x + x^2$.

What conditions are needed for stability?

Exercise 13.7 (Neural Control of a Point Mass) Repeat Example 13.6 using a neural network.

Chapter 14

Decentralized Systems

14.1 Overview

In this chapter, we will study the control of MIMO systems where constraints are placed on the flow of information. In particular we will consider the control of decentralized systems where there are constraints on information exchange between subsystems. Decentralized control systems often arise from either the physical inability of subsystem information exchange or the lack of computing capabilities required for a single central controller. Furthermore, at times it may be more convenient to design a controller in a decentralized framework since each subsystem is often much more simple than the composite MIMO system.

Within a decentralized framework, the overall system is broken into N subsystems each with its own inputs and outputs. A decentralized control law is then defined using local subsystem signals. Thus the ith subsystem does not have access to the signals associated with the jth subsystem when $i \neq j$. Figure 14.1 shows a decentralized system with four subsystems, each influenced by interconnections to one or more of the other subsystems. Here we use the notation S_i to represent the ith subsystem and $I_{i,j}$ is an interconnection which defines how the jth subsystem influences the ith subsystem.

The goal of this chapter will be to familiarize the reader with some of the tools used in the design and analysis of decentralized controllers for nonlinear systems. There has been a large volume of work dedicated to the design of decentralized controllers so it will only be possible to touch on a few decentralized techniques. The reader is urged to see the "For Further Study" section at the end of this book for references to related decentralized approaches.

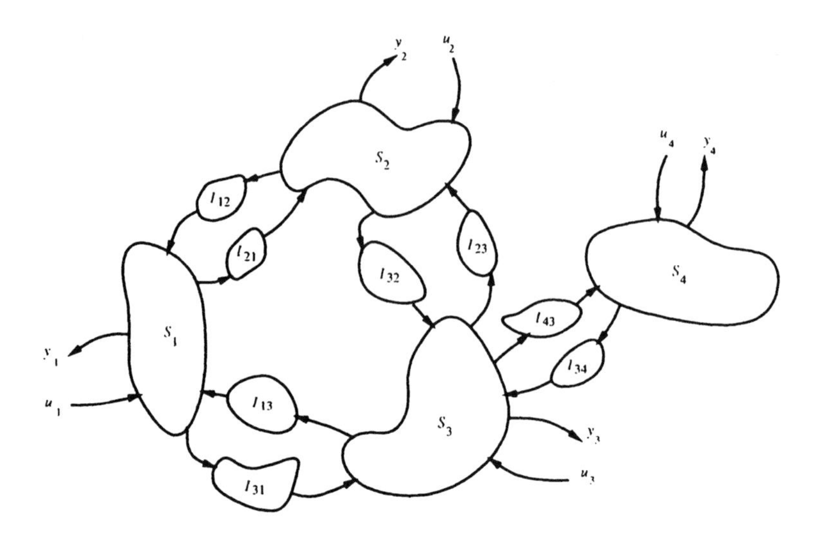

Figure 14.1. A decentralized system.

14.2 Decentralized Systems

Consider the MIMO system defined by

$$\begin{aligned} \dot{\xi} &= f_\xi(\xi) + g_\xi(\xi)u \\ y &= h_\xi(\xi), \end{aligned} \tag{14.1}$$

with $y, u \in \mathbb{R}^N$, and $\xi \in \mathbb{R}^n$, where it is assumed that $n \geq N$. At this point (14.1) is a standard MIMO system and thus control designs presented earlier in this book may be applicable.

We will assume that the decomposition of (14.1) into subsystems is achieved by some means. When communication constraints are placed on the design of a control system, it may be obvious how to define subsystems. When a decentralized approach is chosen for computational or other design reasons, the specification of a subsystem may often be determined due to physical reasons such as when each input has a strong relationship to a particular output. This may occur, for example, if two bodies are connected by a spring. In this case if the spring were removed, the subsystems would be independent from one another. The spring acts as an interconnection between the subsystems. We will assume that the definition of the subsystems is obvious from the problem statement.

Let T_i be a (decentralized) transformation used to define the subsystem S_i states as $x_i = T_i(\xi_i)$ with dynamics defined by

$$S_i : \begin{aligned} \dot{x}_i &= f_i(x_i) + g(x_i)u_i + \Delta_i(t, x) \\ y_i &= h_i(x_i), \end{aligned} \tag{14.2}$$

for $i = 1, \ldots, N$, where y_i is the output of the ith subsystem and the state vector for the composite system is defined by $x = [x_i^{\mathsf{T}}, \ldots, x_N^{\mathsf{T}}]^{\mathsf{T}}$. A decentralized transformation is one in which $x_i = T_i(\xi_i)$ so that only local measurements are used in the transformation.

We will be interested in designing a decentralized controller for systems in which the interconnections do not necessarily satisfy matching conditions, and the strengths of the connections may be bounded by arbitrary smooth functions of the subsystem outputs. To simplify the notation, we will assume that $n_i = n$ for $i = 1, \ldots, N$ so that each isolated subsystem has n states. Consider the subsystem transformed into

$$S_i: \quad \begin{aligned} \dot{x}_{i,1} &= f_{i,1}(\bar{x}_{i,1}) + x_{i,2} + \Delta_{i,1}(t,x) \\ &\vdots \\ \dot{x}_{i,n-1} &= f_{i,n-1}(\bar{x}_{i,n-1}) + x_{i,n} + \Delta_{i,n-1}(t,x) \\ \dot{x}_{i,n} &= f_{i,n}(\bar{x}_{i,n}) + u_i + \Delta_{i,n}(t,x), \end{aligned} \tag{14.3}$$

with output $y_i = x_{i,1}$. Here the notation $\bar{x}_{i,j} = [x_{i,1}, \ldots, x_{i,j}]^{\mathsf{T}}$ is used. Each $\Delta_{i,j}$ represents the interconnection between subsystems. Our goal will be to design a controller which forces $y_i \to 0$ for $i = 1, \ldots, N$ when the interconnections $\Delta_{i,j}$ satisfy certain bounding conditions which will be defined next. Note here that each $\Delta_{i,j}$ is influenced by subsystems $S_1, \ldots, S_N$ and should not be considered to be associated with the specific interconnection $I_{i,j}$ which defines the influence of S_j upon S_i.

We will assume that the interconnections are defined by

$$I_i: \quad \begin{aligned} \dot{q}_i &= \phi_i(t,x) \\ \Delta_i &= s_i(t,q_i,x) \end{aligned} \tag{14.4}$$

with $q_i \in \mathrm{R}^{m_i}$ and $\Delta_i \in \mathrm{R}^{n_i}$. Notice that I_i combines the effects of the individual interconnections $I_{i,1}, \ldots, I_{i,N}$. When there are no dynamics associated with the interconnection, we let $\Delta_i = s_i(t,x)$. When $\Delta_i \equiv 0$, we say that (14.2) represents the isolated dynamics for the ith subsystem.

Example 14.1 Assume that we wish to design a control system for a fixture used to test the functionality of an integrated circuit (IC) before being shipped. The goal of the control system is to accurately position a stage (which holds the IC) in two linear degrees of freedom and a rotational degree of freedom (often referred to as an X-Y-Θ stage).

Restoring springs have been included in the system design so that the stage returns to some nominal position when power is removed. The dynamics of the system are described by

$$\begin{aligned} m\ddot{x} &= F_x - F_1(x,\theta) \\ m\ddot{y} &= F_y - F_2(y,\theta) - F_3(y,\theta) \\ J\ddot{\theta} &= T + c_1 F_1(x,\theta) + c_2 F_2(y,\theta) + c_3 F_3(y,\theta), \end{aligned} \tag{14.5}$$

where x, y and θ are the linear and angular stage degrees of freedom, m is the stage mass, J is the moment of inertia, F_x is the force along the x-axis, F_y is the force along the y-axis, and T is the torque applied to the stage. The term F_i is the force exerted by the ith spring where

$$\begin{aligned} F_1 &= k_1(x - d_1 \sin(\theta) - l_1) \\ F_2 &= k_2(y + d_2 \sin(\theta) - l_2) \\ F_3 &= k_3(y + d_3 \sin(\theta) - l_3), \end{aligned} \tag{14.6}$$

where k_i is the stiffness of the ith spring, l_i is the natural length of the ith spring. The value d_1 is the distance from the x-axis to the spring attachment point, while d_2 and d_3 are the distances from the y-axis to the attachment points for springs 2 and 3. The value of c_i defines the moment arm for the ith spring.

A MIMO control approach may be used to design a control system for this example. Assume, however, that we would like to design decentralized controllers where an individual control system is designed for each degree of freedom of the stage. That is, the controller for the x-axis will not use information about the y or θ states. Thinking of the IC tester as a decentralized system, we may express the dynamics as

$$\begin{aligned} \ddot{x} &= F_x/m + \Delta_x(x, \theta) \\ \ddot{y} &= F_y/m + \Delta_y(y, \theta) \\ \ddot{\theta} &= T/J + \Delta_\theta(x, y, \theta) \end{aligned} \tag{14.7}$$

where

$$\begin{aligned} \Delta_x &= -F_1(x, \theta)/m \\ \Delta_y &= -F_2(y, \theta)/m - F_3(y, \theta)/m \\ \Delta_\theta &= c_1 F_1(x, \theta)/J + c_2 F_2(y, \theta)/J + c_3 F_3(y, \theta)/J. \end{aligned} \tag{14.8}$$

We will find that this form is suitable for various static and adaptive decentralized control approaches. $\triangle$

14.3 Static Controller Design

In this section, we will learn how to develop a static controller for a decentralized system. As with the design of centralized controllers, the properties of a Lyapunov function will then be used to place bounds on states which may be used as inputs to a finite approximator used within the control law.

14.3.1 Diagonal Dominance

When designing controllers for decentralized systems, one typically tries to find robust controllers when the interconnections are ignored. By making each local controller "very robust," the effects of the interconnections are dominated by the inherently robust local controllers. A powerful tool often used in the design of decentralized controllers is the concept of diagonal dominance. The following theorem is one possible application of diagonal dominance.

Theorem 14.1: *The inequality $x^\top(G+D)x \geq 0$ holds for all x if $G, D \in \mathrm{R}^{n\times n}$ where $G = [g_{i,j}]$ and $D = diag(d_1,\ldots,d_n)$ with*

$$d_i \geq n\left(1 + g_{1,i}^2 + \cdots + g_{n,i}^2\right)$$

defined along the columns of G, or

$$d_i \geq n\left(1 + g_{i,1}^2 + \cdots + g_{i,n}^2\right)$$

defined along the rows of G.

Proof: Let $z = x^\top(G+D)x$. Notice that

$$z = \sum_{i=1}^{n} d_i x_i^2 + \sum_{i=1}^{n}\sum_{j=1}^{n} g_{i,j}x_i x_j = \sum_{i=1}^{n}\sum_{j=1}^{n}\left[\frac{d_i}{n}x_i^2 + g_{i,j}x_i x_j\right]. \tag{14.9}$$

Since $p^2 \pm pq \geq p^2/2 - q^2/2$ we may let $p^2 = \frac{d_i}{n}x_i^2$ and $q = \sqrt{\frac{n}{d_i}}g_{i,j}x_j$ to show

$$\begin{aligned} z &\geq \sum_{i=1}^{n}\sum_{j=1}^{n}\left[\frac{d_i}{2n}x_i^2 - \frac{ng_{i,j}^2}{2d_i}x_j^2\right] \\ &= \sum_{i=1}^{n}\frac{d_i}{2}x_i^2 - \frac{n}{2}\sum_{j=1}^{n}\left(\frac{g_{1,j}^2}{d_1} + \cdots + \frac{g_{n,j}^2}{d_n}\right)x_j^2. \end{aligned} \tag{14.10}$$

To get $z \geq 0$ for all x, we should ensure that each d_i is chosen such that

$$d_i \geq n\left(\frac{g_{1,i}^2}{d_1} + \cdots + \frac{g_{n,i}^2}{d_n}\right) = n\sum_{j=1}^{n}\frac{g_{j,i}^2}{d_j} \tag{14.11}$$

for all i.

Now consider the choice

$$d_i \geq n\left(1 + g_{i,1}^2 + \cdots + g_{i,n}^2\right).$$

Starting from the right hand side of (14.11) we find

$$n\sum_{j=1}^{n}\frac{g_{j,i}^2}{d_j} \leq n\sum_{j=1}^{n}\frac{g_{j,i}^2}{n\left(1 + g_{j,1}^2 + \cdots + g_{j,n}^2\right)}. \tag{14.12}$$

Since $g_{j,i}^2 \le 1 + g_{j,1}^2 + \cdots + g_{j,n}^2$ for $i = 1, \ldots, n$

$$n \sum_{j=1}^{n} \frac{g_{j,i}^2}{d_j} \le n \le d_i, \tag{14.13}$$

for each $i = 1, \ldots, n$ so (14.11) is satisfied. Since $z = z^\top = x^\top (G^\top + D)x$, the same steps may be used to find

$$d_i^2 \ge n(1 + g_{1,i}^2 + \cdots + g_{n,i}^2)$$

is also sufficient for $z \ge 0$ for all x, which completes the proof. ∎

We will use Theorem 14.1 to help determine how to stabilize each subsystem so that the effects of the interconnections do not cause closed-loop system instabilities. It should be noted that Theorem 14.1 only provides sufficient, and not necessary conditions on the choice of the diagonal terms. In fact the results of Theorem 14.1 may be rather conservative for certain applications.

14.3.2 State-Feedback Control

We will begin our investigation into decentralized control design by first considering the case where a decentralized controller is to be developed for (14.3) with interconnection dynamics governed by

$$I_i : \begin{array}{rcl} \dot{q}_i & = & \phi_i(t, q_i, x) \\ \Delta_{i,j} & = & s_{i,j}(t, q_i, \bar{x}_{i,j}, y) \end{array}, \tag{14.14}$$

with bound

$$|\Delta_{i,j}| \le \rho + \zeta_{i,j}(\bar{x}_{i,j}) \sum_{k=1}^{N} \psi_{i,k}(|y_k|). \tag{14.15}$$

Here $\zeta_{i,j} : \mathsf{R}^j \to \mathsf{R}^+$ and $\psi_{i,k} : \mathsf{R} \to \mathsf{R}^+$ are smooth nonnegative functions and $\rho \in \mathsf{R}$. It is assumed that the functions $\zeta_{i,j}$ and $\psi_{i,k}$ are bounded for all bounded inputs. Notice that the bound on $\Delta_{i,j}$ is defined in terms of other subsystem outputs due to the $\psi_{i,k}$ elements.

We will additionally require that the I_i dynamics are input-to-state stable so that there exists some $V_{q_i}(t, q_i)$ such that

$$\begin{array}{rcl} \gamma_{q1_i}(q_i) & \le & V_{q_i}(t, q_i) \le \gamma_{q2_i}(q_i) \\ \dot{V}_{q_i} & \le & -\gamma_{q3_i}(q_i) + \psi_{q_i}(x), \end{array} \tag{14.16}$$

where γ_{q1_i}, γ_{q2_i}, and γ_{q3_i} are class-$\mathcal{K}_\infty$. Thus if we may design a control law which forces x to be bounded, then each q_i will also remain bounded.

The following lemma will be useful in the development of our decentralized control scheme:

Lemma 14.1: *Given a nonnegative continuous function $\psi : \mathsf{R} \to \mathsf{R}^+$, there exists a nonnegative continuous function $\phi : \mathsf{R} \to \mathsf{R}^+$ such that $\psi(|x|) \leq |x|^n \phi(|x|) + d$ where $n > 0$ and $d \geq 0$ are finite constants.*

Proof: Choose $\epsilon > 0$ to be a finite constant and $d = \max_{x \in [0,\epsilon]} \psi(x)$. It is sufficient that ϕ be any nonnegative continuous function such that $\phi(|x|) \geq \psi(|x|)/|x|^n$ for all $|x| \geq \epsilon$. Some ϕ is guaranteed to exist (see Example 14.2 for one such choice). When $|x| \leq \epsilon$, we find

$$\psi(|x|) \leq d \leq |x|^n \phi(|x|) + d$$

since $|x|^n \phi(|x|)$ is nonnegative. When $|x| \geq \epsilon$ we find

$$\psi(|x|) = |x|^n \frac{\psi(|x|)}{|x|^n} \leq |x|^n \phi(|x|) \leq d + |x|^n \phi(|x|)$$

since d is positive. Thus $\psi(|x|) \leq d + |x|^n \phi(|x|)$ for all x. ■

The following example provides one possible choice for ϕ in Lemma 14.1.

Example 14.2 Suppose that we are given some ψ and that we wish to find a ϕ such that $\psi(|x|) \leq |x|^n \phi(|x|) + d$ as described in Lemma 14.1. Choose some $\epsilon > 0$ and let $d = \max_{x \in [0,\epsilon]} \psi(x)$. Then $\phi = \psi(|x|)/\epsilon^n$ satisfies the requirements in the proof of Lemma 14.1 since we find $\psi(|x|)/\epsilon^n > \psi(|x|)/|x|^n$ when $|x| > \epsilon$. Thus

$$\psi(|x|) \leq |x|^n \frac{\psi(|x|)}{\epsilon^n} + d$$

for all x. △

Using the results of Lemma 14.1 and Example 14.2 we assume that there exist known constants $d_{i,k}$ and smooth functions $\phi_{i,k}$ such that

$$\psi_{i,k}(|y_k|) \leq d_{i,k} + \phi_{i,k}(|y_k|)\sqrt{|y_k|}. \tag{14.17}$$

for all $y_k \in \mathsf{R}$.

In this chapter we will focus on the decentralized output regulation problem where it is desired that $y_i \to 0$ for $i = 1, \ldots, N$ with $y_i = x_{i,1}$. Now consider the error system

$$e_{i,j} = x_{i,j} - v_{i,j-1}(\bar{x}_{i,j-1}), \tag{14.18}$$

for $i = 1, \ldots, N$ and $j = 1, \ldots, n$ with $v_{i,0} = 0$. The term $v_{i,j}$ will be defined shortly. Taking the derivative of the errors, we find

$$\dot{e}_{i,j} = f_{i,j} + x_{i,j+1} + \Delta_{i,j} - \sum_{k=1}^{j-1} \frac{\partial v_{i,j-1}}{\partial x_{i,k}} \left[f_{i,k} + x_{i,k+1} + \Delta_{i,k} \right]$$

$$= \quad f_{i,j} + e_{i,j+1} + v_{i,j} - \sum_{k=1}^{j-1} \frac{\partial v_{i,j-1}}{\partial x_{i,k}} [f_{i,k} + x_{i,k+1}] \tag{14.19}$$
$$+\Delta_{i,j} - \sum_{k=1}^{j-1} \frac{\partial v_{i,j-1}}{\partial x_{i,k}} \Delta_{i,k}$$

for $i = 1, 2, \ldots, N$ and $j = 1, \ldots, n-1$. Choosing

$$v_{i,j} = -(\kappa_i + \nu_{i,j}(\bar{x}_{i,j}))e_{i,j} - e_{i,j-1} - f_{i,j} + \sum_{k=1}^{j-1} \frac{\partial v_{i,j-1}}{\partial x_{i,k}} [f_{i,k} + x_{i,k+1}],$$

we find

$$\dot{e}_{i,j} = -(\kappa_i + \nu_{i,j})e_{i,j} - e_{i,j-1} + e_{i,j+1} + \Delta_{i,j} - \sum_{k=1}^{j-1} \frac{\partial v_{i,j-1}}{\partial x_{i,k}} \Delta_{i,k} \tag{14.20}$$

for $j = 1, \ldots, n-1$ where $e_{i,0} = 0$ and each $\nu_{i,j}$ will be chosen to account for the interconnections.

The control law is then defined as

$$u_i = -(\kappa_i + \nu_{i,n}(\bar{x}_{i,n}))e_i - e_{i-1} - f_{i,n} + \sum_{k=1}^{n-1} \frac{\partial v_{i,n-1}}{\partial x_{i,k}} [f_{i,k} + x_{i,k+1}] \tag{14.21}$$

so that

$$\dot{e}_{i,n} = -(\kappa_i + \nu_{i,n})e_{i,n} - e_{i,n-1} + \Delta_{i,n} - \sum_{k=1}^{n-1} \frac{\partial v_{i,n-1}}{\partial x_{i,k}} \Delta_{i,k}, \tag{14.22}$$

where $\nu_{i,n}$ will again be chosen to account for the system interconnections. Let $e_i = [e_{i,1}, \ldots, e_{i,n}]^\top$. Then the subsystem error dynamics may be expressed as

$$\dot{e}_i = A_i e_i + \begin{bmatrix} -\nu_{i,1}e_{i,1} + \delta_{i,1} \\ \vdots \\ -\nu_{i,n}e_{i,n} + \delta_{i,n} \end{bmatrix}, \tag{14.23}$$

where

$$A_i = \begin{bmatrix} -\kappa_i & 1 & 0 & \cdots & 0 \\ -1 & -\kappa_i & 1 & & \vdots \\ 0 & -1 & -\kappa_i & & \\ \vdots & & & \ddots & 1 \\ 0 & \cdots & & -1 & -\kappa_i \end{bmatrix}, \tag{14.24}$$

and

$$\delta_{i,j} = \Delta_{i,j} - \sum_{k=1}^{j-1} \frac{\partial v_{i,j-1}}{\partial x_{i,k}} \Delta_{i,k}. \tag{14.25}$$

The following theorem summarizes the resulting closed-loop stability properties of the proposed static controller.

Theorem 14.2: *Given the subsystem (14.3) with interconnections bounded by (14.15), the decentralized control law (14.21) with $\nu_{i,j}$ defined by (14.28) will ensure that each e_i (and thus each y_i) is UUB.*

Proof: Define $V_{s_i} = \frac{1}{2}\sum_{j=1}^{n} e_{i,j}^2$ for each subsystem. Using (14.23) we find

$$\dot{V}_{s_i} = \frac{1}{2} e_i^\top (A_i^\top + A_i) e_i + e_i^\top \begin{bmatrix} -\nu_{i,1} e_{i,1} + \delta_{i,1} \\ \vdots \\ -\nu_{i,n} e_{i,n} + \delta_{i,n} \end{bmatrix}. \tag{14.26}$$

Using the interconnection bound (14.15), we find

$$|\delta_{i,j}| \le \rho + \zeta_{i,j} \sum_{l=1}^{N} \psi_{i,l}(|y_l|) + \sum_{k=1}^{j-1} \left| \frac{\partial v_{i,j-1}}{\partial x_{i,k}} \right| \left(\rho + \zeta_{i,k} \sum_{l=1}^{N} \psi_{i,l}(|y_l|) \right),$$

so that terms may be rearranged to obtain

$$|\delta_{i,j}| \le \rho \left(1 + \sum_{k=1}^{j-1} \left| \frac{\partial v_{i,j-1}}{\partial x_{i,k}} \right| \right) + \left(\zeta_{i,j} + \sum_{k=1}^{j-1} \zeta_{i,k} \left| \frac{\partial v_{i,j-1}}{\partial x_{i,k}} \right| \right) \sum_{l=1}^{N} \psi_{i,l}.$$

Using (14.17) we are guaranteed that $\psi_{i,l} \le d_{i,l} + \phi_{i,l}(|y_l|)\sqrt{|y_l|}$. The inequality $2xq_l \le x^2/\sqrt{c} + \sqrt{c}q_l^2$ with $c > 0$ may now be used with

$$2x = \left(\zeta_{i,j} + \sum_{k=1}^{j-1} \zeta_{i,k} \left| \frac{\partial v_{i,j-1}}{\partial x_{i,k}} \right| \right) \qquad q_l = \phi_{i,l}(|y_l|)\sqrt{|y_l|}$$

so that

$$\begin{aligned} |\delta_{i,j}| \le\ & \rho \left(1 + \sum_{k=1}^{j-1} \left| \frac{\partial v_{i,j-1}}{\partial x_{i,k}} \right| \right) + \left(\zeta_{i,j} + \sum_{k=1}^{j-1} \zeta_{i,k} \left| \frac{\partial v_{i,j-1}}{\partial x_{i,k}} \right| \right) \sum_{l=1}^{N} d_{i,l} \\ & + \frac{N}{4\sqrt{c}} \left(\zeta_{i,j} + \sum_{k=1}^{j-1} \zeta_{i,k} \left| \frac{\partial v_{i,j-1}}{\partial x_{i,k}} \right| \right)^2 + \sum_{l=1}^{N} \sqrt{c} |y_l| \phi_{i,l}^2. \end{aligned}$$

Let

$$\begin{aligned} z_{i,j} =\ & \rho \left(1 + \sum_{k=1}^{j-1} \left| \frac{\partial v_{i,j-1}}{\partial x_{i,k}} \right| \right) + \left(\zeta_{i,j} + \sum_{k=1}^{j-1} \zeta_{i,k} \left| \frac{\partial v_{i,j-1}}{\partial x_{i,k}} \right| \right) \sum_{l=1}^{N} d_{i,l} \\ & + \frac{N}{4\sqrt{c}} \left(\zeta_{i,j} + \sum_{k=1}^{j-1} \zeta_{i,k} \left| \frac{\partial v_{i,j-1}}{\partial x_{i,k}} \right| \right)^2, \end{aligned} \tag{14.27}$$

and

$$\nu_{i,j} = \eta_i z_{i,j}^2 + q_i(y_i), \tag{14.28}$$

where we choose $\eta_i > 0$. The smooth function q_i will be chosen to achieve diagonal dominance in the composite system (this will be defined later in the proof). Then from (14.26)

$$\begin{aligned}
\dot{V}_{s_i} &\leq -\kappa_i e_i^\top e_i + \sum_{j=1}^{n} \left(-\nu_{i,j} e_{i,j}^2 + |e_{i,j}||\delta_{i,j}|\right) \\
&\leq -\kappa_i e_i^\top e_i \qquad (14.29) \\
&\quad + \sum_{j=1}^{n} \left(-\eta_i z_{i,j}^2 e_{i,j}^2 + |z_{i,j}||e_{i,j}| - q_i e_{i,j}^2 + |e_{i,j}| \sum_{l=1}^{N} \sqrt{c}|y_l|\phi_{i,l}^2\right).
\end{aligned}$$

Since $-x^2 \pm 2xs \leq s^2$ we may let $x^2 = \eta_i z_{i,j}^2 e_{i,j}^2$ and $s = \frac{1}{2\sqrt{\eta_i}}$ to show

$$\dot{V}_{s_i} \leq -2\kappa_i V_{s_i} + \frac{n}{4\eta_i} - 2q_i V_{s_i} + \sum_{j=1}^{n} |e_{i,j}| \sum_{l=1}^{N} \sqrt{c}|y_l|\phi_{i,l}^2. \qquad (14.30)$$

Notice that we combined the $q_i e_{i,j}^2$ terms in (14.29) to obtain $2q_i V_{s_i}$ and have used $\sum_{j=1}^{n} \frac{1}{4\eta_i} = \frac{n}{4\eta_i}$ in (14.30).

To choose each $q_i(y_i)$ we will now consider the composite system. Let $V_s = \sum_{i=1}^{N} V_{s_i}$. Since $|y_l| = |e_{l,1}|$ and $|e_{i,j}| \leq \sqrt{2V_{s_i}}$ for all i, j we obtain

$$\begin{aligned}
\dot{V}_s &\leq \sum_{i=1}^{N} \left[-2\kappa_i V_{s_i} + \frac{n}{4\eta_i} - 2q_i V_{s_i} + \sqrt{2c}\sum_{j=1}^{n} \sqrt{V_{s_i}} \sum_{l=1}^{N} \phi_{i,l}^2 \sqrt{V_{s_l}}\right] \\
&= \sum_{i=1}^{N} \left[-2\kappa_i V_{s_i} + \frac{n}{4\eta_i} - 2q_i V_{s_i} + n\sqrt{2c}\sqrt{V_{s_i}} \sum_{l=1}^{N} \phi_{i,l}^2 \sqrt{V_{s_l}}\right]
\end{aligned}$$

It is now possible to rearrange terms to obtain

$$\dot{V}_s \leq -2\bar{\kappa} V_s + \frac{nN}{4\bar{\eta}} - 2\left[\begin{array}{ccc} \sqrt{V_{s_1}} & \cdots & \sqrt{V_{s_N}} \end{array}\right](D+G)\left[\begin{array}{c} \sqrt{V_{s_1}} \\ \vdots \\ \sqrt{V_{s_N}} \end{array}\right],$$

where $\bar{\kappa} = \min_i(\kappa_i)$, $\bar{\eta} = \min_i(\eta_i)$, $D = \mathrm{diag}(q_1, \ldots, q_N)$, and

$$G = -\frac{n\sqrt{c}}{\sqrt{2}} \left[\begin{array}{ccc} \phi_{1,1}^2 & \cdots & \phi_{1,N}^2 \\ \vdots & & \vdots \\ \phi_{N,1}^2 & \cdots & \phi_{N,N}^2 \end{array}\right].$$

The signal $\phi_{j,i}$ is available to the ith subsystem since it is defined in terms of the output y_i. Thus the signals in the ith column of G may be used to define the controller for the ith subsystem.

Using diagonal dominance, we may choose the q_i terms so that $D+G$ is positive definite. In particular, from Theorem 14.1 we choose

$$q_i = \mu_i N \left[1 + \frac{cn^2}{2}\left(\sum_{j=1}^{N} \phi_{j,i}^4(|y_i|)\right)\right], \tag{14.31}$$

where $\mu_i \geq 1$ so that

$$\dot{V}_s \leq -2\bar{\kappa}V_s + \frac{nN}{4\bar{\eta}}. \tag{14.32}$$

Using (14.32) we find V_s is UUB and thus each $|e_{i,j}|$ is UUB. ■

Using (14.32) we are ensured that

$$\lim_{t\to\infty} |V_s| \leq \frac{nN}{8\bar{\eta}\bar{\kappa}},$$

where the values of $\bar{\eta}$ and $\bar{\kappa}$ may be set by the designer. Thus the ultimate bound may be made arbitrarily small by proper choice of the parameters in the control law. Using the properties of diagonal dominance, it is possible to ensure that each subsystem has its output driven to an arbitrarily small ball using only local states even though its dynamics are influenced by the other subsystems. The following example demonstrates how to use the above decentralized control design.

Example 14.3 Consider the system defined by

$$\begin{aligned} \dot{x}_{1,1} &= x_{1,1}^2 x_{2,1} + u_1 \\ \dot{x}_{2,1} &= x_{2,1}^2 x_{1,1} + u_2, \end{aligned} \tag{14.33}$$

where it is desired that $x_{i,1} \to 0$ for $i = 1, 2$. Notice that this fits the form $\dot{x}_{i,1} = \Delta_{i,1} + u_i$, where the interconnections are defined by

$$\Delta_{1,1} = x_{1,1}^2 x_{2,1} \qquad \Delta_{2,1} = x_{2,1}^2 x_{1,1}.$$

Also, each interconnection may be bounded as

$$\begin{aligned} |\Delta_{1,1}| &\leq \rho + \zeta_{1,1}\psi_{1,1} \qquad & \zeta_{1,1} &= x_{1,1}^2 \qquad & \psi_{1,1} &= |x_{2,1}| \\ |\Delta_{2,1}| &\leq \rho + \zeta_{2,1}\psi_{2,1} \qquad & \zeta_{2,1} &= x_{2,1}^2 \qquad & \psi_{2,1} &= |x_{1,1}| \end{aligned} \tag{14.34}$$

with $\rho = 0$. We also have

$$\begin{aligned} \psi_{1,1} &\leq d_{1,2} + \phi_{1,1}\sqrt{|x_{2,1}|} \qquad & \phi_{1,1} &= \sqrt{|x_{2,1}|} \\ \psi_{2,1} &\leq d_{2,1} + \phi_{2,1}\sqrt{|x_{1,1}|} \qquad & \phi_{2,1} &= \sqrt{|x_{1,1}|} \end{aligned} \tag{14.35}$$

with $d_{1,2} = d_{2,1} = 0$.

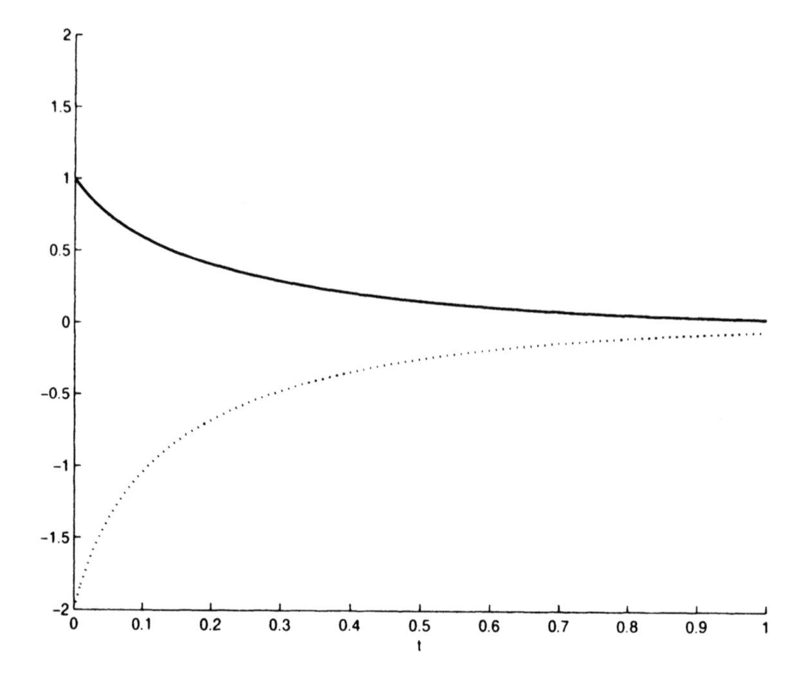

Figure 14.2. Closed-loop trajectory for $x_{1,1}$ (–) and $x_{2,1}$ $(\cdots)$.

The error terms are simply defined by $e_{1,1} = x_{1,1}$ and $e_{2,1} = x_{2,1}$. The controller is now chosen to be

$$u_i = -\left(\kappa + \nu_{i,1}\right) e_{i,1},$$

where according to (14.28) $\nu_{i,1} = \eta z_{i,1} + q_i$. Using (14.27) we choose $z_{i,1} = \frac{1}{2\sqrt{c}}\zeta_{i,1}$, and using (14.31) we choose

$$q_1 = \mu\left[2 + c\phi_{1,2}^4\right] \qquad q_2 = \mu\left[2 + c\phi_{2,1}^4\right].$$

The state trajectory of the resulting closed-loop system is shown in Figure 14.2 when $x_{1,1}(0) = 1, x_{2,1}(0) = -2$, and the controller parameters are chosen to be $\kappa = 1$, $c = 0.2$, $\mu = 1$, and $\eta = 1$. $\triangle$

14.3.3 Using a Finite Approximator

The development of the control laws up to this point have not placed any bounds upon each y_i. If approximators are to be used, for example, to represent the bounding interconnection terms $\phi_{j,i}(y_i)$, then we must know the space in which each y_i may travel. In this section we will use the previous results to develop bounds on the subsystem outputs.

When analyzing a static decentralized controller, the Lyapunov candidate for the composite system is typically defined by

$$V_s = \sum_{i=1}^{N} a_i V_{s_i},$$

where V_{s_i} is the Lyapunov candidate for the ith subsystem and each $a_i > 0$ (in the proof of Theorem 14.2, we let $a_i = 1$). If it is possible to conclude that $\dot{V}_s \leq -k_1 V_s + k_2$, then

$$V_s \leq \frac{k_2}{k_1} + \left(V_s(0) - \frac{k_2}{k_1}\right) e^{-k_1 t}. \tag{14.36}$$

But $a_i V_{s_i} \leq V_s$, so

$$V_{s_i} \leq \frac{k_2}{k_1 a_i} + \left(\frac{V_s(0)}{a_i} - \frac{k_2}{k_1 a_i}\right) e^{-k_1 t} \tag{14.37}$$

for all t. Using (14.37) we see that $V_{s_i} \leq V_{r_i}$ where

$$V_{r_i} = \max\left(\frac{V_s(0)}{a_i}, \frac{k_2}{k_1 a_i}\right).$$

With $y_i^2 \leq 2V_{s_i}$, we find

$$|y_i| \leq b_i = \sqrt{2 \max\left(\frac{V_s(0)}{a_i}, \frac{k_2}{k_1 a_i}\right)} \tag{14.38}$$

for $i = 1, \ldots, N$. If the approximations $\mathcal{F}_{j,i}(y_i, \theta_{j,i}) \approx \phi_{j,i}(y_i)$ are defined on $|y_i| \leq b_i$, then one may use (14.38) to pick the bounds on the initial conditions and controller parameters to ensure that $|y_i(t)| \leq b_i$ for all t when using the decentralized controller.

14.4 Adaptive Controller Design

We are now ready to design adaptive decentralized controllers in which each subsystem will contain its own parameter estimate and update law. We will see that the adaptive controllers may be designed to either account for unknown interconnection bounds or unknown subsystem dynamics.

14.4.1 Unknown Subsystem Dynamics

We will start our study of decentralized adaptive control by considering the subsystem dynamics

$$S_i : \begin{array}{rcl} \dot{x}_{i,1} &=& x_{i,2} + \Delta_{i,1}(t,x) \\ &\vdots& \\ \dot{x}_{i,n-1} &=& x_{i,n} + \Delta_{i,n-1}(t,x) \\ \dot{x}_{i,n} &=& f_i(x_{i,1}) + u_i + \Delta_{i,n}(t,x), \end{array} \tag{14.39}$$

with output $y_i = x_{i,1}$. Here we will assume that the interconnections satisfy (14.15) and that the function f_i is to be approximated. We will assume that there exists some linear in the parameter approximator $\mathcal{F}_i(y_i, \hat{\theta}_i)$ and ideal parameter set θ_i such that $|f_i(x_{i,1}) + \mathcal{F}_i(y_i, \theta_i)| \leq W_i$ when $|y_i| \leq b_i$. Thus $\mathcal{F}(y_i, \theta_i) \approx -f_i(y_i)$ when $|y_i| \leq b_i$. The goal of our control design will be to find a stabilizing adaptive controller which ensures that $|y_i| \leq b_i$ for all t and that $|y_i| \to 0$.

The error system is chosen to be

$$e_{i,j} = x_{i,j} - v_{i,j-1}(\bar{x}_{i,j-1}), \tag{14.40}$$

for $i = 1, \ldots, N$ and $j = 1, \ldots, n$ with $v_{i,0} = 0$. Taking the derivative of the errors, we find

$$\begin{aligned} \dot{e}_{i,j} &= x_{i,j+1} + \Delta_{i,j} - \sum_{k=1}^{j-1} \frac{\partial v_{i,j-1}}{\partial x_{i,k}} [x_{i,k+1} + \Delta_{i,k}] \\ &= e_{i,j+1} + v_{i,j} - \sum_{k=1}^{j-1} \frac{\partial v_{i,j-1}}{\partial x_{i,k}} [x_{i,k+1}] \\ &\quad + \Delta_{i,j} - \sum_{k=1}^{j-1} \frac{\partial v_{i,j-1}}{\partial x_{i,k}} \Delta_{i,k} \end{aligned} \tag{14.41}$$

for $i = 1, 2, \ldots, N$ and $j = 1, \ldots, n-1$. Choosing

$$v_{i,j} = -(\kappa_i + \nu_{i,j}(\bar{x}_{i,j}))e_{i,j} - e_{i,j-1} + \sum_{k=1}^{j-1} \frac{\partial v_{i,j-1}}{\partial x_{i,k}} [x_{i,k+1}],$$

we find

$$\dot{e}_{i,j} = -(\kappa_i + \nu_{i,j})e_{i,j} - e_{i,j-1} + e_{i,j+1} + \Delta_{i,j} - \sum_{k=1}^{j-1} \frac{\partial v_{i,j-1}}{\partial x_{i,k}} \Delta_{i,k} \tag{14.42}$$

for $j = 1, \ldots, n-1$ where $e_{i,0} = 0$ and each $\nu_{i,j}$ will be chosen to account for the interconnections. So far the adaptive controller design follows the steps as for the static controller case.

The control law is then defined as

$$u_i = -(\kappa_i + \nu_{i,n}(\bar{x}_{i,n}) + \bar{\gamma}_i)e_i - e_{i-1} + \mathcal{F}(y_i, \hat{\theta}_i) + \sum_{k=1}^{n-1} \frac{\partial v_{i,n-1}}{\partial x_{i,k}} [x_{i,k+1}] \tag{14.43}$$

so that

$$\begin{aligned} \dot{e}_{i,n} &= -(\kappa_i + \nu_{i,n} + \bar{\gamma}_i)e_{i,n} - e_{i,n-1} + f_i(y_i) + \mathcal{F}(y_i, \hat{\theta}_i) \\ &\quad + \Delta_{i,n} - \sum_{k=1}^{n-1} \frac{\partial v_{i,n-1}}{\partial x_{i,k}} \Delta_{i,k}, \end{aligned} \tag{14.44}$$

where $\bar{\gamma}_i > 0$ will be used to compensate for errors associated with the adaptive feedback, and $\nu_{i,n}$ will again be chosen to account for the system interconnections. The subsystem error dynamics may now be expressed as

$$\dot{e}_i = A_i e_i + \begin{bmatrix} -\nu_{i,1} e_{i,1} + \delta_{i,1} \\ \vdots \\ -\nu_{i,n} e_{i,n} + \delta_{i,n} \end{bmatrix} + b(-\bar{\gamma}_i e_{i,n} + f_i(y_i) + \mathcal{F}(y_i, \hat{\theta}_i)), \quad (14.45)$$

where A_i is defined by (14.24), $b = [0, \dots, 0, 1]^\top$, and

$$\delta_{i,j} = \Delta_{i,j} - \sum_{k=1}^{j-1} \frac{\partial v_{i,j-1}}{\partial x_{i,k}} \Delta_{i,k}. \quad (14.46)$$

The update law is finally chosen to be

$$\dot{\theta}_i = -\Gamma_i \left[e_{i,n} \left(\frac{\partial \mathcal{F}_i}{\partial \theta} \right)^\top + \sigma_i (\hat{\theta}_i - \theta_i^0) \right]. \quad (14.47)$$

The following theorem summarizes the stability of the closed loop system.

Theorem 14.3: *Assume that for a given linear in the parameter approximator $\mathcal{F}_i(y_i, \hat{\theta}_i)$, there exists some $\theta_i \in \mathrm{R}^{p_i}$ such that*

$$|\mathcal{F}_i(y_i, \theta_i) + f_i(y_i)| \leq W_i$$

for all $y_i \in S_{y_i}$, where $S_{y_i} = \{y_i \in \mathrm{R} : |y_i| \leq b_i\}$ with b_i known. Assume that the control parameters are chosen such that $B_{y_i} \subseteq S_{y_i}$, where B_{y_i} is defined by (14.54). Given the system dynamics defined by (14.39) with interconnection bounds satisfying (14.15), the decentralized controller (14.43) with update laws (14.47) will ensure that the errors associated with each subsystem are UUB.

Proof: We will begin the proof by following the same steps used in the case of the static decentralized controller. Following the steps up to (14.30) one may obtain

$$\begin{aligned} \dot{V}_{s_i} \leq\ & -2\kappa_i V_{s_i} + \frac{n}{4\eta_i} - 2q_i V_{s_i} + \sum_{j=1}^{n} |e_{i,j}| \sum_{l=1}^{N} \sqrt{c} |y_l| \phi_{i,l}^2 \quad (14.48) \\ & + e_{i,n} \left(-\bar{\gamma}_i e_{i,n} + f_i(y_i) + \mathcal{F}(y_i, \hat{\theta}_i) \right), \end{aligned}$$

where we have used $e_i^\top b = e_{i,n}$. Define $\nu_{i,j} = \eta_i z_{i,j}^2 + q_i(y_i)$ with $\eta_i > 0$ while $z_{i,j}$ and q_i are defined by (14.27) and (14.31) respectively. Taking the derivative of $V_s = \sum_{i=1}^{N} V_{s_i}$ we now find

$$\dot{V}_s \leq -2\bar{\kappa} V_s + \frac{nN}{4\bar{\eta}} + \sum_{i=1}^{N} e_{i,n} \left(-\bar{\gamma}_i e_{i,n} + f_i(y_i) + \mathcal{F}(y_i, \hat{\theta}_i) \right). \quad (14.49)$$

Since

$$\begin{aligned} f_i(y_i) + \mathcal{F}(y_i, \hat{\theta}_i) &= f_i(y_i) + \mathcal{F}(y_i, \theta_i) - \mathcal{F}(y_i, \theta_i) + \mathcal{F}(y_i, \hat{\theta}_i) \\ &= f_i(y_i) + \mathcal{F}(y_i, \theta_i) + \frac{\partial \mathcal{F}_i}{\partial \hat{\theta}_i}\left(\hat{\theta}_i - \theta_i\right), \end{aligned}$$

one finds

$$\begin{aligned} \dot{V}_s \leq & -2\bar{\kappa} V_s + \frac{nN}{4\bar{\eta}} + \sum_{i=1}^{N} e_{i,n}\left(-\bar{\gamma}_i e_{i,n} + f_i(y_i) + \mathcal{F}(y_i, \theta_i)\right) \\ & + \sum_{i=1}^{N} e_{i,n} \frac{\partial \mathcal{F}_i}{\partial \hat{\theta}_i}\left(\hat{\theta}_i - \theta_i\right). \end{aligned} \tag{14.50}$$

We will now include the effects of the update law.

Consider the composite Lyapunov candidate $V_a = V_s + \sum_{i=1}^{N} \frac{1}{2}\tilde{\theta}_i^\top \Gamma_i^{-1} \tilde{\theta}_i$, where $\tilde{\theta}_i = \hat{\theta}_i - \theta_i$ is the parameter estimate error for the ith subsystem and $\Gamma_i > 0$. Taking the derivative of V_a and substituting the parameter update law (14.47), we find

$$\dot{V}_a \leq -2\bar{\kappa} V_s + \frac{nN}{4\bar{\eta}} + \sum_{i=1}^{N}\left(-\bar{\gamma}_i e_{i,n}^2 + W_i|e_{i,n}|\right) - \sum_{i=1}^{N} \sigma_i \tilde{\theta}_i^\top\left(\hat{\theta}_i - \theta_i^0\right) \tag{14.51}$$

when $|y_i| \leq b_i$ for $i = 1, \ldots, N$. Since $-\tilde{\theta}_i^\top(\hat{\theta}_i - \theta_i^0) = -\tilde{\theta}_i^\top \tilde{\theta}_i - \tilde{\theta}_i^\top(\theta_i - \theta_i^0)$, we find

$$\begin{aligned} -\tilde{\theta}_i^\top(\hat{\theta}_i - \theta_i^0) &\leq -\frac{1}{2}|\tilde{\theta}_i|^2 + \frac{1}{2}|\theta_i - \theta_i^0|^2 \\ &\leq -\frac{1}{2\lambda_{\max}(\Gamma_i^{-1})}\tilde{\theta}_i^\top \Gamma_i^{-1} \tilde{\theta}_i + \frac{1}{2}|\theta_i - \theta_i^0|^2. \end{aligned}$$

Also $-\bar{\gamma}_i e_{i,n}^2 + W_i|e_{i,n}| \leq W_i^2/(4\bar{\gamma}_i)$ so that

$$\dot{V}_a \leq -2\bar{\kappa} V_s - \sum_{i=1}^{N} \frac{\sigma_i \tilde{\theta}_i^\top \Gamma_i^{-1} \tilde{\theta}_i}{2\lambda_{\max}(\Gamma_i^{-1})} + \frac{nN}{4\bar{\eta}} + \sum_{i=1}^{N}\left(\frac{W_i^2}{4\bar{\gamma}_i} + \sigma_i \frac{1}{2}|\theta_i - \theta_i^0|^2\right) \tag{14.52}$$

Let $k_1 = \min\left(2\bar{\kappa}, \frac{\sigma_1}{\lambda_{\max}(\Gamma_i^{-1})}, \ldots, \frac{\sigma_N}{\lambda_{\max}(\Gamma_N^{-1})}\right)$ and

$$k_2 = \frac{nN}{4\bar{\eta}} + \sum_{i=1}^{N}\left(\frac{W_i^2}{4\bar{\gamma}_i} + \frac{\sigma_i}{2}|\theta_i - \theta_i^0|^2\right). \tag{14.53}$$

Then $\dot{V}_a \leq -k_1 V_a + k_2$ so that V_a is bounded by $V_a \leq \max(V_a(0), k_2/k_1)$. Since $y_i \leq \sqrt{2V_s} \leq \sqrt{2V_a}$ we are guaranteed that $y_i \in B_{y_i}$ where

$$B_{y_i} = \left\{ y_i \in \mathsf{R} : |y_i| \leq \sqrt{2\max\left(V_a(0), \frac{k_2}{k_1}\right)} \right\}. \tag{14.54}$$

Since we may make k_2/k_1 arbitrarily small by proper choice of the controller parameters, it is always possible to choose the initial conditions such that $y_i \in S_{y_i}$. ■

Since $\dot{V}_a \leq -k_1 V_a + k_2$, the ultimate bound on V_a is given by k_2/k_1. Since k_2/k_1 may be made arbitrarily small, it is possible to force the ultimate bound on each y_i to also be arbitrarily small since $|y_i| \leq \sqrt{2V_a}$.

14.4.2 Unknown Interconnection Bounds

In the definition of the static decentralized control law, we assumed that the signal

$$q_i = \mu_i N \left[1 + \frac{cn^2}{2} \left(\sum_{j=1}^{N} \phi_{j,i}^4(|y_i|) \right) \right] \tag{14.55}$$

is known (that is, each $\phi_{i,j}$ is known). Recall that the $\phi_{i,j}$'s are based on the subsystem interconnection. Here we will study the control of (14.3) when the functions $\phi_{i,j}$ are not known. One should keep in mind that the $\zeta_{i,j}$ bounding terms defined in (14.15) and the $d_{i,k}$ terms in (14.17) are also based on the interconnections and will be used in the definition of the controller to follow. Thus only part of the interconnection is assumed to be unknown. We chose to first study the case where each $\phi_{i,j}$ is unknown for simplicity.

Define a bounding approximation to q_i as

$$\hat{q}_i = \mu_i N \left[1 + \frac{cn^2}{2} \mathcal{F}(y_i, \hat{\theta}) \right], \tag{14.56}$$

where $\hat{\theta}$ is a vector of adjustable parameters. Assume that there exists some $\theta_i \in \mathrm{R}^{p_i}$ such that

$$\mathcal{F}_i(y_i, \theta_i) \geq \sum_{j=1}^{N} \phi_{j,i}^4(|y_i|), \tag{14.57}$$

when $y_i \in S_{y_i}$. Thus, $q_i(y_i) \leq \hat{q}_i(y_i, \theta_i)$ when $y_i \in S_{y_i}$. We will later show that it is possible to define the adaptive controller parameters such that $y_i \in B_{y_i}$ for all t where we can force $B_{y_i} \subseteq S_{y_i}$.

We will now define the error system for the adaptive problem as

$$e_{i,j} = x_{i,j} - v_{i,j-1}(\bar{x}_{i,j-1}, \hat{\theta}_i), \tag{14.58}$$

for $i = 1, \ldots, n$ with $v_{i,0} = 0$ where now the term $v_{i,j}$ is also dependent upon the current parameter estimate $\hat{\theta}_i$. Taking the derivative of the errors, we find

$$
\begin{aligned}
\dot{e}_{i,j} &= f_{i,j} + x_{i,j+1} + \Delta_{i,j} - \sum_{k=1}^{j-1} \frac{\partial v_{i,j-1}}{\partial x_{i,k}} \left[f_{i,k} + x_{i,k+1} + \Delta_{i,k} \right] - \frac{\partial v_{i,j-1}}{\partial \hat{\theta}_i} \dot{\hat{\theta}} \\
&= f_{i,j} + e_{i,j+1} + v_{i,j} - \sum_{k=1}^{j-1} \frac{\partial v_{i,j-1}}{\partial x_{i,k}} \left[f_{i,k} + x_{i,k+1} \right] - \frac{\partial v_{i,j-1}}{\partial \hat{\theta}_i} \dot{\hat{\theta}}_i \\
&\quad + \Delta_{i,j} - \sum_{k=1}^{j-1} \frac{\partial v_{i,j-1}}{\partial x_{i,k}} \Delta_{i,k}. \qquad (14.59)
\end{aligned}
$$

Choosing

$$
\begin{aligned}
v_{i,j} &= -(\kappa_i + \nu_{i,j}(\bar{x}_{i,j}, \hat{\theta}_i)) e_{i,j} - e_{i,j-1} - f_{i,j} \\
&\quad + \sum_{k=1}^{j-1} \frac{\partial v_{i,j-1}}{\partial x_{i,k}} \left[f_{i,k} + x_{i,k+1} \right] + \tau_{i,j},
\end{aligned}
$$

the error dynamics become

$$
\begin{aligned}
\dot{e}_{i,j} &= -(\kappa_i + \nu_{i,j}) e_{i,j} - e_{i,j-1} + e_{i,j+1} - \frac{\partial v_{i,j-1}}{\partial \hat{\theta}_i} \dot{\hat{\theta}}_i + \tau_{i,j} \\
&\quad + \Delta_{i,j} - \sum_{k=1}^{j-1} \frac{\partial v_{i,j-1}}{\partial x_{i,k}} \Delta_{i,k}, \qquad (14.60)
\end{aligned}
$$

for $j = 1, \ldots, n-1$ where $e_{i,0} = 0$ and each $\nu_{i,j}$ will be chosen to account for the interconnections. The terms $\tau_{i,j}(\bar{x}_{i,j}, \hat{\theta}_i)$ will be defined based on the update law as was done when developing the indirect adaptive controller for strict feedback systems.

The control law is now chosen as

$$
\begin{aligned}
u_i &= -(\kappa_i + \nu_{i,n}(\bar{x}_{i,n}, \hat{\theta}_i)) e_i - e_{i-1} - f_{i,n} \qquad (14.61) \\
&\quad - \sum_{k=1}^{n-1} \frac{\partial v_{i,n-1}}{\partial x_{i,k}} \left[f_{i,k} + x_{i,k+1} \right] + \tau_{i,n}
\end{aligned}
$$

so that

$$
\begin{aligned}
\dot{e}_{i,n} &= -(\kappa_i + \nu_{i,n}) e_{i,n} - e_{i,n-1} - \frac{\partial v_{i,n-1}}{\partial \hat{\theta}_i} \dot{\hat{\theta}}_i + \tau_{i,n} \\
&\quad + \Delta_{i,n} - \sum_{k=1}^{n-1} \frac{\partial v_{i,n-1}}{\partial x_{i,k}} \Delta_{i,k}, \qquad (14.62)
\end{aligned}
$$

where $\nu_{i,n}$ will again be used to account for the system interconnections. Based on the form used in the static control case (defined by (14.28)), we will consider the choice

$$
\nu_{i,j} = \eta_i z_{i,j}^2 + \hat{q}_i(y_i, \hat{\theta}), \qquad (14.63)
$$

where $z_{i,j}$ is defined by (14.27). Note that the subsystem dynamics may be written as

$$\dot{e}_i = A_i e_i + \begin{bmatrix} -\nu_{i,1}e_{i,1} + \tau_{i,1} + \delta_{i,1} \\ \vdots \\ -\nu_{i,n}e_{i,n} + \tau_{i,n} + \delta_{i,n} \end{bmatrix} + \begin{bmatrix} D_{i,1} \\ \vdots \\ D_{i,n} \end{bmatrix} \dot{\hat{\theta}}_i, \tag{14.64}$$

where A_i is defined by (14.24), $D_{i,j} = -\frac{\partial v_{i,j-1}}{\partial \hat{\theta}_i}$, and

$$\delta_{i,j} = \Delta_{i,j} - \sum_{k=1}^{j-1} \frac{\partial v_{i,j-1}}{\partial x_{i,k}} \Delta_{i,k}.$$

Define $V_{s_i} = \frac{1}{2}\sum_{j=1}^{n} e_{i,j}^2$. The update law for the ith subsystem is now chosen to be

$$\dot{\hat{\theta}}_i = \Gamma_i \left[2V_{s_i} \left(\frac{\partial \hat{q}_i}{\partial \hat{\theta}_i} \right)^{\top} - \sigma_i(\hat{\theta}_i - \theta_i^0) \right], \tag{14.65}$$

where $\Gamma_i \in \mathrm{R}^{p \times p}$ is positive definite and symmetric and $\sigma_i > 0$. We will now choose $\tau_{i,j}$ so that

$$\begin{bmatrix} D_{i,1} \\ \vdots \\ D_{i,n} \end{bmatrix} \dot{\hat{\theta}}_i + \begin{bmatrix} \tau_{i,1} \\ \vdots \\ \tau_{i,n} \end{bmatrix} = M_i \bar{e}_{i,n}, \tag{14.66}$$

where M_i is a skew-symmetric matrix (i.e., $M_i^\top + M_i = 0$).

Notice that using the update law (14.65) with

$$V_{s_i} = \frac{1}{2} \begin{bmatrix} e_{i,1} & \cdots & e_{i,n} \end{bmatrix} \begin{bmatrix} e_{i,1} \\ \vdots \\ e_{i,n} \end{bmatrix}$$

we find

$$\begin{aligned} \begin{bmatrix} D_{i,1} \\ \vdots \\ D_{i,n} \end{bmatrix} \dot{\hat{\theta}}_i &= \begin{bmatrix} D_{i,1} \\ \vdots \\ D_{i,n} \end{bmatrix} \Gamma_i \left(\frac{\partial \hat{q}_i}{\partial \hat{\theta}_i} \right)^{\top} \bar{e}_{i,n}^{\top} \bar{e}_{i,n} - \sigma_i \begin{bmatrix} D_{i,1} \\ \vdots \\ D_{i,n} \end{bmatrix} \Gamma_i \left(\hat{\theta}_i - \theta_i^0 \right) \\ &= \begin{bmatrix} 0 & \cdots & 0 \\ m_{i,2,1} & \cdots & m_{i,2,n} \\ \vdots & & \vdots \\ m_{i,n,1} & \cdots & m_{i,n,n} \end{bmatrix} \bar{e}_{i,n} - \sigma_i \begin{bmatrix} D_{i,1} \\ \vdots \\ D_{i,n} \end{bmatrix} \Gamma_i (\hat{\theta}_i - \theta_i^0), \end{aligned}$$

where

$$m_{i,j,k} = D_{i,j} e_{i,k} \Gamma_i \left(\frac{\partial \hat{q}_i}{\partial \hat{\theta}_i} \right)^{\top} = -\frac{\partial v_{i,j-1}}{\partial \hat{\theta}_i} e_{i,k} \Gamma_i \left(\frac{\partial \hat{q}_i}{\partial \hat{\theta}_i} \right)^{\top}.$$

Notice that $m_{i,j,k}$ is defined using $x_{i,1}, \ldots, x_{i,l}$ where $l = \max(j,k)$. Thus $\tau_{i,j}$ may only be defined using the terms $m_{i,j,k}$ with $k \le j$ and $\hat{\theta}_i$. Choosing

$$\bar{\tau}_{i,n} = -\begin{bmatrix} 0 & 0 & \ldots & 0 \\ m_{2,1} & m_{2,2} & \ddots & \vdots \\ \vdots & \vdots & \ddots & 0 \\ m_{n,1} & m_{n,2} & \ldots & m_{n,n} \end{bmatrix} \bar{e}_{i,n} + \sigma_i \begin{bmatrix} D_{i,1} \\ \vdots \\ D_{i,n} \end{bmatrix} \Gamma_i(\hat{\theta}_i - \theta_i^0)$$
$$- \begin{bmatrix} 0 & & \ldots & & 0 \\ 0 & 0 & & & \\ 0 & m_{2,3} & 0 & & \vdots \\ \vdots & \vdots & \ddots & \ddots & \\ 0 & m_{2,n} & \ldots & m_{n-1,n} & 0 \end{bmatrix} \bar{e}_{i,n} \tag{14.67}$$

ensures (14.66) holds with M_i a skew-symmetric matrix. Notice that the first term in the right-hand side of (14.67) is used to cancel as many $m_{i,j,k}$ terms as possible. The last term in the right-hand side of (14.67) is then used to ensure that M_i is skew-symmetric. The error dynamics for the ith subsystem may now be expressed as

$$\dot{e}_i = (A_i + M_i)e_i + \begin{bmatrix} -\nu_{i,1}e_{i,1} + \delta_{i,1} \\ \vdots \\ -\nu_{i,n}e_{i,n} + \delta_{i,n} \end{bmatrix}. \tag{14.68}$$

We are now ready to state the closed-loop properties obtained when using the proposed decentralized adaptive controller.

Theorem 14.4: *Assume that for a given linear in the parameter approximator $\mathcal{F}_i(y_i, \hat{\theta}_i)$, there exists some $\theta_i \in \mathrm{R}^{p_i}$ such that*

$$\mathcal{F}_i(y_i, \theta_i) \ge \sum_{j=1}^{N} \phi_{j,i}^4(|y_i|)$$

for all $y_i \in S_{y_i}$ where each $\phi_{j,i}$ defines the bounds on the interconnections and S_{y_i} contains a ball centered at the origin. Assume that the control parameters are chosen such that $B_{y_i} \subseteq S_{y_i}$ where B_{y_i} is defined by (14.54). Given the system dynamics defined by (14.3) with interconnection bounds satisfying (14.15), the decentralized controller (14.61) with update laws (14.65) will ensure that the errors associated with each subsystem are UUB.

Proof: Define $V_i = V_{s_i} + \frac{1}{2}\tilde{\theta}_i^\top \Gamma_i^{-1} \tilde{\theta}_i$ for each subsystem with $V_{s_i} = \frac{1}{2}\sum_{j=1}^{n} e_{i,j}^2$ and $\tilde{\theta}_i = \hat{\theta}_i - \theta_i$. Then following the steps in the proof of

Theorem 14.2, we find

$$\dot{V}_i \le \frac{1}{2}\bar{e}_{i,n}^\top \left(A_i^\top + G_i^\top + A_i + G_i\right)\bar{e}_{i,n} + \sum_{j=1}^{n}\left(-\nu_{i,j}e_{i,j}^2 + |e_{i,j}||\delta_{i,j}|\right) + \tilde{\theta}_i^\top \Gamma_i^{-1}\dot{\hat{\theta}}_i. \tag{14.69}$$

Since

$$\nu_{i,j} = \eta z_{i,j}^2 + \hat{q}_i(y_i,\hat{\theta}_i) = \eta z_{i,j}^2 + \hat{q}_i(y_i,\theta_i) + \frac{\partial \hat{q}_i}{\partial \hat{\theta}_i}\tilde{\theta}_i,$$

we find

$$\dot{V}_i \le -2\kappa_i V_{s_i} + \sum_{j=1}^{n}\left(-\eta_i z_{i,j}^2 e_{i,j}^2 + |z_{i,j}||e_{i,j}|\right) + \tilde{\theta}_i^\top \Gamma_i^{-1}\dot{\hat{\theta}}_i + \sum_{j=1}^{n}\left(-\hat{q}_i(y_i,\theta_i)e_{i,j}^2 - \frac{\partial \hat{q}_i}{\partial \hat{\theta}_i}\tilde{\theta}_i e_{i,j}^2 + |e_{i,j}|\sum_{l=1}^{N}\sqrt{c}|y_l|\phi_{i,l}^2\right). \tag{14.70}$$

Using the inequality $-x^2 \pm 2xy \le y^2$ and combining terms, we find

$$\dot{V}_i \le -2\kappa_i V_{s_i} + \frac{n}{4\eta_i} - 2\hat{q}_i(y_i,\theta_i)V_{s_i} + \sum_{j=1}^{n}|e_{i,j}|\sum_{l=1}^{N}\sqrt{c}|y_l|\phi_{i,l}^2 - 2V_{s_i}\frac{\partial \hat{q}_i}{\partial \hat{\theta}_i}\tilde{\theta}_i + \tilde{\theta}_i^\top \Gamma_i^{-1}\dot{\hat{\theta}}_i. \tag{14.71}$$

Using the definition of the update law (14.65), the Lyapunov candidate decreases according to

$$\dot{V}_i \le -2\kappa_i V_{s_i} + \frac{n}{4\eta_i} - 2\hat{q}_i(y_i,\theta_i)V_{s_i} + \sum_{j=1}^{n}|e_{i,j}|\sum_{l=1}^{N}\sqrt{c}|y_l|\phi_{i,l}^2 - \frac{\sigma_i}{2}|\tilde{\theta}_i|^2 + \frac{\sigma_i}{2}\left|\theta_i - \theta_i^0\right|^2 \tag{14.72}$$

where we have used $-2\tilde{\theta}_i^\top(\hat{\theta}_i - \theta_i^0) \le -|\tilde{\theta}_i|^2 + |\theta_i - \theta_i^0|^2$.

We will now consider the composite Lyapunov candidate $V = \sum_{i=1}^N V_i$ and also define $V_s = \sum_{i=1}^N V_{s_i}$. Since $|y_l| = |e_{l,1}|$ and $|e_{i,j}| \le \sqrt{2V_{s_i}}$ for all i,j we obtain

$$\dot{V} \le \sum_{i=1}^{N} -2\kappa_i V_{s_i} + \frac{n}{4\eta_i} - 2\hat{q}_i(y_i,\theta_i)V_{s_i} + n\sqrt{2c}\sqrt{V_{s_i}}\sum_{l=1}^{N}\phi_{i,l}^2\sqrt{V_{s_l}} - \frac{\sigma_i}{2}|\tilde{\theta}_i|^2 + \frac{\sigma_i}{2}\left|\theta_i - \theta_i^0\right|^2.$$

It is now possible to rearrange terms to obtain

$$\dot{V} \leq -2\bar{\kappa}V_s + \frac{nN}{4\bar{\eta}} - 2\left[\ \sqrt{V_{s_1}} \ \cdots \ \sqrt{V_{s_N}}\ \right]^\top (D+G)\begin{bmatrix} \sqrt{V_{s_1}} \\ \vdots \\ \sqrt{V_{s_N}} \end{bmatrix} + \sum_{i=1}^{N}\left(-\frac{\sigma_i}{2}|\tilde{\theta}_i|^2 + \frac{\sigma_i}{2}\left|\theta_i - \theta_i^0\right|^2\right) \quad (14.73)$$

where $\bar{\kappa} = \min_i(\kappa_i)$, $\bar{\eta} = \min_i(\eta_i)$, $D = \text{diag}(\hat{q}_1(y_1,\theta_1),\ldots,\hat{q}_N(y_N,\theta_N))$, and

$$G = -\frac{n\sqrt{c}}{\sqrt{2}}\begin{bmatrix} \phi_{1,2}^2 & \cdots & \phi_{1,N}^2 \\ \vdots & & \vdots \\ \phi_{N,2}^2 & \cdots & \phi_{N,N}^2 \end{bmatrix}. \quad (14.74)$$

Since

$$\hat{q}_i(y_i,\theta) \geq \mu_i N\left[1 + \frac{cn^2}{2}\left(\sum_{j=1}^{N}\phi_{j,i}^4(|y_i|)\right)\right], \quad (14.75)$$

we may use Theorem 14.1 to show that

$$\begin{aligned} \dot{V} &= -2\bar{\kappa}V_s + \frac{nN}{4\bar{\eta}} + \sum_{i=1}^{N}\left(-\frac{\sigma_i}{2}|\tilde{\theta}_i|^2 + \frac{\sigma_i}{2}|\theta - \theta_i^0|^2\right) \\ &\leq -2\bar{\kappa}V_s - \sum_{i=1}^{N}\frac{\sigma_i}{2\lambda_{\max}(\Gamma_i^{-1})}\tilde{\theta}_i^\top\Gamma_i^{-1}\tilde{\theta}_i + \frac{nN}{4\bar{\eta}} + \sum_{i=1}^{N}\frac{\sigma_i}{2}|\theta - \theta_i^0|^2. \end{aligned}$$

Choose

$$k_1 = \min\left(2\bar{\kappa}, \frac{\sigma_1}{\lambda_{\max}(\Gamma_1^{-1})}, \ldots, \frac{\sigma_N}{\lambda_{\max}(\Gamma_N^{-1})}\right),$$

and

$$k_2 = \frac{nN}{4\bar{\eta}} + \sum_{i=1}^{N}\frac{\sigma_i}{2}|\theta - \theta_i^0|^2$$

so that $\dot{V} \leq -k_1 V + k_2$.

With this form, we find $V \leq V_r$ for all t with

$$V_r = \max\left(V(0), \frac{k_2}{k_1}\right).$$

Since $V_{s_i} \leq V_r$ for $i = 1,\ldots,N$, and $|y_i|^2 = |e_{i,1}|^2 \leq 2V_{s_i}$ we find $y_i \in B_{y_i}$ for all t where

$$B_{y_i} = \left\{y_i \in \mathsf{R} : |y_i| \leq \sqrt{2V_r}\right\}. \quad (14.76)$$

By properly choosing the control parameters and initial conditions, one can make B_{y_i} arbitrarily small. In particular, we may make k_2 arbitrarily small by picking $\bar{\eta}$ large and each σ_i small. Additionally, k_1 made be made large by picking $\bar{\kappa}$ large and $\lambda_{\max}(\Gamma_i^{-1})$ small. The effects of the parameter estimate initial conditions may be made small by choosing $\lambda_{\max}(\Gamma_i^{-1})$ small since

$$\begin{aligned} V_i(0) &= V_{s_i}(0) + \frac{1}{2}\tilde{\theta}_i^{\top}(0)\Gamma_i^{-1}\tilde{\theta}_i(0) \\ &\leq V_{s_i}(0) + \frac{1}{2}\lambda_{\max}(\Gamma_i^{-1})\left|\tilde{\theta}_i(0)\right|^2. \end{aligned}$$

Thus it is always possible to make $B_{y_i} \subseteq S_{y_i}$ since S_{y_i} contains a ball centered at the origin. In addition, since $\lim_{t\to\infty} V = k_2/k_1$ we find the ultimate bound on each output to be

$$\lim_{t\to\infty} |y_i| \leq \sqrt{\frac{2k_2}{k_1}}. \tag{14.77}$$

■

In the proof of Theorem 14.4, we could have defined θ_i as any value which causes $D+G$ in (14.73) to be positive definite. The value defined by (14.55) is just one such choice.

Ideally the adaptive controller is able to compensate for all the uncertainty associated with the interconnections. This way if the interconnection bounds are approximated on-line, it may be possible to add additional subsystems without redesigning the individual controllers. The approximator just adjusts itself so that the additional subsystems are accounted for.

14.5 Summary

In this chapter we learned how to use the concept of diagonal dominance to develop both static and adaptive decentralized controllers. It was shown that if one ensures that each subsystem is designed to be sufficiently stable using only local signals, then the composite system is stable. The static controller was designed using knowledge of the form of the interconnection between each of the other subsystems.

It was shown that it is possible to design adaptive controllers in which an approximator is adjusted on-line to compensate for uncertainties in the isolated subsystem dynamics. It was also shown that it may be possible to adaptively compensate for interconnection uncertainties. This is particularly appealing since the designer requires less knowledge of the interaction between subsystems, which is often the reason for considering a decentralized framework.

14.6 Exercises and Design Problems

Exercise 14.1 (Diagonal Dominance) Consider the matrices

$$A = \begin{bmatrix} 0 & a \\ a & 0 \end{bmatrix} \qquad D = \begin{bmatrix} d & 0 \\ 0 & d \end{bmatrix}$$

where $a \in \mathsf{R}$ and $d \geq 0$. Find the minimum value for d as a function of a such that $A + D$ is positive definite. Compare this result to the values obtained using Theorem 14.1.

Exercise 14.2 (Scalar Decentralized Stability) Consider the scalar subsystem

$$S_i : \ \dot{x}_i \ = \ \sum_{j=1}^{N} m_{i,j} x_j,$$

for $i = 1, \ldots, N$. Given the composite system Lyapunov candidate $V = \sum_{i=1}^{N} d_i x_i^2$ with $d_i > 0$, show that $x_i = 0$ is an exponentially stable equilibrium point if there exists some $D = \text{diag}(d_1, \ldots, d_N)$ such that $DM + M^\top D < 0$ and $M = [m_{i,j}]$.

Exercise 14.3 (M-Matrix) Consider the case where a local controller $u_i = \nu_i(x_i)$ is defined for the ith subsystem

$$S_i : \ \dot{x}_i \ = \ f_i(x_i) + g(x_i)u_i + \Delta_i(t, x_i)$$

such that given positive definite functions $V_i(x_i)$, we find

$$\begin{aligned} \frac{\partial V_i}{\partial x_i}\left(f_i(x_i) + g(x_i)u_i\right) &\leq -\alpha_i \psi_i^2(x_i) \\ \left|\frac{\partial V_i}{\partial x_i}\right| &\leq \beta_i \psi_i(x_i), \end{aligned} \tag{14.78}$$

where the functions ψ_i are positive definite and continuous. Use the composite Lyapunov function $V = \sum_{i=1}^{N} d_i V_i$ with $d_i > 0$ to show that if there exists some $D = \text{diag}(d_1, \ldots, d_N)$ such that $DS + S^\top D > 0$, where

$$s_{i,j} = \begin{cases} \alpha_i - \beta_i \eta_{i,i} & i = j \\ -\beta_i \eta_{i,j} & i \neq j \end{cases}$$

then the composite system is asymptotically stable if the interconnections may be bounded by

$$|\Delta_i| \leq \sum_{j=1}^{N} \eta_{i,j} \psi_j$$

for some $\eta_{i,j} \geq 0$ [108]. The matrix S is often referred to as a Minkowski matrix or M-matrix.

Exercise 14.4 (Decentralized Control Design) Assume that the interconnections of a decentralized system satisfy

$$|\Delta_{i,j}| \leq \rho + \sum_{k=1}^{N} \psi_{i,k}(y_k)|y_k|$$

where $\rho \geq 0$ and each $\psi_{i,k}$ is nonnegative. Design a decentralized controller that takes advantage of this simplification over (14.15).

Exercise 14.5 (Connected Pendulums) Consider the system defined by two pendulums connected by a spring. The dynamics for this system are given by

$$S_1 : \begin{array}{rcl} \dot{\theta}_1 & = & \omega_1 \\ J_1\dot{\omega}_1 & = & k(\theta_2 - \theta_1) + u_1 \end{array}$$

$$S_2 : \begin{array}{rcl} \dot{\theta}_2 & = & \omega_2 \\ J_2\dot{\omega}_2 & = & k(\theta_1 - \theta_2) + u_2, \end{array}$$

where θ_i and ω_i are the position and angular velocity for the ith pendulum, respectively. Here k is an interconnection constant based on the spring stiffness and lever arm associated with the connection point. Define decentralized controllers for each torque input u_i so that $\theta_i \to 0$ for $i = 1, 2$.

Exercise 14.6 (Decentralized Fuzzy/Neural Control) Consider i^{th} the subsystem defined by

$$S_i : \begin{array}{rcl} \dot{x}_{i,1} & = & \dot{x}_{i,2} \\ \dot{x}_{i,2} & = & u_i + \Delta_i \end{array} \qquad (14.79)$$

for $i = 1, \ldots, N$. Design an adaptive fuzzy or neural controller when $N = 3$ and

- $\Delta_i = x_{i,2}x_{i+1,1}$
- $\Delta_i = x_{i+1,1}\sin(x_{i-1,1})$
- $\Delta_i = x_{i+1,1}^2$,

where the subsystem subscripts are taken modulo-N so that $x_{N+1,1} = x_{1,1}$. Is it possible to define one adaptive decentralized controller that is able to work for each of the interconnections?

Exercise 14.7 (Angular Velocity Control) The angular velocity dynamics for a spacecraft may be described by

$$J\begin{bmatrix} \dot{\omega}_x \\ \dot{\omega}_y \\ \dot{\omega}_z \end{bmatrix} = \begin{bmatrix} 0 & \omega_z & -\omega_y \\ -\omega_z & 0 & \omega_x \\ \omega_y & -\omega_x & 0 \end{bmatrix} J \begin{bmatrix} \omega_x \\ \omega_y \\ \omega_z \end{bmatrix} + \begin{bmatrix} u_x \\ u_y \\ u_z \end{bmatrix}, \qquad (14.80)$$

where J is a diagonal inertial matrix with positive elements. Use a decentralized control approach to design controllers $u_x = \nu_x(\omega_x)$, $u_y = \nu_y(\omega_y)$, $u_z = \nu_z(\omega_z)$ so that $\omega_x \to \Omega$, and $\omega_y, \omega_z \to 0$.

Exercise 14.8 (Stage Control) Develop a decentralized controller for the 3-axis stage of Example 14.1 so that $x, y, \theta \to 0$.

Chapter 15

Perspectives on Intelligent Adaptive Systems

15.1 Overview

Adaptive fuzzy and neural systems do not exist as an isolated topic devoid of relationships to other fields. It is important to understand how they relate to other fields in order to strengthen your understanding of them, and to see how general the ideas on adaptation are.

We have emphasized that the methods of this book have their foundations in conventional adaptive control and that there are many relationships to techniques, ideas, and methodologies there. Adaptive fuzzy and neural systems are also an "intelligent control" technique, and hence there are certain relationships between them and other intelligent control methods. In this chapter we will provide a brief overview of some of the basic relationships between adaptive fuzzy and neural control systems and other control methods. This will give the reader who has a good understanding of adaptive fuzzy and neural control systems a glimpse of related topics in other areas. Moreover, it will give the reader who has a good understanding of other areas of control a better understanding with what the field of adaptive fuzzy and neural control systems is concerned.

We begin the chapter by providing a conventional control engineering perspective on adaptive fuzzy and neural control systems. Following this, we briefly discuss "foraging strategies" for adaptive control, genetic adaptive control strategies and some relationships to adaptive fuzzy and neural control systems. Next, we explain how expert systems can be used as controllers, and particularly how they can be used to supervise adaptive controllers. Then we briefly discuss adaptive planning systems. Finally, we provide an overview of the general area of (hierarchical) intelligent and autonomous control where we briefly explain the possible roles of adaptive fuzzy and neural systems in such general controllers.

The reader interested only in adaptive fuzzy and neural control systems can certainly ignore this chapter; we do not, however, advise this as the relationships to other fields often suggest ideas on how to expand the basic methods and may provide key ideas on how to solve a control problem for a particular application.

15.2 Relations to Conventional Adaptive Control

There are intimate relationships between conventional adaptive systems and adaptive fuzzy and neural control systems. For instance, the type of mathematics used here is what is typically required as background for conventional adaptive control. The approaches borrow their basic structures (e.g., direct, indirect) from conventional approaches and in fact the proofs extend those from conventional adaptive control. Often the control problems that we study may be redefined so that appropriate conventional approaches may be used. Though the adaptive routines are the same and the stability proof follows with some extensions from conventional adaptive control, we do not place the same restrictions upon the plant dynamics so the methods also have some key differences.

The basic differences between conventional adaptive control and adaptive fuzzy and neural control systems stem from the use of the tuning of general nonlinear structures (i.e., the fuzzy or neural systems) to match some unknown nonlinearity (even if only the parameters that enter linearly are tuned). Compared with conventional adaptive control of linear systems with unknown constant coefficients, here our concern is with tuning functions to match unknown functions, rather than parameter tuning to find unknown parameters. Of course we parameterize our nonlinearities and hence tune functions via the tuning of parameters. Really, in the linear case, we simply assume that the unknown function is linear and we use a linear approximator structure and tune it; in this sense conventional adaptive control of linear systems is a special case of adaptive fuzzy and neural control systems.

Conventional adaptive approaches also often assume that the unknown portion of the system dynamics are in the form of unknown coefficients which multiply *known* functions, such as the system defined by $\dot{x} = ax^2 + u$, where a is an unknown constant. If the system were defined by $\dot{x} = \sin(ax)x + u$ with a unknown, however, most of the conventional approaches would not be applicable.

As we pointed out, conventional approximation structures such as polynominals can be used, as well as fuzzy or neural structures. However, the perspective of on-line function approximation grew out of the neural and fuzzy research communities and the fuzzy and neural structures are easy to use, and enjoy good approximation properties, so they are often employed. Moreover, it is sometimes found that the fuzzy system is easy to initialize with heuristic knowledge that can shape the nonlinearity so that it is closer to its optimal shape. Regardless, there is no reason that all the approaches in this book cannot work for conventional approximation structures and it is for this reason that we can take the view that the approaches used in this book are all quite conventional! Indeed, adaptive fuzzy and neural control systems can be considered to be an application of "on-line function approximation" strategies to control as we have emphasized throughout this book.

To summarize, for adaptation, the approaches in this book use optimization methods (e.g., gradient or least squares) to reduce an error between two variables (e.g., estimation or tracking error) by adjusting a fuzzy or neural system to match an unknown nonlinear function (e.g., the input-output mapping of a physical system, or an ideal stabilizing control function). Intuitively, stability is ensured by adjusting the parameters in the fuzzy or neural systems in a direction within the parameter space in an attempt to reduce the approximation error. As the approximation error improves, the errors within the adaptive systems tend to also improve by the way we defined the adaptive systems. The ability of the approaches to handle general classes of systems arises from the generality of the nonlinear approximator structures that fuzzy or neural systems and others offer.

Broadly speaking, the general objective of the approaches is to achieve robust adaptive control for more general classes of nonlinear systems, an objective that is not unlike that of conventional adaptive control. Next, we discuss several other "intelligent" adaptive control approaches, ones that are not nearly as mature (mathematically speaking, and in some cases for actual implementations), as the methods discussed in this book. You could then view the outline below as providing a research agenda in the sense that adaptive control methods are outlined that have seen little if any mathematical analysis (e.g., stability analysis).

15.3 Genetic Adaptive Systems

A genetic algorithm (GA) uses the principles of evolution, natural selection, and genetics from natural biological systems in a computer algorithm to simulate evolution. Essentially, the genetic algorithm is an optimization technique that performs a parallel, stochastic, but directed search to evolve the most fit population. In particular, it provides a stochastic optimization method where if it "gets stuck" at a local optimum, it tries to simultaneously find other parts of the search space and "jump out" of the local optimum to a global one.

Traditionally, genetic algorithms have been used for off-line design, search, and optimization. There are ways, however, to evolve controllers or estimators (fuzzy, neural or conventional) while the system is operating, rather than in off-line design. Consider, for instance, a "genetic model reference adaptive controller" (GMRAC). For the GMRAC there is a genetic algorithm that maintains a population of strings, each of which is a descriptor (or chromosome) that represents a candidate controller. This genetic algorithm uses a process model (e.g., a linear model of the process) and data from the process to evaluate the fitness of each controller in the population. It does this evaluation at each time step by simulating future plant outputs (using the process model) with each candidate controller and forming a fitness function based on the error between the predicted output for each controller and that of the reference model (which characterizes the desired performance). Using this fitness evaluation, the genetic algorithm propagates controllers into the next generation via the standard genetic operators (e.g., selection, cross-over, and mutation). The controller that is the most fit one in the population at each time step is used to control the system. This allows the GMRAC to automatically evolve a controller from generation to generation (i.e., from one time step to the next, but of course multiple generations could occur between time steps) and hence to tune a controller in response to changes in the process, or due to user change of the specifications in the reference model. Overall, the GMRAC provides unique features where alternative controllers can be quickly applied to the problem if they appear useful (e.g., the process (re)enters a new operating condition).

It is also possible to use the genetic algorithm in on-line tuning of estimators. The closest analogy to such an approach is the use of the gradient method for on-line estimator tuning as we have done at several places in this book. Basically, all you need to do is to solve the optimization problem via the genetic algorithm rather than the gradient method. The challenge, is, however, to formulate the problem so that the GA can be used for the optimization. For the GA you have to be able to evaluate a finite number of solutions and rank them. It is easy to do this in identification since you can simply run N identifier models with the parameters of the identifier model loaded into a chromosome in the GA (then use a certainty equivalence approach to controller construction where the estimated parameters at time k are taken to be the ones that best estimate the current system output). For the direct adaptive strategies you have to take an approach like we do for the GMRAC (which is a direct adaptive strategy). A general genetic adaptive strategy is shown in Figure 15.1. There, we have a genetic adaptive estimator that identifies the plant model that is used in the GMRAC to evaluate the fitness of candidate controllers in its population (some may recognize this as a type of stochastic adaptive model predictive control strategy).

Although is not too difficult to construct genetic adaptive systems, and in some applications they seem to work quite well, proving that they possess stability properties may be quite challenging, if not impossible. The

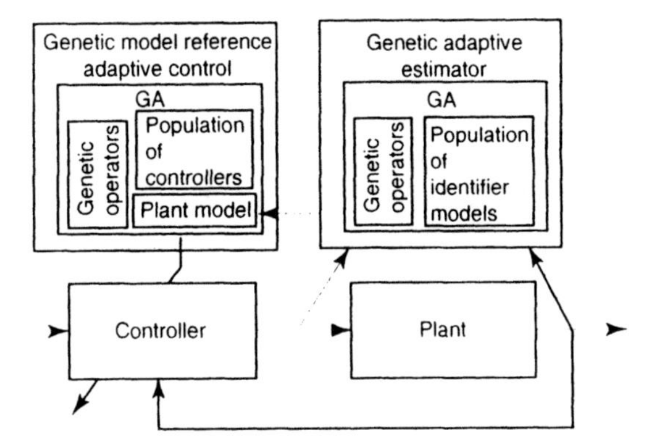

Figure 15.1. General genetic adaptive control.

challenge arises due to plant nonlinearities, stochastic influences that could exist in the nonlinear plant, and the nonlinear stochastic nature of the controller.

Finally, note that conventional nongradient optimization methods (e.g., pattern search, direct search, and some stochastic approximation methods) and optimization methods that model how animals forage for food can also be used in a similar manner to genetic algorithms for adaptive control.

15.4 Expert Control for Adaptive Systems

Initially, for the sake of our discussion, we will simply view the expert system that is used here as a controller for a dynamic system, in the same way that we can view a fuzzy system as a controller for a system without adaptation. Suppose that the expert system serving as a feedback controller has a reference input r and feedback variable y. It uses the information in its knowledge-base and its inference mechanism to decide what command input u to generate for the plant. Conceptually, then, we see that the expert controller is closely related to the direct (nonadaptive) fuzzy controller. There are, however, several differences. First, the knowledge-base in the expert controller could be a rule-base but is not necessarily so. It could be developed using other knowledge-representation structures, such as frames, semantic nets, causal diagrams, and so on. Second, the inference mechanism in the expert controller is more general than that of the fuzzy controller. It can use more sophisticated matching strategies to determine which rules should be allowed to fire. It can use more elaborate inference strategies. For instance, some expert systems use "refraction," where if a rule has fired recently it may not be allowed back into the "conflict set" (i.e., the set of rules that are allowed to fire) or it may use "recency," where rules that were fired most recently are given priority in being fired again, among various other priority schemes.

Up till now we have only discussed the direct (nonadaptive) expert controller. It may be possible to develop adaptation (learning) strategies for expert systems in an analogous manner to how we have in this book for fuzzy systems. Moreover, it is also possible to use an expert system in a "supervisory" role for conventional controllers or for the supervision of fuzzy or neural, estimators or controllers. This is shown in Figure 15.2. Such high-level adaptive controller tuning can be useful for practical implementations of adaptive control systems. For instance, an expert controller may be useful for supervision of the adaptation mechanism (e.g., slowing the adaptation when it appears that the system may go unstable, or speeding up adaptation when the responses are sluggish). Alternatively, the expert controller may perform on-line tuning of a reference model that the adaptive controller is trying to get the closed-loop system to behave like; then, the overall system will be "performance adaptive" (i.e., it will try to adapt to improve performance when possible and reduce performance requirements when they are not possible to achieve).

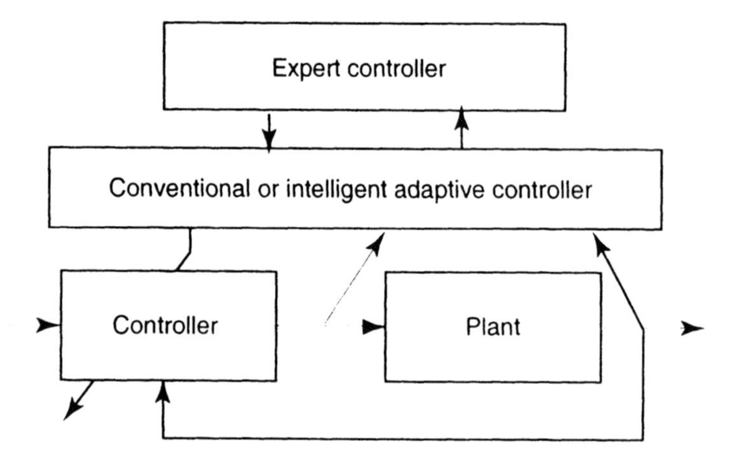

Figure 15.2. Expert system for supervising an adaptive controller.

15.5 Planning Systems for Adaptive Control

Artificially intelligent planning systems (computer programs that emulate the way experts plan) have been used in path planning and to make high-level decisions about control tasks for robots. A generic planning system can be configured in the architecture of a standard control system with the word "planner" replacing the word "controller" and "problem domain" replacing "plant." Here, the "problem domain" is the environment that the planner operates in. There are measured outputs y_k at step k (variables of the problem domain that can be sensed in real time), control actions u_k (the ways in which we can affect the problem domain), disturbances d_k (which represent random events that can affect the problem domain and hence the measured variable y_k), and goals g_k (what we would like to achieve in the problem domain, analogous to reference inputs). There are closed-loop specifications that quantify performance and stability requirements.

It is the task of the planner to monitor the measured outputs and goals so that it may generate control actions that will counteract the effects of the disturbances and result in the goals and the closed-loop specifications being achieved. To do this, the planner performs "plan generation," where it projects into the future (usually a finite number of steps, and often using a model of the problem domain) and tries to determine a set of candidate plans. Next, this set of plans is pruned to one plan that is the best one to apply at the current time (e.g., one plan is selected that minimizes consumption of resources). The plan is then executed, and during execution the performance resulting from the plan is monitored and evaluated. Often, due to disturbances, plans will fail, and hence the planner must generate a new set of candidate plans, select one, then execute that one (or "tweak" the existing plan). Some planning systems use "situation assessment" to try to estimate the state of the problem domain (this can be useful in execution monitoring and plan generation).

Advanced planning systems such as the one shown in Figure 15.3 perform "world modeling," where a model of the problem domain is developed in an on-line fashion (similar to on-line system identification, but here it could be for an automata model), and "planner design" that uses information from the world modeler to tune the planner (so that it makes the right plans for the current problem domain). Clearly, there is a close connection in this strategy to conventional indirect adaptive control schemes and adaptive model predictive control.

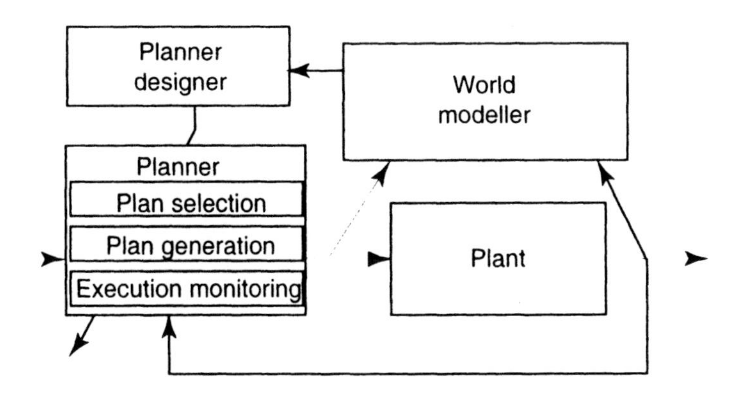

Figure 15.3. Adaptive planning system.

While it is not difficult to conceptualize how to construct an adaptive planning strategy, it can be very difficult to analyze the stability and performance properties of practical adaptive planning systems as there is often a need for the use of models that contain automata-like characteristics, and standard ordinary differential equations. Such general "hybrid" models, plus the need to consider the effects of disturbances on the planning strategies and nonlinear optimization in plan construction, generally make the analysis of stability, robustness, and performance properties very difficult.

15.6 Intelligent and Autonomous Control

Highly autonomous systems[1] have the capability to independently perform complex tasks with a high degree of success (e.g. human beings and animals are highly autonomous systems). It seems unlikely that any system can be considered to be highly autonomous without an ability to learn about its environment (i.e., be self-adapting) and improve its performance based on what it has learned. Hence, concepts of learning and adaptation, topics central to the focus of this book, are essential for autonomous systems.

It is possible to construct what some would call "artificial" (as contrasted with biological) autonomous systems where we seek to automate functions not normally performed on machines. In fact, consumer and governmental demands for such systems are common today. For instance, in the emerging area of intelligent vehicle and highway systems (IVHS), engineers are designing vehicles and highways that can fully automate vehicle route selection, collision avoidance, steering, braking, and throttle control to reduce congestion and improve safety. In avionic systems, a "pilot's associate" computer program has been designed to emulate the functions of mission and tactical planning that in the past may have been performed by the copilot. In manufacturing systems, efficiency optimization and flow control are being automated, and robots are routinely replacing humans in performing relatively complex tasks. Moreover, there are significant efforts underway to implement autonomous land vehicles and autonomous underwater vehicles which, for example, operate in a hazardous environment.

From a broad historical perspective, each of these applications began at a low level of automation, and through the years each has evolved into a more autonomous system. For example, today's automotive cruise controllers are the ancestors of the controllers that achieve coordinated control of steering, braking, and speed for autonomous automobile driving or collision avoidance. The terrain following, terrain avoidance control systems for low-altitude flight are ancestors of an artificial pilot's associate that can

[1] Autonomous systems to be discussed in this subsection are not to be confused with nonlinear ordinary differential equations $\dot{x}(t) = f(t,x)$ that do not have a dependence on time so that $f(t,x) = f(x)$.

integrate mission and tactical planning activities. The general trend has been for engineers to incrementally "add more intelligence" in response to consumer, industrial, and government demands and thereby create systems with increased levels of autonomy.

In this process of enhancing autonomy by adding intelligence, engineers often study how humans solve problems, then try to directly automate their knowledge and techniques to achieve high levels of automation. Other times, engineers study how intelligent biological systems perform complex tasks (e.g., neural networks), then seek to automate "nature's approach" in a computer algorithm or circuit implementation to solve a practical technological problem. Such approaches, where we seek to emulate the functionality of an intelligent biological system to solve a technological problem (sometimes called "biomimicry" or "bio-inspiration"), can be collectively named "intelligent systems and control techniques." It is by using such techniques that some engineers are trying to create highly autonomous systems such as those listed above.

It is important to note that the fuzzy and neural systems used in this book are only poor approximations of their biological counterparts. The (nonadaptive) fuzzy system only crudely models deduction as it occurs in humans and the adaptive fuzzy controller only crudely models the human induction process. Moreover, the neural network models that we use are only crude representations of their biological counterparts. Hence, from a broad perspective, the approaches in this book, while termed "intelligent," are only beginning to exploit the mechanisms of intelligence that have evolved in biological systems. Control theory and technology are, however, themselves evolutionary and as the complexity of the models and methods increase, it is hoped that we will be able to achieve higher and higher levels of autonomy.

How does a truly autonomous controller operate? Figure 15.4 shows a functional architecture for an intelligent autonomous controller with an interface to the process involving sensing (e.g., via conventional sensing technology, vision, touch, smell, etc.), actuation (e.g., via hydraulics, robotics, motors, etc.), and an interface to humans (e.g., a driver, pilot, crew, etc.) and other systems.

The "execution level" has low-level numeric signal processing and control algorithms (e.g., PID, optimal, adaptive, or intelligent control; parameter estimators, failure detection and identification (FDI) algorithms). The "coordination level" provides for tuning, scheduling, supervision, and redesign of the execution-level algorithms, crisis management, planning and learning capabilities for the coordination of execution-level tasks, and higher-level symbolic decision making for FDI and control algorithm management. The "management level" provides for the supervision of lower-level functions and for managing the interface to the human(s) and other systems. In particular, the management level will interact with the users

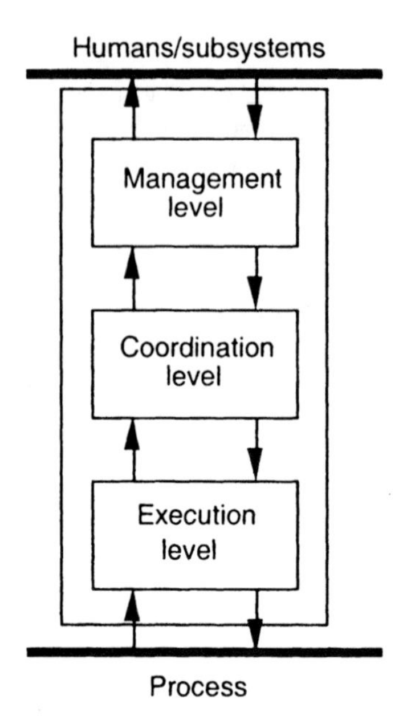

Figure 15.4. Intelligent autonomous controller.

in generating goals for the controller and in assessing the capabilities of the system. The management level also monitors performance of the lower-level systems, plans activities at the highest level (and in cooperation with humans), and performs high-level learning about the user and the lower-level algorithms.

Adaptive intelligent systems (e.g., fuzzy, neural, genetic, expert, and planning) can be employed as appropriate in the implementation of various functions at the three levels of the intelligent autonomous controller. For example, adaptive fuzzy or neural control may be used at the execution level for adaptation, genetic algorithms may be used in the coordination level to pick an optimal coordination strategy, and adaptive planning systems may be used at the management level for sequencing operations. Hierarchical controllers composed of a hybrid mix of intelligent and conventional systems are commonly used in the intelligent control of complex dynamic systems. This is because to achieve high levels of autonomy, we often need high levels of intelligence, which calls for incorporating a diversity of decision-making approaches for complex dynamic learning and reasoning.

In summary, we see that there can be many roles for the intelligent adaptive estimation and control systems in the larger context of intelligent autonomous control for complex systems. It is a challenging problem to develop general design and analysis strategies for complex hierarchical intelligent autonomous control systems as the underlying dynamics are often "hybrid" in the sense that they involve dynamics that can be represented by a combination of automata and ordinary differential equations (e.g., we may use automata to represent the problem domain for a planner, but the planner may supervise an adaptive controller for a plant that is most conveniently represented by ordinary differential equations). Hence, while it is relatively easy to construct a hierarchical interconnection of intelligent adaptive estimation and control strategies, it is a challenge to prove that the overall hierarchical adaptive system achieves some type of stable robust operation, or achieves a certain performance level.

15.7 Summary

Within this chapter we have presented a brief introduction of the role of fuzzy systems and neural networks in the area of intelligent adaptive systems. In particular, we have overviewed the relationships between adaptive fuzzy and neural systems and the following topics:

- Conventional adaptive control.
- Genetic adaptive control.
- Expert control for adaptive controller supervision.
- Adaptive planning systems.
- Intelligent autonomous controllers.

This chapter is meant to expose you in other related topics so we encourage you to study the next section and pursue some of the relationships that we have highlighted.

For Further Study

The material presented here is intended to help the reader gain a better understanding of the origins of many of the concepts used throughout this book. A number of references to related material are also provided to help expand the applicability of the concepts developed in the previous chapters.

Stability Analysis: Chapter 2

The stability proofs in this book have largely been developed using principles of Lyapunov stability. A nice introduction to differential equations and stability may be found in [70, 154, 153], while the use of Lyapunov analysis for both continuous and discrete-time systems is presented in [97]. Other references on stability theory include [68, 111, 166]. Good references for nonlinear control, which cover many of the mathematical topics presented in Chapter 2, include [108, 230, 205]. Other books that provide similar background material are books on conventional adaptive control [79, 191].

In this book we have typically assumed that the system $\dot{x} = f(t, x, u)$ and controller $u(t, x)$ are defined such that f is piecewise continuous in t and Lipschitz continuous in x. When controllers are defined using the principles of variable structure control [227, 204] (sliding mode control) or projection, one often finds that the resulting f is discontinuous in x. Since the traditional existence and uniqueness conditions do not hold under this case, one may use the approaches in [51, 123, 174, 199].

The concept of input-to-state stability was introduced by Sontag [207]. For more details on properties of input-to-state stability, see e.g., [210, 211]. Extensions to input-to-state practical stability may be found in [92] and integral input-to-state stability may be found in [5, 209]. Other stability analysis techniques appropriate for nonlinear systems include the use of small gain theorems [93, 91, 222], passivity [20, 164], and the describing function method [108].

Throughout this book we have discussed stability in terms of a Lyapunov function associated with a closed-loop system. Another way to view stabilization is via a control Lyapunov function. Given the error system

$$\dot{e} = \alpha(t,x) + \beta(x)u, \tag{.1}$$

a control Lyapunov function is a smooth positive definite radially unbounded function $V(e)$ such that

$$\inf_{u \in \mathbf{R}^m} \left\{ \frac{\partial V}{\partial e}\alpha + \frac{\partial V}{\partial e}\beta u \right\} < 0. \tag{.2}$$

Just as a Lyapunov function implies stability for a system with the control defined, a control Lyapunov function implies stabilizability of a system with an undefined input. See [114, 115] for more details.

Neural Networks and Fuzzy Systems: Chapters 3–5

There are many good books on neural networks. The reader may want to consider [67, 72]. Using a gradient descent approach to minimize a cost function based on the difference between the true and desired neural network output, the backpropagation technique was developed [186]. With this tool, researchers showed that neural networks are able to approximate a large variety of nonlinear functions. This later led to the work [17, 33, 57, 75, 126] in which it was shown that neural networks are indeed universal approximators. Various properties of approximators are studied in [13, 59, 61, 171]. In [208] it is shown that for certain control problems it may be better to use a two layer neural network rather than a single layer. Early applications of neural networks for use in control and identification was studied in [160]. This led to a large volume of applications and theory related to the use of neural networks in feedback systems [6, 31, 49, 156, 194].

L. Zadeh [250, 251] introduced the field of fuzzy sets and logic, while E. Mamdani [145, 144] suggested that fuzzy systems be used in feedback controllers. The universal approximation property for fuzzy systems is discussed in [24, 231, 233, 235]. For more details on fuzzy logic see, for instance, [110], and for a general treatment of how to use fuzzy logic in engineering see [184]. For more details on fuzzy systems and control see [170, 42, 213, 233, 235].

An introduction to the topical area of approximation is given in [201, 95], where the authors also cover wavelets, and other approximators and properties in some detail. The idea of linearizing a nonlinear approximator was studied in [29] in their stability proofs. For more details on linear least squares methods, especially for system identification, see [135]. For a more detailed treatment of optimization theory see the introductory book [137]. For a more advanced treatment, see [16, 181].

There are also a number of journals dedicated to neural networks and fuzzy systems. See, for example, IEEE Transactions on Neural Networks, IEEE Transactions on Fuzzy Systems, Neural Networks, IEEE Transactions on Systems, Man, and Cybernetics, and Fuzzy Sets and Systems. Along with these publications are conferences such as the IEEE International Symposium on Intelligent Control.

State Feedback Control: Chapters 6–8

There are a number of good references for general nonlinear control design including [81, 108, 128, 205]. For more details on feedback linearization and diffeomorphisms, see [81, 149, 236]. A discussion of pole placement may be found in most basic control texts dealing with state-space techniques including [27, 54]. Robust stability has been addressed in the literature using a number of techniques including nonlinear damping [96] and sliding mode control [40, 204, 227]. For a more detailed discussion of robust stability, see the work [114, 133]. Also see [92, 79] for a discussion of dynamic normalization.

Integrator backstepping was introduced as a control design tool in [104, 187]. Robust extensions to integrator backstepping were developed in [148, 203, 182] for systems with uncertainties, while extensions to systems with unmodeled dynamics were given in [118, 179]. The backstepping approach is covered in detail in [115].

The concepts of zero dynamics and static state feedback of single-input single-output (SISO) nonlinear systems can be readily generalized to multivariable nonlinear systems, but in general two cases arise, static and dynamic state feedback. The reader may consult [191, 193, 81] for a review of this material.

For a given controller structure, there may be a number of different choices available for the controller parameters that results in a stable closed-loop system. To help choose a particular controller, one may define a cost function for the closed-loop system that is to be minimized. This is the objective of optimal control [19, 109]. For example, a linear quadratic regulator (LQR) is defined as the controller $u = Kx$ (where K is to be chosen) that minimizes a cost function defined to keep the states and/or control input small [128]. Kalman has shown that one of the features of an optimal design is that the closed-loop system is inherently robust. Optimal design techniques do not readily carry over to the nonlinear world, but there has been progress in the development of inverse optimal control techniques for nonlinear systems that result in a robust closed-loop system [195].

There are a number of good references that deal with direct and indirect adaptive control of linear systems. See, for example, [11, 63, 128, 159, 191]. The field of adaptive control of nonlinear systems has continued to see a great deal of research [50, 103, 100, 116, 161, 162, 177, 220]. The work in [115, 149] discusses a number of approaches to the adaptive control of nonlinear systems, while in [219] the authors consider the case where the system contains actuator and sensor nonlinearities. The reader interested in more discussion on robust adaptive control is urged to see [79, 92, 90, 172, 175, 221]. For a discussion on adaptive control with nonlinear parameterizations, see the work [18, 136].

The papers [160, 173, 189] started an important branch within the field of nonlinear adaptive control, that of nonlinear neural and fuzzy adaptive control. The papers [30, 45, 47, 48, 130, 134, 160, 172, 173, 176, 185, 188, 189, 242, 247, 244] make use of neural networks as approximators of nonlinear functions, whereas [26, 76, 124, 217, 232, 234] use fuzzy systems for the same purpose and [160, 185] use dynamical neural networks. An interesting study on issues related to the use of local (finite support) approximators in adaptive control can be found in [46]. The neural and fuzzy approaches are most of the time equivalent, differing between each other only by the structure of the approximator chosen.

Among those works in which tunable parameterized functions are used, a major difference can be devised in the choice of the parameterization: linear in [26, 22, 47, 48, 45, 76, 172, 173, 189, 188, 217, 233, 248] and non-linear in [30, 134, 130, 129, 160, 176, 242, 244]. Most of the papers deal with indirect adaptive control, trying first to identify the dynamics of the systems and eventually generating a control input according to the certainty equivalence principle (with some modification to add robustness to the control law), whereas very few authors face the direct approach. Finally, the most frequent classes of systems considered in these papers are the SISO affine ones, that is, systems of the type $\dot{x} = f(x) + g(x)u$ with $u \in \mathsf{R}$.

The reader interested in using neural networks and/or fuzzy systems in identification may wish to read [95, 122, 127, 160, 173, 188, 190, 201, 218]. Other properties of parameter identification are given in [131, 254].

Output Feedback Control: Chapters 10 and 11

The class of output-feedback systems considered in this book (see (10.5)) was introduced by Marino and Tomei in [146]. The control design presented here, which essentially relies on the backstepping methodology, goes along the line of the solution presented in the book [115]. Additional details on the stabilization of output-feedback systems can be found in the book [149].

The general output feedback stabilization problem considered in Chapter 10 and specifically the development of a separation principle for nonlinear systems has been the object of intense research since the beginning of the nineties. The work by Esfandiari and Khalil, in [44], was the first to highlight the potentially destabilizing effect of high-gain observers employed in closed-loop (i.e., the peaking phenomenon), which we have seen to be one of the two major obstacles to the achievement of a separation principle, and to suggest the employment of control input saturation as a cure to the peaking phenomenon. Even though the applicability of the results obtained in that paper was limited to minimum phase input-output feedback linearizable systems, and the region of attraction of the closed-loop system employing the output feedback controller could not be made arbitrarily large, the ideas contained in the paper had a major influence in all the literature that followed.

The second obstacle to the development of a separation principle, namely the fact that the observability mapping of the plant depends in general on the control input and its time derivatives, i.e., $y_e = \mathcal{H}(x, u, \ldots, u^{(n_u-1)})$, and hence in order to estimate the state of the system one needs to know $n_u - 1$ time derivatives of the control input, was removed by Tornambé in [225]. The main idea contained in [225] is that of employing a chain of integrators at the input side of the plant and use the state feedback controller for the original system to design a state feedback controller for the augmented system (the plant plus a chain of integrators). This key idea removes the obstacle mentioned above since the time derivatives of the control input are now simply given by the states of the integrators and hence the state of the system can be estimated by inverting the mapping $\mathcal{H}$ and using a high-gain observer to estimate the vector y_e containing the output and its $n-1$ time derivatives. This also allows for the removal of the minimum phase assumption required in [44] but the stability results obtained in this paper are only local since control input saturation is not employed and hence the peaking phenomenon affects the systems states. The fundamental work of Teel and Praly in [223] merged the ideas of Tornambé and those of Esfandiari and Khalil in [44] to prove that a generic nonlinear SISO system which is globally stabilizable and uniformly completely observable is semiglobally stabilizable by output feedback, thus developing a separation principle for a rather general class of nonlinear systems. The main ingredients of the controller developed in [223] are the chain of integrators at the input side of the system and saturation. The practical limitation of these results (as well as all the results cited above) is the need to know the explicit inverse of the observability mapping $\mathcal{H}$ to reconstruct the state of the plant. See also the work by Atassi and Khalil in [12] for further extensions and a more general separation principle.

The separation principle proved in Theorem 10.2 is taken from [143] and contains similarities to the results cited above (e.g., a chain of integrators is added at the input side of the system) with two important differences. First, the standard high-gain observer employed to estimate the output derivatives is replaced by the nonlinear observer (10.39) with the dynamic projection (10.63). Second, saturation is not needed since the dynamic projection (10.63) *eliminates* the peaking phenomenon in the observer states. As we have seen in Chapter 10, this approach has the practical advantage of avoiding the need to calculate the explicit inverse of the observability mapping $\mathcal{H}$. Furthermore, an advantage of using dynamic projection to confine the observer states to within a prespecified compact set is that the assumption that the plant be *uniformly completely observable* can be considerably relaxed, see [143] for the details and Chapter 12 for an example.

The idea of using two cascaded observers to estimate the state of the plant without adding integrators is taken from [140] and is inspired by the work by Tornambé in [224], where the author employs a parallel connection of two high-gain observers to estimate the output and input derivatives, respectively. The estimates obtained from the observers are then employed to calculate an estimate of the state of the plant by inverting the observability mapping $\mathcal{H}$. The observer employed in this paper, (10.88), (10.89) is different from that found in [224] and has once again the advantage of directly estimating the state of the plant, avoiding the inversion of the mapping $\mathcal{H}$. Tornambé did not investigate the employment of his estimation scheme in closed-loop and hence the result in Theorem 10.3 is new to the output feedback literature.

Tracking is one of the important problems in control theory and researchers are still working at developing a general theory in the nonlinear setting using only output feedback (in a partial information setting). For output-feedback systems, its global solution is well-known (see [149, 115]) while, for nonminimum-phase systems, achieving asymptotic tracking is considered to be a challenging problem even when the state of the plant is known. When the plant is *slightly nonminimum-phase*, it may be approximated by a minimum-phase system, and a tracking controller for the approximated model may be employed to achieve s small tracking error (see [71]). Another way to achieve approximate tracking entails finding an output function with respect to which the system is minimum-phase, and modifying the original reference to create an appropriate desired trajectory for the new output (see the works in [64] and [15] among others). A more general way to address tracking problems is to use differential flatness, introduced by Fliess *et al.* in [52], which is related to the concept of dynamic feedback linearization (see [25]). It is well-known (e.g., [52], [150], [229]) that for differentially flat systems the state feedback tracking problem can be easily solved since the state and the control input are completely characterized by the output and a finite number of its derivatives. Hence, given a reference trajectory, one can uniquely determine the corresponding state and control trajectories reproducing the desired output, and use this information to solve the tracking problem. A second major approach (which has some relationship to differential flatness) to solving tracking problems was introduced in [41] by Devasia *et al.*, and involves calculating a stable inverse of the plant (see Chapter 6). If the plant is differentially flat, a stable inverse must exist. In general, the stable inverse can be found as the solution of an integral equation which can be approximated iteratively by means of a Picard-like iteration. The stable inverse so obtained is employed as a feedforward term in a regulation scheme and, being in general non-causal, it may require pre-actuation. The other major approach to solving tracking problems is the theory of output regulation (also referred to as the servomechanism theory, see [37]), originally introduced by Davison, Francis, and Wonham in [37, 53] for linear systems, and extended to nonlinear systems by Isidori and Byrnes in [80]. The output regulation problem entails finding a dynamic controller that makes the output track a reference trajectory and reject a time-varying disturbance, both of which are generated by a neutrally stable exosystem. In order to do that one seeks the solution of a set of nonlinear partial differential equations referred to as the regulator equations which, as shown in [77], is identical to the solution of the integral equation in the stable inversion approach. In the nonlinear full information setting one implements a control law which depends on the state of the exosystem (see [81]) while, in the error feedback setting (analogous to what we defined to be the partial information setting in Chapter 10) one does not need the state of the exosystem, but rather needs to find a suitable internal model which is employed to asymptotically generate the tracking control action. The solution to the error feedback output regulation problem for general classes of nonlinear systems is presently an open research topic.

The methodology to solve the tracking problem introduced in Chapter 10 can be found in [141] and has some similarities to both the stable inversion approach of Devasia *et al.* and output regulation theory. The results presented in this book rely on the existence of a *practical internal model* (see Definition 10.2). It can be shown that a practical internal model always exists when the plant is differentially flat and the flat output is measurable. Analogously, it can be shown that the methodology presented in this book can be employed in the output regulation framework, see [142] and [141] and for the details.

For systems in adaptive tracking form (11.1) (originally defined by Marino and Tomei) the reader may refer to [147, 103, 115, 117] for global stabilization results and to [180] for some generalizations. The adaptive controllers for systems in adaptive tracking form presented in this book follows [103]. For more general nonlinear systems with constant unknown parameters, Khalil in [107] developed adaptive output feedback controllers based on a separation principle, see also [3]. The class of systems considered in [107, 3] is similar to that considered in Example 11.1 and the tools developed in Section 11.3.2 can be employed to develop stable adaptive controllers for such systems.

Discrete-Time Control: Chapter 13

In [157] the authors introduce the notion of zero dynamics and minimum phase for discrete-time systems. There has been research on the stabilization of nonlinear discrete-time systems using state feedback control [106, 23, 21], adaptive output feedback control (for certain classes of systems) [206, 245], and the MIMO case has been studied for a Hammerstein model in [252]. The work in [120] focuses on stability analysis of a feedback linearizing controller for a class of nonlinear discrete-time systems that have constraints on the inputs. Development of controllers for discrete-time nonlinear systems with certain types of input and output nonlinearities is studied in [219]. The work in [132] focuses on H_∞-control of discrete-time nonlinear systems; in particular, the problem of disturbance attenuation is studied under different assumptions about the type of feedback information that is used.

The book [63] is a good starting point for results concerning adaptive control of discrete-time systems that are linear in a set of unknown parameters. A scheme was presented in [101, 102] for the adaptive control of the nonlinear system $y(k+1) = a^*\zeta(y(k)) + u(k)$, where a^* is an unknown constant and ζ is a nonlinear function which is not necessarily sector bounded; stability results are established that are not dependent upon parameter initialization. The work in [255] focuses on the development of stable adaptive controllers for discrete-time strict-feedback nonlinear systems. Here, a very different approach is taken from the work in this book since rather than a traditional on-line estimation and control approach, the authors separate the problem into a nonlinearity basis identification phase and look-ahead control phase. They obtain global stability and tracking results. In [139] the authors provide necessary and sufficient conditions for the equivalence between a general class of discrete-time systems and discrete-time systems in strict feedback form. They use a backstepping algorithm and obtain stability results. In [241] the authors show that it may not always be possible to use adaptive feedback to stabilize a discrete-time system.

Within [28], an indirect adaptive control routine is presented using a class of neural networks. Stability is established assuming that if the neural networks are properly adjusted, they are able to perfectly model the plant. These ideal modeling requirements are relaxed within [29], while ensuring tracking to an ϵ-neighborhood of zero. Stability for these routines is established assuming that the initial estimates of the nonlinear system are chosen to be sufficiently accurate, and that $f(x(k), u(k))$ vanishes at the origin. In [83] the authors consider a discrete-time MIMO nonsquare feedback linearizable system without zero dynamics, but with a bounded state disturbance. They perform indirect adaptive control using linear in the parameter neural networks, and on-line tuning without the need of initialization conditions, and obtain uniform ultimate boundedness of the tracking error. In [84] similar results to [83] are obtained under the same assumptions, but using multilayered neural networks (a nonlinear in the parameter tunable function), by showing passivity of the neural network. Within [172], a stable neural network control methodology is presented for continuous-time strict feedback systems using the backstepping method [103]. See [1, 129, 85, 82, 86, 94] for other fuzzy/neural approaches.

Decentralized Control: Chapter 14

Decentralized control has long been used in the control of power systems and spacecraft since the problem may often be viewed as a number of interconnected systems. See [155] for general stability theory related to decentralized systems. An early form of diagonal dominance led to the concept of an M-matrix [78, 108, 243].

Model reference adaptive control (MRAC) based designs for decentralized systems have been studied in [73, 78, 58, 36, 238] for the continuous-time case and in [183, 165] for the discrete-time case. These approaches, however, are limited to decentralized systems with linear subsystems and possibly nonlinear interconnections. Decentralized adaptive controllers for robotic manipulators were presented in [56, 32, 196], while a scheme for nonlinear subsystems with a special class of interconnections was presented in [198].

Integrator backstepping is used in [237] to help relax relative degree requirements on the subsystems. Polynomial bounded interconnections were developed in [200], while general nonlinear interconnections were studied in [69]. The matching conditions were relaxed in [87, 88] when the interconnections could be bounded by polynomials. [65, 89] relax both matching conditions and polynomial bounding of the interconnections. These results consider subsystems that are linear in a set of unknown parameters, or consider the uncertainties to be contained within the dynamics describing the subsystem interconnections which are bounded. Decentralized techniques using neural networks and fuzzy systems were developed in [34, 105, 214, 246, 253] that allowed for more general classes of isolated subsystem dynamics.

In this book we have allowed interconnections bounded by functions that are not necessarily polynomials and have relaxed the matching condition requirement. In the adaptive case, it was shown that it is possible to adaptively approximate the bounds on the interconnections (not just polynomial coefficients). The approach considered in this book, however, does assume that the interconnections are bounded by functions defined in terms of the subsystem outputs and not necessarily the full states.

Applications: Chapters 9 and 12

Good books on dynamics of mechanical systems include [60, 98, 197]. The reader interested in the control of specific types of applications may wish to read [9, 212] for robotic systems; [99, 151] for aerospace applications; [112, 125, 202] for electric machines; and [4, 35, 38, 43, 55] for systems with friction. Other books that contain a number of traditional control problems (such as the inverted pendulum) include [108, 128, 149, 205]. Within this text, the satellite dynamics of Example 6.3 may be found in [149], the ball and beam dynamics of Exercise 6.12 in [195], the M-link robot of Exercise 6.13 and robot with a flexible joint of Exercise 6.20 in [212], the inverted pendulum of Exercise 6.14 in [108, 149], a modified version of the magnetic levitation example of Chapter 12 in [108], and the surge tank of Exersise 7.11 in [169]. The electric machine models throughout this text may be found in the following: the field-controlled DC motor of Exercise 6.19 in [202], stepping motor of Exercise 8.7 in [119],

For recent implementations of various control techniques, one may wish to look through IEEE Transactions on Control Systems Technology and IEEE Control Systems Magazine.

Perspectives on Intelligent Adaptive Systems: Chapter 15

For more details on genetic algorithms, see the books [62, 152] or article [215]. The use of genetic algorithms in (indirect) adaptive control was first studied in [113]. The genetic model reference adaptive controller (a direct adaptive scheme) was first introduced in [178]. For more details on (direct) expert control, see [168], or [138] for a control-engineering analysis of the feedback loop that is inherent in the expert system inference process. The idea of using expert systems to supervise adaptive control systems was first introduced in [10] and is also investigated in [8]. The section on planning systems is based on [167]. For an artificial intelligence perspective on planning systems, that attempts to relate planning ideas to control theory, see [39]. For a general introduction to intelligent control, see the books [8, 226, 66, 228, 239] or articles [7, 2, 216].

Bibliography

[1] M. S. Ahmed and M. F. Anjum, *Neural-net based self-tuning control of nonlinear plants*, Int. J. Control **66** (1997), no. 1, 85–104.

[2] J. S. Albus, *Outline for a theory of intelligence*, IEEE Trans. on Systems, Man, and Cybernetics **21** (1991), no. 3, 473–509.

[3] B. Aloliwi and H. K. Khalil, *Robust adaptive output feedback control of nonlinear systems without persistence of excitation*, Automatica **33** (1997), no. 11, 2025–2032.

[4] J. Amin, B. Friedland, and A. Harnoy, *Implementation of a friction estimation and compensation technique*, IEEE Control Systems Magazine (1997), 71–76.

[5] D. Angeli, E. D. Sontag, and Y. Wang, *A characterization of integral input-to-state stability*, IEEE Trans. Automat. Contr. **45** (2000), no. 6, 1082–1097.

[6] P. J. Antsaklis, *Neural networks in control systems*, IEEE Control Systems Magazine (1990), 3–5.

[7] P. J. Antsaklis and K. M. Passino, *Towards intelligent autonomous control systems: Architecture and fundamental issues*, Journal of Intelligent and Robotic Systems **1** (1989), 315–342.

[8] P. J. Antsaklis and K. M. Passino (eds.), *An introduction to intelligent and autonomous control*, Kluwer Academic Publishers, Norwell, MA, 1993.

[9] H. Asada and J. J. E. Slotine, *Robot analysis and control*, John Wiley and Sons, New York, 1986.

[10] K. J. Åström, J. J. Anton, and K. E. Arzen, *Expert control*, Automatica **22** (1986), no. 3, 277–286.

[11] K. J. Åström and B. Wittenmark, *Adaptive control*, Addison-Wesley Publishing Company, Reading, MA, 1995.

[12] A. Atassi and H. K. Khalil, *A separation principle for the stabilization of a class of nonlinear systems*, IEEE Trans. Automat. Contr. **44** (1999), no. 9, 1672–1687.

[13] A. R. Barron, *Universal approximation bounds for superpositions of a sigmoid function*, IEEE Trans. Inform. Theory **39** (1993), no. 3, 930–945.

[14] R. G. Bartle, *The elements of real analysis*, John Wiley and Sons, New York, 1976.

[15] L. Benvenuti, M. D. Di Benedetto, and J. W. Grizzle, *Approximate output tracking for nonlinear non-minimum phase systems with an application to flight control*, International Journal of Robust and Nonlinear Control **4** (1994), no. 3, 397–414.

[16] D. P. Bertsekas, *Nonlinear programming*, Athena Scientific Press, Belmont, MA, 1996.

[17] E. K. Blum and L. K. Li, *Approximation theory and feedforward networks*, Neural Networks **4** (1991), 511–515.

[18] J. D. Bošković, *Adaptive control of a class of nonlinearly parameterized plants*, IEEE Trans. Automat. Contr. **43** (1998), no. 7, 930–934.

[19] A. E. Bryson and Y. C. Ho, *Applied optimal control*, Halsted Press, Washington, D.C., 1975.

[20] C. Byrnes, A. Isidori, and J. Willems, *Passivity, feedback equivalence, and the global stabilization of minimum phase nonlinear systems*, IEEE Trans. Automat. Contr. **36** (1991), 1228–1240.

[21] C. Byrnes, W. Lin, and B. Ghosh, *Stabilization of discrete-time nonlinear systems by smooth state feedback*, Systems Control Lett. **21** (1993), 255–263.

[22] R. Carelli, E. F. Camacho, and D. Patino, *A neural network based feedforward adaptive controller for robots*, IEEE Transactions on Systems, Man, and Cybernetics **25** (1995), no. 9, 1281–1288.

[23] B. Castillo, S. D. Gennaro, S. Monaco, and D. Normand-Cyrot, *Nonlinear regulation for a class of discrete-time systems*, Syst. and Control Letters **20** (1993), 57–65.

[24] J. L. Castro, *Fuzzy logic controllers are universal approximators*, IEEE Trans. Syst. Man, Cybern. **25** (1995), no. 4, 629–635.

[25] B. Charlet, J. Lévine, and R. Marino, *On dynamic feedback linearization*, Systems Control Lett. (1989), no. 13, 143–151.

[26] B.-S. Chen, C.-H. Lee, and Y.-C. Chang, *H^∞ tracking design of uncertain nonlinear SISO systems: Adaptive fuzzy approach*, IEEE Transactions on Fuzzy Systems **4** (1996), no. 1, 32–43.

[27] C.-T. Chen, *Linear system theory and design*, Holt, Rinehart and Winston, Inc, New York, 1984.

[28] F.-C. Chen and H. K. Khalil, *Adaptive control of nonlinear systems using neural networks*, Int. J. Control **55** (1992), 1299–1317.

[29] ______, *Adaptive control of a class of nonlinear discrete-time systems using neural networks*, IEEE Trans. Automat. Contr. **40** (1995), no. 5, 791–801.

[30] F.-C. Chen and C.-C. Liu, *Adaptively controlling nonlinear continuous-time systems using multilayer neural networks*, IEEE Trans. Automat. Contr. **39** (1994), no. 6, 1306–1310.

[31] J. Y. Choi and J. A. Farrell, *Nonlinear adaptive control using networks of piecewise linear approximators*, Proc. of 38th Conf. Decision Contr. (Phoenix, Az), December 1999, pp. 1671–1676.

[32] Y. K. Choi and Z. Bien, *Decentralized adaptive control scheme for control of a multi-arm-type robot*, Int. J. Control **48** (1988), no. 4, 1715–1722.

[33] G. Cybenko, *Approximations by superpositions of a sigmoidal function*, Math. Contr. Signals, Syst. **2** (1989), 303–314.

[34] F. Da, *Decentralized sliding mode adaptive controller design based on fuzzy neural networks for interconnected uncertain nonlinear systems*, IEEE Trans. Neural Networks **11** (2000), no. 6, 1471–1480.

[35] P. Dahl, *Measurement of solid friction parameters of ball bearings*, Proceedings of the 6th Annual Symposium on Incremental Motion, Control Systems and Devices (University of Illinois), 1977, pp. 49–60.

[36] A. Datta, *Performance improvement in decentralized adaptive control: A modified model reference scheme*, IEEE Trans. Automat. Contr. **38** (1993), no. 11, 1717–1722.

[37] E. J. Davison, *The robust control of a servomechnism problem for linear time-invariant multivariable systems*, IEEE Trans. Automat. Contr. **21** (1976), 25–34.

[38] C. C. de Wit, H. Olsen, K. Åström, and P. Lischinsky, *A new model for control of systems with friction*, IEEE Trans. Automat. Contr. **40** (1995), no. 3, 419–425.

[39] T. Dean and M. P. Wellman, *Planning and control*, Morgan Kaufman, San Mateo, CA, 1991.

[40] R. A. DeCarlo, S. H. Żak, and G. P. Matthews, *Variable structure control of nonlinear multivariable systems: A tutorial*, Proc IEEE **76** (1988), no. 3, 212–232.

[41] S. Devasia, D. Chen, and B. Paden, *Nonlinear inversion-based output tracking*, IEEE Trans. Automat. Contr. **41** (1996), no. 7, 930–942.

[42] D. Driankov, H. Hellendoorn, and M. Reinfrank, *An introduction to fuzzy control*, 2nd ed., Springer-Verlag, New York, 1996.

[43] H. Du and S. S. Nair, *Low velocity friction compensation*, IEEE Control Systems Magazine (1998), 61–69.

[44] F. Esfandiari and H. K. Khalil, *Output feedback stabilization of fully linearizable systems*, Int. J. Control **56** (1992), no. 5, 1007–1037.

[45] S. Fabri and V. Kadirkamanathan, *Dynamic structure neural networks for stable adaptive control of nonlinear systems*, IEEE Transactions on Neural Networks **7** (1996), no. 5, 1151–1167.

[46] J. A. Farrell, *Motivations for local approximators in passive learning control*, J. of Intelligent Sys. and Contr. **1** (1996), no. 2, 195–210.

[47] ———, *Persistence of excitation conditions in passive learning control*, IFAC World Congress (San Francisco, CA), 1996, pp. 313–318.

[48] ———, *Persistence of excitation conditions in passive learning control*, Automatica **33** (1997), no. 4, 699–703.

[49] ———, *Stability and approximator convergence in nonparametric nonlinear adaptive control*, IEEE Trans. Neural Networks **9** (1998), no. 5, 1008–1020.

[50] G. Feng, *Analysis of a new algorithm for continuous-time robust adaptive control*, IEEE Trans. Automat. Contr. **44** (1999), no. 9, 1764–1768.

[51] A. F. Filippov, *Differential equations with discontinuous right-hand side*, Amer. Math. Soc. Translations **42** (1964), no. 2, 199–231.

[52] M. Fliess, M. Lévin, P. Martin, and P. Rouchon, *Flatness and defect of non-linear systems: Introductory theory and examples*, Int. J. Control **61** (1995), no. 6, 1327–1361.

[53] B. A. Francis and W. M. Wonham, *The internal model principle of control theory*, Automatica **12** (1976), 457–465.

[54] G. F. Franklin, J. D. Powell, and A. Emami-Naeini, *Feedback control of dynamic systems*, 2nd ed., Addison-Wesley, Reading, MA, 1991.

[55] B. Friedland and Y.-J. Park, *On adaptive friction compensation*, IEEE Trans. Automat. Contr. (1992), no. 10, 1609–1612.

[56] L.-C. Fu, *Robust adaptive decentralized control of robot manipulators*, IEEE Trans. Automat. Contr. **37** (1992), no. 1, 106–110.

[57] K. Funahashi, *On the approximate realization of continuous mappings by neural networks*, Neural Networks **2** (1989), 183–192.

[58] D. T. Gavel and D. D. Šiljak, *Decentralized adaptive control: Structural conditions for stability*, IEEE Trans. Automat. Contr. **34** (1989), no. 4, 413–426.

[59] S. Geva and J. Sitte, *A constructive method for multivariate function approximation by multilayer perceptrons*, IEEE Trans. Neural Networks **3** (1992), no. 4, 621–624.

[60] J. H. Ginsberg, *Advanced engineering dynamics*, Harper & Row, New York, 1988.

[61] F. Girosi and G. Anzellotti, *Convergence rates of approximation by translates*, Art. Intell. Lab. Tech. Rep. 1288, Mass. Inst. Technol., 1992.

[62] D. Goldberg, *Genetic algorithms in search, optimization and machine learning*, Addison-Wesley, Reading, MA, 1989.

[63] G. C. Goodwin and K. S. Sin, *Adaptive filtering prediction and control*, Prentice Hall, Englewood Cliffs, NJ, 1984.

[64] S. Gopalswamy and K. Hedrick, *Tracking nonlinear nonminimum-phase systems using sliding control*, Int. J. Control **57** (1993), no. 5, 1141–1158.

[65] Y. Guo, Z.-P. Jiang, and D. J. Hill, *Decentralized robust disturbance attenuation for a class of large-scale nonlinear systems*, Systems Control Lett. **37** (1999), 71–85.

[66] M. Gupta and N. Sinha (eds.), *Intelligent control: Theory and practice*, IEEE Press, Piscataway, NJ, 1995.

[67] M. Hagan, H. Demuth, and M. Beale (eds.), *Neural network design*, PWS Publishing, Boston, MA, 1996.

[68] W. Hahn, *Stability of motion*, Springer-Verlag, Berlin, 1967.

[69] M. C. Han and Y. H. Chen, *Decentralized control design: Uncertain systems with strong interconnections*, Int. J. Control **61** (1995), no. 6, 1363–1385.

[70] P. Hartmann, *Ordinary differential equations*, 2nd ed., Birkhäuser, Boston, 1982.

[71] J. Hauser, S. Sastry, and G. Meyer, *Nonlinear control design for slightly nonminimum phase systems: Application to V/STOL aircraft*, Automatica **28** (1992), no. 4, 665–679.

[72] J. Hertz, A. Krogh, and R. G. Palmer, *Introduction to the theory of neural computation*, Addison-Wesley Publishing Company, 1991.

[73] A. Hmamed and L. Radouane, *Decentralized nonlinear adaptive feedback stabilization of large-scale interconnected systems*, IEE Proc. D, Control Theory & Appl. **130** (1983), no. 2, 57–62.

[74] R. A. Horn and C. R. Johnson, *Matrix analysis*, Cambridge University Press, Cambridge (Cambridgeshire); New York, 1985.

[75] K. Hornik, M. Stinchcombe, and H. White, *Multilayered feedforward networks are universal approximators*, Neural Networks **2** (1989), 359–366.

[76] F.-Y. Hsu and L.-C. Fu, *A new design of adaptive robust controller for nonlinear systems*, Proceedings of the American Control Conference, June 1995.

[77] L. R. Hunt and G. Meyer, *Stable inversion for nonlinear systems*, Automatica **33** (1997), no. 8, 1549–1554.

[78] P. A. Ioannou, *Decentralized adaptive control of interconnected systems*, IEEE Trans. Automat. Contr. **31** (1986), no. 4, 291–298.

[79] P. A. Ioannou and J. Sun, *Robust adaptive control*, Prentice Hall, Englewood Cliffs, New Jersey, 1996.

[80] A. Isidori and C. I. Byrnes, *Output regulation of nonlinear systems*, IEEE Trans. Automat. Contr. **35** (1990), 131–140.

[81] A. Isidori, *Nonlinear control systems*, 3rd ed., Springer-Verlag, London, 1995.

[82] S. Jagannathan, *Adaptive fuzzy logic control of feedback linearizable discrete-time dynamical systems under persistence of excitation*, Automatica **34** (1998), no. 11, 1295–1310.

[83] S. Jagannathan and F. L. Lewis, *Discrete-time neural net controler for a class of nonlinear dynamical systems*, IEEE Trans. Automat. Contr. **41** (1996), no. 11, 1693–1699.

[84] ———, *Multilayer discrete-time neural-net controller with guaranteed performance*, IEEE Trans. Neural Networks **7** (1996), no. 1, 107–130.

[85] S. Jagannathan, F. L. Lewis, and O. Pastravanu, *Discrete-time model reference adaptive control of nonlinear dynamical systems using neural networks*, Int. Journal of Control **64** (1996), no. 2, 217–239.

[86] S. Jagannathan, M. W. Vandegrift, and F. L. Lewis, *Adaptive fuzzy logic control of discrete-time dynamical systems*, Automatica **36** (2000), 229–241.

[87] S. Jain and F. Khorrami, *Decentralized adaptive control of a class of large-scale interconnected nonlinear systems*, IEEE Trans. Automat. Contr. **42** (1997), no. 2, 136–154.

[88] ———, *Decentralized adaptive output feedback design for large-scale nonlinear systems*, IEEE Trans. Automat. Contr. **42** (1997), no. 5, 729–735.

[89] Z.-P. Jiang, *Decentralized and adaptive nonlinear tracking of large-scale systems via output feedback*, IEEE Trans. Automat. Contr. **45** (2000), no. 11, 2122–2128.

[90] Z.-P. Jiang and D. J. Hill, *A robust adaptive backstepping scheme for nonlinear systems with unmodeled dynamics*, IEEE Trans. Automat. Contr. **44** (1999), no. 9, 1705–1711.

[91] Z.-P. Jiang and M. Y. Mareels, *A small-gain control method for nonlinear cascaded systems with dynamic uncertainties*, IEEE Trans. Automat. Contr. **42** (1997), no. 3, 292–308.

[92] Z.-P. Jiang and L. Praly, *Design of robust adaptive controllers for nonlinear systems with dynamic uncertainties*, Automatica **34** (1998), no. 7, 825–840.

[93] Z.-P. Jiang, A. R. Teel, and L. Praly, *Small-gain theorem for iss systems and applications*, Math. Contr. Lett. **7** (1994), 95–120.

[94] T. A. Johansen, *Fuzzy model based control: Stability, robustness, and performance issues*, IEEE Trans. Fuzzy Systems **2** (1994), no. 3, 221–234.

[95] A. Juditsky, H. Hjalmarsson, A. Benveniste, B. Deylon, L. Ljung, J. Sjoberg, and Q. Zhang, *Nonlinear black-box modeling in system identification: Mathematical foundations*, Automatica **31** (1995), no. 12, 1691–1724.

[96] V. Jurdjevic and J. P. Quinn, *Controllability and stability*, J. Differential Equations **28** (1978), 381–389.

[97] R. E. Kalman and J. E. Bertram, *Control system analysis and design via the "second method" of Lyapunov, Parts I and II*, Journal of Basic Engineering **82** (1960), 371–400.

[98] T. R. Kane, *Dynamics*, Holt, Rinehart, and Willson, New York, 1968.

[99] T. R. Kane, P. W. Likins, and D. A. Levinson, *Spacecraft dynamics*, McGraw-Hill, New York, 1983.

[100] I. Kanellakopoulos, *Passive adaptive control of nonlinear systems*, Int. J. Adapt. Control Signal Process. **7** (1993), 339–352.

[101] ______, *A discrete-time adaptive nonlinear system*, Proc. 1994 American Control Conf. (Baltimore, MD), June 1994, pp. 867–869.

[102] ______, *A discrete-time adaptive nonlinear system*, IEEE Trans. Automat. Contr. **39** (1994), no. 11, 2362–2365.

[103] I. Kanellakpoulos, P. V. Kokotović, and A. S. Morse, *Systematic design of adaptive controllers for feedback linearizable systems*, IEEE Trans. Automat. Contr. **36** (1991), no. 11, 1241–1253.

[104] ______, *A toolkit for nonlinear feedback design*, Systems & Control Letters **18** (1992), 83–92.

[105] A. Karakaşoğlu, S. I. Sudharsanan, and M. K. Sudareshan, *Identification and decentralized adaptive control using dynamical neural networks with application to robotic manipulators*, IEEE Trans. Neural Networks **4** (1993), no. 6, 919–930.

[106] D. Kazakos and J. Tsinias, *Stabilization of nonlinear discrete-time systems using state detection*, IEEE Trans. Automat. Contr. **38** (1993), 1398–1400.

[107] H. K. Khalil, *Adaptive output feedback control of nonlinear systems represented by input-output models*, IEEE Trans. Automat. Contr. **41** (1996), no. 2, 177–188.

[108] ______, *Nonlinear systems*, 2nd ed., Macmillan Publishing Company, New York, New York, 1996.

[109] D. E. Kirk, *Optimal control theory: An introduction*, Prentice Hall, Englewood Cliffs, N.J., 1970.

[110] G. J. Klir and B. Yuan, *Fuzzy sets and fuzzy logic: Theory and applications*, Prentice-Hall, Englewood Cliffs, NJ, 1995.

[111] N. N. Krasowsky, *Problems of the theory of stability of motion*, Stanford Univ. Press, Stanford, CA, 1963.

[112] P. C. Krause and O. Wasynczuk, *Electromechanical motion devices*, McGraw-Hill, New York, 1989.

[113] K. Kristinsson and G. Dumont, *System identification and control using genetic algorithms*, IEEE Transactions on Systems, Man, and Cybernetics **22** (1992), no. 5, 1033–1046.

[114] M. Krstić and H. Deng, *Stabilization of nonlinear uncertain systems*, Springer, London, 1998.

[115] M. Krstić, I. Kanellakopoulos, and P. Kokotović, *Nonlinear and adpative control design*, Wiley, New York, 1995.

[116] M. Krstić and P. V. Kokotović, *Adaptive nonlinear design with controller-identifier separation and swapping*, IEEE Trans. Automat. Contr. **40** (1995), no. 3, 426–440.

[117] ———, *Adaptive nonlinear output-feedback schemes with Marino-Tomei controller*, IEEE Trans. Automat. Contr. **41** (1996), no. 2, 274–280.

[118] M. Krstić, J. Sun, and P. V. Kokotović, *Robust control of nonlinear systems with input unmodeled dynamics*, IEEE Trans. Automat. Contr. **41** (1996), 913–920.

[119] B. C. Kuo, *Automatic control systems*, Prentice Hall, Englewood Cliffs, New Jersey, 1987.

[120] M. Kurtz and M. Henson, *Stability analysis of a feedback linearizing control strategy for constrained nonlinear systems*, Proc. of the American Control Conference (Albuquerque, NM), June 1997, pp. 3480–3484.

[121] J. Kurzweil, *On the inversion of Ljapunov's second theorem on stability of motion*, American Mathematical Society Translations, Series 2 **24** (1956), 19–77.

[122] E. G. Laukonen and S. Yurkovich, *A ball and beam testbed for fuzzy identification and control design*, June 1993, American Control Conference, San Francisco, CA, pp. 665–669.

[123] Y. S. Ledyaev and E. D. Sontag, *A lyapunov characterization of robust stability*, Nonlinear Analysis **37** (1999), 813–840.

[124] C.-H. Lee and S.-D. Wang, *A self-organizing adaptive fuzzy controller*, Fuzzy Sets and Systems (1996), 295–313.

[125] W. Leonhard, *Control of electrical drives*, Spriner-Verlag, Berlin, 1985.

[126] M. Leshno, V. Y. Lin, A. Pinkus, and S. Schocken, *Multilayer feedforward networks with a nonpolynomial activation function can approximate any function*, Neural Networks **6** (1993), 861–867.

[127] A. U. Levin and K. S. Narendra, *Control of nonlinear dynamical systems using neural networks - part ii: Observability, identification, and control*, IEEE Trans. Neural Networks **7** (1996), no. 1, 30–42.

[128] W. Levine (ed.), *The control handbook*, CRC Press, Boca Raton, FL, 1996.

[129] F. L. Lewis, S. Jagannathan, and A. Yeşildirek, *Neural network control of robot manipulators and nonlinear systems*, Taylor & Francis, Philadelphia, PA, 1999.

[130] F. L. Lewis, A. Yeşildirek, and K. Liu, *Multilayer neural-net robot controller with guaranteed tracking performance*, IEEE Trans. Neural Networks **7** (1996), no. 2, 1–12.

[131] J.-S. Lin and I. Kanellakopoulos, *Nonlinearities enhance parameter convergence in output-feedback systems*, IEEE Trans. Automat. Contr. **43** (1998), no. 2, 204–222.

[132] W. Lin and C. Byrnes, *H^∞-control of discrete time nonlinear systems*, IEEE Trans. Automat. Contr. **41** (1996), no. 4, 494–510.

[133] Y. Lin, E. Sontag, and Y. Wang, *A smooth converse Lyapunov theorem for robust stability*, SIAM Journal on Control and Optimization **34** (1996), no. 1, 124–160.

[134] C.-C. Liu and F.-C. Chen, *Adaptive control of non-linear continuous-time systems using neural networks - general relative degree and MIMO cases*, International Journal of Control **58** (1993), no. 2, 317–355.

[135] L. Ljung, *System identification: Theory for the user*, Prentice Hall, Englewood Cliffs, New Jersey, 1987.

[136] A.-P. Loh, A. M. Annaswamy, and F. P. Skantze, *Adaptation in the presence of a general nonlinear parameterization: An error model approach*, IEEE Trans. Automat. Contr. **44** (1999), no. 9, 1634–1652.

[137] D. G. Luenberger, *Linear and nonlinear programming*, Addison-Wesley, Reading, MA, 1984.

[138] A. D. Lunardhi and K. M. Passino, *Verification of qualitative properties of rule-based expert systems*, Int. Journal of Applied Artificial Intelligence **9** (1995), no. 6, 587–621.

[139] A. Madani, S. Monaco, and D. Normand-Cyrot, *Adaptive control of discrete-time dynamics in parametric stric-feedback form*, Proceedings of the 35th Conference on Decision and Control, December 1996, pp. 2659–2664.

[140] M. Maggiore and K. Passino, *Robust output feedback control of incompletely observable nonlinear systems without input dynamic extension*, Proc. of 39th Conf. Decision Contr. (Sydney, Australia), 2000.

[141] ———, *Practical internal models for output feedback tracking in nonlinear systems*, Proc. 2001 American Control Conf., 2001, pp. 2803–2808.

[142] ———, *Sufficient conditions for the solution of the semiglobal output tracking problem using practical internal models*, Proocedings of the IFAC Symposium on Nonlinear Control Systems Design (NOLCOS), St. Petersburg, Russia, 2001, pp. 1572–1577.

[143] ———, *Output feedback control of stabilizable and incompletely observable systems: Theory*, Proocedings of the 2000 American Control Conference (Chicago, IL), June 2000, pp. 3641–3645.

[144] E. Mamdani, *Advances in the linguistic synthesis of fuzzy controllers*, Intl. Journal of Man-Machine Studies **8** (1976), no. 6, 669–678.

[145] E. Mamdani and S. Assilian, *An experiment in linguistic synthesis with a fuzzy logic controller*, Intl. Journal of Man-Machine Studies **7** (1975), no. 1, 1–13.

[146] R. Marino and P. Tomei, *Dynamic output-feedback stabilization and global stabilization*, Systems Control Lett. **17** (1991), 115–121.

[147] ———, *Global exponential tracking control of nonlinear systems by output feedback. part I: Linear parametrization*, IEEE Trans. Automat. Contr. **38** (1993), no. 1, 17–32.

[148] ———, *Robust stabilization of feedback linearizable time-varying uncertain nonlinear systems*, Automatica **29** (1993), 181–189.

[149] ———, *Nonlinear control design: Geometric, adaptive & robust*, NJ: Prentice Hall, 1995.

[150] P. Martin, S. Devasia, and B. Paden, *A different look at output tracking: Control of a VTOL aircraft*, Automatica **32** (1996), no. 1, 101–107.

[151] D. McLean, *Automatic flight control systems*, Prentice Hall, New York, 1990.

[152] Z. Michalewicz, *Genetic algorithms + data structure = evolution programs*, Springer-Verlag, Berlin, 1992.

[153] A. N. Michel, *Differential equations, ordinary*, Encyclopedia of Physical Science and Technology (R. A. Meyers, ed.), vol. 4, Academic Press, 1987, pp. 283–316.

[154] A. N. Michel and R. K. Miller, *Ordinary differential equations*, Academic Press, New York, 1982.

[155] A. N. Michel and R. K. Miller, *Qualitative analysis of large scale dynamical systems*, Academic Press, New York, 1977.

[156] W. T. Miller, R. S. Sutton, and P. J. Werbos (eds.), *Neural networks for control*, The MIT Press, Cambridge, MA, 1990.

[157] S. Monaco and D. Normand-Cyrot, *Minimum-phase nonlinear discrete-time systems and feedback stabilization*, Proc. of 26th Conf. Decision Contr., 1987, pp. 979–986.

[158] F. K. Moore and E. M. Greitzer, *A theory of post-stall transients in axial compression systems-part I: Development of equations*, Journal of Turbomachinery **108** (1986), 68–76.

[159] K. S. Narendra and A. M. Annaswamy, *Stable adaptive systems*, Prentice-Hall, Englewood Cliffs, NJ, 1989.

[160] K. S. Narendra and K. Parthasarathy, *Identification and control of dynamical systems using neural networks*, IEEE Trans. Neural Networks **1** (1990), no. 1, 4–27.

[161] R. Ordóñez and K. M. Passino, *Adaptive control for a class of nonlinear systems with a time-varying structure*, IEEE Trans. Automat. Contr. **46** (2001), no. 1, 152–155.

[162] ———, *Indirect adaptive control for a class of nonlinear systems with a time-varying structure*, International Journal of Control **74** (2001), no. 7, 701–717.

[163] R. Ordóñez, J. Zumberge, J. T. Spooner, and K. M. Passino, *Experiments and comparative analyses in adaptive fuzzy control*, IEEE Transactions on Fuzzy Systems **5** (1997), no. 2, 167–188.

[164] R. Ortega, *Passivity properties for the stabilization of cascaded nonlinear systems*, Automatica **27** (1991), no. 2, 423–424.

[165] R. Ortega and A. Herrera, *A solution to the decentralized adaptive stabilization problem*, Systems Control Lett. **20** (1993), 299–306.

[166] P. C. Parks and V. Hahn, *Stability theory*, Prentice Hall, New York, 1992.

[167] K. M. Passino and P. J. Antsaklis, *A system and control theoretic perspective on artificial intelligence planning systems*, Int. Journal of Applied Artificial Intelligence **3** (1989), 1–32.

[168] K. M. Passino and A. D. Lunardhi, *Qualitative analysis of expert control systems*, Intelligent Control: Theory and Practice (M. Gupta and N. Sinha, eds.), IEEE Press, Piscataway, NJ, 1996, pp. 404–442.

[169] K. M. Passino and S. Yurkovich, *Fuzzy control*, The Control Handbook (W. Levine, ed.), CRC Press, Boca Raton, FL, 1996, pp. 1001–1017.

[170] ———, *Fuzzy control*, Addison Wesley Longman, Menlo Park, CA, 1998.

[171] T. Poggio and F. Girosi, *Networks for approximation and learning*, Proc. of the IEEE **78** (1990), 1481–1497.

[172] M. M. Polycarpou, *Stable adaptive neural control scheme for nonlinear systems*, IEEE Trans. Automat. Contr. **41** (1996), no. 3, 447–451.

[173] M. M. Polycarpou and P. A. Ioannou, *Identification and control of nonlinear systems using neural network models: Design and stability analysis*, Electrical Engineering - Systems Report 91-09-01, University of Southern California, September 1991.

[174] ———, *On the existance and uniqueness of solutions in adaptive control systems*, IEEE Trans. Automat. Contr. **38** (1993), no. 3, 474–479.

[175] ———, *A robust adaptive nonlinear control design*, Proc. 1993 American Control Conf., June 1993, pp. 1365–1369.

[176] M. M. Polycarpou and M. J. Mears, *Stable adaptive tracking of uncertain systems using nonlinearly parametrized on-line approximators*, Int. J. Control **70** (1998), no. 3, 363–384.

[177] J.-B. Pomet and L. Praly, *Adaptive nonlinear regulation: Estimation from the lyapunov equation*, IEEE Trans. Automat. Contr. **37** (1992), no. 6, 729–740.

[178] L. Porter and K. M. Passino, *Genetic model reference adaptive control*, Proc. of the IEEE Int. Symp. on Intelligent Control (Chicago), August 1994, pp. 219–224.

[179] L. Praly and Z.-P. Jiang, *Stabilization by output-feedback for systems with ISS inverse dynamics*, Systems and Control Letters **21** (1993), 19–33.

[180] ———, *Stabilization by output-feedback for systems with ISS inverse dynamics*, Systems Control Lett. **21** (1993), no. 1, 19–33.

[181] W. H. Press, S. A. Teukolsky, W. T. Vetterling, and B. P. Flannery, *Numerical recipies in c*, 2nd ed., Cambridge University Press, 1992.

[182] Z. Qu, *Robust control of nonlinear uncertain systems under generalized matching conditions*, Automatica **29** (1993), 985–998.

[183] J. S. Reed and P. A. Ioannou, *Discrete-time decentralized adaptive control*, Automatica **24** (1988), no. 3, 419–421.

[184] T. Ross, *Fuzzy logic in engineering applications*, McGraw-Hill, New York, 1995.

[185] G. A. Rovithakis and M. A. Christodoulou, *Adaptive control of unknown plants using dynamical neural networks*, IEEE Trans. Syst. Man, Cybern. **24** (1994), no. 3, 400–412.

[186] D. Rumelhart, G. E. Hinton, and R. J. Williams, *Learning representations by back-propagating errors*, Nature **323** (1986), 533–536.

[187] A. Saberi, P. V. Kokotoviċ, and H. J. Sussmann, *Global stabilization of partially linear composite systems*, SIAM J. Control Opt. **28** (1990), 1491–1503.

[188] N. Sadegh, *A perceptron network for functional identification and control of nonlinear systems*, IEEE Trans. Neural Networks **4** (1993), no. 6, 982–988.

[189] R. M. Sanner and J.-J. E. Slotine, *Gaussian networks for direct adaptive control*, IEEE Trans. Neural Networks **3** (1992), no. 6, 837–863.

[190] P. Sastry, G. Santharam, and K. Unnikrishnan, *Memory neuron networks for identification and control of dynamical systems*, IEEE Trans. Neural Networks **5** (1994), no. 2, 306–319.

[191] S. S. Sastry and M. Bodson, *Adaptive control: Stability, convergence, and robustness*, Prentice Hall, Englewood Cliffs, New Jersey, 1989.

[192] S. S. Sastry and A. Isidori, *Adaptive control of linearizable systems*, IEEE Trans. Automat. Contr. **34** (1989), no. 11, 1123–1131.

[193] ———, *Adaptive control of linearizable systems*, IEEE Trans. Automat. Contr. **34** (1989), no. 11, 1123–1131.

[194] A. V. Sebald and J. Schlenzig, *Minimax design of neural net controllers for highly uncertain plants*, IEEE Trans. Neural Networks **5** (1994), no. 1, 73–82.

[195] R. Sepulchre, M. Janković, and P. V. Kokotović, *Constructive nonlinear control*, Springer-Verlag, Berlin, 1997.

[196] H. Seraji, *Decentralized adaptive control of manipulators: Theory, simulation, and experimentation*, IEEE Trans. Robotics and Automation **5** (1989), no. 2, 183–201.

[197] A. A. Shabana, *Dynamics of multibody systems*, 2nd ed., Cambridge University Press, New York, 1998.

[198] S. Sheikholeslam and C. A. Desoer, *Indirect adaptive control of a class of interconnected nonlinear dynamical systems*, Int. J. Control **57** (1993), no. 3, 743–765.

[199] D. Shevitz and B. Paden, *Lyapunov stability theory of nonsmooth systems*, Proc. of 32nd Conf. Decision Contr., December 1993, pp. 416–421.

[200] L. Shi and S. K. Singh, *Decentralized adaptive controller design for large-scale systems with higher order interconnections*, IEEE Trans. Automat. Contr. **37** (1992), no. 8, 1106–1118.

[201] J. Sjoberg, Q. Zhang, L. Ljung, A. Benveniste, B. Deylon, P. Glorennec, H. Hjalmarsson, and A. Juditsky, *Nonlinear black-box modeling in system identification: A unified overview*, Automatica **31** (1995), no. 12, 1725–1750.

[202] G. R. Slemon and A. Straughen, *Electric machines*, Addison-Wesley, Reading, MA, 1980.

[203] J.-J. E. Slotine and J. K. Hedrick, *Robust input-output feedback linearization*, Internation Journal of Control **57** (1993), 1133–1139.

[204] J.-J. E. Slotine and S. S. Sastry, *Tracking control of nonlinear systems using sliding surfaces, with application to robot manipulators*, Int. J. Control **38** (1983), no. 2, 465–492.

[205] J.-J. E. Slotine and W. Li, *Applied nonlinear control*, Prentice Hall, Inc., Englewood Cliffs, New Jersey, 1991.

[206] Y. Song and J. Grizzle, *Adaptive output-feedback control of a class of discrete-time nonlinear systems*, Proc. of the American Control Conference (San Francisco), June 1993, pp. 1359–1364.

[207] E. D. Sontag, *Smooth stabilization implies coprime factorization*, IEEE Trans. Automat. Contr. **34** (1989), no. 4, 435–443.

[208] ———, *Feedback stabilization using two-hidden-layer nets*, IEEE Trans. Neural Networks **3** (1992), no. 6, 981–990.

[209] ———, *Comments on integral variants of ISS*, Systems Control Lett. **34** (1998), 93–100.

[210] E. D. Sontag and Y. Wang, *On characterizations of the input-to-state stability property*, Systems Contr. Lett. **24** (1995), 351–359.

[211] ———, *New characterizations of input-to-state stability*, IEEE Trans. Automat. Contr. **41** (1996), no. 9, 1283–1294.

[212] M. W. Spong and M. Vidyasagar, *Robot dynamics and control*, Wiley, New York, 1988.

[213] J. T. Spooner and K. M. Passino, *Stable adaptive control using fuzzy systems and neural networks*, IEEE Trans. Fuzzy Systems **4** (1996), no. 3, 339–359.

[214] ———, *Decentralized adaptive control of nonlinear systems using radial basis neural networks*, IEEE Trans. Automat. Contr. **44** (1999), no. 11, 2050–2057.

[215] M. Srinivas and L. M. Patnaik, *Genetic algorithms: A survey*, IEEE Computer Magazine (1994), 17–26.

[216] R. F. Stengel, *Toward intelligent flight control*, IEEE Trans. on Systems, Man, and Cybernetics **23** (1993), no. 6, 1699–1717.

[217] C.-Y. Su and Y. Stepanenko, *Adaptive control of a class of nonlinear systems with fuzzy logic*, IEEE Trans. Fuzzy Systems **2** (1994), no. 4, 285–294.

[218] T. Takagi and M. Sugeno, *Fuzzy identification of systems and its applications to modeling and control*, IEEE Trans. on systems, man, and cybernetics **15** (1985), no. 1, 116–132.

[219] G. Tao and P. Kokotović, *Adaptive control of systems with actuator and sensor nonlinearities*, Wiley, New York, 1996.

[220] D. G. Taylor, P. V. Kokotović, R. Marino, and I. Kanellakopoulos, *Adaptive regulation of nonlinear systems with unmodeled dynamics*, IEEE Trans. Automat. Contr. **34** (1989), no. 4, 405–412.

[221] A. R. Teel, *Adaptive tracking with robust stability*, Proc. of 32nd Conf. Decision Contr., December 1993, pp. 570–575.

[222] ———, *A nonlinear small gain theorem for the analysis of control systems with saturation*, IEEE Trans. Automat. Contr. **41** (1996), no. 9, 1256–1270.

[223] A. R. Teel and L. Praly, *Global stabilizability and observability imply semi-global stabilizability by output feedback*, Systems Control Lett. **22** (1994), 313–325.

[224] A. Tornambè, *High-gain observers for non-linear systems*, International Journal of Systems Science **23** (1992), no. 9, 1475–1489.

[225] ———, *Output feedback stabilization of a class of non-minimum phase nonlinear systems*, Systems & Control Letters **19** (1992), 193–204.

[226] S. Tzafestas (ed.), *Methods and applications of intelligent control*, Kluwer Academic Publishers, Norwell, MA, 1997.

[227] V. I. Utkin, *Variable structure systems with sliding modes*, IEEE Trans. Automat. Contr. **AC-22** (1977), no. 2, 212–221.

[228] K. Valavanis and G. Saridis, *Intelligent robotic systems: Theory, design, and applications*, Kluwer Academic Publishers, Norwell, MA, 1992.

[229] M. Van Nieuwstadt, M. Rathinam, and R. M. Murray, *Differential flatness and absolute equivalence of nonlinear control systems*, SIAM Journal on Control and Optimization **36** (1998), no. 4, 1225–1239.

[230] M. Vidyasagar, *Nonlinear systems analysis*, Prentice Hall, Inc., Englewood Cliffs, New Jersey, 1993.

[231] L.-X. Wang, *Fuzzy systems are universal approximators*, 1st IEEE conference on fuzzy systems, March 1992, pp. 1163–1170.

[232] ———, *Stable adaptive fuzzy control of nonlinear systems*, IEEE Trans. Fuzzy Systems **1** (1993), no. 2, 146–155.

[233] ———, *Adaptive fuzzy systems and control: Design and stability analysis*, Prentice Hall, Englewood Cliffs, NJ, 1994.

[234] ———, *A supervisory controller for fuzzy control systems that guarantees stability*, IEEE Trans. Automat. Contr. **39** (1994), no. 9, 1845–1847.

[235] ———, *A course in fuzzy systems and control*, Prentice-Hall, Englewood Cliffs, NJ, 1997.

[236] F. W. Warner, *Foundations of differential manifolds and lie groups*, Springer-Verlag, New York, 1983.

[237] C. Wen and Y. C. Soh, *Decentralized adaptive control using integrator backstepping*, Automatica **33** (1997), no. 9, 1719–1724.

[238] ———, *Decentralized model reference adaptive control without restrictions on subsystem relative degree*, IEEE Trans. Automat. Contr. **44** (1999), no. 7, 1464–1469.

[239] D. White and D. Sofge (eds.), *Handbook of intelligent control: Neural, fuzzy and adaptive approaches*, Van Nostrand Reinhold, New York, 1992.

[240] M. Widjaja and S. Yurkovich, *Intelligent control for swing up and balancing of an inverted pendulum system*, Proc. of the 4th IEEE Int. Conf. on Control Applications, Albany, NY, September 1995, pp. 534–542.

[241] L.-L. Xie and L. Guo, *Fundamental limitations of discrete-time adaptive nonlinear control*, IEEE Trans. Automat. Contr. **44** (1999), no. 9, 1777–1782.

[242] T. Yabuta and T. Yamada, *Neural network controller characteristics with regard to adaptive control*, IEEE Trans. on Systems, Man, and Cybernetics **22** (1992), no. 1, 170–177.

[243] T. C. Yang, *An algorithm to find the smallest possible values in a matrix to satisfy the M-matrix condition*, IEEE Trans. Automat. Contr. **42** (1997), no. 12, 1738–1740.

[244] A. Yeşildirek and F. L. Lewis, *Feedback linearization using neural networks*, Automatica **31** (1995), no. 11, 1659–1664.

[245] P.-C. Yeh and P. Kokotović, *Adaptive output-feedback design for a class of nonlinear discrete-time systems*, IEEE Trans. Automat. Contr. **40** (1995), 1663–1668.

[246] G. G. Yen, *Decentralized neural controller design for space structural platforms*, IEEE International Conf. on Sys., Man, and Cybernetics, October 1994, pp. 2126–2131.

[247] A. Yeşildirek and F. L. Lewis, *A neural network controller for feedback linearization*, Proc. of 33rd Conf. Decision Contr. (Lake Buena Vista, FL), December 1994, pp. 2494–2499.

[248] T.-K. Yin and C. S. G. Lee, *Fuzzy model-reference adaptive control*, IEEE Transactions on Systems, Man, and Cybernetics **25** (1995), no. 12, 1606–1615.

[249] S. Yurkovich and M. Widjaja, *Fuzzy controller synthesis for an inverted pendulum system*, IFAC Control Engineering Practice **4** (1996), no. 4, 445–469.

[250] L. A. Zadeh, *Fuzzy sets*, Informat. Control **8** (1965), 338–353.

[251] ———, *Outline of a new approach to the analysis of complex systems and decision processes*, IEEE Trans. on Systems, Man. and Cybernetics **3** (1973), no. 1, 28–44.

[252] J. Zhang and S. Lang, *Adaptive control of a class of multivariable nonlinear systems and convergence analysis*, IEEE Trans. Automat. Contr. **34** (1989), 787–791.

[253] T.-P. Zhang and C.-B. Feng, *Decentralized adaptive fuzzy control for large-scale nonlinear systems*, Fuzzy Sets and Systems **92** (1997), 61–70.

[254] Y. Zhang, P. A. Ioannou, and C.-C. Chien, *Parameter convergence of a new class of adaptive controllers*, IEEE Trans. Automat. Contr. **41** (1996), no. 10, 1489–1493.

[255] J. Zhao and I. Kanellakopoulos, *Adaptive control of discrete-time strict-feedback nonlinear systems*, Proc. of the American Control Conference (Albuquerque, NM), June 1997, pp. 828–832.

Index

CPSIA information can be obtained at www.ICGtesting.com
Printed in the USA
BVOW030219270112

281416BV00007B/44/P

9 780471 415466